terra australis 27

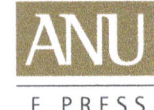

© 2008 ANU E Press

Published by ANU E Press
The Australian National University
Canberra ACT 0200 Australia
Email: anuepress@anu.edu.au
Web: http://epress.anu.edu.au

National Library of Australia Cataloguing-in-Publication entry

Author:	McDonald, Josephine.
Title:	Dreamtime superhighway : an analysis of Sydney Basin rock art and prehistoric information exchange / Jo McDonald.
ISBN:	9781921536168 (pbk.) 9781921536175 (pdf)
Series:	Terra Australis ; 27
Notes:	Bibliography.
Subjects:	Rock paintings--New South Wales--Sydney Basin.
	Petroglyphs--New South Wales--Sydney Basin.
	Visual communication in art--New South Wales--Sydney Basin.
	Art, Aboriginal Australian--New South Wales--Sydney Basin.
	Aboriginal Australians--New South Wales--Sydney Basin--Antiquities.
Dewey Number:	709.011309944

Copyright of the text remains with the contributors/authors, 2006. This book is copyright in all countries subscribing to the Berne convention. Apart from any fair dealing for the purpose of private study, research, criticism or review, as permitted under the *Copyright Act*, no part may be reproduced by any process without written permission. Inquiries should be made to the publisher.

Series Editor: Sue O'Connor

Typesetting and design: Silvano Jung

Cover photograph by Jo McDonalnd
Back cover map: *Hollandia Nova*. Thevenot 1663 by courtesy of the National Library of Australia.
Reprinted with permission of the National Library of Australia.

Terra Australis Editorial Board: Sue O'Connor, Jack Golson, Simon Haberle, Sally Brockwell, Geoffrey Clark

Terra Australis reports the results of archaeological and related research within the south and east of Asia, though mainly Australia, New Guinea and island Melanesia — lands that remained *terra australis incognita* to generations of prehistorians. Its subject is the settlement of the diverse environments in this isolated quarter of the globe by peoples who have maintained their discrete and traditional ways of life into the recent recorded or remembered past and at times into the observable present.

List of volumes in *Terra Australis*

Volume 1: Burrill Lake and Currarong: Coastal Sites in Southern New South Wales. R.J. Lampert (1971)

Volume 2: Ol Tumbuna: Archaeological Excavations in the Eastern Central Highlands, Papua New Guinea. J.P. White (1972)

Volume 3: New Guinea Stone Age Trade: The Geography and Ecology of Traffic in the Interior. I. Hughes (1977)

Volume 4: Recent Prehistory in Southeast Papua. B. Egloff (1979)

Volume 5: The Great Kartan Mystery. R. Lampert (1981)

Volume 6: Early Man in North Queensland: Art and Archaeology in the Laura Area. A. Rosenfeld, D. Horton and J. Winter (1981)

Volume 7: The Alligator Rivers: Prehistory and Ecology in Western Arnhem Land. C. Schrire (1982)

Volume 8: Hunter Hill, Hunter Island: Archaeological Investigations of a Prehistoric Tasmanian Site. S. Bowdler (1984)

Volume 9: Coastal South-West Tasmania: The Prehistory of Louisa Bay and Maatsuyker Island. R. Vanderwal and D. Horton (1984)

Volume 10: The Emergence of Mailu. G. Irwin (1985)

Volume 11: Archaeology in Eastern Timor, 1966–67. I. Glover (1986)

Volume 12: Early Tongan Prehistory: The Lapita Period on Tongatapu and its Relationships. J. Poulsen (1987)

Volume 13: Coobool Creek. P. Brown (1989)

Volume 14: 30,000 Years of Aboriginal Occupation: Kimberley, North-West Australia. S. O'Connor (1999)

Volume 15: Lapita Interaction. G. Summerhayes (2000)

Volume 16: The Prehistory of Buka: A Stepping Stone Island in the Northern Solomons. S. Wickler (2001)

Volume 17: The Archaeology of Lapita Dispersal in Oceania. G.R. Clark, A.J. Anderson and T. Vunidilo (2001)

Volume 18: An Archaeology of West Polynesian Prehistory. A. Smith (2002)

Volume 19: Phytolith and Starch Research in the Australian-Pacific-Asian Regions: The State of the Art. D. Hart and L. Wallis (2003)

Volume 20: The Sea People: Late-Holocene Maritime Specialisation in the Whitsunday Islands, Central Queensland. B. Barker (2004)

Volume 21: What's Changing: Population Size or Land-Use Patterns? The Archaeology of Upper Mangrove Creek, Sydney Basin. V. Attenbrow (2004)

Volume 22: The Archaeology of the Aru Islands, Eastern Indonesia. S. O'Connor, M. Spriggs and P. Veth (2005)

Volume 23: Pieces of the Vanuatu Puzzle: Archaeology of the North, South and Centre. S. Bedford (2006)

Volume 24: Coastal Themes: An Archaeology of the Southern Curtis Coast, Quuensland. S. Ulm (2006)

Volume 25: Lithics in the Land of the Lightning Brothers: The Archaeology of Wardaman Country, Northern Territory. C. Clarkson (2007)

Volume 26: Oceanic Explorations: Lapita and Western Pacific Settlement. S. Bedford, C. Sand and S.P. Connaughton (2007)

Volume 27: Dreamtime Superhighway: Sydney Basin Rock Art and Prehistoric Information Exchange. J. McDonald (2008)

Volume 28: New Directions in Archaeological Science. A. Fairbairn, S. O'Connor and Ben Marwick (2008)

Volume 29: Islands of Inquiry: Colonisation, Seafaring and the Archaeology of Maritime Landscapes. G. Clark, F. Leach and S. O'Connor (2008)

terra australis 27

Dreamtime Superhighway

Sydney Basin Rock Art and Prehistoric Information Exchange

Jo McDonald

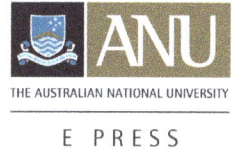

E PRESS

FOREWORD

I remember the visit with clarity: it was 1991 and my first trip to Australia. One day was spent at Sydney's West Head with the young and enthusiastic Jo McDonald, where we toured some of the rock engraving sites and the painted Great Mackerel rock shelter. I can still see the large flat curving rocks at West Head on which animals had been engraved - fish, kangaroos, a goanna. And, as is so often the case when I visit rock art sites, my imagination gets going as to the 'how's' and 'why's'. I try to run some sort of a show in my head as to the people who produced these, viewed these, and what kinds of significations the images and their makings played in the lives of past peoples. At the time of this visit, Jo McDonald was well into her 1994 dissertation—the work upon which this monograph is based. Already, in that PhD dissertation, she was able to touch on some of the things that my imagination was searching for. Thus, it is with great pleasure and even more enthusiasm that I am writing this Foreword to a revised and thoroughly up-dated monograph based on that initial dissertation research.

At the time of the 1994 dissertation, the anthropological and archaeological study of 'rock art' was really emerging into new trends and new prominence. Surely Australia was one of the leaders in the training of students and in the research that contextualized the images and 'art', thanks to such scholars as Andrée Rosenfeld, Robert Layton, and John Clegg among others. By this time, the work of David Lewis-Williams and colleagues in southern Africa had set off numerous studies world-wide into the relationship between the production of rock art images and altered states of consciousness and the role of shamans in image-making practices. But this was not the direction or focus of rock art studies in Australia, and McDonald's original work would not be tempted either. Rather, she carried out an extensive project of contextualizing the rock art in question, and in two different ways. First, she wanted to give us an understanding of the rock art in its archaeological contexts—sheltered or open, dates and chronologies, site types and, in general, what the archaeology could inform on in terms of social and cultural lives associated with the image-making. This dimension is thoroughly expanded in this present monograph, a genuine testimony to precisely why so many insist on a true archaeology of 'rock art'. Second, she wanted to try out a model that suggested the ways in which visual culture—such as rock art-making and its images and forms—could perhaps be understood as a system of communication, as a way of signaling, so–to–speak, among and between various social factions and groups. Could we gain some insight into the 'why's' and 'how's' by stretching our notion of context, meaning, and function? Was the image making itself part of the stretching—and marking—of social relations, which we knew were so crucial to the on-going-ness of Aboriginal society? These were solid and provocative questions of the mid 1990's and McDonald rose to the challenge with an admirable array of data and a compelling conceptual framework that drew on the social communication/information exchange models of the time.

It is now nearly fifteen years later. Much has changed in archaeology and the study of the so-called rock arts; the landscape is 'in', the spirit world and shamanic practices are still with us, rock art has been attributed to even earlier time ranges of the Pleistocene in many parts of the globe, and more theoretical frameworks circulate widely—structuration, practice theory, agency, social memory, costly signaling, post-colonialism, to name but a few. Even the term 'rock art' itself has been subjected to critique and scrutiny, and it can be heard, from time to time, that there is not much to be gained from an archaeology of rock art: what, after all, might stone tools have to 'say' about the making and meanings of rock imagery?

It is refreshing to read here that McDonald has not decided to publish her monograph by picking one from the list of the 'new' approaches. Nor has she tried out, as Chris Tilley has done in a unique and innovative comparative study of Scandinavian rock art, how the materials might be understood using several different interpretive frameworks to find out which one works the best–at least for now! Rather, this is a revised and updated study that draws forcefully on much richer data and interpretive source materials because so much fine and extensive new research has been carried out and made available—much of it by McDonald herself. Suddenly, the archaeological contexts become more, not less, important and useful. At this point, the more reflective ethnohistories and documentary sources are brought to bear in new ways. Locational scrutiny–with many new known and excavated sites and new dating techniques—allows McDonald to further support the hypothesis that the greater stylistic homogeneity in the engraved medium demonstrates larger scale group cohesion. And the more stylistically heterogeneous pigment sites demonstrate localised group identifying behaviour.

We are well on our way to grappling intimately with the materiality of image making and the resultant images and forms. We are well on our way to accessing some of the 'why's' and 'how's'. We can feel that we have a better grasp on what these cultural practices might have been about—as best as one can from a strikingly different cultural vantage point and one that cannot really divorce itself from being that of the colonizer, at that. But none the less, when one brings to bear—as McDonald does here—such an array of well thought through sets of information, such a detailed documentation of place, natural worlds, variations and yet patterns in cultural practices, settlement histories, and specific images and their patterning, the reader is rewarded with what a 'deep archaeology' can do for getting somewhat closer to the sensibilities of the art makers, to their worlds of social and cultural negotiations and habits. And to the archaeologist of art, to the anthropologist of art, to those of us who try to think about past worlds and their continuations with transformations into present lives, McDonald has done us a real service. I would like to think that this work will bring some satisfaction to not only to those of us who 'analyze' and 'study' the archaeology of rock art, but also to Aboriginal people who might find in it not only an enormous amount of information about historical practices and places, but also a source for discussion, reflection, debate and inspiration. Just how we relate to the material worlds we imagine and produce is a question for all times. McDonald shows here that it is these material and visual 'interventions' into social life—this image–making—may well have been precisely about how differing dimensions of social life were both instantiated and maintained. Not many of us can say that for our own work, sites and regions. That we see here the very evolution and expansion of a project that has flowered with time and with an enormous amount of new research and methods, suggests we should never let those old theses, reports and projects sit on the shelf. We have a new vision here, yet one that does not stray too much from its original incarnation. This is both heartening and admirable. Especially for researchers who might still doubt the efficacy of wedding substantive archaeology with 'rock art', this monograph is a must read.

Margaret W. Conkey

University of California, Berkeley

TABLE OF CONTENTS

1. INTRODUCTION 1
Previous work on the Sydney rock art — the research context 3
Sampling issues 4
The analysis — rock art as stylistic messaging 5

2. THE SYDNEY BASIN - ENVIRONMENTAL CONTEXT 6
Geology of the Sydney Basin 6
Geomorphology: structure and terrain 6
Climatic and eustatic change 11
Vegetation 13
Ku-ring-gai Chase National Park 13
Mangrove Creek 13
Yengo National Park 14
Conclusions 15

3. SOCIAL CONTEXT 16
Social organisation and language boundaries 17
Linguistic Evidence 20
Land use strategies and habitations 22
Living sites 23
Material culture 24
Shields 25
Ceremonial behaviour and initiation 25
Culture Heroes 27
Art 28
Body marking and personal attire 29
The impact of European contact on Aboriginal society 31
The relevance of ethnohistoric evidence to patterning in the art 32

4. ARCHAEOLOGICAL CONTEXT 34
A behavioural model for the Sydney region 39
Summary of the model with likely art correlates 41

5. THE ROCK ART OF THE SYDNEY BASIN 43
Sydney Art in the general scheme of Australian rock art 43
Previous work in the Sydney region — the growth of a data base 45
The Sydney Basin Art Style: a current definition 46
Schema 46
Chronology 46
The Samples 51
The Open Engraving sites 51
Assemblage Size 51

Site Nature and Topographic Location 51

Site Associations 52

Motifs 53

The Georges River style boundary 53

THE SHELTER ART SITES 55

Assemblage Size 55

Site Nature and Topographic Location 55

Site Associations 56

Motifs 56

Technique 58

COMPARISON OF THE TWO ART CONTEXTS: SHELTERED AND OPEN ART SITES 65

Assemblage Size 65

Topographic location 66

Motif 66

Composition 67

6. EXCAVATIONS AT YENGO 1 AND YENGO 2 69

ENVIRONMENTAL CONTEXT 70

THE SITES 71

Yengo 1 (AHIMS #37-5-1) 71

The Art 71

The Hand Stencils 74

Left or right? 79

Size 79

The Grinding Grooves 81

Yengo 2 (AHIMS #37-5-2) 83

Stencils 84

EXCAVATION PROCEDURES 88

Yengo 1 88

Yengo 2 90

RESULTS - YENGO 1 90

Stratigraphy 90

Stratigraphic Layers 91

Dates 98

The Excavated Lithic Assemblage 99

    Artefact Density 100

    Assemblage Characteristics 100

    Artefacts with Retouch/Usewear 100

    Cores 101

    Backed Artefacts 101

    Ground Fragments 102

    Other Implement Types 102

- Hafting Residue ... 102
- The Analysis Squares ... 102
- Assemblage Characteristics ... 102
- Raw Material ... 102
- Size ... 103
- Artefact Types ... 105
- Vertical Distribution of Material ... 107
  - Artefact Density and Percentage frequency ... 108
  - Artefact deposition rates ... 109
  - Raw Materials ... 112
  - Size ... 113
  - Artefact types ... 113
  - Amorphously retouched flakes and flaked pieces ... 114
  - Backed Artefacts ... 115
  - Ground Fragments ... 116
  - Cores ... 116
  - Summary ... 118
  - Faunal Remains ... 120
  - Pigment ... 125
- Activity areas: spatial patterning ... 125
  - Artefact Density ... 125
  - Implements and other artefact types ... 126
- Other Economic Remains ... 127
  - Shell ... 127
  - Pigment ... 128
  - Seeds ... 128
  - Faunal material ... 128

Results - Yengo 2 ... 129
- Stratigraphy ... 130
- Cultural Material ... 130

Discussion and Interpretation ... 130
- Yengo 1 ... 130
- Artefact accumulation rates ... 132
- The Lithic Assemblage ... 132
- The Faunal Remains ... 134
- The Engravings ... 134
- The Pigment Art ... 136
- Stencils ... 137
- The Grinding Grooves ... 138
- What Type of Site? ... 138
- Yengo 2 ... 140

The Yengo Sites in a Regional Context ... 141

7. EXCAVATION AT THE GREAT MACKEREL ROCKSHELTER 143

    ENVIRONMENTAL CONTEXT 143

    THE SITE 143

    The Art 144

        Stencils 145

        The Deposit 145

    Field methods 149

    Results 149

        Cultural Material 149

            Stratigraphy 150

            Dates 152

            Shell 153

    Proportions of Shellfish Species 154

    Temporal Variation in Shellfish Exploitation 156

        Size 157

    The Bone 161

    Dietary Estimates 162

    Stone Artefacts 163

        Raw Material 163

        Size 163

        Artefact Types 163

        Change through time 164

    Shell Artefacts 166

    The Pigment 168

    DISCUSSION AND INTERPRETATION 169

8. EXCAVATIONS AT UPSIDE-DOWN-MAN 171

    ENVIRONMENTAL CONTEXT 172

    THE SITE 172

        The Art 172

            Stencils 172

            Engravings 173

            Change over time 176

            Art Dates 181

    THE EXCAVATION 183

        Aims 183

        Methods 183

        Stratigraphy 184

        Correlation of Excavation Layers with Stratigraphic Layers 186

        Dates 186

        Cultural Material 187

            Stone Artefacts, Charcoal and Unsorted Residue 188

Shell and Bone	189
Plant Remains and Pigment	189
Change over time	189
ARTEFACT ANALYSIS	190
Raw Material	191
Size	192
Artefact Types	192
Vertical Distribution	192
Artefact Density	193
Raw Material	196
Size	199
Artefact Type	200
Discussion	204
The Archaeology of Mangrove Creek	205
UMCC	205
Sampling Issues	206
Discussion	208
Interpretation of the Upside-Down-Man site	210
Assemblage Size and Characteristics	211
Artefact Accumulation Rates	211
Hatchets and grinding technology	212
The art and the occupation evidence	213
The AMS samples	214
Upside-Down Man in the local context	215

9. THE CONTEMPORANEITY OF ART AND DEPOSIT 217

THE PROBLEM, PREVIOUS APPROACHES, CURRENT AIMS	217
REGIONAL PATTERNS FOR THE SYDNEY REGION	220
Henry Lawson Drive	223
Mill Creek	223
Barden's Creek	223
Bull Cave	224
Mangrove Creek	224
Dingo and Horned Anthropomorph	224
Emu Tracks 2	225
Angophora Reserve	225
Audley	225
Daley's Point (Milligan's)	226
CONCLUSIONS	226

10. DIACHRONIC VARIATION IN THE ART OF THE SYDNEY BASIN 229

A PREVIOUS REGIONAL APPROACH TO DIACHRONIC VARIABILITY	230
AN EARLIER MODEL FOR DIACHRONIC CHANGE IN THE SHELTER ART	231

Diachronic variability in the art of Mangrove Creek	232
The Sample and the Technique variables	232
Superimposition Analysis	236
Changes in Motif Preference over Time	237
Multivariate Analyses	241
Small sites	243
Medium sites	244
Large and complex sites	244
Conclusions	245
Mangrove Creek Art Sequence	246

11. SYNCHRONIC VARIATION: SYDNEY BASIN ENGRAVED ART 251

Introduction	251
Defining a regional style: methodology	252
Motif Variables	252
Counting	253
Analyses	253
Correspondence Analysis (CA): data, results and interpretation	253
Regional Analysis	255
Language Areas. Searching for boundaries and between-group distinctiveness	255
Motif Assemblage Differences across the Basin	258
Darkingung	259
Darug	259
Guringai	259
Sydney (Eora)	259
Tharawal	259
Summary	262
Correspondence Analysis and Language Areas	263
Drainage Basins	263
CA according to drainage basins	263
1) Darkingung language group (drainage basins 1, 5 and 6)	263
2) the Guringai/Darug language boundary (drainage basins 10 - 13)	266
3) The Tharawal language area (drainage basins 18 – 21)	270
Conclusions	271
Ridge top vs vertical engraving sites	272
Rare Motifs	275
Composition	278
Shields	278
Culture Heroes	286
Engraving sites: conclusions	289

12. REGIONAL SYNCHRONIC VARIATION: SHELTER ART 292

Introduction	292

Correspondence Analysis: regional data, results and interpretation	292
Motif	292
Technique	294
Regional Comparison	295
Language Areas	296
Motif and Technical Variation across the Basin	298
Darkingung Language Area	298
Darug Language Area	300
Guringai Language Area	302
Sydney (Eora) Language Area	304
Tharawal Language Area	305
Summary	307
Correspondence Analysis, Language Areas and Drainage Basins	308
Shelter Art Motifs	308
1) Darkingung Language Area (drainage basins 1, 5 and 6)	308
2) East-west patterning Guringai/Darug language boundary (drainage basins 10 - 13)	311
3) The Tharawal language area (drainage basins 18 - 21)	315
Summary	319
Shelter Art Technique	319
1) Darkingung language group (drainage basins 1, 5 and 6)	319
2) East-west patterning Guringai/Darug language boundary (drainage basins 10 - 13)	321
3) Tharawal and Darug language areas (Basins 18 - 21)	325
Summary	328
Rare Motifs	329
Composition	331
Shields	331
Culture Heroes	336
Shelter art sites: conclusions	338
13. DREAMTIME SUPERHIGHWAY: MODELLING A REGIONAL STYLE	**340**
Art characteristics in two contexts	340
Engraving sites	340
Shelter art sites	340
General comparisons of the two contexts	341
Contemporaneity of art and deposit	342
Diachronic change in the shelter art	342
Synchronic variability: both art components	343
Engravings	343
Shelter Art Sites	344
Social context and stylistic information	345
Shelters vs engravings	345

	Public vs Private engravings	347
A MODEL FOR SOCIAL AND TERRITORIAL INTERACTION ACROSS THE REGION		347
	Pre Bondaian 30,000 years ago to 8,000BP	349
	Early Bondaian 8,000 years to c.4,000 years BP	349
	Middle Bondaian c.4,000 years to c.1,000 years BP	349
	Late Bondaian c.1,000 years to European contact	349

14. REFERENCES 352

Tables

Table 4.1:	Open Site excavations on the Cumberland Plain since 1993, showing areas excavated (testing and open area), total lithics and artefacts retrieved.	34
Table 4.2:	The Eastern Regional Sequence (dates after JMcD CHM 2005e).	39
Table 5.1:	The Sydney Basin Engraving component: motif frequency and %f.	54
Table 5.2:	The Sydney Basin Engraved Component: clumped motif frequency, %frequency of totals and identifiable motifs.	54
Table 5.3:	Shelter Art component: summary motif frequency information.	57
Table 5.4:	Shelter Art component: clumped motif frequency and % frequency.	57
Table 5.5:	Shelter Art component: summary technique frequency information.	65
Table 6.1:	Yengo 1. Art Assemblage. Motif and technique information.	74
Table 6.2:	Yengo 1. Left and right handedness of the coloured stencils.	79
Table 6.3	Yengo 1. Size ranges of hand stencils (group n≥3).	80
Table 6.4:	Yengo 1. The grinding grooves (measurements in centimetres).	81
Table 6.5:	Yengo 2. Art Assemblage. Motif and technique information.	83
Table 6.6:	Yengo 2. Handedness (left or right) of the coloured stencils.	84
Table 6.7:	Yengo 2. Size ranges of hand stencils (where n >3).	85
Table 6.8:	Yengo 1. Excavated Pit Dimensions.	88
Table 6.9:	Yengo 1. Excavated Depth, Spits and Stratigraphic Layers. All excavated squares.	96
Table 6.10:	Yengo 1. Correlation of spits with analytical units and stratigraphic layers: Analysis squares (1B, 4A, 3S and 6) only.	97
Table 6.11:	Yengo 1. Radiocarbon determinations.	97
Table 6.12:	Yengo 1. Features dated by the charcoal samples. Identification number is that used on stratigraphic sections (Figure 6.30).	98
Table 6.13:	Yengo 1. Average artefact densities (density^1 = artefacts/m^3; density^2 = artefacts/kg).	99
Table 6.14:	Yengo 1. Cores.	100
Table 6.15:	Yengo 1. Distribution of retouched and broken backed blades across the site.	100
Table 6.16:	Yengo 1. Artefact Totals per Spit both Analysis Squares.	102
Table 6.17:	Artefact Totals per Spit; Squares 1B, 4A, 3S, 6. Artefacts >1cm.	102
Table 6.18:	Yengo 1. Raw Material Percentages in Squares 1B-4 and 3S-6.	103
Table 6.19:	Yengo 1. Size ranges, non-modified artefacts, Squares 1B-4 and 3S-6.	103
Table 6.20:	Size ranges: modified artefact and core assemblage; Squares 1B-4 and 3S-6.	103

Table 6.21:	Yengo 1. Age-depth calculations. Artefact totals and bone weights per 100 years (refer Figure 6.37, Table 6.9).	110
Table 6.22:	Yengo 1. Size ranges of cores in the analysis squares (per stratigraphic layer).	113
Table 6.23:	Yengo 1. Size ranges of artefacts with R/U in each stratigraphic layer. Includes ground fragments and backed blades.	113
Table 6.24:	Yengo 1. Core platform characteristics per stratigraphic layer.	117
Table 6.25:	Taxonomic composition of the Yengo1 faunal assemblage.	123
Table 6.26:	Breakdown of animal varieties identified at Yengo 1. Categories of Small, Medium and Large by NISP.	124
Table 6.27:	Yengo 1. Distribution of pipeclay and pigment across the site and throughout the sequence.	125
Table 6.28:	Yengo 1. Distribution of implements, R/U and micro-debitage (artefacts <1cm) across the site.	129
Table 6.29:	Yengo 2. Location of artefacts and charcoal throughout the two test pits.	130
Table 7.1:	Art Assemblage: Motif and Technique Information.	144
Table 7.2:	Proportions of shell, bone, artefacts, charcoal and pigment in cultural material (all squares).	149
Table 7.3:	Distribution of Cultural Material by Square. Weights in grams.	149
Table 7.4:	Correlation of excavated spits (Squares 6, 6A, 6B and 8), analytical units and stratigraphic layers.	152
Table 7.5:	Radiocarbon dates, depth below surface and stratigraphic associations.	153
Table 7.6:	Age depth calculations. Artefacts accumulating per 100 year.	153
Table 7.7:	Shell species identified at the Great Mackerel site (after Dakin 1980).	154
Table 7.8:	Estimated Number of Individuals (MNI) in Squares 6 and 8.	157
Table 7.9:	Weight (in g.) of shellfish species in Squares 6 and 8.	158
Table 7.10:	Square 6; Minimum numbers: change through time by analytical unit.	159
Table 7.11:	Square 8. Minimum numbers: Change through time by analytical unit.	159
Table 7.12:	Square 6. Shellfish species and other cultural material; change over time by analytical unit. Weight in grams except where indicated.	160
Table 7.13:	Great Mackerel Square 8. Shellfish species and other cultural components through time per analytical unit. Weights in grams except where indicated.	161
Table 7.14:	Squares 6 and 8. Dominant shell species - peak midden units.	161
Table 7.15:	Great Mackerel, all squares. Raw materials and artefact totals.	163
Table 7.16:	Artefacts with Retouch/usewear. Raw Material vertical distribution.	164
Table 7.17:	Squares 6. Changes in Raw Material, per cultural layer.	165
Table 7.18:	Square 8. Changes in Raw Material, per cultural layer.	166
Table 7.19:	Great Mackerel, Square 6. Distribution of shell artefacts.	168
Table 7.20:	Great Mackerel. Distribution of pipeclay throughout the archaeological sequence (weight in grams).	168
Table 8.1:	UDM Shelter. Motif and Technique Information.	175
Table 8.2:	UDM Shelter. Technique variables. Variables in identified superimposition relationships (Table 8.4) highlighted in bold.	176
Table 8.3:	Proposed Diachronic Sequence; Mangrove Creek Shelter Site Phases 1-3. Those elements observed at UDM in bold.	180
Table 8.4:	UDM shelter. Recorded superimposition information.	180

Table 8.5:	UDM Shelter. AMS dates from the two macropod motifs.	181
Table 8.6:	UDM Shelter. Correlation of analytical units with excavation spits and stratigraphic layers (Squares 4, 6 and 8).	186
Table 8.7:	Radiocarbon dates, depth below surface and stratigraphic associations.	187
Table 8.8:	Proportions of lithic, charcoal, shell, bone, residue and 'other' from the excavated squares. Weights in grams.	187
Table 8.9:	Proportions of cultural material retrieved from the excavated squares. Weights in grams except where specified.	188
Table 8.10:	UDM Shelter. Artefact Types according to raw material and size. The three analysed squares (6B, 8B and 8C).	193
Table 8.11:	UDM Shelter. Modified artefacts and raw material types.	193
Table 8.12:	UDM Shelter. R/U artefacts and cores and raw material types.	195
Table 8.13:	UDM Shelter. Age-depth calculations: artefacts per year.	195
Table 8.14:	UDM Shelter. Size ranges for backed blades and artefacts with retouch/ usewear per stratigraphic layer.	199
Table 8.15:	UDM Shelter. Size ranges for cores per stratigraphic layer.	199
Table 8.16:	UDM Shelter. Raw material per analytical unit. All Artefacts.	199
Table 8.17:	UDM Shelter. Raw material per analytical unit. Artefacts >1cm.	200
Table 8.18:	Platform characteristics over time. Frequency per analytical unit.	201
Table 8.19:	UDM Shelter. Artefacts with R/U per stratigraphic layer. Platform and flaked piece types.	202
Table 8.20:	Colour distribution of ground FGB artefacts; as per stratigraphic unit.	203
Table 8.21:	UMCC Sites with >100 excavated artefacts. Reworked estimated artefacts totals in successive periods of time (Attenbrow 2004: Table 6.7).	207
Table 8.22:	UMCC Sites with >100 excavated artefacts. Reworked estimated totals of artefacts in successive periods of time (Attenbrow 2004: Table 6.13).	208
Table 8.23:	UDM: Rates of Artefact accumulations in successive phases.	211
Table 9.1:	Excavated shelters with art in the Sydney region.	221
Table 10.1:	Techniques and/or colour. Artistic variables in the diachronic analysis.	234
Table 10.2:	Mangrove Creek valley: different sized assemblages located in Upper Mangrove Creek compared with the midstream Warre Warren sites.	234
Table 10.3:	Mangrove Creek shelter art sites. Number and % frequency of motifs for each technique variable.	236
Table 10.4:	Mangrove Creek shelter art sites: Sites at which different technique variables are present.	237
Table 10.5:	Mangrove Creek Shelter Art sites. Superimposition sequence.	241
Table 10.6:	Mangrove Creek shelter art: Motif totals in the proposed art phases.	242
Table 10.7:	Proposed Diachronic Sequence in the Mangrove Creek sites. Mangrove Creek Pigment Phases 1-3.	242
Table 10.8:	Phased Sequence: Mangrove Creek Shelter Art sites.	243
Table 10.9:	Techniques and/or colour variables used in Correspondence Analysis.	245
Table 10.10:	Excavated shelters with radiocarbon determinations, estimated archaeological phases and designated art phases.	248
Table 11.1:	Engraving Sites. Clumped motif variables in Correspondence Analysis.	253
Table 11.2:	Analytical grouping of engraving sites according to AHIMS numbers. These groups were used in the regional interpretation of the CA results.	254

Table 11.3:	Language areas, codes and sample sizes.	258
Table 11.4:	Drainage Basins, Language Areas and Sample sizes.	264
Table 11.5:	Engraving sites. Motif total, maximum incidence at any particular site, number of sites in the region with motif present, and % of sites with motif.	276
Table 11.6:	Rare Engraving Motifs. Distribution per Language Area.	277
Table 11.7:	Engraving sites. Significant values for rare motifs per language areas.	278
Table 11.8:	Engraved sites. Shield Design Types according to language areas and drainage basins. Sites with single design types only.	283
Table 11.9:	Engraving sites. Shields Distribution of mixed design types according to language areas and drainage basins. Type 2C, 4 and 5 varieties.	285
Table 11.10:	Engraving sites. Culture heroes: compositional details.	286
Table 11.11:	Daramulan and Biaime types, language areas and drainage basins.	288
Table 12.1:	Shelter Art sites: Clumped motif variables in Correspondence Analysis.	293
Table 12.2:	Analytical grouping of shelter art sites used in regional analysis.	293
Table 12.3:	Shelter Art sites: technique variables used in the CA.	295
Table 12.4:	Shelter Art sites (motif): Language areas, codes and sample sizes.	308
Table 12.5:	Shelter Art sites (motif): Drainage Basins, Language Areas and Samples.	309
Table 12.6:	Shelter Art Motif totals. Maximum motif incidence, number of sites in the region with motif present, and %f of sites with motif.	329
Table 12.7:	Rare Shelter Art Motifs: Distribution per Language Area.	331
Table 12.8:	Shelter art motifs. Significant values of rare motifs per language area.	331
Table 12.9:	Shelter Art sites. Culture heroes: compositional details.	337
Table 13.1:	Statistics for regional assemblage sizes, both art components.	346
Table 13.2:	Statistics for total number of motif categories recorded in both art components across the region.	346

Illustrations

Figure 1.1:	The study area. The Sydney Basin showing the extent of Hawkesbury sandstone, cities and major rivers.	2
Figure 2.1:	A typical upper tributary gully with a ridgeline in background. Six engraving sites were located in this gully.	7
Figure 2.2:	The Georges River to the south of the region. View to the west from Alford's Point.	7
Figure 2.3:	Engraving site on the ridge top overlooking Berowra Creek. This vast expanse of sandstone only had a single engraved motif on it.	8
Figure 2.4:	A relatively small boulder with engravings near Maroota, south of the Hawkesbury River. Mundoes and anthropomorphs are on this sloping surface.	8
Figure 2.5:	Large shelter with honeycomb weathering near Warre Warren Creek. Pigment art is located on a smooth surface at one side of this shelter.	9
Figure 2.6:	This small overhang near Wheelbarrow Ridge forms a lip in an extensive sandstone platform. Despite its small size, this shelter has a moderately large pigment art assemblage on its ceiling.	9
Figure 2.7:	Large free-standing boulders such as this one near Howes Valley provide ideal shelters for a range of assemblage sizes.	10
Figure 2.8:	This is the smallest recorded shelter in the region, near Warre Warren Creek. Three pigment motifs were found in this overhang.	10

Figure 3.1:	The four language areas defined for the region (after Capell 1970). 21
Figure 4.1:	Radiocarbon determinations from Sydney region, millennial sequence..............41
Figure 5.1:	Engraving Sites. Percentage frequency of different assemblage sizes................51
Figure 5.2:	Engravings in Ku-ring-gai Chase National Park. Male and female figures and fish positioned in a tessellated pavement....................47
Figure 5.3:	Engravings in Ku-ring-gai Chase. Fish and bird motifs positioned within and across natural tessellations. 47
Figure 5.4:	Large macropod engraving at Maroota, south of the Hawkesbury River. This macropod is being struck by two boomerangs. The row of pits crossing this motif traverses the entire assemblage. 48
Figure 5.5	Engraved shield motif near Berowra Creek. The internal design on this shield matches early ethnohistoric descriptions of the St George Cross................48
Figure 5.6:	Devils Rock Maroota. Recording engravings in daylight using large acetate mirrors. In foreground is a Biaime culture hero. 49
Figure 5.7:	Large emu with clutch of eggs. The emu tracks are in outline form unlike the pecked intaglio version often found in the region's rockshelters. 49
Figure 5.8:	Macropod motif, superimposed by a post-contact sailing ship............................50
Figure 5.9:	Engraved snake motif consisting of five parallel zig-zag lines. The individual peck marks are clearly visible. 50
Figure 5.10:	Shelter Art Sites: assemblage size information. 55
Figure 5.11:	Frieze of charcoal women and men beneath a red outlined wombat; hand stencils, white paintings and drawings and engraved macropod tracks 59
Figure 5.12:	Black outlined and infilled turtle with linear infill positioned in a scalloped alcove with a faded, infilled snake motif on the surface below 59
Figure 5.13:	Black infilled and white outlined dingoes and macropods on Yatala Creek near the Colo River. These motifs are large and affected by water seepage 60
Figure 5.14:	Swinton's shelter, with its heavily superimposed art. Stencils and drawings in red, black and white and a red horned anthropomorph........................60
Figure 5.15:	Black and white bichrome goannas (Phase 3) over red painted goannas (Phase 2) at the northern end of Swinton's shelter. 61
Figure 5.16:	Three Birds Site: Engraved bird and macropod tracks on shelf at back of shelter just above floor level. 61
Figure 5.17:	Cafe's Cave, Putty: near the northern limit of the study area. White painted complex-non-figuratives (CXNF), hand and axe stencils. 62
Figure 5.18:	Tic Alley: white painted anthropomorph and shield with internal design (2B). These motifs are painted over black drawings. 62
Figure 5.19:	Gunyah Beach, near the confluence of Cowan Creek and the Hawkesbury River. Stencilled fish tails. 63
Figure 5.20:	Boorai Creek, Wollemi National Park. Red stencilled kangaroo tails. This site also has stencils of boomerangs, hands and a kangaroo skin bag. 63
Figure 5.21:	Colo Heights. Panel of stencilled material objects including boomerangs, parrying shield, axe, woomera and large hand stencils........................64
Figure 5.22:	Swinton's shelter. Pink and bichrome stencils of boomerangs, and hand stencils...........64
Figure 5.23:	Motif preferences for the two art contexts. Motif classification (excluding unidentified motifs). 67
Figure 5.24:	Subject preference for the two art contexts. Clumped motif classes. 68
Figure 6.1:	Locality Map showing site context. 69
Figure 6.2:	Site Plan, Yengo 1. 72

Figure 6.3:	Cross-sections, Yengo 1.	73
Figure 6.4:	Engraved panel at the front of Yengo 1, showing the engravings and their (excavated) depth below the surface level.	73
Figure 6.5:	Yengo 1: grinding grooves - pecked and pounded areas.	75
Figure 6.6:	View upslope towards sites Yengo 1 and Yengo 2 (#'s 37-5-1 and 37-5-2). Sites are arrowed - Yengo 1 is on the left.	76
Figure 6.7:	Yengo 1. The engraved panel at front of shelter prior to excavation.	76
Figure 6.8:	Yengo 1. Interior view of shelter showing relationship between engraved boulders stencilled art on ceiling and deposit and square (1A).	77
Figure 6.9:	Yengo 1. Detail of stencils and white painted motifs on rear wall.	77
Figure 6.10:	Yengo 1. White and red hand stencils on rear wall of shelter.	78
Figure 6.11:	Yengo 1. White stencils and painted complex-non-figurative design on back wall of shelter.	78
Figure 6.12:	Yengo 1. Stencils. The range, median and variance (in cm).	80
Figure 6.13:	Yengo 1. Grinding grooves on sloping back wall.	82
Figure 6.14:	Yengo 1. Detail of axe found in undergrowth outside the shelter.	82
Figure 6.15:	Yengo 2. Site Plan.	85
Figure 6.16:	Yengo 2. Cross-sections.	86
Figure 6.17:	Yengo 2. Black and white possums and goannas over red motifs.	86
Figure 6.18:	Yengo 2. Black and white goanna, black drawn and white painted non-figurative motifs and red solid figure on southern side of shelter.	87
Figure 6.19:	Yengo 2. White hand stencil variations - wrist-wrist combination on ceiling of shelter, superimposed over red outline macropod.	87
Figure 6.20:	Yengo 1. Squares 1B, 4A and 5A, showing excavated units.	88
Figure 6.21:	Yengo 1. Squares 1A, 1B and 1R, showing excavated units.	89
Figure 6.22:	Yengo 1. Squares 2A, 3S, 3N and 6, showing excavated units.	89
Figure 6.23:	Yengo 1. Squares 2A, 2B, 3S and 6E, showing excavated units.	90
Figure 6.24:	Yengo 2. Completed excavation.	91
Figure 6.25:	Yengo 2. Squares 4B and 4C. Excavated spits and stratigraphy.	92
Figure 6.26	Stratigraphic section. Northern baulk of Squares 4A and 5A.	95
Figure 6.27:	Trench aligned with Square 1, showing site's internal stratigraphy.	95
Figure 6.28:	Yengo 1. Southern baulk, square 1C, showing buried boulder and engraved boulder.	96
Figure 6.29:	Yengo 1 Squares 2A, 2B, 3S and 6. Stratigraphic sections.	97
Figure 6.30:	Schematic section showing engraved boulder and charcoal samples.	99
Figure 6.31:	Yengo 1. Proportions of raw materials in the two analysis squares.	105
Figure 6.32:	Yengo 1. Proportions of raw materials in the two analysis squares. Artefacts >1cm.	106
Figure 6.33:	Yengo 1. Raw material proportions for backed blades, artefacts with R/U and cores. Both analysis squares.	107
Figure 6.34:	Yengo 1 Square 1B-4. Artefact Distribution (%f) by analytical unit.	108
Figure 6.35:	Yengo 1. Square 1B/4. Artefact density (/kg) through time.	108
Figure 6.36:	Yengo 1. Square 3S-6 Artefact Distribution.	109
Figure 6.37:	Yengo 1. Age depth curve for square 4A. Upper date from square 2B. Analytical units shown on vertical axis.	110
Figure 6.38:	Yengo 1. Artefacts accumulation per 100 years: square 1B-4.	111

Figure 6.39:	Correlation between sample size and raw materials; analytical units.	111
Figure 6.40:	Yengo 1. Vertical trends in raw material proportions. All artefacts.	112
Figure 6.41:	Vertical distribution of raw materials. Artefacts >1cm.	114
Figure 6.42:	Yengo 1. Vertical distribution of raw material per stratigraphic layer.	115
Figure 6.43:	Yengo 1. Proportions of size categories over time per stratigraphic layer. Cores and artefacts with R/U.	116
Figure 6.44:	Yengo 1. Vertical distribution of artefacts with R/U.	117
Figure 6.45:	Yengo 1. Changing raw material preferences over time. Artefacts with retouch/usewear. Both squares. Ground fragments excluded.	118
Figure 6.46:	Yengo 1. Vertical distribution of backed blades and ground fragments in both analysis squares. Stratigraphic layers indicated.	119
Figure 6.47:	Yengo 1. Raw Materials, backed artefacts through vertical sequence.	120
Figure 6.48:	Yengo 1. Vertical distribution of cores per analytical unit.	121
Figure 6.49:	Yengo 1. Changing platform characteristics of cores through time. Both analysis squares.	122
Figure 6.50:	Yengo 1. Vertical distribution of bone. Based on bone weights %f. Squares 3S, 3N, 4A and 6 (data from Steele 1994: Table A1.4).	123
Figure 6.51:	Yengo 1. Bone accumulation rates. Based on bone weights %f. Squares 3S, 3N, 4A and 6 (data from Steele 1994: Table A1.4; see Table A1.22).	124
Figure 6.52:	Yengo 1. The excavation grid layout and artefacts retrieved per pit.	126
Figure 6.53:	Yengo 1. Artefact densities and percentage frequency distributions of lithic material across the site.	127
Figure 6.54:	Yengo 1. Spatial distribution of backed artefacts and flakes and flaked pieces with retouch/usewear.	131
Figure 6.55:	Yengo 1. Spatial distribution of ground pieces and micro-debitage.	132
Figure 6.56:	Yengo 1. Spatial distribution of bone fragments and distribution of bone densities.	133
Figure 7.1:	Locality Map showing Great Mackerel rockshelter and its context.	143
Figure 7.2:	The Great Mackerel rockshelter. View from the north-east.	144
Figure 7.3:	The Great Mackerel rockshelter. Art assemblage. Faint white hand stencil in varying sizes.	146
Figure 7.4:	The Great Mackerel rockshelter. Stencil composition with feet, hands and digging sticks.	146
Figure 7.5:	The Great Mackerel rockshelter, Panel A. Older exfoliated red hand stencils are superimposed by white stencils.	147
Figure 7.6:	The Great Mackerel rockshelter. Site Plan.	147
Figure 7.7:	The Great Mackerel rockshelter. Cross-sections.	148
Figure 7.8:	Excavating in square 6, provenancing shell and collecting samples.	148
Figure 7.9:	The Great Mackerel rockshelter. Squares 6, 6A and 6B at the completion of excavation. View to the western baulk from the east-north-east.	150
Figure 7.10:	The Great Mackerel rockshelter. Square 8 excavated to bedrock. View of southern baulk.	151
Figure 7.11:	Excavated spits, Squares 6 and 8.	151
Figure 7.12:	Stratigraphic sections, Squares 6 and 8.	155
Figure 7.13:	Age depth curve based on square 6 spit depths.	156
Figure 7.14:	Artefact and cultural material accumulation rates per 100 years	157
Figure 7.15:	Dominant shellfish species in the midden layer. Squares 6 and 8.	160

terra australis 27

Figure 7.16:	Great Mackerel: Artefact with retouch/usewear. Raw materials.	165
Figure 7.17:	Raw Material proportions in the Midden and Artefact Layers.	166
Figure 7.18:	The Great Mackerel rockshelter. Excavated complete fishhooks.	167
Figure 7.19:	Great Mackerel shell scrapers. Examples shown front and back.	167
Figure 8.1:	Locality Map. UDM Shelter in its local context.	171
Figure 8.2:	UDM Shelter. Site Plan.	173
Figure 8.3:	UDM Shelter. Site Cross-sections A - D.	174
Figure 8.4:	View of the internal space of the rockshelter, facing north-east.	174
Figure 8.5:	UDM Shelter. The three engraved male human figures.	175
Figure 8.6:	Panel 2 (above) and Panel 3. Incised macropods on far right of Panel 2 (detail Figure 8.10). The current floor level is c. 1m below Panel 3.	177
Figure 8.7:	Pecked Upside-down-man beneath Panel 1. Note the charcoal infill in the hands and feet.	177
Figure 8.8:	UDM Panel 4. White painted blobs, bird tracks and white hand stencils.	178
Figure 8.9:	UDM Panel 5. Red anthropomorphs (one with white outline) beneath white outlined macropods.	178
Figure 8.10:	UDM Panel 2. Detail of incised macropods over black charcoal macropods.	179
Figure 8.11:	Panel 3 showing context of incised panel and the macropod #2 in the alcove (arrowed right) which was sampled for AMS dating.	182
Figure 8.12:	UDM Shelter. Macropod motif from which AMS samples were collected.	182
Figure 8.13:	UDM Shelter. Excavated spits – all squares.	185
Figure 8.14:	Square 4, excavation in progress. Note the marked colour contrast between the reddened ashy lens below Layer 1.	186
Figure 8.15:	Square 8, excavation to bedrock. The colour differences in the different strata are quite clear.	187
Figure 8.16:	UDM Shelter. Stratigraphic sections, Squares 4 and 8.	188
Figure 8.17:	UDM Shelter. Stratigraphic Section, Square 6.	188
Figure 8.18:	UDM. Cultural components in the three analysed squares over time.	190
Figure 8.19:	The bifacially flaked basalt axe pre-form collected by Pat Vinnicombe	191
Figure 8.20:	UDM. Raw material proportions in the three analysed squares.	194
Figure 8.21:	UDM. Raw material proportions in the three analysed squares.	194
Figure 8.22:	Artefact densities in the three analysed squares showing locations of the four radiocarbon dates.	195
Figure 8.23:	UDM shelter. Age depth curve and artefact accumulation rates by analytical and stratigraphic units.	196
Figure 8.24:	Correlation between sample size and number of raw material types per analytical unit.	196
Figure 8.25:	UDM Shelter. Vertical distribution of raw material. All artefacts.	197
Figure 8.26:	UDM Shelter. Vertical distribution of raw material per analytical unit.	198
Figure 8.27:	UDM. Raw material proportions plotted against sample size.	198
Figure 8.28:	UDM Shelter. Platform, flaked piece and core types over time.	202
Figure 8.29:	UDM Shelter. Vertical distribution of artefacts with retouch usewear, grinding and backed blades.	203
Figure 8.30:	UDM Shelter. Modified artefacts and raw materials: analytical units.	204
Figure 8.31:	UDM Shelter. Proportions of modified material per analytical unit.	204

Figure 8.32: Reworked UMCC data. Local rates of artefact accumulation in successive periods........207

Figure 8.33: UMCC reworked data. Rates of habitation establishment and habitation use in the seven sites with >100 excavated artefacts. 209

Figure 8.34: Mangrove Creek Valley. Rates of artefact accumulation in successive periods. Attenbrow's seven large sites plus UDM 212

Figure 8.35: Mangrove Creek Valley. Rates of habitation establishment and habitation use in the eight sites (including UDM) with >100 excavated artefacts 213

Figure 9.1: The four shelter art sites excavated for this research. 219

Figure 9.2: Dated shelter art sites showing length of occupation and period of most intensive artefact accumulation. 227

Figure 10.1: Results of the superimposition analysis. 239

Figure 10.2: Mangrove Creek shelter sites. Motif preferences in the three art phases.........................240

Figure 10.3: Number of sites demonstrating the three identified phases and cumulative frequency of sites with particular phases in use. 240

Figure 10.4: Mangrove Creek shelter CA results. Scree slope plot of latent roots...............................241

Figure 10.5: Mangrove Creek CA Results. Bivariate plot of component scores. Technical variables..243

Figure 10.6: Mangrove Creek CA Results. Bivariate plot of component scores. Technical variables..246

Figure 10.7: Bivariate plot of component scores, Mangrove Creek. Small, medium and large-very large sites plotted separately. 247

Figure 11.1: CA Scores: entire Engraving component (705 sites). Each dot represents many sites. 254

Figure 11.2: Engraving CA results: Plot of the latent roots indicating that the variance in the data set is well accounted for by the first two components. 255

Figure 11.3: CA results: Engraving Sites. Bivariate Plot of Variable Scores. 256

Figure 11.4: Quadrant labels used in the following discussion of the CA results. 256

Figure 11.5: Percentage of homogeneous engraving sites in each analytical Group. 257

Figure 11.6: Bivariate plot of sample size and number of motifs recorded per sample area. The five language areas and two randomly generated *Guringai* samples..........................258

Figure 11.7: Darkingung Language Area. Motif Assemblage. 260

Figure 11.8: Darug Language Area. Motif Assemblage. 260

Figure 11.9: Guringai Language Area. Motif assemblage..261

Figure 11.10: Sydney (*Eora*) Area: Motif Assemblage. 262

Figure 11.11: Tharawal Area: Motif Assemblage..262

Figure 11.12: The 25 drainage basins defined across the Sydney region. 265

Figure 11.13: Bivariate plots, CA results. Darkingung language area - Upper Macdonald, central Macdonald and Mangrove Creek drainage basins. 266

Figure 11.14: Cattai Drainage Basin. Bivariate plot of CA results..268

Figure 11.15: Berowra Drainage Basin (*Darug* and *Guringai* Language Areas). Bivariate plot of CA results...268

Figure 11.16: Cowan Drainage Basin. Bivariate plot of CA results..269

Figure 11.17: Pittwater Drainage Basin. Bivariate plot of CA scores. 269

Figure 11.18: *Tharawal* Language Area (Woronora and Port Hacking Drainage Basins). Bivariate Plot of CA Scores. 271

Figure 11.19: Two examples of vertical engraving sites from Berowra Creek (top) and Cowan Creek (bottom). 272

Figure 11.20: Smith's Creek, Ku-Ring-Gai Chase National Park. The red stencils (arrowed and inset) in this shelter must have been produced by artists standing in a canoe at high tide. Photo taken at low tide. 273

Figure 11.21: Shelter on Cowan Creek, the floor of which is in the littoral zone. An engraved fish (inset) is located on the interior floor surface (arrowed). 273

Figure 11.22: Motif histogram for engraving sites in ridgetop, hillslope and vertical engravings in estuarine valley bottom..274

Figure 11.23: CA bivariate plots according to topographic location. *Guringai* language area: Berowra and Cowan drainage basins. 276

Figure 11.24: Engraving component. Sites at which particular motifs appear. 277

Figure 11.25: Distribution of sites with engraved women and profile people..........................279

Figure 11.26: Distribution of engraving sites with culture heroes and shield motifs...............280

Figure 11.27: Distribution of sites with engraved axe and macropod tracks. 281

Figure 11.28: Distribution of engraving sites with CXNF and contact motifs.........................282

Figure 11.29: Site 575 with 15 shield motifs and three boomerangs. 283

Figure 11.30: Range of shield designs present in the engraved component.............................284

Figure 11.31: Distribution map of main shield designs..285

Figure 11.32: Distribution of Biaime and Daramulan type engraved motifs. 288

Figure 11.33: Engraved Culture Heroes. Daramulan motifs which appears to have been altered over time. 289

Figure 11.34: Paired and/or transitional *Daramulan* and *Biaime* motifs. Note that *Daramulan* below (on #45-2-16) has an altered outline also...............................290

Figure 12.1: CA results. Scree slope plot of the latent roots showing that the variance is well accounted for by the first two components...293

Figure 12.2: CA results, shelter motifs. Component scores a) motifs and b) sites. 294

Figure 12.3: Quadrant identification used in interpreting the shelter art CA results. 294

Figure 12.4: CA results Technique. Plot of the latent roots demonstrating that the variance in the data set is well accounted for by the first two components....................295

Figure 12.5: CA Results for Technique. Plot of component and eigen scores. 296

Figure 12.6: Shelter Art, Motif. Homogeneous shelter sites in each analysis group.............297

Figure 12.7: Shelter Art, Technique. Homogeneous shelter sites in each group. 297

Figure 12.8: *Darkingung* Language Area, Motif Assemblage..298

Figure 12.9: *Darkingung* Language Area: depictive motifs. 299

Figure 12.10: *Darkingung* Language Area. Techniques employed. 299

Figure 12.11: *Darkingung* Language Area depictive motifs. Form..299

Figure 12.12: *Darkingung* Language Area. Colour usage...299

Figure 12.13: Darug (North) Language Area. Motif assemblage...300

Figure 12.14: *Darug* (North) Language Area. Depictive Motifs...300

Figure 12.15: *Darug* (North) Language Area. Technical options employed. 301

Figure 12.16: *Darug* (North) Language Area depictive motifs. Form..................................301

Figure 12.17: *Darug* (North) Language Area. Colour preferences......................................301

Figure 12.18: *Darug* (South) Language Area. Motif Assemblage..301

Figure 12.19: *Darug* (South) Language Area. Depictive Motifs..302

Figure 12.20: *Darug* (South) Language Area. Technical options employed. 302

Figure 12.21: *Darug* (South) Language Area. Form. 302

Figure 12.22: *Darug* (South) Language Area. Colour preferences...302

Figure 12.23: *Guringai* Language Area. Motif Assemblage. 303

Figure 12.24: *Guringai* Language Area. Depictive Motifs...303

Figure 12.25: *Guringai* Language Area. Technical options employed. 303

Figure 12.26: *Guringai* Language Area depictive motifs. Form. 304

Figure 12.27: *Guringai* Language Area. Colour preferences. 304

Figure 12.28: *Eora* Language Area. Motif Assemblage. 304

Figure 12.29: *Eora* Language Area. Depictive Motifs. 305

Figure 12.30: *Eora* Language Area. Technical options employed. 305

Figure 12.31: *Eora* Language Area depictive motifs. Form. 305

Figure 12.32: *Eora* Language Area. Colour preferences. 305

Figure 12.33: *Tharawal* Language Area. Motif Assemblage..306

Figure 12.34: *Tharawal* Language Area. Depictive Motifs..306

Figure 12.35: *Tharawal* Language Area. Technical options employed. 306

Figure 12.36: *Tharawal* Language Area depictive motifs. Form. 307

Figure 12.37: *Tharawal* Language Area. Colour preferences. 307

Figure 12.38: *Darkingung* Language Area. Bivariate plot of CA scores: motifs. 309

Figure 12.39: *Darkingung* Language Area: Motif. Bivariate plots for the three drainage basin groupings...310

Figure 12.40: *Darug* and *Guringai* Language Areas: Motif. CA scores..312

Figure 12.41: *Darug* drainage basins: motifs. Bivariate plot of CA scores..313

Figure 12.42: *Guringai* drainage basins: motif. Bivariate plot of CA scores. 314

Figure 12.43: *Tharawal* and *Darug* language areas: motif. Bivariate plot - CA scores. 316

Figure 12.44: *Tharawal* and *Darug* drainage basins: motif. Bivariate plot - CA scores..........................317

Figure 12.45: *Tharawal* drainage basins: motif. Bivariate plot of CA scores. 318

Figure 12.46: *Darkingung* language area: technique. Bivariate plot of CA scores. 320

Figure 12.47: Darkingung drainage basins: technique. Bivariate plot of CA scores.................................320

Figure 12.48: *Darug* and *Guringai* language areas: technique. Plot - CA scores....................................322

Figure 12.49: Cattai and Berowra drainage basins: technique. Plot - CA scores. 323

Figure 12.50: Pittwater and Cowan Drainage basins: technique. Plot of CA scores................................324

Figure 12.51: *Darug* and *Tharawal* Language Areas: technique. CA scores. 326

Figure 12.52: *Darug* Georges and Mill/Williams drainage basins. Plot of CA scores.............................327

Figure 12.53: *Tharawal* drainage basins: technique. Bivariate plot of CA scores. 328

Figure 12.54: Occurrences of sites with particular motifs..330

Figure 12.55: Pigment art sites. Shelters with women and profile people motifs. 332

Figure 12.56: Pigment art sites. Shelters with culture heroes and shield motifs. 333

Figure 12.57: Pigment art sites. Shelters with human feet and macropod tracks.....................................334

Figure 12.58: Pigment art sites. Shelters with axes and contact motifs...335

Figure 13.1: Model for territorial organisation and interaction across the Sydney region......................351

Abstract

This monograph is based on PhD research completed at the Australian National University in 1994. The research examines prehistoric rock art from the Sydney region in coastal south-eastern Australia. The rock art occurs in two distinct contexts provided by the sandstone bedrock which defines this region. Engraving (or petroglyph) sites occur in open locations on horizontal platforms. In rockshelter locations there is pigment art (drawings, stencils and paintings) and occasionally engravings.

The principal aim of this research was to define a model for cultural interaction to describe a prehistoric art system. Information exchange theory provided the basis for this proposed model. By perceiving 'style' from a functional perspective the region's art was seen as a conduit for the expression of social affiliations. The concept of social context, e.g. public versus private, has been extremely important in developing this argument. So has the notion that style is a means of non-verbal communication used to negotiate identity.

Varying levels of stylistic heterogeneity reveal different types of social information. Higher levels of stylistic homogeneity in prehistoric art can be interpreted in terms of larger-group cohesion. Higher levels of heterogeneity are interpreted as demonstrating local-group identifying behaviour. In the Sydney region, complex patterns of variability in both art contexts demonstrate the nature of the contacts between language groups, as well as areas where the stresses resulting from these contacts may have been the greatest.

Patterns in stylistic variability were explored with the effects of medium, diachronic change and synchronic variability all considered. The contemporaneity of art and occupation evidence was explored in three decorated rockshelters excavated for this research. General regional patterns were also investigated.

The rock art in the Sydney region functioned as a prehistoric information superhighway. Through stylistic behaviour, groups around the region who were not in constant verbal contact with each other were able to communicate important social messages and demonstrate both broad-scale group cohesion and within-group distinctiveness. Throughout the Sydney region people have signalled information about themselves making interaction more predictable during a period of substantial social change.

Acknowledgements and Preface

'No (wo)man is an island'.

This research would not have been possible without the assistance of many people. I would particularly like to thank my PhD supervisor, Andrée Rosenfeld, for her support, friendship and inspiration over the years. My advisers, Isabel McBryde and (the late) Anthony Forge also generously imparted experience and wisdom.

A large amount of excavation and rock art recording fieldwork was undertaken, assisted by many of my professional colleagues and post-graduate peers. My thanks go to all of these people.

The Great Mackerel site was excavated with the help of Terri Bonhomme, Neville Baker, Dominic Steel and Jon Barlow. Bronwyn Conyers (then NSW NPWS) and Harry Mumbulla (then Metropolitan Local Aboriginal Land Council) visited the site during excavation. Sharon Sullivan (then NPWS Regional Director) granted permission to excavate within a National Park: Geoff Spencer (then Superintendent, Ku-ring-gai Chase NP) and Alex Kneer (then KCNP Ranger) allowed us to camp at The Basin and to use fire-trails not otherwise open for general use.

During the two field seasons at *Big Yango*, I was assisted by John Appleton, Val Attenbrow, Sally Brockwell, John Clegg, Hilary du Cros, Sarah Forbes, Neil Hanson, Margrit Koettig, Judy McDonald, Andrew McWilliam, Helen Marshall, Kelvin Officer, Tony Ogden, Ann O'Gorman, Andrée Rosenfeld, Annie Ross, Somsuda Rutnin, Jonathan Saunders, Laurajane Smith, Kate Sullivan, Paul Taçon, Marina Walkington, and Vivienne Wood. Helen Brayshaw, Ann Conway, Dave Lambert, Ian Webb and Richard, Sonia and Ben Wright also trekked along the Boree Track to visit. Excavation at *Big Yango* would not have been possible without the permission of John Bowen. Thanks to the Bowen family and to property managers, Steve and Janine Marsh for their hospitality, assistance and good humoured tolerance.

Val Attenbrow, Tessa Corkill, Denise Donlon, John Edgar and Judith Field assisted in the excavation at Upside-Down-Man. Dave Lambert assisted in the collection of the art charcoal samples from this site for AMS radiocarbon dating.

Rock art recording work along the Boree Track was undertaken with the assistance of Mary Dallas and David Bell. Frying Pan Rock will never be the same!

Much art recording work was undertaken during the Sydney Basin Rock Art Project, funded by three National Estate Grants. Thanks to Laurajane Smith, Annie Ross, Bronwyn Conyers and Warren Bluff (Stage II) and Mary Dallas, Paul Taçon, Steve King, Jim Hatfield and Ken Cutmore of Metropolitan Local Aboriginal Land Council (Stage III). The NSW NPWS provided a helicopter for recording work in the remote north-west of the region, and a boat for work along Berowra and Cowan Creeks. Mary Dallas provided accommodation aboard the *Cruisin' Bokkus Club* during the Berowra Waters stint.

Many art sites were recorded in the Warre Warren Aboriginal Place, funded by the NSW Forestry Commission. Assistance here was provided by Kelvin Officer, Jan Klaver, Dave Lambert, Ingereth McFarlane and Kerry Navin.

David Macgregor (ANU) assisted tirelessly and cheerfully in the preparation of the many field trips. Field funding for the excavations at Mount Yengo and Great Mackerel was provided by the Australian National University.

Ingereth McFarlane and Beth White assisted with the sorting of some lithics (Yengo 1 and UDM). Phil Boot and Huw Barton commented on some artefacts identified with resin/blood/plant residue. Dominic Steele analysed the faunal remains from Yengo 1. Seeds from UDM and Yengo 1 were identified by Nena Panich of the Seeds Laboratory, NSW Agriculture Department.

Conventional radiocarbon dates were provided by the ANU Quaternary Research Centre. Some AMS dates were provided by the Arizona University AMS Facility, others (UDM) were provided by the new facility at ANSTO, Lucas Heights. An Australian Museum Grant-in-Aid paid for two of the UDM shelter AMS dates. John Head (ANU Quaternary Research Centre) provided advice in all matters charcoal.

Ross Cunningham and Christine Donnelly (Faculties' Statistics Department ANU) provided statistical guidance. The advice of Richard Wright in this regard was (as usual) invaluable. Figures 1.1, 2.1, and 3.1 were drawn by Win Mumford; Figures 6.15, 6.16, 7.6 and 7.7 were drawn by Joan Goodrum. All other illustrations and photographs were created by the author.

My years in residence in Canberra were a stimulating time. Discussions over cappuccinos, beers or the Wash-and-Sort Lab sink with Kim Akerman, Debbie Bird-Rose, Sally Brockwell, Denis Byrne, Sarah Colley, Annie Clarke, Barry Cundy, Sarah Forbes, Darrell Lewis, Ingereth Macfarlane, Kelvin Officer, Colin Pardoe, Somsuda Rutnin, Paul Taçon and Vivienne Wood enriched the intellectual exercise and enlivened the social sphere.

Various friends and colleagues generously read and commented on parts of the original thesis: Val Attenbrow, Helen Brayshaw, Peter Douglas, Ian Lilley, Beth White and Annie Ross. Thank you also to Peter Veth for commenting on parts of this monograph, for reactivating and encouraging my participation in rock art research and for always being stimulating! *Gros Bises*!

The original thesis was originally dedicated to my family (by descent and marriage) - Neil, Oliver, Jesse, Judy, Laurie, Simon and Katie. The conversion of this thesis into a monograph has again required the encouragement, support and co-operation of my (now transformed) residence group. This version is dedicated to Peter, Oliver, Jesse and Zoë – with additional thanks to Sue O'Connor and Andrée Rosenfeld for making me do it!!

Jo McDonald
Canberra

A note on the title and dates:

'Information superhighway' is a now-obsolete term that was used to describe the future of what existed up until the mid-1990s as 'the Internet' (//www.Wikipedia.org/). 'Dreamtime Internet' doesn't have quite the same implication as I had originally intended – so I have retained this now out-dated phraseology in the title of this work.

All the radiocarbon dates returned during this research were uncalibrated. Although more recent research in the region, including that undertaken under the auspices of consulting archaeology, has been calibrated, I have not undertaken a recalibration exercise (McCormac et al. 2004). This is justified by the fact that Attenbrow's (2004) publication of Mangrove Creek was not calibrated and many dating aspects discussed here relate to her data.

1

INTRODUCTION

This monograph derives from doctoral research undertaken at the Australian National University, completed in 1994. This work focused on the prehistoric rock art of the Sydney Basin, with particular focus on placing this in its social context. The art in this region occurs as two quite separate media in different physical locations. Engraving (or petroglyph) sites are found on open sandstone platform while in rock shelter locations the art is predominantly pigment (pictograph) art – which has been drawn, painted and stencilled. The art is generally described as being of the Simple Figurative Style (after Maynard 1977). The aim of the thesis was to explore the sources of stylistic variability in the region's two art media. Diachronic and synchronic variation was investigated, as were the effects of medium and site context. The interrelatedness of art and archaeological evidence was also considered, as was the effect of linguistic boundaries across the region. The contemporaneity of art and other archaeological evidence in shelter sites was explored as a means of testing assumptions about the age of the art. Some direct dating of charcoal motifs was also undertaken. This work was also important for developing models about how the two art components may have functioned across this region. A regional model founded in Information Exchange Theory was proposed.

Two underlying assumptions directed this research. These were, that;

1) the majority of the art production coincides with the archaeologically most visible period of late Holocene occupation, known as the Bondaian. Most of it, therefore, is younger than c.5,000 years old. Art was produced up until European contact, but production ceased at or soon thereafter;   and,

2) the two art components were practiced contemporaneously across the region.

The Sydney Basin is defined geologically and the study area for this research was restricted to the Hawkesbury sandstone formation (Figure 1.1). The homogeneity of the sandstone medium across this formation means that the region's boundaries can be defined objectively, if arbitrarily from a cultural perspective.

For this research, 'style' was defined as the particular way of doing or producing material culture which signals the activity of a particular group of people who distinguish themselves from other, similarly constituted groups. The preferred application of this definition was that used by advocates of information exchange theory whereby style is 'that part of the formal variability in material culture that can be related to the participation of artefacts in the processes of information exchange' (Wobst 1977:321). Style is non-verbal communication which negotiates identity (Wiessner 1990:107).

It is generally recognised that hunter-gatherer behaviour is regionally embedded (Wobst 1983:222). The rock art from the Sydney Basin is recognised as being a distinctive regional style. What makes it distinctive and what distinguishes it from other regional styles has been addressed previously (e.g. Franklin 2004). The apparent variations within this region and possible causes for these was the focus of this research. However, the very presence of this regional style and the fact that it has a recognisable extent is of interest.

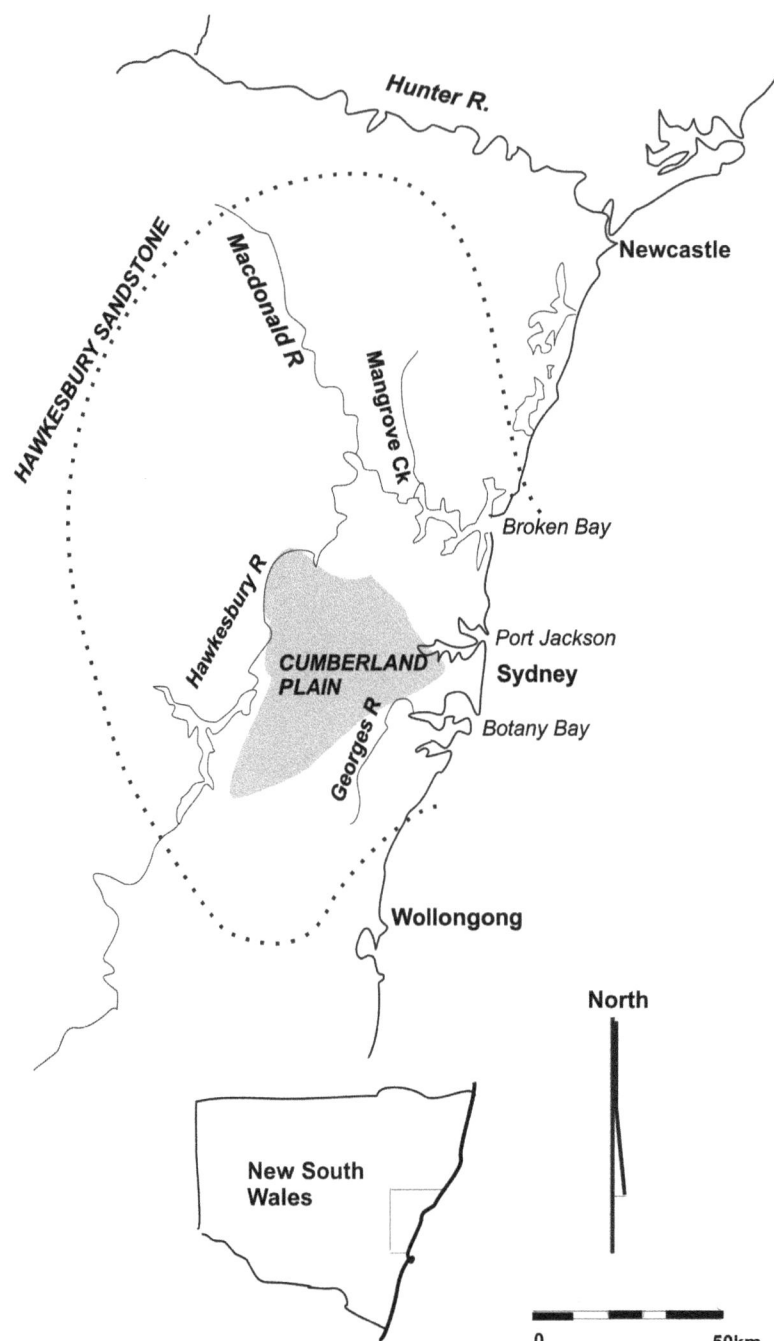

Figure 1.1: The study area. The Sydney Basin showing the extent of Hawkesbury sandstone, cities and major rivers.

What was it about the socio-political culture in the Sydney Basin that led to the proliferation of artistic behaviour? Was the region so resource rich, that the proliferation of artistic activity signals an extraordinary amount of leisure time (Jones 1977)? Was its resource richness the cause of an increased population making it necessary to mediate social interaction across this area, particularly in the late Holocene, post sea-level rise? Why is this regional rock art style an 'island' surrounded by a comparative artistic vacuum? Does this geographically extensive and yet relatively homogeneous art body demonstrate the widespread operation of social mechanisms particularly social identity signalling behaviour?

The Sydney region has a long history of archaeological practice and there is art and archaeological evidence available at many scales of analysis (e.g. Attenbrow 2002, 2004; Hiscock 2003; McCarthy 1979; McMah 1965). There is a proliferation of material with which one can address regional questions (Conkey 1987, Wobst 1983). Small scaled intra and inter site analyses and more detailed local catchment patterns (e.g. Attenbrow 1987, 2004; Attenbrow and Negerevich 1984; McDonald 1986a, 1988a; Officer 1984; Smith 1983) and a proliferation of cultural heritage management project on Sydney's Cumberland Plain in the decade since this dissertation was completed (e.g. Bonhomme Craib and Associates 1999; McDonald and Rich 1993; JMcD CHM 2005a, 2005b, 2006) provide evidence for localised variability, and complement the approach used in this research, which was generally relatively coarse grained and aimed at identifying broader patterning. By identifying localised variability regional patterning was able to be quantified and better understood. Environmental evidence for the Sydney region throughout the Holocene indicates that conditions were fertile and relatively stable. The ethnohistoric literature suggests that the population densities at contact were high. Archaeological evidence suggests these have probably been high throughout the period that most of the art was being produced, i.e. for the last 3,000 years.

It is not known how rock art functioned across this region at European contact. It is known that stylistic behaviours demonstrating social group affiliation was evident amongst the populations in the region at contact. Many of these social practices have not survived in the archaeological record - body scarification and/or painting, tooth avulsion, shield designs and other geographically specific items of (organic) material culture. The presence of these 'portable' forms of stylistic behaviour in the region at contact demonstrate how important group identifying behaviour probably was – which is of course equally clear in the region's rock art.

Previous work on the Sydney rock art – the research context
While the presence of rock art in the Sydney region was recorded by Governor Phillip soon after the First Fleet arrived in 1788, it was another hundred years before systematic recording of the art began. There is no record of questions being asked during the contact period of the Sydney inhabitants either about the function or meaning of the art that was observed. This has resulted in a regional art body without anthropological or social context: one for which the meaning cannot be interpreted except by archaeological methods. The first major publication of rock art recording the Sydney region was by the surveyor W.D. Campbell (1899), and in the first half of the twentieth century more intensive art recording activity was undertaken, with Fred McCarthy being the most prolific publisher (see bibliography) over a forty year period. Ian Sim was another prolific Sydney rock art recorder, as was John Lough. Interpretations of the Sydney engravings throughout this time were based loosely on borrowed ethnographic material from other regions (Elkin 1949; McCarthy 1956, 1961). No systematic analysis of this body of art was undertaken until Lesley Maynard (then McMah) undertook her Honours research in the Sydney region (McMah 1965). This was the first quantitative analysis of any rock art province in Australia, with 285 engraving sites being analysed with the aim '(to) produce, first a typology of the engravings, and second, a spatial distribution of traits, based on the typology' (McMah 1965:7).

A number of Sydney University Bachelor's degree projects, under the supervision of John Clegg, have been directed at more specific stylistic questions. Tania Konecny's (1981) work looked at both engraved and pigment art. The aim of this research was to investigate the possibility of aggregation locales (viz. Conkey 1980) within the Sydney area. Laurajane Smith's work was concerned with identifying archaeological patterning across ethnographically reported tribal boundaries (Smith 1983). Her analysis focused on the Mangrove Creek/Macdonald River area and involved both art components. Kelvin Officer's (1984 ANU) research explored the art south of the Georges River (SPG 1983). The main aim of this research was to explore and describe the formal variability at a local scale. Natalie Franklin's (1984) work explored Maynard's (1976) definition of the Simple Figurative styles, and thus compared the Sydney region with regional assemblages from Port Hedland, Cobar, south-east Cape York and the Grampians. Subsequent to this doctoral research there have been several other rock art projects in the region. Samantha Higgs (2003) looked at the distribution and morphology of 'culture heroes' in the engraved art; while the art of the Illawarra (at the south extent of the region) is the subject of current doctoral research (Julie

Dibden, in prep.). Another ANU project focussed on the geochemical characterisation of pigment samples from the southern Woronora plateau rock art (Ford 2006); while doctoral research on religious sites in the Blue Mountains has also been completed (Kelleher 2003). The ongoing investigation of rock art in the Blue Mountains World Heritage area (Taçon *et al*. 2005, 2006) is an exciting rock art project that has also commenced more recently. This work has demonstrated that the *Darkingung* pigment style continues further west into the Blue Mountains than was previously known.

The Sydney Basin Rock Art Project (SRAP), funded by the NSW National Parks and Wildlife Service and the National Estate Grant Project (McDonald 1985a, 1987, 1990a) formed the basis for the regional recording achieved by this research, and indeed created the impetus for this research project, which commenced in 1987. Other than this management project, little art research in the region has been achieved through cultural heritage management work. This is partly because of overriding time constraints and partly because the majority of urban development in the Sydney region has occurred on the shale-based Cumberland Plain (see Figure 1.1) where there is no rock art. In several instances, consultancy projects have achieved a research oriented outcome with rock art sites (McDonald and Smith 1984, McDonald 1988a). Several specific management plans have also been prepared for shelter art sites: Blackfella's Hands Cave (JMcD CHM 1999) and Whale Cave (JMcD CHM 2002a).

Sampling issues
The data used throughout this research were based on archival materials lodged in the Sites Register of the NSW National Parks and Wildlife Service (now known as AHIMS in the NSW Department of Environment and Climate Change). These records were initially generated both by early publications (particularly by Campbell, McCarthy and Sim) and updated by rock art management effort. The data set was audited during the initial stage of the SRAP (McDonald 1985a) and groundtruthed in the subsequent fieldwork stages (Stages II and III: McDonald 1987, 1990a). The samples comprise either the art assemblages of sites which had been recorded in sufficient detail (i.e. their entire assemblages had been drawn) by other recorders, or those which have been recorded by me using the classification system devised for the Rock Art Project. These are not random samples: they have inherent biases and inconsistencies.

The samples used for these analyses comprised 717 engraving sites and 546 shelter art sites. These represented 39.5% and 32.7% (respectively) of the known sites in the region at that time. Of the samples used, 167 shelter art sites (30.6%) and 61 engraving sites (8.5%) were recorded by me. Many biases in this sample were identified early in the Rock Art Project (McDonald 1985a) and these were addressed by subsequent Stage's fieldwork, where the reliability of the recordings made by previous recorders was assessed.

Most of the problems inherent in these samples result from the way that the NPWS Sites Register accrued over time, and particularly during its formative years. The unreliability of many site locations is one problem which can often only be resolved by field relocation. Variation in recorder competence and consistency over time is another problem. Also, over the years, the definition of a site has undergone considerable change. What McCarthy described as a 'group' in the 1940's may have stretched over several kilometres of ridgeline - and today is described as ten sites.

More than twelve weeks of rock art recording fieldwork was completed in the course of this research to record sites and to assess these problems. A detailed description of procedures used can be found elsewhere (McDonald 1990b). Only the drawn recordings of other recorders were used here: written descriptions alone were excluded. This was justified on the basis of the Stage II results (McDonald 1987) which demonstrated that written descriptions often underestimated assemblage size by as much as 400%.

The unsystematic way that the samples were originally collected is one problem could not be addressed. Only a relatively small number of sites has resulted from the few (small) areas systematically surveyed (e.g. Attenbrow 1981, 1987; Attenbrow and Negerevich 1981; Gunn 1979; Sefton 1988; McDonald 1986a, 1988a; McDonald and Smith 1984). These biases cannot be quantified and this is a recognised shortfall in the data base.

The analysis – rock art as stylistic messaging
The pigment and engraved art components appear to have functioned in different ways and to have presented different opportunities for projecting personal and group identity. The social context of the art's production is considered to account for these differences. Based on the precepts of Information Exchange Theory (Wobst 1977), it is anticipated that the art which was viewed by a larger proportion of the population and which is situated in a public context, would be expected to contain the best potential for stylistic messaging of an *emblemic* kind. Art produced in a domestic context appears to have allowed the opportunity for individual stylistic expressions (Wiessner 1990), resulting in the development of localised stylistic trends. The shelter art sites most of which have concomitant evidence for occupation are located within valley systems. These are demonstrably also living sites. All members of a local territorial group (men, women and children) will have used these sites. Local groups operating in their own territories are least likely to be communicating a politically motivated need for large scale group cohesion. The art in this context, thus, while not defining the locations of boundaries *per se*, contains the best potential for stylistic bounding information.

Most of the region's large engraving sites are on ridgelines, away from the economic resource areas and often not in the centre of any particular group's foraging territory (see also Layton 1989, 1992b). They are often on access routes around the region. They are rarely associated with other forms of occupation evidence (although notably, few such locations have been subject to subsurface investigation). It has been argued that the engraving sites in Sydney fulfilled a ritual function (Elkin 1949, McCarthy 1956, 1959a), and McCarthy (1961) also suggested that shelter art sites fulfilled a similar role. This interpretation was based on the presence of particular motifs at art sites, and by ethnographic analogy with Bora ceremonies from the north coast of NSW[1]. While there is no ethnographic evidence for this interpretation, it is suggested that engraving sites may have fulfilled a regional bonding function and provided the opportunity for large scale group cohesion. This may in part have been achieved by a ritual function.

There is regional evidence for gender exclusivity in economic and ritual behaviour and for inter-group participation in some ceremonies, particularly for male initiation. If any particular group, i.e. men or women, used these sites for ceremonial purposes, it would be expected that these sites would be viewed by particular and more limited section of the population. The audience was restricted, not public. Stylistic patterning with this art form then should be clinal within and between social units. Inter-group aggregation for the purposes of ceremony would similarly have resulted in a mixing of social messages. Given that these sites are often on recorded access routes, it is possible that generic social group identity messaging was further reinforced in these locations. Such a usage would contribute to a homogenizing of boundary information in this component.

---

[1] It is notable that none of the Sydney ethnohistoric or later ethnographic literature describes local initiation ceremonies including the production of rock art (see below).

2

THE SYDNEY BASIN - ENVIRONMENTAL CONTEXT

The Sydney region is located on the south-east coast of Australia between the coastline and the Great Dividing Range (Figure 1.1). The Sydney Basin is defined geologically and the study area for this research is restricted to the Hawkesbury sandstone formation which is the surface bedrock in the centre of the Greater Sydney Basin. Sydney is located towards the centre of the study area, and the cities of Newcastle and Wollongong roughly define its northern and southern extents. The Hawkesbury sandstone covers an area of approximately 190km x 90km - 17,100 square kilometres.

Throughout this work the study area will be called the Sydney region or the Sydney Basin interchangeably, this meaning the geographic extent of the Hawkesbury sandstone within the Greater Sydney Basin.

Geology of the Sydney Basin

The Greater Sydney Basin resulted from a marine transgression at the end of the Late Palaeozoic glaciation, followed by a marine regression during Late Permian and Triassic times. Three stratigraphic divisions have been defined within this region. The lowest division is the Narrabeen group, followed by the overlying Hawkesbury sandstone. The Wianamatta group is uppermost (Branagan et al. 1976: 28). The Wianamatta shale predominates in the centre of the region, forming the Cumberland Plain.

The Hawkesbury sandstone is a fairly friable medium with relatively homogenous grain size. It weathers cavernously to form overhangs (shelters) which occur in a range of topographic locations (see Figure 2.1-2.8). It also occurs as flat topped outcrops (platforms of varying sizes) and boulders, mainly on ridge tops but also along the sides of gullies and in valley bottoms.

For the purposes of this work, geological homogeneity was significant because the sandstone provides a homogeneous medium of a relatively restricted extent. In other parts of the Greater Sydney Basin, there are different surface sandstones (e.g. the Nowra sandstone, Narrabeen sandstone etc.), some of which also contain different styles of rock art (e.g. Officer 1992, 1993). The boundaries of this study area were selected on the basis of geology rather than on a presumed extent of the Sydney Basin style. Thus it was possible to impose a natural (and neutral) boundary and not pre-empt any conclusions about the region's outer limits.

Geomorphology: structure and terrain

Much of the earlier archaeological literature for the Region describes the area between the coast and the Great Dividing Range as the coastal strip (Lawrence 1968, Lampert 1971a, 1971b; Poiner 1976, Ross 1976, Attenbrow 1987). In the Sydney Region the 'coastal strip' is almost 90km wide and this descriptor conceals important geographic/environmental divisions. There is a maritime zone (which includes estuarine influences) on the eastern side of the coastal strip, while the coastal hinterland (or 'inland': Lampert and Hughes 1974, Poiner 1976) forms the western portion of the study area. Further subdivisions based on specific ecosystems can also be identified. These are relevant to our understanding and interpretation of prehistoric Aboriginal land use patterns in the Region.

Chapter 2: the Sydney Basin- environmental context

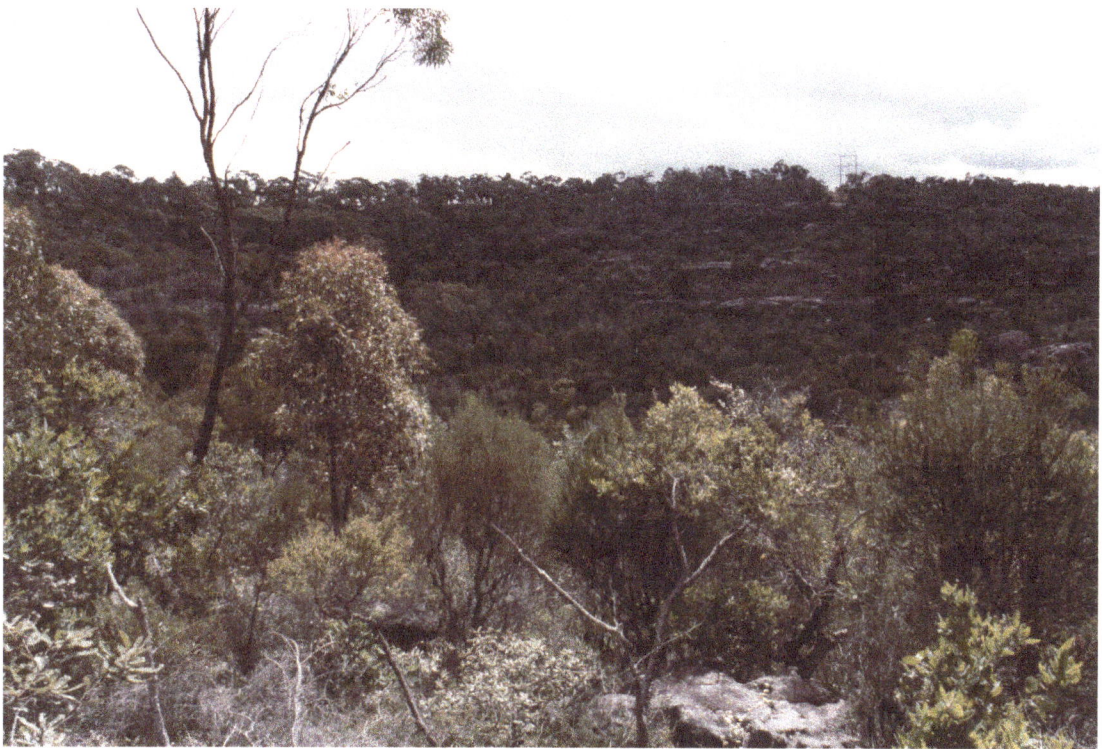

Figure 2.1: A typical upper tributary gully with a ridgeline in background. Six engraving sites were located in this gully on various sandstone shelves and platforms visible.

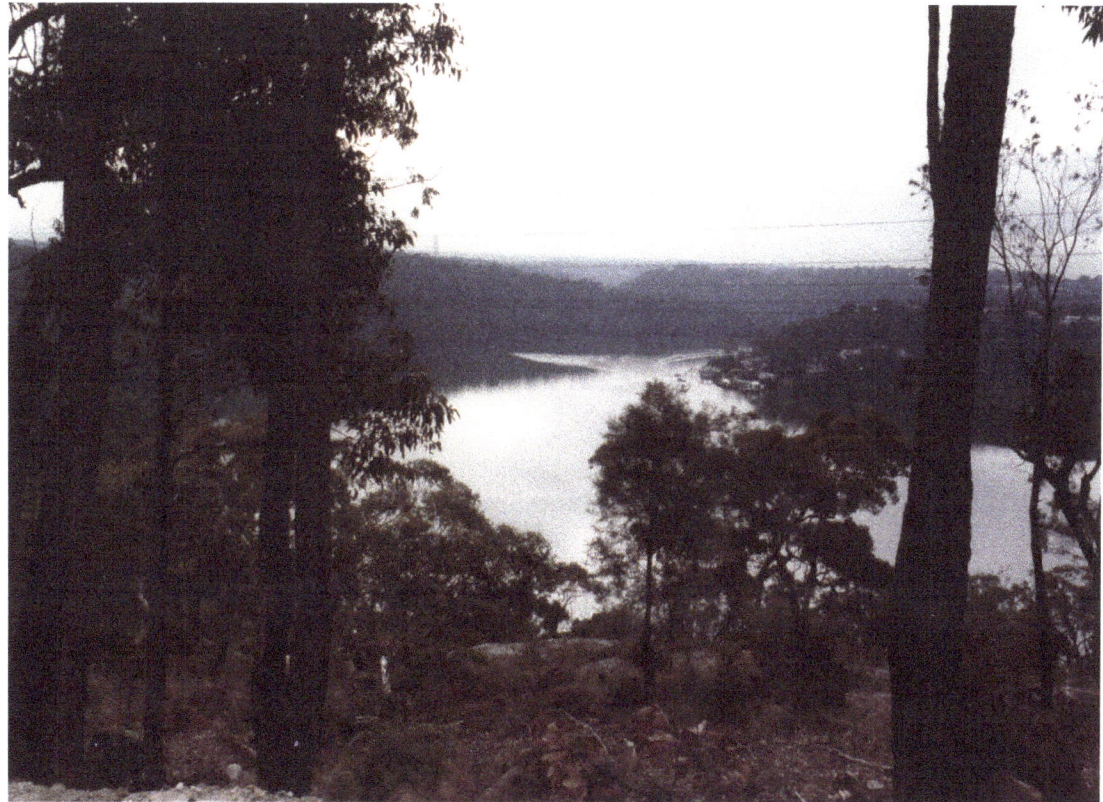

Figure 2.2: The Georges River to the south of the region. View to the west from Alford's Point.

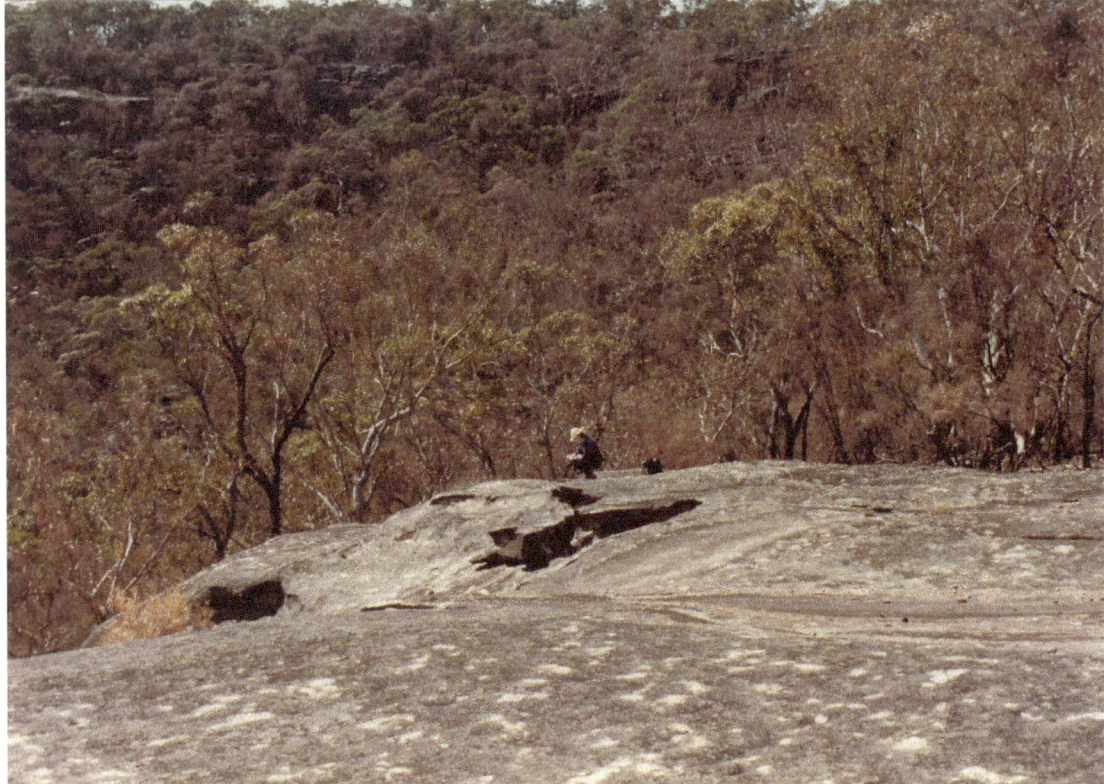

Figure 2.3: Engraving site on the ridge top overlooking Berowra Creek. This vast expanse of sandstone only had a single engraved motif on it.

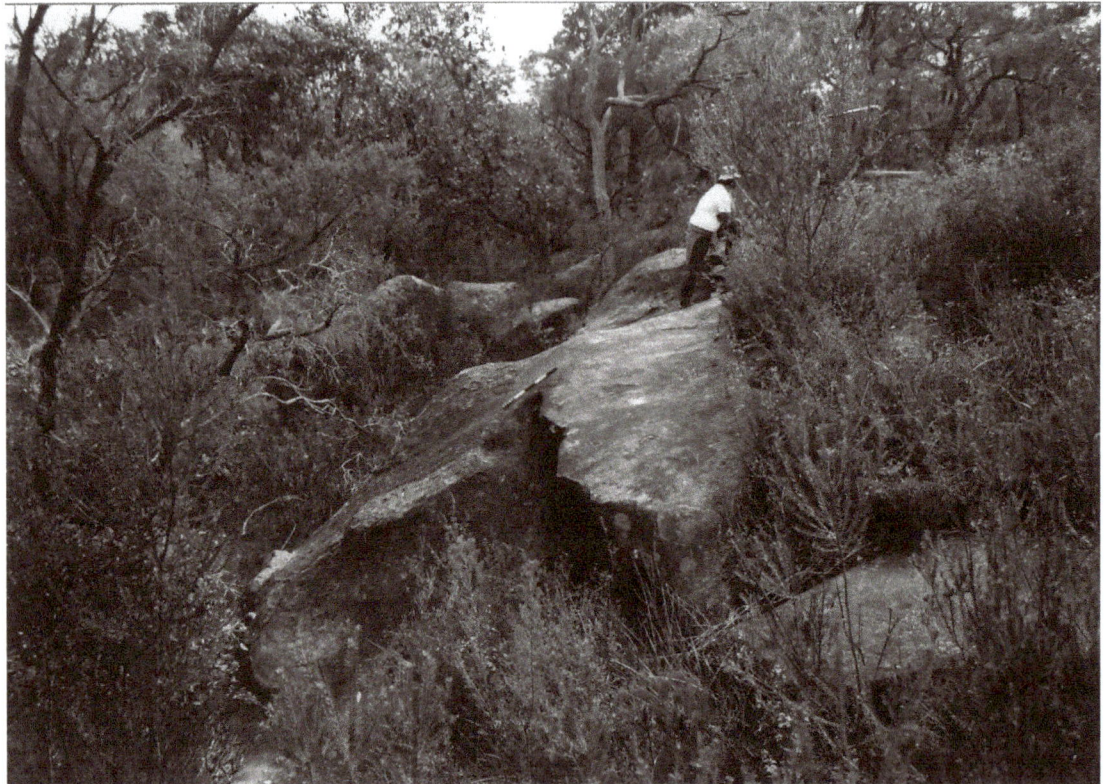

Figure 2.4: A relatively small boulder with engravings near Maroota, south of the Hawkesbury River. Engraved mundoes and anthropomorphs are located on this sloping top surface.

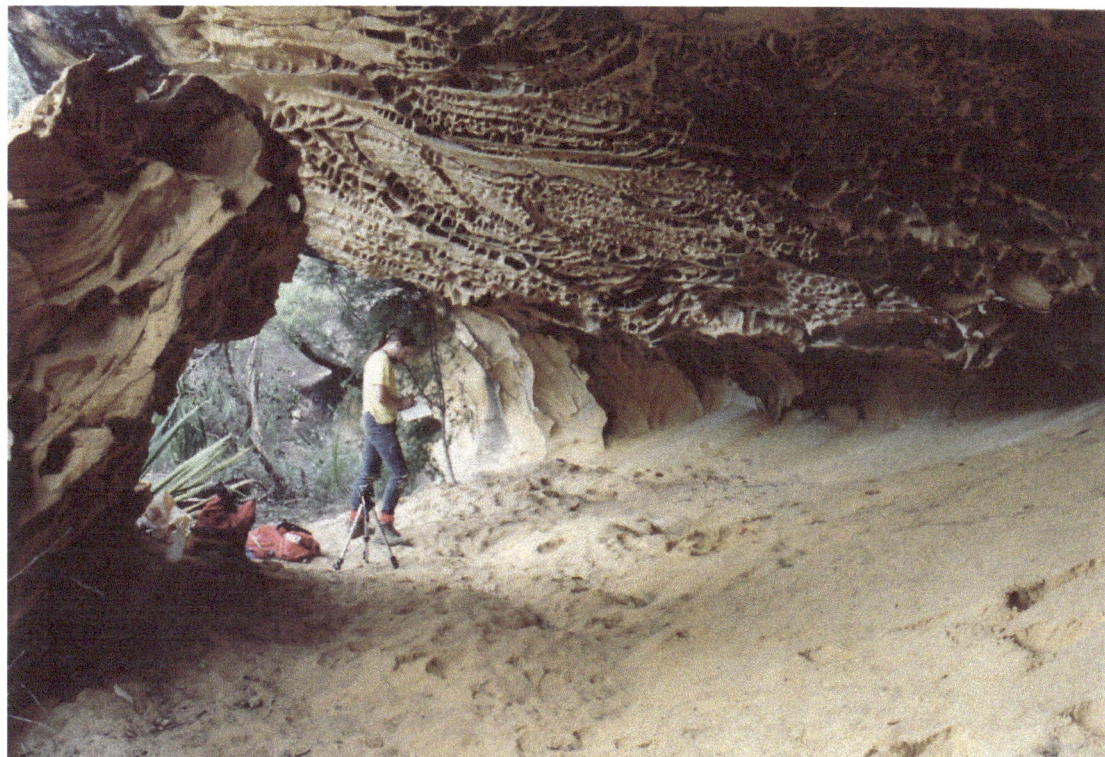

Figure 2.5: Large shelter with honeycomb weathering near Warre Warren Creek. Pigment art is located on a smooth surface at one side of this shelter.

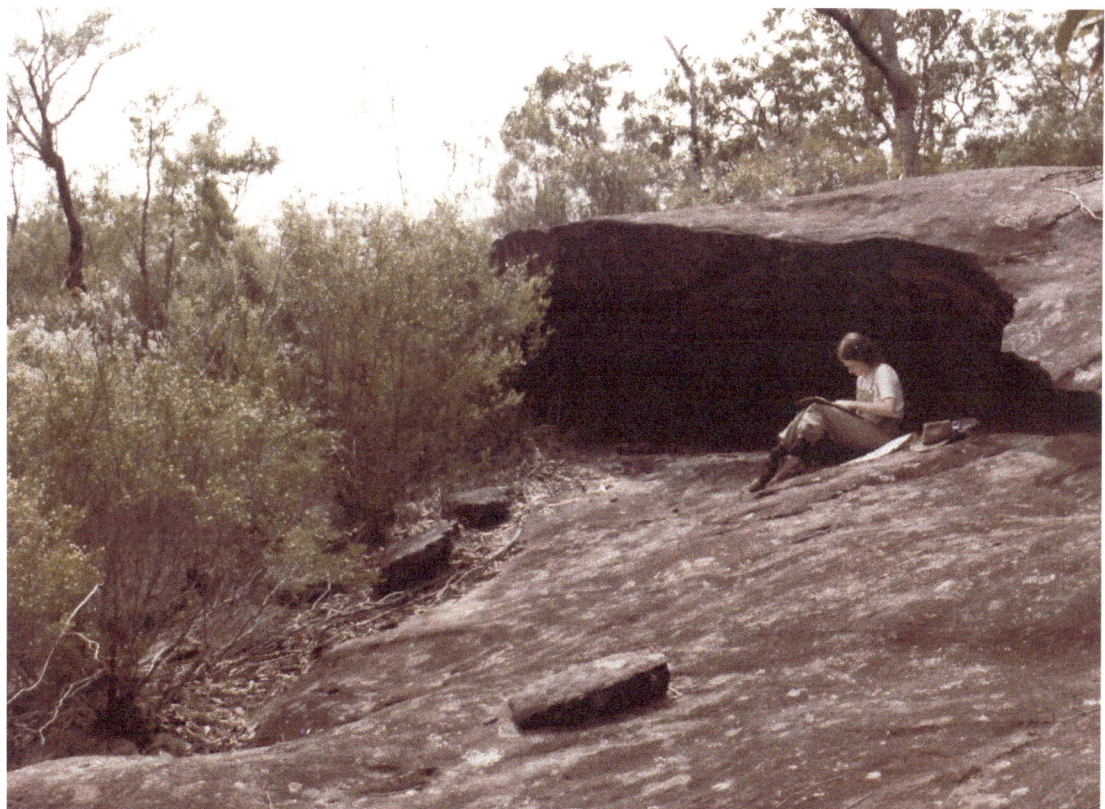

Figure 2.6: This small overhang near Wheelbarrow Ridge forms a lip in an extensive sandstone platform. Despite its small size, this shelter has a moderately large pigment art assemblage on its ceiling.

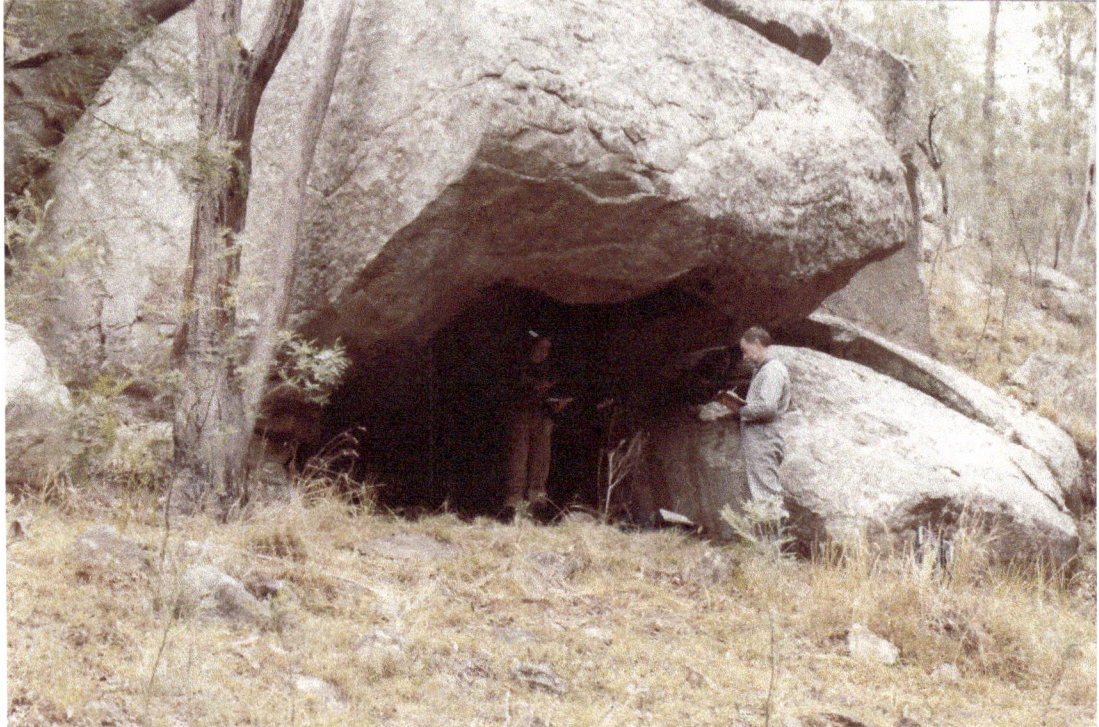

Figure 2.7: Large free-standing boulders such as this one near Howes Valley provide ideal shelters for a range of assemblage sizes.

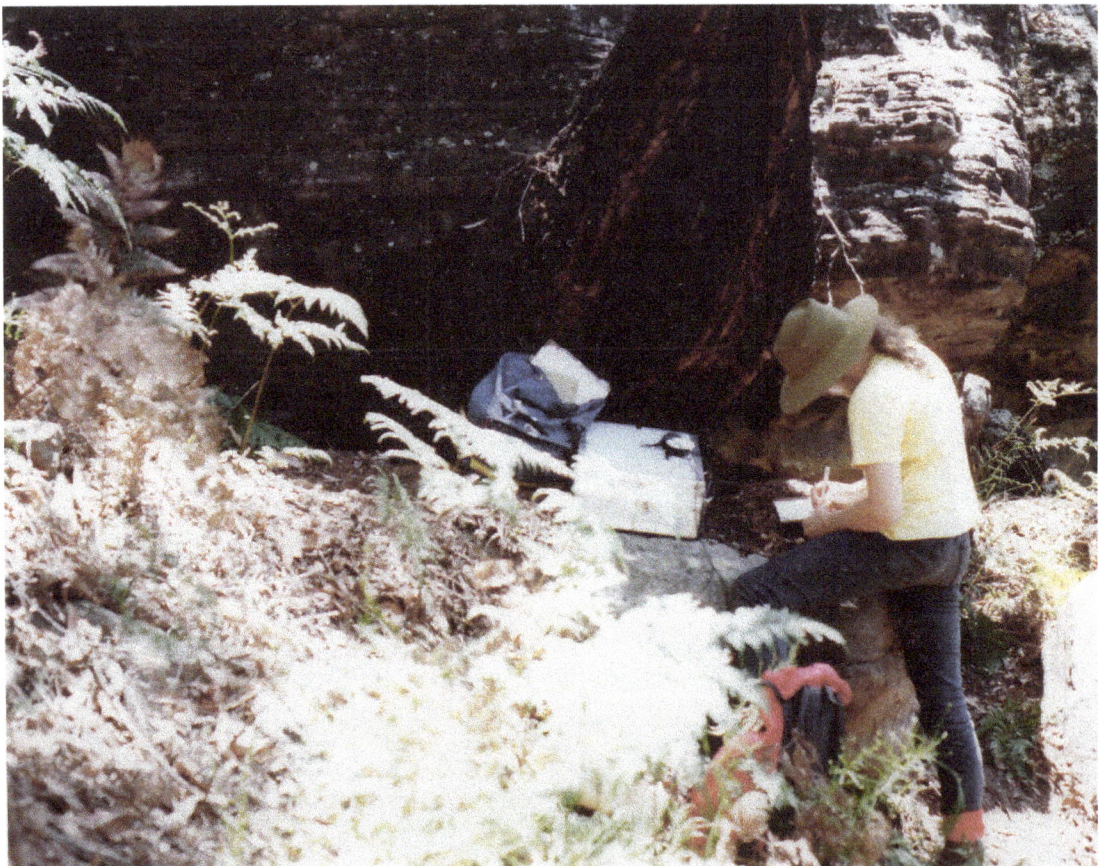

Figure 2.8: This is the smallest recorded shelter in the region, near Warre Warren Creek. To enter this shelter you have to crawl in behind the tree. Three pigment motifs were found in this overhang.

The maritime zone includes the open coastal margin, coastal heath and estuaries. This zone extends roughly 15km inland from the coastline, with estuarine conditions along major waterways (e.g. the Hawkesbury, Parramatta and Georges Rivers). The hinterland zone includes riverine, forest and woodland environments. It is the strip of country between the maritime zone and the Blue Mountains. Throughout this work 'maritime' and 'coastal' will be used interchangeably to describe that zone, while 'hinterland' and 'inland' will be used to describe the westerly portion of the region. In both zones and all localised environments, the characteristics of the underlying sandstone geology dominate: with the landscape being unrelentingly dissected and fairly rugged.

In the centre of the Sydney Basin is the Cumberland Plain. This consists of open plain woodland on shale geology. Relief here is low and gently undulating. Surrounding the Cumberland Plain are the Hornsby, Woronora and Blue Mountains Plateaux. The boundaries between the Cumberland Plain and the adjacent north (Hornsby) and south (Woronora) sandstone plateaux are poorly defined. To the west of the Plain, however, is the distinctive Lapstone Monocline and Nepean Fault system (Branagan *et al.* 1976: 45). This research was confined mostly to the Hornsby and Woronora Plateaux, although some sites (on Hawkesbury sandstone) were known on the Blue Mountains Plateau.

Altitude varies across the region. The Hornsby Plateau has a maximum altitude of 250m AHD (Australian Height Datum) between Hornsby and the Hawkesbury River. Between the Hawkesbury River and the Hunter Range, elevation is consistently between 250-350m AHD. Mt Yengo, in the north-west of the region, has the highest elevation at 386m. Closer to the coast, the elevation is between 100-150m AHD.

The Blue Mountains Plateau, which is part of the Great Dividing Range, rises steeply from the Nepean to 250m AHD in the first line of ridges around Blaxland. The maximum elevation in the region is recorded at Blackheath in the Blue Mountains (1,087m AHD).

The Nepean-Hawkesbury River is a significant drainage catchment in this region. This system drains the three major plateaux, as well as part of the southern Tablelands. The Nepean River rises in the south-west of the Sydney Basin, and drains the Avon, Cordeaux, Cataract Rivers to its east, and the Nattai, Wollondilly and Cox's River from the west. It becomes the Hawkesbury River downstream of Richmond. The major river systems of the Colo and the Macdonald are also drained by the Hawkesbury River. Downstream of Wiseman's Ferry, the point at which the river becomes estuarine, there are several important creeks. Major catchments include Mangrove, Berowra, Cowan and Mooney Mooney Creeks. Pittwater and Brisbane Water are major embayments north and south of Broken Bay, wherein the Hawkesbury River meets the Tasman Sea.

Other major river systems occur within the central and south-eastern parts of the region. These include Port Jackson (Middle Harbour and the Lane Cove and Parramatta Rivers), Botany Bay (the Cook's and Georges Rivers) and Bates Bay (Port Hacking).

The regional or 'culture-area' population (Peterson 1976: 51) is described well by this set of catchments. The geographic extent of Hawkesbury sandstone is smaller than the outer Hawkesbury-Nepean watershed boundaries, particularly in the south and west. The northern boundary of the catchment is at the Hunter Range, which also delimits the northern extent of the Hawkesbury sandstone.

Climatic and eustatic change

The Sydney region presently possesses a generally warm climate with uniform annual rainfall. In the Blue Mountains there is a long, mild summer, while elsewhere there is a hot summer. Temperatures are generally dependent upon season, aspect, distance from sea and elevation. Proximity to the sea is the major factor in maintaining temperate conditions, namely mild winter minimum temperatures and summer maximum temperatures. On the coast the summer temperature is commonly less than 26°C, while that on the Cumberland Plain is greater than 29°C. Winter minimums, on the other hand are commonly between 8-9°C on the coast and 5-6°C inland (http://www.bom.gov.au/climate/averages/tables).

The average annual rainfall also varies with distance from the coast. The coastal annual fall is in the order of 1,210mm, with the heaviest falls in autumn. In the hinterland, the rainfall is lower (around 670 mm/year) and the heaviest falls are in summer (Bureau of Meteorology: http://www.bom.gov.au/climate/averages/tables).

At the height of the Last Interglacial period, between 120,000 to 130,000 years ago, sea levels were up to 5 metres higher than today. Throughout the Late Pleistocene temperatures in south-eastern Australia are known to have fluctuated as the region experienced glacial and periglacial conditions. Glaciation commenced around 31-34,000 years ago, and at this time the mean temperature was probably only 3°C below present. During the period between 25-15,000 years ago, however, mean annual temperatures would have been between 6°C to 10°C below present (Kershaw et al. 2000: 490). Throughout the late Pleistocene, conditions were generally drier than they are today (Dodson and Thom 1992: 121, 132) – with rainfall possibly at 50% of today's values (Kershaw et al. 2000: 491). By 15,000 BP, temperatures were rising and deglaciation was in progress. The colder conditions lasted until c.10,000 BP. Lands to the east of the coastal Ranges would have experience a gradual increase in rainfall and vegetation growth between 15–10,000 BP (Dodson and Hope 1983: 75).

By 10,000 years ago (at the beginning of the Holocene), conditions in many coastal areas would have been similar to what they are today. Between 8-4,000 BP, it was warmer and wetter with both temperature and rainfall higher than at present. Around 5,000 years ago, a return to a slightly cooler climate is thought to have occurred, while after 3,000BP the El Niño-Southern Oscillation (ENSO) began to operate as it does now – with an increase in seasonality for rainfall, and more marked winter-summer patterns in some areas (Markgraf and Diaz 2000: 475). Temperatures continued to drop so that in the period from between 3,500 and 2,000 BP the temperatures were 2-3°C lower than today and it was drier (Harrison and Dodson 1993: 279-80). In the last 1,000 years the temperature and rainfall have been relatively stable.

The main effect of these temperature and rainfall variations would have been in terms of the vegetation regimes likely to be available to Aboriginal people. With moister conditions, rainforests and wet sclerophyll would have increased (at the expense of grasslands and dry sclerophyll) and swamps would have increased in size. Species diversity and abundance within communities may also have changed. With drier conditions, the reverse of these characteristics would be expected – and more open conditions would have prevailed. Changes in water availability and vegetation patterns would have affected available animal species distributions – and potentially fire regimes as well (Attenbrow 2004, Dodson and Mooney 2002, Markgraf and Diaz 2000).

During the last Glacial Maximum (LGM) between 30,000-15,000BP (Chappell, in prep.) the sea level was between 110-130m below its present level (Roy 1998). The Sydney coastline, now at Sydney Heads, would have been 25-35 kilometres further east. The major river systems – which are now estuarine - would have been fresh water. As the sea level rose (between ca 18,000 – 6,500 BP), the catchments of these rivers would have slowly become inundated with estuarine conditions eventually being established. The current coastline has only existed for the last 6,500 years (Chappell 1982:71, 1983b: 121; Chappell and Thom 1977:278-80). Sydney has a drowned embayed coast, with prior bedrock valleys partially infilled by sandy barriers, tidal flats, and deltaic plains (Roy and Thom 1981:471). Rocky headlands alternate with bays which incorporate bay head beaches, barrier beaches and lagoons. There is no broad coastal lowland (Branagan et al. 1976: 50). The coastal zone morphology is due primarily to the Holocene eustatic sea level rise. Many of the present coastal forms, however, were initiated during Pleistocene stages of higher sea level. Following the Holocene transgression, these have been reworked and extended.

While generally stable from c.6,500 years ago, there was another sea level rise (between 1-2m in height) sometime between 4-3,000 BP (Baker and Haworth 2000). It has been argued that the marine coastal ecology probably did not stabilise until c.4,000 BP, meaning that shellfish and estuarine resources probably did not become important to coastal/estuarine inhabitants until this time (Callaghan 1980, White and O'Connell 1982).

The drowning of the coast affected the region in several ways. It reduced the available land area for habitation, altered the configuration of the coastline and (eventually) substantially increased the estuarine conditions along the present shorelines. While the sea was encroaching on the land mass, it has been argued that there would have been an immature coastal morphology with no, or fewer, lagoons, less established tidal rock platforms and generally less shallow water, compared with periods of stable sea level. An immature coast is less diverse and also poorer in resources. Coastal productivity would not have been as great during the period of rising sea levels in the late Pleistocene/early Holocene, as in the mid- to late Holocene (Callaghan 1980; Lampert and Hughes 1974; Walters 1992b; White and O'Connell 1982). The implications of these changing climatic conditions and resource opportunities for human populations living in the region are obvious.

Vegetation

The vegetation in the Sydney Region is diverse and dependant primarily on geology, aspect and topography. Proximity to the coast and water (sea, rivers, creeks etc.), rainfall and elevation are also determining factors. Vegetation surveys have been undertaken in various ecological zones within the region. The following three are summarised here to demonstrate the regional variation, because these locales contain large numbers of art sites, including those excavated for this research.

Ku-ring-gai Chase National Park

Maritime conditions prevail here with estuarine and major tidal creek system, coastal heath, open woodland. Ku-ring-gai Chase National Park is bounded on the north by the Hawkesbury River and to the east by Pittwater. The ridge along the Cowan-Berowra peninsula marks its western boundary, and Cowan Creek flows south-north through its centre.

A vegetation analysis for the park (NPWS 1988:7-9), covering more than 15,000ha, identified 24 plant communities, several of which are of limited size and associated with unusual geological and topographic features (e.g. remnant rainforest communities along western Pittwater, diatreme vegetation communities at Campbell's and Smith's Craters and dyke vegetation communities on West Head).

Predominant vegetation communities include coastal heath, swamp sedge, littoral vegetation, low woodland and open forest (Beadle *et al*. 1986).

Mangrove Creek

Mangrove Creek drains north-to-south into the Hawkesbury River and is c.36 kilometres long. It is the second major northern tributary valley along the Hawkesbury River west of the coastline, and has a catchment area of 420 sq km (Vinnicombe 1980). It is at the hinterland/maritime transition with estuarine conditions present at its confluence with the Hawkesbury.

The Judge Dowling Range forms its western watershed boundary while Peat's Ridge forms its eastern boundary. This catchment's northernmost tributaries start in the Hunter Range.

The climate of this area can be generally categorised as fringe temperate maritime. While the summers are hot and warm and the winters cool to mild, these are more extreme than on the truly temperate maritime coastal strip. Rainfall on the coast and on the Somersby Plateau averages 1,200–1,600mm per year (Bureau of Meteorology 2007: http://www.bom.gov.au/climate/averages/).

A vegetation survey of Upper Mangrove Creek (Benson 1979) identified four communities in this area: Woodland, Open Forest (associated with Hawkesbury sandstone), Open Forest/Tall Open Forest (associated with the Narrabeen Group on the lower drier aspect of the hillslopes) and Tall Open Forest.

The various topographic locations defined in this catchment by Attenbrow (2004) were used in this research. These are broadly ridge tops, hill sides and valley bottoms. The cut-off for distinguishing these topographic locations was 5m in elevation below ridge tops and above valley bottoms (respectively).

Yengo National Park

This area is true hinterland at the north-west edge of Hornsby Plateau at the most northern extent of the Hawkesbury sandstone. In the upper reaches of the Macdonald River this area has higher average elevation and relief. Most deeply incised creek lines and rivers in this area have Narrabeen sandstone exposed in the lowest valley levels. Extensive areas of Quaternary alluvium occur here along the valley bottoms of major creek and river systems, and along the western ridgeline boundary, e.g. Mellong Swamp.

The annual average rainfall varies considerably here. This results from the area's distance from the coast, and the counteracting orographic effects of the elevated landscape. At Howes Valley the annual rainfall is 737mm (Bureau of Meteorology 1975). Most of this area has little standing water and there is rapid runoff of rainwater. Numerous springs, however, are known to exist across the area (Sanders *et al.* 1988: 15) these being of variable reliability (John Bowen, *Big Yango*, pers. comm. 1988).

While thirteen vegetation communities were identified here, more than 75% of the vegetation can be characterised by four communities. The most common vegetation community across the area is the Hawkesbury sandstone woodland, followed by the Narrabeen-Hawkesbury ironbark forest, the Sheltered Hawkesbury Sandstone forest and the Complex Hawkesbury sandstone sheltered forest. On richer and moister soils are pockets of rainforest and sheltered forest. Paperbark swamps are also recorded in poorly drained, alluvial areas (Sanders *et al.* 1988: 64-81).

The vegetation regimes across the hinterland area are extremely complex and variable. Geology and landform are major determining factors in the distribution of vegetation communities. The ridges capped by Hawkesbury sandstone are much less varied geologically than the gullies. The gullies cut through a range of geological strata as well as colluvium and alluvium (Sanders *et al.* 1988: 82-6), providing a complex mosaic of soil types and vegetation communities in these locations.

While the economic resources of this true hinterland are significantly reduced in comparison with the maritime or estuarine environments, it can be expected that the resources of this area were not marginal since the complex distribution of the vegetable resources would have ensured ecological variability and localised resource rich high-value patches (cf. Fletcher-Jones 1985).

Conclusions

The Sydney region is within the fertile coastal zone. Environmental conditions across the region vary in terms of localised biomass. The two main economic zones defined for this work are the coastal and hinterland zones.

Localised resources could be expected within these broadly defined environmental zones. Geological variation is known to delimit the extent of particular lithic resources (e.g. silcrete is found in cobbles of the St Mary's formation on the Cumberland Plain; the gravel beds of the Nepean are a localised source of cherts and silicified tuff[2]; quartz is ubiquitous across the Hawkesbury sandstone formation (although good quality sources are not common); axe quality

[2]Previously known as 'indurated mudstone'

basalt occurs across the region in localised areas (e.g. Kulnura, Mount Yengo, Prospect, Barranjoey headland).

Food resources would not have been dispersed evenly across the region. The maritime/estuarine influence extends approximately 10km inland along major river systems. Fish, water birds and eels would have contributed significantly to the diets of those living further inland. However, even in close proximity to major waterways, there could not have been the same degree of aquatic specialisation afforded by the coastal zone. These variations in resources are likely to have affected the social organisation of groups across the region.

3

SOCIAL CONTEXT

> While anthropology can help elucidate the complexity of cultural systems at particular points in time, archaeology can best document long-term processes of change (Layton 1992b:9)

In the original thesis one chapter was dedicated to the ethnohistoric and early sources for the Sydney region and another to previous regional archaeological research. Since 1994, Valerie Attenbrow has published both her PhD thesis (1987, 2004) and the results of her Port Jackson Archaeological Project (Attenbrow 2002). More recent, extensive open area excavations on the Cumberland Plain done as cultural heritage management mitigations (e.g. JMcD CHM 2005a, 2005b, 2006) have also altered our understanding of the region's prehistory. As Attenbrow's *Sydney's Aboriginal Past* (2002) deals extensively with ethnohistoric evidence from the First Fleet and early days of the colony the ethnohistoric and historic sources explored for this thesis have been condensed to provide the rudiments for the behavioral model developed for prehistoric Sydney rock art.

The British First Fleet sailed through Sydney Heads on 26 January 1788. Within two years an epidemic of (probably) smallpox had reduced the local Aboriginal population significantly – in Farm Cove the group which was originally 35 people in size was reduced to just three people (Phillip 1791; Tench 1793; Collins 1798; Butlin 1983; Curson 1985; although see Campbell 2002). This epidemic immediately and irreparably changed the traditional social organisation of the region.

The Aboriginal society around Port Jackson was not studied systematically, in the anthropological sense, by those who arrived on the First Fleet. Numerous accounts were made of the more obvious aspects of Aboriginal culture (e.g. Bradley 1786-92; Collins 1798, 1802, 1804; Dawes 1790; Hunter 1793; Phillip 1789, 1791; Tench 1793; Watling 1794; White 1790). Over the next 50 years a number of detailed references to Aboriginal life in the region were made (e.g. Barrallier 1802, Angas 1847, Threlkeld 1824-1859 in Gunson 1974, the Russians between 1814-1822 in Barratt 1981), but explicitly anthropological work was not undertaken in the region until the late 19th century, when R.H. Mathews (1896d, 1897c, 1897e, 1898b, 1900, 1901, 1908; Mathews and Everitt 1900) studied the languages and social organisation of various tribes in south-eastern Australia. These studies, however, took place in already devastated communities living in much altered circumstances more than 100 years after contact, and after a second epidemic of smallpox in the 1830's (Butlin 1983, Curson 1985).

While the Sydney ethnohistory is disjointed and its interpretation requires many 'leaps of faith', the First Fleet journalists were 'trained military observers' and the very survival of the colony depended on 'their observations and assessments' (Isabel McBryde, pers. comm., 1990). Also, in the early days of the colony, there was an atmosphere of philanthropy. Governor Phillip was intent on carrying out orders to 'establish with th[e native inhabitants] a strict amity and alliance' and to treat them 'with the utmost kindness' (Phillip 1789[1970]:36). Many of the descriptive accounts are useful in establishing daily activities of the Port Jackson Aborigines at 1788. The attendant interpretations and conclusions, however, particularly about the more abstract qualities of Aboriginal life, must be treated carefully (and see Clendinnen 2003).

In 1788 certain obvious differences were observed between groups of people living in the region. These included tribal (family and language) groupings, economic and social divisions (coastal/inland, gender divisions, social prohibitions, ceremonies etc.). While this evidence is not sufficient to develop a firm behavioural model, it does provide material 'clues' for the interpretation of the stylistic patterning encountered in the art (and see Attenbrow 1988, 2002; Kohen 1986, 1988, 1993; Kohen and Lampert 1988; Lawrence 1968; Ross 1976, 1988; Vinnicombe 1980). Previous archaeological models based on these observations provide contradictory interpretations on social organisation (cf. Poiner 1971, Ross 1976, Kohen 1986).

This review establishes the behavioural parameters relevant to an interpretation of the art produced by the prehistoric inhabitants of the Sydney region. Of particular importance, are the social divisions which were recognised across the region and the types of social 'boundaries' which might have existed. Stylistic behaviour depends not only on social cohesion, and the maintenance of social ties, but also on social exclusivity, and the maintenance of boundaries between groups of people (Wiessner 1983, 1990; Wobst 1977).

Social organisation and language boundaries

> [They are] divided into families. Each family has a particular place of residence, from which is derived its distinguishing name. This is formed by adding the monosyllable *Gal* to the name of the place: thus the southern shore of Botany Bay is called *Gwea*, and the people who inhabit it style themselves *Gweagal*. (Collins 1798[1975]:453)

The Sydney region falls within the south-east coast cultural-area group (Peterson 1976). While people would usually have subsisted in smaller localised groups, these groups would have formed part of a larger population. The social organisation of the Aboriginal people around Port Jackson was observed as named groups associated with designated tracts of land. It was generally recognised that the basic economic unit in the region was the family group - one or two adult males, their wives and dependants (young and old) - several of which usually teamed together to forage in a fairly restricted area (Lawrence 1968:171).

Within a year of arriving in Sydney a number of named groups had been recognised by the recorders on the First Fleets; the *Cadigal, Cammerragal, Wannegal, Wallumedigal, Gweagal, Boromedegal, Noronggerragel, Borogegal* and the *Gomerrigal* (Phillip to Lord Sydney, 13 Feb, 1790; HRA 1,1 [1914]:160). Only the first six of these were provenanced. Linguistic information (Capell 1970, Dawes 1790) has been used to supplement this list (Kohen 1988: Figure 2).

The complexity of kin relationship within and between family groups was not recorded – and nor is the relationship between family groups in the larger social structure understood. The presence of linguistic sub-groups was indicated by different dialects although the connections and social mechanisms which enabled the larger 'culture' group to function were not understood. The division of territory was still less clear, although most early writers observed that smaller groups had specific connections with particular locations (e.g. Collins 1798[1975]:453, Hunter 1793[1968]:62).

When considering the nature of territories and boundaries, particularly in terms of how these may be reflected in rock art, it is recognised that 'boundaries' are not entirely useful concepts when it comes to viewing Australian Aboriginal territorial organisation. Aboriginal territorial organisation is much more complex than any Eurocentric concepts of individual, or even corporate, ownership (Rigsby and Sutton 1982; Sutton 1995). While the anthropological literature is divided over the basic units of territoriality, definitions of 'the tribe' and the bases for territorial organisation, certain substantive matters are basic to a consideration of territoriality in contact Sydney. By necessity these are projected from ethnographic material from elsewhere on the continent. The following generalisations regarding Aboriginal territorial organisation are relevant:

i) the demarcation of defined tracts of territory and sets of totemic sites was usual in Aboriginal society. The extent of this area can be defined as the extent of the resident population's ecosystem (Peterson 1976:58). While the periphery of this may have been vague, the 'heartland' of any group's estate gave that group its habitation and name;

ii) a kinship defined 'residence group' occupied a defined tract of land (its range). This small scale exploitative population combined for various economic reasons: conservation of effort, population pressure, seasonality and scheduling, a mixed diet and the division of labour (Peterson 1976:59). Band size varied considerably by region across Australia (Peterson 1985:50);

iii) residence groups comprised not only the members of the local descent (and therefore language) group, but wives, and also others of 'foreign agnatic stock' (Stanner 1965:11). There was fairly fluid movement of individuals between groups. Band size has been recorded as being remarkably consistent amongst hunter gatherers generally and Australia in particular (Peterson 1985:38). These are cited as variously being between 25-50 people with the optimum being 30 (Davidson 1938, Tindale 1974);

iv) in many instances, 'zones of indeterminacy' exist between adjacent groups 'without clash over title' (Stanner 1965:12). These and indeterminate tracts (used, for instance, as access routes) were not the exclusive possession of any one band;

v) the local group population was usually part of a larger regional or 'culture-area' population (Peterson 1976:51) and as such, shared a number of cultural traits and/or characteristics (Kroeber 1939, McCarthy 1940, Lampert 1971b). Visits between groups within a region were ruled by convention. 'Constant interaction of th[is] kind characterised both the religious and the secular life. There was a real interest in mixing with neighbours, and a strong moral requirement to share life-supports with them' (Stanner 1965:2). The evidence in the Sydney region is sparse [e.g. Collins (1798) and Howitt (1904)] but would appear to indicate that as many as 200 men or 600-800 people might congregate for the final sequences of initiation ceremonies;

vi) the Aboriginal life pattern had a marked polarity; the population aggregated and dispersed in successive phases. Periodicity and length of the phases varied considerably, but generally, the local group clustered in good times and dispersed in the bad;

vii) the degree of interaction between groups depended on general ecological conditions and good seasons, on population density and a tendency towards the conservation of effort; e.g. by division of labour and the ethic of reciprocity (Peterson 1976:58; Stanner 1965:3-7). Localised seasonal abundances, such as evidenced by the Bogong moth feasts (Flood 1980), the Bunya nut festivals (Morwood 1987), the annual mutton bird migration (Threlkeld in Gunson 1974:65, Lampert 1966:97), would also have encouraged interaction, as would have the (less predictable) 'windfall' resources such as beached whales reported in the Sydney region (Tench 1793[1961]:176, Bradley 1786-92:120, Collins 1798 [1975]:490, Threlkeld in Gunson 1974:55);

On these bases, certain parameters for the current analysis are clear.

- Distinct bands, speaking separate languages or dialects would have been identifiable, and have inhabited discrete tracts of land. Given the relatively rich ecological conditions of the Sydney coastal region, each of these groups could have maintained a degree of economic independence, since their ranges would have provided for the basics of survival (Stanner 1965, Godwin 1990).

- These bands would have been part of a larger clan group, whose estate would have represented the extent of the land inhabited by that larger group of people with economic and ritual rights. This larger group is assumed here to be the language group.
- Interaction between the clan groups would have occurred on the periphery of these estates for economic activities. Shared ritual responsibilities would have required interaction between these sub-groups of the larger culture area, and social conventions would have controlled such visits. Ritual and economic relationships between people and land (Hiatt 1962:284, Peterson 1985:24), and the distinction between land-using and right-holding groups would have regulated this interaction.

- Interaction between clan (or language) groups to hold ceremonies indicates larger scale group cohesion of the culture area population, and there is evidence that considerable stability in social activities is thus achieved; e.g. 'correct' designs are maintained on Pukamani Poles amongst the Tiwi by supervision from neighbouring (totemically related) individuals (Hart 1970). Hiatt (1965) records similar 'owner/manager' interaction in Gidgingali ceremony, as they have been made to bark painting production (Morphy 1977, 1989; Taylor 1987, 1989). Collins commented (1798 [1975]:467) on an initiation ceremony which could not commence in the absence of elders from adjacent 'tribes'. Similar social customs (as manifested by ceremonial behaviour) were observed across the Sydney region, and these appear to have relied on ritual relationships with neighbouring groups.

Specifically anthropological descriptions of social organisation in the Sydney region were made by R.H. Mathews (although see comments above). He states that all members of the *Darkingung* community (in the north-west of the region) were segregated into two moieties (phratries) *Dilbi* and *Kuparthin*, whose names correspond with the Kamilaroi (Mathews 1897e:161, 170). Each moiety was further subdivided into two sections, for the *Darkingung* these were named *Bya* and *Kubbi* (for the former moiety) and *Ippai* and *Kumbo* (for the latter). On the basis of these moieties and sections, totemic affiliation and marriage relations were determined. While totemic affiliation controlled many social interactions, it did not dictate general economic activities; 'members of each group, and consequently of the totems also, [were] found in all the local divisions of the tribe' (Mathews 1897e:159).

Mathews notes that the *Darkingung* tribe had 'uterine [matrilineal] descent' (Mathews 1897e:170), with a dispersal of male and female members of the same totem. Conversely, Mathews records that the groups in the south-west of the region – the *Gandangara* speaking tribes (which by implication include the *Darug*: see below) had patrilineal descent, but that the women married out of the group into which they had been born (Mathews and Everitt 1900:264). Here, while the females of the same totem were dispersed by marriage, the males of the same totem, would have co-resided: there was patrilineal descent and virilocal residence.

Peterson (1986:17) indicates that these post-marriage arrangements would have resulted in two different types of residential groups, with the *Darkingung* exhibiting characteristics of the mixed group and the *Gandangara* adhering to the characteristics of a kin group. Peterson argues that the latter situation was the culturally prescribed ideal, both anthropologically (e.g. Radcliffe-Brown 1931) and by individual adult men. The former, however, was probably more common. Most groups probably oscillated between the two, depending on economic obligations and ritual (Peterson 1986:26).

Several early commentators noted an inequality amongst the 'tribes' around Port Jackson, stating that the *Cammaraygal* (from the North Shore) were the largest and strongest group by the 'influence of their numbers and muscular appearance' and that 'there is no doubt of their decided superiority over all tribes with whom we were acquainted' (Collins 1798[1975]:453). These comments were recorded after the 1789 epidemic, which (as mentioned above) greatly affected the groups on the southern shore of Port Jackson. This relative 'superiority' may have reflected the survival of an intact band of initiated people upon whom the relict bands depended for continuing social activities and cohesion. However, the possibility of one group in the region exerting overriding control in the social sphere may be of relevance.

Linguistic Evidence

The linguistic evidence for the region indicates the presence of discrete language groups (Capell 1970; Dawes 1790; Mathews 1897c, 1901; Mathews and Everitt 1900; Threlkeld in Fraser 1892; Tindale 1974; Troy 1990). This evidence is sketchy, and there are conflicting views on how it can be interpreted (Kohen 1986, 1988; Kohen and Lampert 1988; Ross 1976, 1988). The boundaries between these different language groups, as well as inter-relationships between these create the greatest disagreement in archaeological interpretation.

The geographic distribution of linguistic groups within the region relies heavily on late nineteenth century research into relict groups and regions, i.e. *Kamilaroi* and *Wiradjiri* (e.g. Mathews 1897c, 1897e, 1903, 1904). Linguistic evidence collected at contact was largely in the form of unprovenanced word lists (Collins 1798: Appendix XII, Hunter 1793[1968]:523, Tench 1793[1961]:291-3). The Dawes manuscript provides a detailed and comprehensive analysis of the Sydney language - and details interaction between several notable Sydney Aboriginal people such as *Barangaroo* and *Benalong* and their wives (e.g. Dawes 1790:14).

Lancelot Threlkeld arrived in Sydney in the 1820's and completed a detailed and provenanced grammar of the *Guringai*[3] language sometime thereafter (Capell 1970). This agrees with a vocabulary by J.F. Mann (Capell 1970), completed in the 1870's and based on the information of Long Dick, son of *Bongaree*[4]. The Dawes, Threlkeld and Mann manuscripts, while giving detailed vocabularies, do not indicate the geographic distributions of these languages.

Mathews' earlier work in the region defined three distinctive languages, the *Darkingung*, *Gandangara* and *Tharawal*. *Darug* was defined as a dialect of *Gandangara* (Mathews and Everitt 1900:265). Mathew's definition provided an incomplete coverage of the region, specifically not including the coastal area north the Hawkesbury and possibly not north of Port Jackson (Capell 1970). Mathews placed the *Darkingung* to north of the Hawkesbury River in the drainage basins of the Macdonald and Colo Rivers, Putty Creek and Wollombi Brook (Mathews 1897c:1). The *Gandangara* were said to have existed in 'the coastal district ... from the Hawkesbury River to Cape Howe, extending inland to the Blue Mountains, and thence southerly ...' (Mathews and Everitt 1900:262). The *Tharawal* speaking people were spread over the coast from Port Hacking to Jervis Bay ...extend[ing] inland for a considerable distance (Mathews 1901:127). Mathews recorded *Darug* dialect being spoken at 'Campbelltown, Liverpool, Camden, Penrith, and possibly as far east as Sydney, where it merged with *Thurrawal*' (Mathews and Everitt 1900:265).

Capell linguistic analysis (1970) concluded that four language groups (each with varying numbers of dialects) existed in the region at contact. These were *Guringai, Darkingung, Darug* and *Tharawal* (Figure 3.1). The presence and location of the *Gandangara* language is not discussed in Capell's paper.

The *Guringai* inhabited the coast between Port Jackson and somewhere north of Wyong, where it met the *Awaba* language (Threlkeld in Gunson 1974, Mann 1885).

The *Darkingung* speakers occupied land to the west of the *Guringai*, north of the Hawkesbury River (following Mathews 1897c). The boundary between the *Guringai* and *Darkingung* is along Mooney Mooney Creek (cf. Kohen 1986, Smith 1983) and thence northerly (Capell 1970:22).

---
[3]The orthography for these languages varies between sources. Unless being directly quoted the spellings used throughout this thesis will be *Guringai, Darkingung, Darug, Tharawal* and *Eora*.
[4]*Bongaree* was 'king ... of the Pittwater tribe' (Macquarie 1822:258) or 'Broken Bay tribe' (Barratt 1981: Plate III).

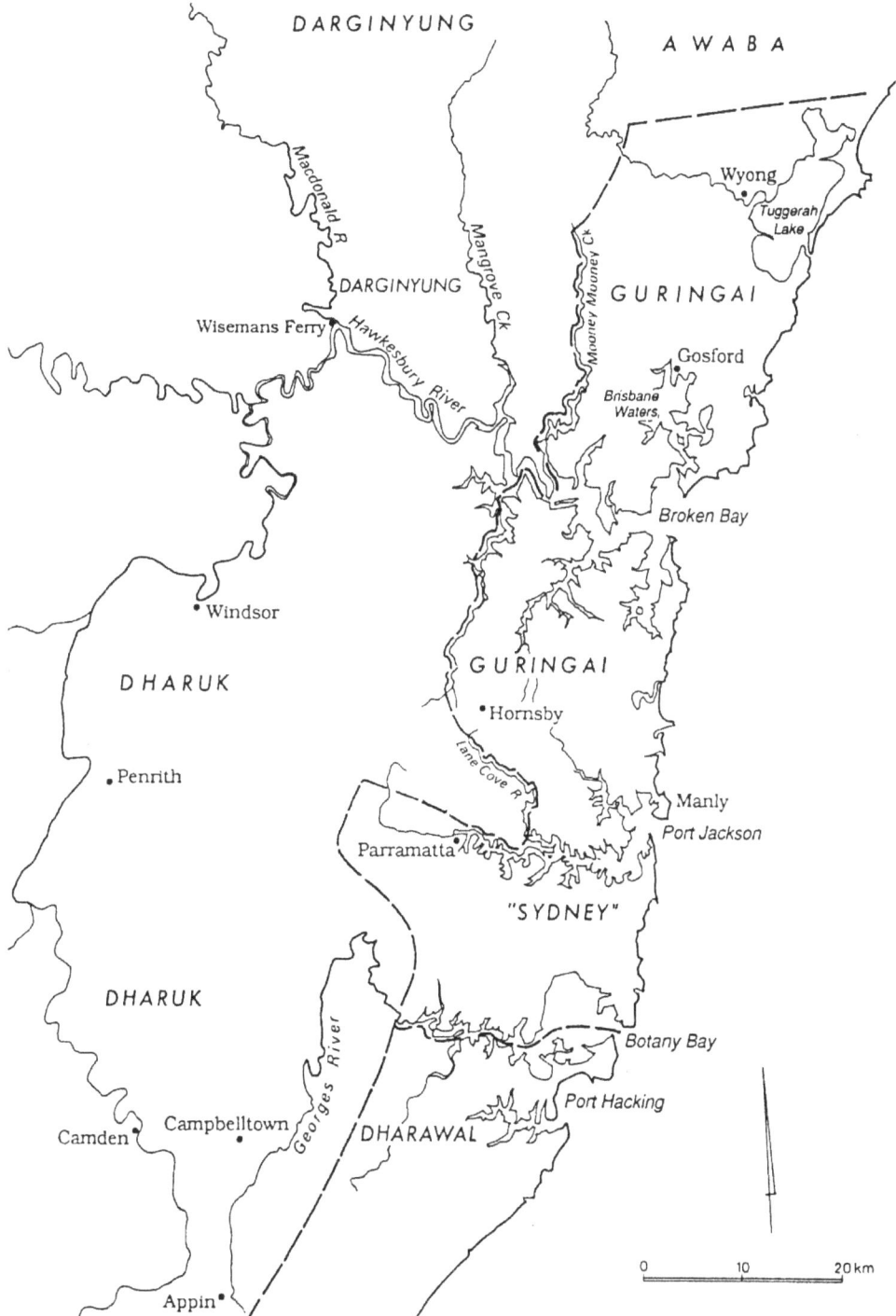

Figure 3.1: The four language areas defined for the region (after Capell 1970: Figure 1).

The *Guringai's* western neighbours south of the Hawkesbury were the *Darug* speakers. This boundary is placed along Berowra Creek and the Lane Cove River to the northern shore of Port Jackson.

Darug speakers inhabited the area south of the Hawkesbury River covering the Cumberland Plain and including the upper reaches of the Georges and Nepean Rivers. South and/or east of the Georges River between the *Darug* and the coast were the *Tharawal* speakers, their southern boundary occurring well south of this study area (cf. Tindale 1940).

Between the *Guringai* and *Tharawal* on the coast, Capell places a dialect 'or even sub-dialect' (1970:22) of *Darug*, which he calls the 'Sydney' language, described elsewhere as the *Eora* (Tindale 1974, Troy 1990). The extent of this dialect was 'limited to the peninsula on which

Sydney now stands' (Capell 1970:22). This designation is based primarily on the fact that both *Tharawal* and *Guringai* are 'affix-transferring' languages, but that *Darug* is not.

A problem with Capell's organisation is the nature of several of his boundaries: specifically smaller water courses - such as Mooney Mooney and Berowra Creeks. It many coastal areas of Australia, tribal boundaries occur at the edges of catchments (viz. Peterson 1976, Tindale 1974) rather than dissecting catchments and logically - band ranges. Mathews' (1897c) description of *Darkingung* territory supports a watershed model, and certainly in the rugged and drier area occupied by the *Darkingung*, it makes sense that the 'heartlands' of these clan estates has ridgelines as their boundaries.

Many of the ridgelines around the Sydney region, e.g. the Boree Track and Kulnura/Peat's Ridge, were documented access routes (and see Ross 1976: Map 2.1). Due to the rugged sandstone landscape, movement around the region would have been mostly by way of ridgelines. It makes sense that major access routes would have occurred in 'zones of intermediacy' (Stanner 1965:12) - on the periphery of a clan's estate (and see Layton 1989:2, 1992b:9). It would seem unlikely that boundaries between clan estates and language groups would have been creeklines, unless these were major river systems (such as the Hawkesbury or Part Jackson) whereby access and crossing, required a canoe.

Conversely, the linguistic evidence indicates that Broken Bay, a significant physical barrier, was not a language boundary. The Hawkesbury River near Wiseman's Ferry, does appear to have been a language boundary (between *Darkingung* and *Darug* speakers), while ethnohistoric evidence suggests that the people on the Hawkesbury in the vicinity of Richmond Hill lived on both sides of the river - '[o]n the opposite bank of the river they had *left* their wives and children' (Tench 1793[1961]:230, *my emphasis*). In the south of the region, the Georges River appears to have been the boundary between the *Eora* and *Darug*, and the *Darug* and *Tharawal*.

There is no dispute that distinctive linguistic divisions existed within the region at European contact. The geographic distribution of these groups is less clear.

To the north and south of the region there is firm linguistic evidence for a separation of coastal and hinterland groups; the *Darkingung* and *Guringai* in the north and the *Darug/Gandangara* and *Tharawal* in the latter. In the centre of the region, primarily covering the Cumberland Plain and Sydney peninsula, there is a suggestion of linguistic continuity between the coast and Cumberland Plain. For the purposes of this study, the distribution of the four language groups (Figure 3.1) present at contact, the *Darkingung, Guringai, Darug* and *Tharawal*, will be tested.

Land use strategies and habitations

Early sources suggested that there was little contact between the coastal and inland tribes (e.g. Tench 1793, Collins 1798[1975]). This was based on differences in economic behaviour as well as on findings made during early explorations that the Port Jackson Aborigines had no knowledge of the country north or west of Parramatta, nor south of the Georges River (Phillip 1792, Tench 1793, Barrallier 1802). A complete separation of 'hunters and fishers' was reported, presumably in terms of coastal and inland groups, and some archaeological interpretations agree with this (Ross 1976, Kohen 1986, Kohen and Lampert 1988). Numerous references indicate that specific adaptation to different resources existed.

Along the coast, the protein portion of the diet was seen as being entirely based upon seafood. Captain Cook noted that 'shellfish is their chief support yet they catch other sorts of fish' (Cook [Beaglehole ed.] 1955:312). Collins gained some insight into the range of foods eaten;

> Fish is their chief support ... the woods, exclusive of the animals which they occasionally find in their neighbourhood, affords them little sustenance; a few berries, the yam and fern root, the flowers of the different Banksia, and at times some honey, make up the whole vegetable catalogue. ... The wood natives also make a paste formed of the fern-root and the large and small ant bruised together; in the season they also add the eggs of this insect.
> (Collins 1798[1975]:461-2)

Archaeological research indicates that this seafood bias in the coastal diet has been overstated (McDonald 1992a, Megaw 1968a). At the Angophora Reserve site, maritime resources (i.e. fish and shellfish) contributed to less than 8% of the calorific content of food remains (Wood 1989:82), while Attenbrow's Port Jackson work has similarly identified the importance of terrestrial and avian fauna in the dietary remains of coastal rockshelter sites, particularly at Balmoral Beach (Attenbrow 2002: Table 7.2).

Ethnohistoric reports indicate considerable diversity in adaptations to environmental conditions. Barrallier in his expedition through *Darug* and *Gandangara* territory in the early nineteenth century, describes the swamps in the Nepean River area as being excellent sources of fish, shellfish and 'enormous' eels and he states that;

> the people from this area usually fed upon opossum and squirrels, which are abundant in that country, and also upon kangaroo rats and kangaroo, but they can only catch this last one with the greatest trouble, and they are obliged to unite in great numbers to hunt it. (Barrallier 1802 [1975]:2-3)

Such a kangaroo hunt, with a large group using fire, spears and 'tomahawks' was described near a Menangle Swamp. The participants were spaced at '30 paces ... [and] formed a circle [covering] an area of 1 or 2 miles' (Barrallier 1802[1975]:3). In the order of 100 people appear to have been involved in this hunt, suggesting co-operation between several bands.

Lizards and grubs, 'particularly those which are found in the trunks of trees' (Barrallier 1802[1975]:6, Collins 1798[1975]:462), were also documented as part of the diet. For the purpose of collecting these grubs (*Cahbrogal*) a specific utensil was used, this being described as:

> a switch about twelve inches long and of the thickness of a fowl's feather ... One of the extremities of this stick is provided with a hook. ... [which is used upon finding evidence of these grubs in the bark of trees having] widen[ed] the hole ... with their axe ... dip their switch into the hole, and, by means of the hook, draw it out, and eat it greedily. (Barrallier 1802[1975]:6)

Other specialised inland adaptations to localised resources include 'squirrel traps' in hollow trees and 'decoys for the purpose of ensnaring birds' (Tench 1793[1961]:154-5). These decoys were assessed as having great utility as they were full of quail feathers. These structures were described as complex (see also Phillip in *HAR*, 1:156 and Collins 1798[1975]:462) and were made of reeds and 'underwood'. They were described as being 'long and narrow, shaped like a mound raised over a grave; with a small aperture at one end for admission of the prey; and a grate made of sticks at the other' (Tench 1793[1961]:154-5). One such structure described by Collins 'was between 40-50 feet long' (1798[1975]:462). He also describes animal and bird traps near inland lagoons as consisting of excavated holes with camouflaged tops.

Early accounts remarked on the tree climbing facility of the inland tribes' men (Hunter 1793, Tench 1793, Collins 1798, Barrallier 1802). This was done for the purpose of obtaining possums (usually with assistance of smoke) and was achieved by cutting notches for toeholds 'with a stone hatchet' (Hunter 1793[1968]:430), Tench 1793[1961]:233). Kohen (1986:46) argues that possums and other tree dwelling animals were the woodland tribes' staple, and that edge-ground hatchets were the dominant subsistence item in the inland toolkit. With the exception of hatchet heads, evidence for these types of resources procurement is not preserved the archaeological record. Bones (terrestrial or maritime) are rarely recovered from Cumberland Plain open sites contexts (e.g. JMcD CHM 2005a, b).

Living sites

Most early references focus on the bark huts used as Aboriginal dwellings across the region. The coastal versions of these were described as being larger than the inland ones, being 'formed of pieces of bark from several trees put together in the form of an oven with an entrance ... large enough to hold six to eight people' (Collins [1975]:460). Worn-out canoes were often recycled for this purpose. Tench described a group of five such huts on the northern arm of Botany Bay as a village (1793[1961]:210). Given the above estimate of the holding capacity of these, groups of up to 40 people could have been so accommodated. There are other references to 'villages' on the sea coast around Botany Bay and Pittwater (e.g. Collins 1798[1975]:47, Worgan 1788[1978]:26).

The huts of the 'woodsman' (Collins 1798[1975]:460), were described as being made of the bark of a single tree, bent in the middle and placed on its two ends on the ground 'exactly resembling two cards, set up to form an acute angle' (Tench 1793[1961]:154; and see Phillip 1789[1970]:55-57) and 'affording shelter to only one miserable tenant' (Collins 1798[1975]:460). These shelters (*gunyahs*) would be grouped together, up to a total of nine (Barrington 1802:20).

Observers also noted the use of rockshelters:

> They appear to live chiefly in the caves and hollows of the rocks, which nature has supplied them with, the rocks about the shore being mostly shelving and overhanging so as to afford a tolerable retreat. (Barrington 1802:20)

Collins also commented on the occupation deposit found in shelters, stating:

> these proved a valuable resource to us, and many loads of shells were burnt into lime, while other parts were wheeled into gardens. (Collins 1804[1910]:306)

The bark constructions in the open have not survived in the archaeological record. The fact that the majority of Aboriginal occupation evidence for the region derived from shelter sites at the time this thesis was written (cf. chapter 4) is significant, given that shelters may not have been the focus for habitation at contact – or indeed the last millennium (JMcD CHM 2005a, b, c).

Material culture

The material culture of the Sydney region is poorly represented in major museum collections[5] (Lampert and Konecny 1989). Drawings and descriptions in journals (e.g. Watling 1794, White 1790[1962] and see Barratt 1981) are thus the best source of this information. These have been discussed and illustrated extensively in previous research (Attenbrow 2002; Kohen and Lampert 1988; Lampert 1988; Lawrence 1968; McBryde 1979, 1989; Megaw 1967, 1969, 1993; Ross 1976) and only relevant aspects will be described here.

Of interest here are regional and localised differences noted by early diarists, in particular, those aspects which may be depicted in the art.

Items of material culture include those used for fishing, hunting and collecting as well as weapons. The men's repertoire included spears (hunting, fishing and ritual), spear throwers, clubs, 'swords' or boomerangs, shields (bark and wooden), and stone hatchets (Bradley 1786-92[1969]:121-8; Collins 1798[1975]:487; Hunter 1793[1968]:55; Tench 1789[1961]:50, 1793[1961]: 184,191,200; White 1790[1962]; 152). Women's belongings included fishing hook and line, digging sticks, various bark items (e.g. dilly bags, fishing tackle, water baskets), wooden bowls and large shells used as containers (Hunter 1793[1968]:63; Tench 1793[1961]:143-6, 186; White 1790[1962]:157, 201; Threlkeld in Gunson 1974:54, 66-8; Bellinghausen in Barratt 1981:35).

---

[5] There are sizeable Australian collections in the British and Pitt Rivers Museums, but the number of items deriving from Sydney - this point of first contact with European settlement, is not large. Scattered items have been discovered in obscure museums around the world (Coates 1999).

Shields

Two types of shields were described in contact Sydney. The first was quite light, being made of bark. This type was used by children in practice combat, in defence against sharpened reeds '[with] which they are soon expert' (Collins 1798[1975]:466).

Threlkeld describes most comprehensively the construction of the wooden shields, albeit from around the Lake Macquarie area (*Awabagal/Guringai* tribes). These were 'three feet long by eighteen inches ... lozenge shaped, pointed at top and bottom, and pigeon breasted rather than flat. ... The shields are always painted with white pipeclay and are generally ornamented with a St George's Cross, formed by two bands two or three inches wide, one vertical the other horizontal, coloured red ...' (in Gunson 1974:68). Rossiyisky describes the wooden shields from the Sydney area very similarly, although noting that 'they are daubed with *various* red and white figures' (in Barratt 1981:23; *my emphasis*). Bellinghausen adds that these shields had a 'dry white colouring substance over which was painted red stripes' (in Barratt 1981:41).

No mention was made in any of the early references to shield designs varying across the region, although Campbell's illustration of a *Yoo-lang* ceremony (1798[1975]: Appendix 6; Plates 4, 5 and 6) shows a variety of shield designs, none matching that of Threlkeld (and see shield designs and other material objects depicted in drawing of Bennelong in McBryde 1989: Plate 19).

Further reference is made to this item of material culture in the synchronic analysis of the art. The shield motif, particularly in the engraved assemblage, is dispersed widely along the coastal strip.

Ceremonial behaviour and initiation

The evidence for local ceremonial behaviour is of primary importance in assessing the nature of social interaction across the region. Linguistic differences indicate a separation of groups across the landscape. The ceremonial activities observed indicate that there was an overriding similarity and larger scale group cohesion amongst the peoples of this culture area. These people, however, spent most of their time in distinctive areas, speaking different languages and/or dialects.

Corroborees were observed in the first years at the settlement, and there is evidence that both genders took part in these at varying stages (Collins 1798[1975]:486). Dancing and singing played an important part in ceremonial behaviour as did body painting.

All groups from the hinterland were described as having very similar *bungung* (e.g. Mathews 1897c, 1897e). The main reason cited for these ceremonies was the initiation of young men. A subsidiary reasons was the resolving of 'tribal wrongs which may have been perpetrated since the last initiation gathering' (Mathews 1897c:10).

The ethnohistoric and later reports all describe the initiation ceremonies as involving a number of neighbouring 'tribes' and large gatherings of people (Collins 1804[1910]:311; Mathews 1897c:1-2; Mathews and Everitt 1900:276). Men only took part in the initiation of young men. The women remained in a separate camp (Mathews 1897c, Mathews and Everitt 1900) where it could be presumed (Bell 1983) that they undertook ceremonies of their own. Indeed Mathew's description of the women during the *bungung* ceremony, suggests just this:

> Every morning the mothers of the novices, accompanied by all the old women of the tribes present, repair to the *watyoor* [a specially prepared area], and light one or more fires in the cleared space, around which they sit and sing songs which have reference to the novices. ... These women are collectively known as the *yanniwa*, and the young women or children, or any of the men, are not permitted to go near them when assembled at the *watyoor*. (Mathews 1897c:7)

Mathews describes *Darkingung* and *Gandangara* ceremonies as involving the construction of bora rings and ground sculptures, the latter usually involving large earthen sculptures of *Daramulan* the 'sky god' (Mathews 1904:see below). There is no suggestion amongst any of these sources that rock engravings of such figures served this same purpose.

Collins' description of a *Yoo-lang* tooth avulsion ceremony (1789[1975]:466-86; Plates 1 - 8) is based on an eyewitness account made on the 25th January 1795. The ceremony was held in Farm Cove in *Cadigal* territory, and was presided over by the *Cam-mer-ray* elders. Also present was '*Pemulwoy* - a wood native' (Collins 1798[1975]:466) and leader of the Botany Bay *Bidiagal* tribe (Bridges 1970, Dawes 1790). No bora ring was constructed, or earth sculptures built, but an ovoid area was cleared.

Mathews described tooth avulsion for all initiated males in *Darkingung, Gandangara* and *Tharawal* speaking areas (Mathews 1897c, Mathews and Everitt 1900). Collins (1798[1975]) also described the practice in the Sydney *Eora* area. Phillip described the *Guringai* Pittwater people as also 'missing a front tooth' (HR NSW 1[2] 1893[1978]:131). However, there is conflict in the ethnographic literature and ethnohistoric documentation. Mathews is unequivocal in his description of tooth avulsion taking place, specifically amongst the *Darkingung* and the *Gandangara*. He says of the former:

> the time spent at [these ceremonies] occupies about a fortnight. ... About the middle of this period, preparations are made for the extraction of one of the novice's upper incisor teeth. (Mathews 1897c:7)   and;

> Early the next day the boys ... are shown a colossal horizontal image of *Dharamulan* ... After that, one of the front upper incisors is punched out of each novice in succession. (Mathews and Everitt 1900:279)

On the other hand, one of the major stated differences between the coastal and inland people was that the latter did not practice tooth avulsion (Ross 1988:48, 1990:3-4). The statement is based on comments by Tench (and Phillip *HR NSW* 1[2] 1893[1978]:131), made on an early journey to the Nepean/Hawkesbury, in which two natives met with on the River had not 'suffered the extraction of a front tooth' (Tench 1793[1961]:230).

The ethnohistoric sources are equivocal. Most sources only commented on the rare *presence* of both teeth, rather than the more common absence. Tench, for instance, comments that 'the deficiency of one of the fore teeth of the upper jaw ... we have seen in *almost* the whole of the men' (Tench 1793[1961]:46; *my emphasis*). Phillip also noted (1789[1970]:42) that 'several old men were seen [in an excursion to Broken Bay] who had not lost the tooth nor had their noses prepared to receive that gross appendage [long bone or stick]'.

This evidence is inconclusive and could be interpreted in three ways. Either;

> i) the people located in Port Jackson with both their front teeth were visiting the coast from inland tribes; or,

> ii) tooth avulsion was not universal on the coast or inland; or,

> iii) the men seen on the coast with both upper incisors had not yet been initiated [see Collins' comment about a novitiate who fled in fright during his *Yoo-lang* when it came time to remove his tooth (Collins 1798[1975]:481)].

A gender and location specific trait was described for the coastal women in Sydney. There are no conflicting reports on this practice. This form of social identification involved the removal of the first two joints of the little finger on the left hand. There is no evidence that this was initiatory, in the normal sense of the word, since by all accounts it was performed very soon after birth;

> the operation is performed when they are very young, and is done with a hair, or some other slight ligature. This being tied around at the joint, the flesh soon swells, and in a few days ... the finger mortifies and drops off. (Collins 1798[1975]:458)

Collins interprets this as a practical solution: 'these joints of the little finger were supposed to be in the way when they wound their fishing lines over their hands' (Collins 1798[1975]:458). It also been interpreted 'fishing magic' (Leroi-Gourhan 1968, Marshack 1972):

> [the mortified section] was taken out into the bay, and with great solemnity, committed to the deep. The belief was that the fish would eat this part of the girl's finger, and would ever, thereafter, be attracted to the rest of the hand from which it had come. Thus [she] would always have success at fishing because of the peculiar lure in her fingers. (Scott 1929[1982]:4)

Whatever the cause, this form of mutilation readily identified the women from the coastal tribes, and is firm evidence for the maintenance of group identification at a personal level. There is a shelter site near Mackerel Beach (Pittwater) in which the (left) hand stencils have been identified as female, based on the fact that the little finger is extremely short.

These issues were discussed in some detail in the original research since the practices of tooth avulsion, bodily mutilation and body decoration are excellent examples of a public information system (Wobst 1977) operating in contact Sydney. Indeed, in terms of boundary maintenance and group identification, these means would have been extremely effective. If tooth avulsion did occur in one part of the region and not in another, one might assume that a significant cultural boundary, or at least group affiliation, existed and was being demonstrated. Such a boundary might be reflected by an art style boundary.

Culture Heroes

Mathews prodigiously documented ceremonial events practiced in the late nineteenth century, which he described as the Bora Religion.

Biaime is the principal hero in the Bora mythology (Mathews 1904:340) and is said to have had his home in an outcrop of granite near the town of Byrock, approx 740km from Sydney. The outcrop was called *Bai* by the Aborigines, this word 'signifying the semen of men and animals' (Mathews 1904:340). *Dharamulan* was:

> a sort of half brother or near relative of *Baiame's*. His name is made up from *dhurra*, thigh, and *mulan*, one side, the whole name meaning leg-on-one-side, as he is said to possess one leg only. ... He has a voice like rumbling distant thunder. It fell to his lot to separate the youths from their mothers and teach them the *Burbung* ceremonies. ... A bullroarer is also frequently called *Dharamulan*, its humming sound ... represents his voice. ... he has the magical power of changing his shape, and making his body smaller and larger at pleasure ... (Mathews 1904:343-5)

Radcliffe-Brown (1930) describes the myth of the rainbow serpent as being 'the belief in a gigantic serpent which has his home in deep and permanent water holes and represents the element of water which is of such vital importance to man in all parts of Australia' (Radcliffe-Brown 1930:343). In New South Wales he states:

> [the] cult of the *karia* (rainbow serpent) was often an element of the Bora or initiation ceremonies of the NSW tribes. ... Many of the sacred Bora grounds had representations of the serpent in the form of a sinuous mound of earth up to 40 feet or more ... beliefs about the rainbow serpent were explained to the younger men, [as they were for] *Biaime* [whose

representation in earth occurred at] practically every Bora ground ... (Radcliffe-Brown 1930:345)

Mathews (1904) also describes a *Darkingung* story with a 'mythical malevolent creature resembling a man whose body had a red glow like burning coals, who had his abode in rocky places on the sides and tops of mountains. ... His name was *Ghindaring* (Mathews 1904:345). The *Gandangara* and *Tharawal* had an aquatic monster called *Gurungaty*, who 'resided in deep waterholes, and would drown and eat strange blacks, but would not harm his own people'. The *Tharawal* had another fabled monster *Mumuga*, who possessed great strength and resided in caves in mountainous country. He had very short arms and legs, with hair all over his body but none on his head. While being unable to run very fast, 'he evacuates all the time as he runs, and the abominable smell of his ordure overcomes the individual, so that he is easily captured' (Mathews 1904:345).

Art

The early accounts of rock art in the region are minimal. None of the early writers sought informed opinion about the art they observed, and the conclusion most often drawn was that these were the doodles of children. One reference (Angas 1847 [1969]) suggests that these sites were the domain of the local 'priests' (see below), but this was an observation made well after the impact of contact on social organisation within the region. It was only much later (e.g. Elkin 1949, McCarthy 1947b, 1956, 1959a) that ceremonial significance was attributed, by ethnographic analogy, to the engraved art in the region.

Arthur Phillip recorded the earliest reference [on 22nd April 1788] to the widespread distribution of engraved Aboriginal art:

> In all the excursions of Governor Phillip, and in the neighbourhood of Botany Bay and Port Jackson, the figures of animals, of shields, and weapons, and even of men have been carved upon the rocks, roughly indeed, but sufficiently well to ascertain very fully what was the object intended. Fish were often represented, and in one place the form of a large lizard was sketched out with tolerable accuracy. On the top of one of the hills the figure of a man, in the attitude usually assumed by them when they begin to dance, was executed in a still superior style. (Phillip 1789[1970]:58)

Phillip expressed surprise that the local inhabitants had developed an art form prior to other aspects of 'civilisation', illustrating the settlers' total incomprehension of the nature of the society confronting them:

> That the arts of imitation and amusement should thus in any degree precede those of necessity seems an exception to the rules laid down by theory for the progress of invention. But perhaps it may better be considered proof that the climate is never so severe as to make provision of covering and shelter of absolute necessity. Had these men been exposed to a colder atmosphere, they would doubtless have had clothes and houses, before they attempted to become sculptors. (Phillip 1789[1970]:58)

Other early commentators also noted the presence of the engravings around the settlement. In Botany Bay 'on many of the rocks are to be found delineations of the figures of men and birds very poorly cut' (Tench 1789[1961]:79). In Port Jackson:

> various figures [are] cut on the smooth surfaces of large stones. They consisted chiefly of representations of the natives in different attitudes; of their canoes; of several sorts of fish and animals, and considering the rudeness of the instruments with which the figures must have been executed, they seemed to exhibit tolerably strong likenesses. (White 1790[1962]:141)

George Angas 'discovered' the engravings in the 1840's (Angas 1847[1969]:201). He and one of his friends took 'Old Queen Gooseberry' (*Bungaree's* wife)' as a guide to visit numerous groups of carvings on North Head, and to tell them 'what she knew' about them (Angas 1847[1969]:202). Angas says '[a]t first the old woman objected, saying that such places were all *koradji* or 'priests'' grounds, and that she must not visit them; but at length, becoming more communicative, she told us all she knew and all that she had heard her father saying about them' (Angas 1847[1969]:202). Unfortunately, Angas does not record the stories he was told.

The systematic recording of engraving sites (including their geographic locations) did not commence until the late 19th century, when R.H. Mathews (e.g. 1895a, 1895b, 1895c, 1895d, 1896b, 1896c, 1897a, 1897b, 1897d, 1898a, 1899) and W.D. Campbell (1899) became interested in the task.

A comparison of Mathew's sketches of earthen sculptures (cf. Mathews 1896a: Plate 1) with the Sydney style engravings does reveal some similarities: both are simple outline depictions, mostly large, of anthropomorphs, birds and animals. Elkin (1949) and McCarthy (1961) made much of these similarities in their interpretations of the Sydney art. In particular, Mathews' and Howitt's documentation of the Bora Religion and its principal mythical 'culture heroes', the rainbow serpent, *Dharamulan* and *Biaime* (Howitt 1904; Mathews 1904, 1908; Radcliffe-Brown 1930) were pivotal to McCarthy's interpretations of the Sydney engravings (e.g. 1956, 1959a, 1961).

Elkin states that rock engravings 'were cut to serve as records and symbols of historical, moral and totemic import which could be and were interpreted' (Elkin 1949:32). The work of Elkin (1949), McCarthy (1961) and Sim (1966a) has been interpreted as indicating that the second part of male initiation ceremonies took place in rock engraving galleries (Morris 1978:50), specifically those assemblages containing depictions of mythical beings (Elkin 1949:135, McCarthy 1959a:213, Morris 1978:43).

Morris explains Collins' 1795 eyewitness account of such a ceremony (which does not mention the use of rock engravings) as being only a partial record of the overall event with the more secret-sacred aspect taking place in the absence of women and Europeans, presumably on a nearby engraving site. Collins account, however, describes only men and the novitiates as being present at the tooth avulsion ceremony that he witnessed (1798[1975]:467-85). Interestingly, Mathews also speculates that Collins only witnessed part of the entire ceremony, on the basis that no bora ring or earthen sculptures were mentioned in Collin's account (Mathews and Everitt 1900:281).

An alternative but equally appropriate interpretation is that while the ceremonies within the region were analogous, they were not identical.

The unsubstantiated ethnographic analogies such as those made by Elkin (1949), McCarthy (1961), Sim (1966a) and Morris (1978) about engravings in the region must be considered as tenuous. It is possible that the engravings were connected with ritual behaviour, but this was not documented. The fact that eyewitness accounts of ceremonies (e.g. Collins 1798[1975]) and ethnographic descriptions (Mathews and Everitt 1900) do not mention the art as having a role in these ceremonies tends to weigh against it - at least at European contact.

Body marking and personal attire

> Ambition must have its badges, and where cloathes (sic) are not worn, the body itself must be compelled to bear them. (Phillip 1789[1970]:42)

It is obvious from early accounts that the Sydney Aborigines practiced, and had a considerable sense of, body decoration.

> Notwithstanding the disregard they have shewn for all the finery we could deck them with, they are fond of adorning themselves with scars, ... It is hardly possible to see anything in

> human shape more ugly, than one of these savages thus scarified, and farther ornamented with a fish bone stuck through the gristle of the nose. The custom of daubing themselves with white earth is also frequent among both sexes ... (Tench 1789[1961]:47)

The references to body painting indicate that women and men had equal access to pigment materials. It may be assumed, then, that both genders indulged in other forms of artistic behaviour - such as the pigment art - using these materials.

While there is some question that the cicatrices observed were purely for decoration (cf. Mathews 1898b, 1904), other aspects of personal adornment can have had no other purpose:

> To their hair by the means of yellow gum, they fasten the front teeth of kangaroo, and the jawbone of a large fish, human teeth, pieces of wood, feathers of birds, the tail of the dog, and certain bones taken out of the head of the fish, not unlike human teeth [otoliths?]. The natives who inhabit the south shore of Botany Bay divide their hair into small parcels, each of which they mat together with gum, and form them into lengths like the thrums of a mop. (Collins 1798[1975]:457)

Collins further described the use of pigments for body decoration 'on particular occasions', stating that red signified fighting, while white was used for the more 'peaceful amusement' of dancing (Collins 1798[1975]: 457). He also gave the fullest description of the nature of body decoration, and indicates that there was considerable variation present (see also illustrations by the anonymous Port Jackson painter, in McBryde 1989: Plates 9, 13, 31, 32). The descriptions indicate that body painting designs may have personal (i.e. not a controlled group schema) although this is not certain. He says:

> The fashion of these ornaments was left to each person's taste; and some, when decorated in their best manner, look perfectly horrible ... In general waved lines were marked down each arm, thigh and leg; and in some the cheeks were daubed; and lines drawn over each rib, presented to the beholder a truly spectre-like figure. ... Both sexes are ornamented with scars upon the breast, arm and back ... in some instances these ... resemble the feet of animals. (Collins 1798[1975]:457-8)

There is extensive evidence that body scarification was widespread in the Sydney region (see illustration by the anonymous Port Jackson painter in McBryde 1989: Plates 9, 11, 31; and Megaw 1993: Figure 14a).

Mathews' (1904) work around the Upper Lachlan River indicated that body scarification was related to totems and food avoidance. The position and extent of the scarring was regulated by the customs of the tribe to which the novice belongs. Mathews indicated that scarification was an ongoing process for identifying which foods were no longer taboo. The position of the scars signified the type of animal being released from prohibition. The goanna was marked on the shoulders below another row of scars which denoted the emu, tree grubs were marked by vertical cuts on the left arm, the carpet snake on the chest, below collar bone, and so on (Mathews 1904:262-269).

Mathews does not describe whether the order is dictated by the individual's totem, although it could be assumed that this would be the case. It is likely then, that different moieties/sections would be differently scarred, and that this would have acted as information to the general group about an individual's totemic affiliations, as well as level of initiation. It too would have been a facet of social cohesiveness.

Items of apparel observed around the region included various bands worn around the head, neck and waist (Collins 1798[1975]:459, 465-6; Tench 1793[1961]:186) and Collins also described the ritual gear associated with tooth avulsion. Each novitiate, after tooth removal, was given 'a girdle around his waist in which was stuck a wooden sword; a ligature was put around his head, in which were stuck slips of the grass-gum tree: which, being white, had a curious and not unpleasing effect' (1798[1975]:484-5, Appendix 6; plate 8).

There were rare reports of possum skin rugs being used in the region. Phillip described one such cloak found near the Hawkesbury River as being 'made of the skins of the opossum and flying squirrel, very neatly sewed together, the inside ornamented in diamonds of curved lines, by raising the skin with the point of a small bone, which is made sharp for the purpose' (HR NSW 1[2] 1893[1978]:310). This example was apparently much smaller than those observed on the southern Tablelands (Flood 1980), being described by Bradley as of the size 'to cover a child' (1786-1792[1969]:167-8). The paucity of references to these (e.g. also White 1790[1962]:156; Collins 1798[1975]:486) suggests that the items were rare (cf. the number of references stating that the Aboriginal people wore nothing e.g. HR NSW 1[2] 1893[1978]:129, 132, 222; Hunter 1793[1968]: 59; Stockdale 1789[1950]:44; Tench 1789[1961]:36,47); Worgan 1788[1978]: 13, 18).

Some items of apparel and body markings depicted on anthropomorphic figures in the art include:

- bands (on head, neck and waist);
- body painting designs;
- designs representing cicatrices;
- hair 'styles' and headdresses;
- (possibly) nose bones; and,
- (possibly) possum-skin cloaks.

The impact of European contact on Aboriginal society

While the European settlement in Port Jackson affected most severely the Aboriginal groups living south of the Harbour, the Aboriginal occupants of Broken Bay and the Hawkesbury River were also in contact with Europeans in 1788. Governor Phillip and a party of men in a long boat and cutter explored this northern waterway only a few weeks after the First Fleet arrived. First contact here was friendly and Phillip commented on the large numbers of people in Brisbane Water and Pittwater (Phillip 1789[1970]:40).

The 1789 epidemic caused major cultural upheaval that was not restricted to the immediate environs of the settlement. On visiting Broken Bay, Collins noted:

> The pox has not confined its effects to Port Jackson, for on many places our path was covered with skeletons, and the same spectacles were to be met with in the hollows of most of the rocks in the harbour. ... (1798[1975]:496)

As European settlement expanded, the Aboriginal population dwindled rapidly. In 1840, only 35 Aboriginal names were listed in the Gosford census: and by 1841, only 16 years after the opening of the Lake Macquarie Mission, Threlkeld had to close down because he ran out of Aborigines to teach. Government blanket returns also illustrate the decline in the Aboriginal populations. The evidence suggests that traditional life continued for a short time after contact. The diabolical 1789 epidemic, however, must have severely affected social organisation. An indication that the fabric of society quickly disintegrated is in the very low frequency and restricted distribution of contact art, i.e. motifs of European subject completed in the traditional style.

Relatively few contact motifs have been recorded and these are restricted to the environs of Broken Bay and the Hawkesbury River. None of these occur close to or south of Port Jackson. Many of these motifs are in areas from where Europeans would have been first sighted from a distance and usually across the water, in boats (McDonald 2008).

Mathews observed 'blacks in the Wollombi district execut[ing] paintings in caves up till 1843 ' (Mathews 1895c:56). While isolated artistic incidents may have occurred up until this time, it is clear, that within 50 years of white settlement the population had been drastically reduced and the culture which had previously thrived no longer existed in its pre-contact form.

The art of the region, and its associated archaeological evidence, survives without the traditions which created, explained and gave it purpose. Ethnohistory provides evidence for the complexity of the social system operating here at contact. The archaeological record supports some early accounts and often provides supplementary evidence. There are instances where the archaeological record refutes early accounts (e.g. dietary focus, the distribution of tooth avulsion and some aspects of mortuary behaviour). The 1789 smallpox epidemic means that social traits and organisation observed after this time must be viewed with extreme caution. Archaeological evidence for long term changes preceding the arrival of white settlers also indicates that projecting these interpretations into the past must be done cautiously.

The relevance of ethnohistoric evidence to patterning in the art
From the evidence gleaned it is possible to propose a behavioural model for the region. It is also possible to suggest how certain features of traditional life might affect, or be reflected in, the art of the region. The following aspects are relevant:

1) four languages are recognised as being spoken across the study area at contact. These are the *Darkingung, Guringai, Darug* and *Tharawal.*

    The assumed geographic distribution of the four language groups will be based on Capell's model (Figure 3.1). Certain of the boundaries will be tested in an effort to resolve existing conflicts in interpretation (viz. Kohen 1986, Ross 1976). Testing should also consider a watershed model (viz. Peterson 1976) contra a creekline equals boundary model;

2) residence groups 'bands' in the region consisted of named economic units with designated tracts of land. 'Tribes' are perceived as having comprised a number of these smaller residence groups, speaking dialects of a common language. Within the range of any one linguistic group or tribe, there would have been a number of smaller localised bands (maybe as many as fifteen) who would have had kin and/or totemic links with people in other groups and therefore modes of access to resources;

3) considerable social interaction within and across linguistic boundaries occurred. Organised social events (initiation ceremonies, dances etc.), as well as the exploitation of windfall resources (such as whale feasts) resulted in aggregations of large numbers of people of mixed language groups. Ritual behaviour in the region required the participation, and possibly consent, of neighbouring tribes;

4) there is no evidence for a rigid demarcation of territorial boundaries, although many of the initial observations did occur on the resource rich coastal strip and possibly within one linguistic group. The evidence suggests that the maintenance of clearly defined territorial boundaries was an unlikely behavioural trait. The spatial organisation of art traits may not demonstrate characteristics of smaller scale boundary maintenance (Wobst 1977), particularly at the band level;

5) the presence of distinguishable, localised bands as well as broader language boundaries suggests that there may have been a highly complex pattern of artistic behaviour and signatures within and across tribal (linguistic) 'boundaries';

6) there is no direct evidence that art played a primary role in ceremonial behaviour, nor that it had any mortuary significance;

7) food resources, economic options and adaptive material culture and modes of personal adornment varied across the region. This could be reflected in the different emphasis on maritime and land animals on the coast and inland, as well as a differential distribution of certain material culture items and body decoration;

8) to the north and south of the region, economic differences (east - west) may be reinforced by cultural difference. In both of these areas linguistic boundaries existed at contact between a coastal and hinterland peoples. The absence of sandstone (art sites) on the Cumberland Plain makes resolution of the *Darug/Eora* debate beyond analysis in this context;

9) contact motifs occur in the art in areas where the first contact/sighting was made and where the production of art retained its cultural milieu. The 1789 epidemic means that the time scale for such motifs within the fully functioning artistic system was extremely limited.

The fact that so few post-contact motifs do occur suggests that the cultural destabilisation of the region should not affect the integrity of cultural traits in the prehistoric art assemblage.

4

ARCHAEOLOGICAL CONTEXT

Archaeological evidence demonstrates that the Aboriginal people observed in Sydney at European settlement had not remained unchanged throughout the Holocene. The social dynamic identified by the relatively sparse European observations must be seen in terms of an increasingly complex understanding of the region's late Holocene record.

Temporal and spatial trends in occupation patterns are pertinent to the rock art analyses undertaken here, and these are summarised. In the original thesis most of the regional temporal trends came from rockshelter sites. Val Attenbrow's work in Upper Mangrove Creek catchment (UMCC - 1987, 2004) formed the focus of this, since occupation trends in rockshelter use contextualize the diachronic art analyses.

Attenbrow's 1987 UMCC work focused on 31 habitations. In 1994 this was the most comprehensive data set for the region, demonstrating as it did, the variability within a local catchment using sites from a range of landscapes. Since 1994, more than 50 open sites on (mainly) the Cumberland Plain have been excavated as a result of cultural heritage management projects. These open area excavations currently target c. 100 square metres (following systematic testing) in representative landscapes. The number and variety of sites investigated, and the sheer size of the assemblages retrieved from these Cumberland Plain sites (Table 4.1) provides a stark contrast to the 19,400 artefacts from 16.5 square metres excavated in UMCC (Attenbrow 2004: Tables 3.4 and 4.6).

Table 4.1: Open Site excavations on the Cumberland Plain since 1993, showing areas excavated (testing and open area), total lithics and artefacts retrieved

Location	Site	Total area m^2	Total artefacts	Total lithics	Reference
RHDA*	OWR2	104	743	n/a	JMcD CHM 2005d
Second Ponds Creek	RH/SP12 South	387	19,280	22,860	
	RH/SP12 North	148	1,522	2,004	
	RH/SP13C	120	175	251	
	RH/SP13G	151.5	1,954	2,427	
	RH/SP20	76	27	38	
	RH/SP21	153	194	209	
	RH/SP22	175	3,491	4,453	
	RH/SP13J	35	184	189	JMcD CHM 2004a
	RH/SP7	97	83	88	JMcD CHM 2006c
	RH/AC2	32	695	942	S. Garling 2000
	RH/SP9	162	10,376	12,300	JMcD CHM 1999, 2005a
Caddies Creek	RH/CD5	153	10,777	12,080	JMcD CHM 2007
	RH/CD10	122	2,004	2,549	
	RH/CD7	270	5,482	7,415	
	Mungerie Park	211	5,504	n/a	AMBS 2000
	RH/CD12	248	15,409	19,252	JMcD CHM 2002a, 2005a
	RH/CC2	111	6,705	8,554	JMcD CHM 1999a, 2005a

Location	Site	Total area m²	Total artefacts	Total lithics	Reference
	RH/CC2 mech.	100	1,073	1,167	JMcD CHM 2001, 2005a
	KV1	50	125	141	JMcD CHM 2001, 2005a
	RH/OC1	13	635	678	
	RH/SCT1	21	87	224	
	RH/SC5	54	821	1,099	JMcD CHM 2002b
Total for RHDA	21 sites	2,993.5	87,346	98,920	
Parramatta sand body	CG1	214.5	n/a	6,376	JMcD CHM 2005b
	RTA-G1	132	4,181	4,789	JMcD CHM 2005e
	CG3	123.5	487	1,198	JMcD CHM 2006b
Parramatta	Smith St	77	171	198	JMcD CHM 2004b
Fairfield	OSC1	12	358	358	Mebberson 2002
Colebee release area, Schofields	SA20	148	8,872	n/a	JMcD CHM 2006d
	SA21	112	6,646	n/a	
	SA22	106	2,944	n/a	
	SA23	86	48,873	n/a	
	SA24	114	1,695	n/a	
	SA25	69	9,493	n/a	
	SA26	52	1,835	n/a	
Greystanes	CSIRO2	25	73	131	JMcD CHM 1997a
	PH2+3	192	4,765	4,996	JMcD CHM 2004c
Eastern Ck	Power St bridge	4.5	246	272	Brayshaw McDonald 1993
Eastern Ck	Aus 1 Reedy Ck	121	1,502	2,019	JMcD CHM 2004d
	Aus 4/M7 Hub	125	118	140	JMcD CHM 2005e
	EC3	394	1,419	1,550	JMcD CHM 2006e
South Ck	LEC6-10	27	153	153	Steele 2001
Londonderry	LY2	24	398	1,067	JMcD CHM 1999b
Richmond	RM1	114	2,504	12,226	JMcD CHM 1997b
ADI St Marys	SA1	10	447	1,246	JMcD CHM 1997c
	SA2	17	458	1,163	
	SA3	14	88	188	
	SA4	37	284	539	
	SA5	33	139	325	
	ADI 47+48	193	3,981	4,956	JMcD CHM 2003b
	EP1	85	6,167	n/a	JMcD CHM 2006f
Rossmore	McCann Rd 1	25	11	11	White 2003
West Hoxton	WH3 areas	18.5	3,292	3,344	Rich and McDonald 1995
Regentville	RS1	290	n/a	18,854	Koettig and Hughes 1995; McDonald et al. 1996; Craib et al. 1999
Wattle Grove	WGO3 – 2	54.25	n/a	1,708	JMcD CHM 1998
Total	32	3,061.75	>115,000	>197,842	

*RHDA = Rouse Hill Development Area

These open site excavations have identified an increasing number of sites with stratified deposits, from which the earliest date for human occupation in the region has now been obtained. The Cumberland Plain is of relevance to this research because of its centrality in the region geographically and within the identified culture bloc. While no sandstone shelters or platforms, and therefore no art, occurs across this area, the Plain was inhabited at European contact by *Darug* speakers who also inhabited – and produced art – in the sandstone country to its north and south. Stylistic cohesion within the *Darug* art assemblages is investigated in Chapters 8 and 9 and thus the dynamics of populations living across this area requires comment.

Excavations of shelter and open locations across the Sydney region[6] have yielded 121 radiocarbon determinations (Figure 4.1). There are thousands of known occupation sites and the available radiocarbon ages are thus only indicative of the rates of occupation for each millennium.

These dates reveal a slightly different pattern to that described for UMCC - where habitation establishment rates and numbers of habitations used per millennium increased slowly until the 2nd millennium, and local artefact accumulation rates suggest that this increase commenced in the 3rd millennium (see Attenbrow 2004: Table 10.4; Figure 9.1). The regional dates reveal a much earlier start for occupation - earlier in the late Pleistocene - with occupation becoming archaeologically visible at the beginning of the Holocene. An early spike in artefact accumulation rates in UMCC in the 9th and 10th millennia is replicated with regionally dated assemblages – in the 9th and 8th millennia.

There is a steady increase in sites being used from around 6,000 years ago and almost 80% of the region's determinations date to the last 5,000 years. The number of dated sites peaks in the 2nd millennium - and 28% of the regional dates fall between 1,000-2,000ka BP. There is no indication that the Cumberland Plain was occupied any later than the surrounding Hawkesbury sandstone country. The idea that the Plain was a distinct cultural unit (*contra* Kohen 1987) has no credence or utility, and the movement of people north and south of the Plain would be expected. This open Plain landscape provides complementary material to the Mangrove Creek evidence which focuses on settlement behaviour in sheltered locations. Excavation of more open sites in sandstone country is needed to test whether there was a shift in settlement behaviour towards occupation of open locations here, particularly in the last millennium. Most of the older open sites (inferred on the basis of their lithic assemblages) in the region have still not been dated (JMcD CHM 2005a, 2007) and it is likely that the 121 determinations under-estimate the number of assemblages greater than 5,000 years old. And rising sea levels have no doubt drowned a substantial proportion of the earliest occupation evidence in the region. It is also possible that many of the most recent sites in the region are also omitted from this graph. Most dating exercises focus on establishing the earliest occupation phase, and few lodge dates from the uppermost (potentially more disturbed) layers. Many Cumberland Plain open sites have not been dated because of the absence of reliably associated charcoal features.

The archaeological sequence in Sydney matches patterning found elsewhere. Human arrival on the Australian continent is now generally accepted as c.43-45ka (O'Connell and Allen 2004), with semi-arid south-western New South Wales yielding the oldest human remains known from the continent (Bowler *et al.* 2003). It has been argued that the Cranebrook Terrace, on the Hawkesbury River, has evidence of occupation dating to >40,000 years (Nanson *et al.* 1987), although the authenticity of these artefacts and the security of their context has since been queried by many researchers. Until fairly recently, archaeological shelters in the Blue Mountains [at Shaw's Creek KII: 14,700 yrs BP (Nanson *et al.* 1987)] on the south coast [at Burrill Lake: 20,000 yrs BP (Lampert 1971a)] provided the earliest securely dated evidence for occupation in the region. Open sites on the Cumberland Plain and on the coastal strip now provide good evidence for Pleistocene and early Holocene occupation in Sydney.

A Pleistocene sand body on the Parramatta River, excavated in three different development contexts as sites CG1, RTA-G1 and GG3 (JMcD CHM 2005b, 2005e and 2006a) has returned the oldest date for the region (30,735 ± 407 BP Wk-17435). This extensive sand body was first occupied during the Late Pleistocene at which time an assemblage dominated by silicified tuff artefacts was found. The upper limit for the silicified tuff assemblage is bracketed by age determinations of c.6,000-8,000 BP. Ground stone hatchet heads here are dated to c.3ka, consistent with most age estimations for the earliest appearance of this artefact type in the region[7]. Heat treatment

---

[6]Some sites have multiple determinations, so the number of sites included in this list is less than 121. Dates as reported in Attenbrow 2002, 2004; Balme *et al.* 2001; JMcD CHM 2005d, e, f and 2006c; Mary Dallas Consulting Archaeologists 2002; McDonald 1994; Rich 1993.

[7]An earlier date from Discovery Point (see below) may need to be corrected for the marine reservoir effect (Gillespie 1991, Head 1991) while the earlier date from Jamison's Creek was based on the association of a hatchet with a dated

of silcrete, and backed artefact production occurs in the uppermost units of the sand body - the top 2-3,000 years of which had been truncated by modern buildings. The Parramatta sand sheet provides significant new information about timing and patterns of Aboriginal occupation in region. The three salvage programmes completed here (Table 4.1) provide evidence for distinct and clear changes in the archaeological record through time.

Three other early sites are in open Pleistocene dune contexts in coastal Sydney [at Kurnell, Prince of Wales and at Discovery Point (formerly Tempe House): respectively, Smith *et al.* 1990, Godden Mackay and Austral Archaeology 1997 and JMcD CHM 2005c]. The geomorphic contexts of these are significant both in terms of regional models of occupation and because they have provided a deep matrix which has survived 200 years of European impact.

The site at Discovery Point, as with other stratified open sites across the region, had silcrete in its more recent and intensive occupation phase(s). A calibrated date of 10,700BP relates to the earlier silicified tuff assemblage, which is characterised by relatively sparse deposition rates, non-blade technology and stone rationing behaviour (JMcD CHM 2005c). The most intensive period of artefact production on this sand sheet is characterised by concentrated backed artefact production: the frequency of backed artefacts in the Discovery Point assemblage (6.7%) is higher than in any other reported assemblage in the region (JMcD CHM 2005c: Table 23). Subsequent excavations (JMcD CHM 2006c) were dated using shell samples, which provide an age range between 3,500 - 5,000 BP. The midden deposits were associated with heat-treated silcrete-dominated lithic assemblages. The spatial and vertical configuration of dated shellfish remains and artefact-only assemblages strongly suggests that the extensive distribution of cultural deposits across the sand sheet was created through multiple, structured, short-term occupations: e.g. tool making does not occur in prime food consumption locations. The presence of this technology in association with estuarine shellfish gathering has not been documented previously – and certainly more recent (i.e. in the last millennium) midden assemblages reveal a paucity of associated stone artefacts (e.g. McDonald 1992a, 1992b).

The high frequencies and bulk on-site production of backed artefacts place the Discovery Point assemblages in the Middle Bondaian phase, thought to occur between 1,000-4,000 BP for the greater Sydney region (JMcD CHM 2005a:13; although see Attenbrow 2004). One of the age determinations (3,860–4,218 Cal BP) is at the early end of this previously accepted chronology; the other age determination (4,547–4,940 Cal BP) is earlier, even allowing for possible marine reservoir effect (Gillespie 1991, Head 1991). This site suggests that the Middle Bondaian phase could be pushed back to c.5,000 BP or that 'bulk backed artefact production' in the Sydney area should be re-defined.

Falling within this age bracket are Sydney's oldest dated human remains with its associated backed artefacts – which provide the first archaeological evidence for death by ritual killing in Australia (JMcD CHM 2005f; McDonald *et al.* 2007). Seventeen stone pieces were recovered with these excavated human remains. Two artefacts were embedded in the spine of the skeleton, and another two were found within or adjacent to vertebrae. Three small tips were conjoined to larger pieces making a total of 14 backed artefacts. Usewear on three conjoin sets indicates breakage resulting from hard impact. A fragment of hip bone submitted for dating[8] returned an age determination of 3,480 ± 30 radiocarbon years (CAMS-120202). The calibrated range is 3,630-3,721 cal BP (92% probability).

Punishment by spearing in a ritual fashion is documented in many parts of Australia (e.g. Backhouse 1843, Roth 1909), and is referred to in Sydney's ethnohistoric literature in the spearing of Governor Phillip's Gamekeeper, MacIntyre (e.g. Tench 1793). The Narrabeen man documents the use of backed artefacts as barbs in death spears. It suggests that some social practices in Sydney during the ethnohistoric present can be traced back for almost four millennia.

Hiscock and Attenbrow have argued (2005:142; and see Attenbrow 2004; Hiscock 2002) that the dramatic increase in artefact accumulation rates between 3,000 and 1,000 years ago, was hearth (Kohen 1986).

[8]This bone fragment was submitted for dating after consultation and full informed consent from the Metropolitan Local Aboriginal Land Council.

a response to climatic change. The onset of ENSO dominated climate initiated a trend to drier and more variable rainfall. Hiscock and Attenbrow hypothesize that those conditions stimulated change in foraging practice, perhaps with a shift to higher mobility.

The spearing of the Narrabeen man c.3,700 years ago could indicate that alterations to social organisation and group interaction may have commenced earlier than 3,000 cal BP. The Narrabeen man was slain at a time of high sea stand (Haworth et al., 2002) and may well signal inter-tribal conflict due to increased territoriality and social pressure in this context. The bulk backed artefact production at Discovery Point between 3,500-5,000 cal BP supports this earlier timing.

And the archaeological signature for coastal backed artefacts – at both Narrabeen and Discovery Point - is very similar to that found across the silcrete-dominated Cumberland Plain. Technological similarities, particularly those demonstrated between 3,500-5,000 BP at Parramatta and Discovery Point indicate regional connections – or population movement – in the Middle Bondaian may well have been more fluid than during the last millennium (as observed at contact).

The first three stages of the Rouse Hill Infrastructure Project (McDonald et al. 1994; JMcD CHM 2005a, 2005d: see Table 1) have challenged many preconceptions about Cumberland Plain archaeology. A number of lithic reduction strategies have now been defined, most of which are geared towards microblade production. Stream order and distance to lithic sources have found to be important indicators of site size and complexity. By testing sites in an array of landscapes we can now predict how lithic densities might vary in association with increased stream order (i.e. water permanence) and with distance from streams (JMcD CHM 2007). We have also begun to identify how this pattern of site usage has changed over time – Pre-Bondaian assemblages on Shale appear to be located closer to major creeks at the sandstone – shale interface (preserved in deeper deposits) while Bondaian assemblages are distributed more widely across the landscape.

The processing of plant and animal material has been documented, including the processing of plant material using backed blades (McDonald et al. 1994:283-5; and see Robertson 2005). Quartz artefacts are found on many sites (in small percentages) demonstrating the movement of raw materials from Hawkesbury Sandstone country onto the Plain.

The open site archaeology of the Cumberland Plain has produced extensive evidence for technological production and living sites in the open – and we are continuing to increase our understanding of these assemblages and how they have varied through time. The extensive and unprecedented scale of investigation continues to alter the way we interpret the Aboriginal prehistory of the Sydney Basin. Just as the UMCC work (Attenbrow 2004) demonstrated that regional prehistories should not be written based on a single or few shelters – similarly the Cumberland Plain work indicates that regional prehistories must embrace open site archaeology to explain changing settlement patterns.

Our understanding of the mosaic of habitation indices across the region has evolved since Fred McCarthy (1948, 1964) first characterised the Eastern Regional Sequence (ERS). The currently used terminology in the Sydney region for phases within the ERS are Pre Bondaian (previously Capertian), followed by Early, Middle and Late Bondaian (Hiscock and Attenbrow 2005; JMcD CHM 2005a, b, e). Recent dates allow us to refine the timing of these (Table 4.2).

The change to the Small Tool Tradition is a continent-wide phenomenon of the mid-Holocene and is generally assumed to have been associated with widely ramified social changes (although see Hiscock and Attenbrow 2005: 143). Phases within the Bondaian are based on the introduction and subsequent decline of backed implements, and the increasing predominance of the bipolar technique and the use of quartz. A change in the proportions of raw materials throughout these phases is considered a factor, arguably related to changes in access to and acquisition of supplies (cf. McNiven 1999). The introduction of ground implements around 4,000 BP and shell fishhooks in the last 1,000 years were major technological innovations.

The social import of changes in stone artefacts throughout the Bondaian has been extensively debated (e.g. Attenbrow 2004; Beaton 1985; Hiscock 1986, 1993; Hiscock and Attenbrow 2003,

Table 4.2: The Eastern Regional Sequence (dates after JMcD CHM 2005e)

Period	Approximate age estimate	Description
Pre-Bondaian	30,000-c.8,000 BP	Preference for silicified tuff. At great distances from sources this material was augmented with quartz and unheated silcrete (coarse-grained raw materials). Cores and tools vary widely in size. No backed artefacts, elouera or ground stone. Unifacial flaking was the predominant technique, bipolar flaking was rare. A date of 30,000 BP indicates the earliest identified time frame for this Phase.
Early Bondaian	8,000 - 4,000 BP	There is a decline in silicified tuff as preferred stone and more use of local raw materials, especially at sites occupied for first time. Backed artefacts were uncommon until the later stages of this phase, bipolar flaking occurs widely although relatively rarely at individual sites. Unifacial and bifacial flaking were the dominant technique.
Middle Bondaian	4,000 - 1,000 BP	Stone raw materials vary between and within sites over time. Main phase of backed artefact production. Asymmetric flaking with platform faceting was adopted. Smaller cores and tools, bipolar flaking increases, ground stone artefacts appear infrequently (at less than half of the dated sites). Elouera are rare.
Late Bondaian	1,000 BP to European Contact	Stone raw materials continue to diversify. Backed artefacts possibly decline becoming rare or absent particularly in coastal sites. Bipolar flaking became a little more common. Ground stone is found in low frequencies at the small number of dated sites – but was identified as the major tool type at contact. Elouera a little more frequent.

2005; Johnson 1979; Lourandos 1985; Ross 1985; M.A. Smith 1982; Williams 1985). A dual social system is generally thought to have become dominant along the south-eastern seaboard with the introduction of fishhooks in the last millennium (Walters 1988). The role of women within the economic productions of this may have been radically different from that in place beforehand, where it is assumed that male hegemony was stronger (Hamilton 1980, Walters 1988). The significance of such social change is discussed below.

A behavioural model for the Sydney region
Two previous behavioral models for Aboriginal land use in the Sydney region were those of Ross (1976, 1988) and Kohen (1986, 1988; Kohen and Lampert 1988). Both models favoured a coastal: hinterland social division with coastal and inland groups operating independently within culturally prescribed areas. Seasonal movement inland by coastal tribes was not envisaged as the cultural boundaries were seen as a barrier to such movement. Ceremonial activities were seen by Ross as the critical element in social interaction between the linguistic groups in the area. The main differences between these two models were the nature of regional contacts and interaction, and the location of the designated linguistic boundaries.

These earlier behavioural models are not completely overturned by the vast amount of excavated evidence collected since their inception (Table 4.1), although it is now possible to describe these in terms of residential mobility, stone tool logistics and rock art correlates. The model[9] used in this work proposes the following:

During the late Pleistocene/early Holocene, people had high residential mobility and travelled considerable distances between base camps. They camped near the resources they were exploiting. Residential mobility was high but logistical mobility was not: people did not engage in extensive preparation for specialised foraging. Groups moved within large territorial areas, and

---
[9] I would like to acknowledge Beth White for her contribution in developing this model over the last 10 years - and for analysing the many hundreds of thousands of artefacts we have salvaged in the numerous JMcD CHM open site projects.

the preferred raw material was silicified tuff. We know that this is now sourced primarily from the Nepean River gravels: prior to sea level rise it is possible that the Hawkesbury River would have been a source for this material along its entire length. At greater distances from the Nepean River, local stone was also relied upon. Transported silicified tuff was carried as large cores and tools to make and maintain wooden implements and to butcher animals. While cores and tools were quite large, stone was used sparingly, and few artefacts were discarded. Cores were continually transported as portable raw material supplies. Backed artefacts were rarely made.

Sea levels rose and stabilised after 6,500 BP. Groups that previously occupied the now drowned coastal strip were forced inland. Bays and estuaries formed within areas that were previously low lying valleys and flats. The region's population continued to increase. After 5,000 BP technological strategies underwent substantial change, particularly evident in the emphasis placed on the use of locally available stone. By 4,000 years ago people appear to have occupied smaller territories and on a more permanent basis. Some groups lived full-time on the Cumberland Plain, some full-time in the surrounding sandstone country. Groups moved between these two biogeographic zones. Residential mobility decreased, but logistical pre-equipping (*sensu* Veth 1993) increased. While it is difficult to know what the foraging and logistical ranges of the Cumberland Plain people were – we assume that they used residential bases and defined foraging ranges on annual and extended cycles (after Binford 1982). We have been able to identify a number of sites which qualify as base camps (e.g. RH/CD12, RH/SP12 South, RH/CC2, SA23) while others contain single isolated or several short term knapping events (e.g. RH/SP13J, EC3, Aus1, WH3).

Throughout the last 4-5,000 years on the Cumberland Plain people mainly used locally available silcrete. It varied in quality but was used for a wide variety of tasks, with time and effort spent on heat-treating. The development of heat treatment technology may have occurred before this shift in resource focus. The RTA-G1 site indicates glossy silcrete was used between c.6,700 to 5,050 cal BP (Wk-17436 and Wk-17432). Glossy silcrete artefacts also dominate assemblages at Second Ponds Creek [dated between 3,640 - 3,440 cal BP and 470-290 cal BP (Wk-16227, Wk–16226: JMcD CHM 2005d:73; 135)].

Most artefacts, including cores and tools, were small. People prepared stone at local sources and transported smaller pieces of selected materials back to their residential sites. They often undertook further processing of this raw material to improve its flakeability. On some sites (e.g. RH/CD5, RH/CD12, RH/SP9) backed artefact production was practiced at an industrial scale. These tools have been described as standardized, multifunctional, reliable, maintainable and portable with the production geared to raw material conservatism: characteristics related to risk minimization in peoples' initial survival in arid zone sites (Hiscock 1994: 287, 287). Our studies in the fertile coastal Sydney region have shown that there is considerable variability in the morphology of these implements both within sites and across the Cumberland Plain and that they are not standardized. Further, the quantities of stone and numbers of artefacts left behind at production locations belie the idea of stone conservation. An arid/fertile dichotomy and the effects of this on social dynamics may well explain this finding. Conversely, the massive samples that the current open site salvages have generated may well be describing the full range of variation possible with this tool type. Microscopic usewear and residue studies indicate that these were probably multi-purpose tools (McDonald *et al.* 1994, Robertson 2005). As well as being hafted in spears (McDonald *et al.* 2007), they appear to have been hand-held and used to process plant materials and small game (e.g. birds). More intensive rounded usewear and scarring on the chords of some examples also suggests the processing of a range of soft and harder materials.

Technological change continued to take place in the most recent phase of the sequence. Bipolar flaking increased, probably reflecting even more intensive use of local resources (Hiscock 1994). Ground stone implements were ubiquitous at contact, possibly replacing larger flaked stone tools which had been dropped from the tool kit. Backed artefacts may have declined in frequency during the last millennium (Hiscock 2002, Attenbrow 2004), although there is only good evidence that backed artefact production declined on the coastal strip – where fish hooks and organic (wood, shell and bone) tools do seem to have dominated in this time period. While fewer sites have been dated in this time frame (Figure 4.1), there are various examples of backed artefact production in the last millennium and indeed within the last 500 years [e.g. Loggers shelter 780±80 BP (SUA-1124): Attenbrow 2004; RH/SP12 South at 337±37 BP (Wk-16226): JMcD CHM 2005d; Ropesend Creek rockshelter 230±50 BP (Beta-65747): Rich 1993].

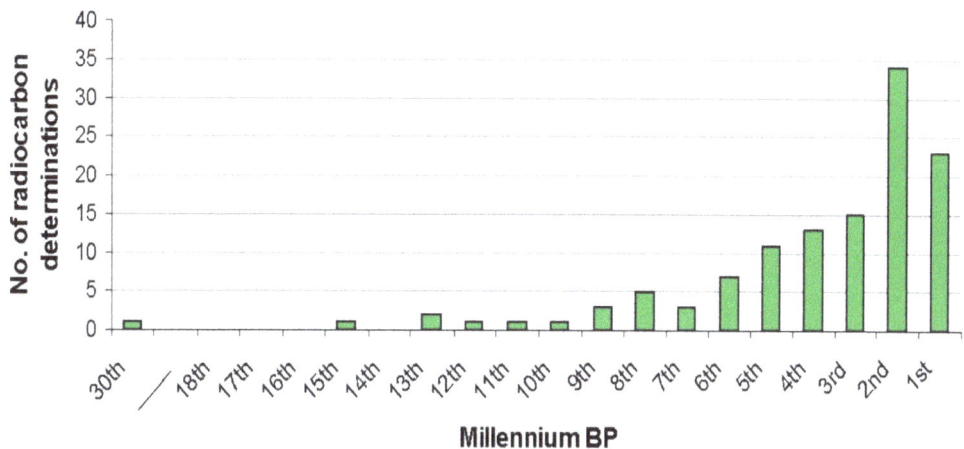

Figure 4.1: Radiocarbon determinations from the Sydney region, millennial sequence.

It is not known whether the gender divide observed on the coast (exemplified by use of fishhooks by women and fishing spears by men) was replicated on the Cumberland Plain. The starch evidence that plant material was processed using backed artefacts (McDonald *et al.* 1994) may hint that women used these tools to process their collected contribution to the diet. It certainly seems unlikely that the dual social system in place in the last millennium was restricted to the coastal strip. Woman's roles in the social system at this time are likely to be demonstrated by gender-sensitive aspects of the archaeological record (e.g. in art and material culture).

Summary of the model with likely art correlates

The Sydney Basin has been occupied since c.30,000 years ago. The earliest stone assemblages, as with Pleistocene evidence from elsewhere across the continent (O'Connor *et al.* 1998; Smith 1987, Thorley 1998, Veth 2005) represent sparse – but real - populations. The technology being used at this time was flexible and settlement appears to be focussed on riverine resources - now preserved in deep sand bodies. The settlement pattern appears to have been one of high residential and low logistical mobility.

Social networks during the Pleistocene were probably more open and far reaching than recorded at contact. Rock art used in this social context would have demonstrated widespread group cohesion (McDonald 2005). The visually homogenous, multivalent engraved (petroglyph) art graphic which is present across vast tracts of Australia – particularly in the arid zone – exemplifies the type of art used at this time. There is some evidence for this type of art functioning as a graphic system across the Sydney region.

In the late Holocene with the advent of larger and more frequent Pre-Bondaian assemblages, we begin to observe indices of social complexity – emerging territoriality and changing levels of mobility. Regionalised style provinces are likely to have evolved with this increasing social complexity, although it seems likely that it was only in areas where high populations and intense social exchanges occurred – combined with stable art matrices (e.g. the Pilbara, Arnhem Land and possibly the Western Desert) that extensive art bodies are likely to survive from this phase. Much of this early evidence in Sydney may have existed on the (now) drowned coastal plain.

The Sydney region's Bondaian assemblages demonstrate a range of archaeological indices with behavioural and resource structure correlates (Ambrose and Lorenz 1990). Residential mobility decreases, there is increased territoriality, group sizes increase and spatial organisation becomes more structured. There is an increase in the use of localised as well as exotic raw material and assemblage diversity burgeons.

Rock art, and symbolic behaviour generally, is seen as an important facilitator and component of increasing and continuing social complexity across the region throughout the late Holocene. The larger culture group (Peterson 1974) would be likely to distinguish itself from other culture groups: local group identity within the culture group would be increasingly demonstrated. Local group identity may be correlated with language boundaries. The dual social system of the last millennium may be identifiable in aspects of the rock art as with items of material culture.

This model forms the basis for the following rock art analyses.

5

THE ROCK ART OF THE SYDNEY BASIN

There are two art contexts in the Sydney Basin: in rockshelters and on open engraving sites. Shelter sites contain mostly pigment art (pictographs) but occasionally also engravings (petroglyphs). Medium is generally defined as 'the physical materials of which the artefact consists; or the techniques employed to produce the artefact' (Clegg 1977:260). Medium here is defined as art context - for pragmatic and theoretical reasons. A shelter art site is defined as all the art which occurs within the dripline boundaries of a single sandstone rockshelter. An engraving site is defined as all the art which is located across the limits of an open sandstone boulder/platform which is usually surrounded by a soil matrix and/or vegetation.

The location or context, in which the art is produced i.e. in the open or in a shelter, may indicate something about the social context of the art (and the site). For the purposes of this study the social context of the art – and whether it was located within a rockshelter or in the open - formed the basis for the initial separation of the two media.

'Art' is defined loosely as all humanly made marks which occur in repeatable identifiable forms. In Sydney, art results from either the application of material (coloured, black or white pigments) for pigment art or from the removal of the sandstone matrix by a variety of techniques (for petroglyphs or engravings[10]). Both art assemblages were classified using motifs, the majority of which have recognisable forms - human, animal or inanimate objects, or can be categorised as geometric shapes. A large proportion the art consists of unrecognisable or incomplete motifs - particularly in the shelter art assemblage. These motifs were included in the initial analysis of both art assemblages, since they provide a more accurate census of the two assemblages and their general technique information. The motif classification used was based on a taxonomy of visually recognisable figurative forms. These have been given the names of the forms which they most closely resemble but these terms are analytical labels (see Clegg 1977, 1981). This classification is a heuristic devise to facilitate the analysis of the data. Given that Sydney Basin rock art is largely figurative, it was pragmatic to give these motifs names which are descriptive.

Sydney Art in the general scheme of Australian rock art
There is continuing debate regarding the geographic and chronological organisation of rock art styles across Australia (McCarthy 1988; Bednarik 1988; Morwood 2002; Franklin 2004) although there has been little empirically based progress made on Maynard's (1976) tripartite model. Maynard's model saw a pan-continental, relatively homogenous engraving style (mainly tracks and circles) - the Panaramitee - replaced by a series of more regionally diverse but simple (engraving and pigment) art styles, the Simple Figurative. The Complex-Figurative was thought to then follow, this being a more complex (pigment and engraving) group of styles restricted mainly to the north-west of the continent.

Increasing difficulties are encountered by this model as the result of the proliferation of data (and dating evidence) from around the country. Some complex figurative art (e.g. Bradshaw figures in the Kimberley) has been found to have significant antiquity (Watchman

---

[10] 'Petroglyphs' and 'Engravings' are used interchangeably throughout this monograph to refer to the rock art images of the Sydney Basin. Both terms imply that various techniques have been applied by the artists, removing the stone matrix to form the resultant image. 'Petroglyph' as a term has been long-used in the Australian art literature (e.g. Worms 1954, McCarthy 1962, Maynard 1977), but it is generally not as popular amongst Australian archaeologists as 'engraving'. The NPWS (now AHIMS) Site Register refers to this type of art as 'Engravings'.

1993b; Roberts *et al*. 1997). The dynamic figures of Kakadu have also been argued to have considerable antiquity (Chaloupka 1977, 1985, 1994; Haskovec 1992; Taçon and Chippindale 1993) and indeed some have older than established Panaramitee dates (e.g. Watchman 1993b). Complex engraved anthropomorphic motifs of great antiquity have now been recorded in the Calvert Ranges (McDonald and Veth 2006) as have ancient figurative forms i.e., embellishments of archaic faces in a number of arid contexts (McDonald 2005b). This creates the most serious challenge to Maynard's tripartite sequence. Increasingly the Panaramitee has been found to demonstrate significant regional heterogeneity (Morwood 1979; Rosenfeld 1991; Rosenfeld *et al.* 1981; Franklin 1988, 1991, 2004) and localised style graphics are argued to date to the earliest production of engraved art throughout the arid zone (McDonald 2005b). The dating evidence does support a late Pleistocene antiquity for the Panaramitee (Rosenfeld 1993). But there is increasing evidence for continuity of this graphic tradition over time – particularly throughout the Western Desert in engraved art (McDonald and Veth 2007), in central Australian engraved and pigment art and body painting (McDonald 2005c, Rosenfeld 2002) and in several eastern regional assemblages, including the Sydney Basin.

The real problem for the Maynard model with the early dates for complex figurative art is that it is obvious that there is not clear diachronic change from an early, highly structured and non-iconic tradition to the later proliferation of figurative (iconic) styles, with increasing design complexity. More recent research (e.g. Rosenfeld and Smith 2002, McDonald 2005c, McDonald and Veth 2006, Ross 2005) has attempted to explore the social mechanisms likely to have been in place for this to occur – and it seems clear that further dating evidence is likely to demonstrate that an early production of highly complex signalling behaviour accompanied the earliest uses of the arid zone – much as has been seen with the use of art throughout Europe during the Palaeolithic.

A vast number of regional Simple Figurative art bodies have now been documented across the country (Cole and Trezise 1992; David 1992, 1994; David and Cole 1990; Flood 1987; Gunn 1983, 1995, 2000, 2003; Hatte 1992; Layton 1992a; McDonald 1998a, 1999; McDonald and Veth 2006; Morwood 1984, 1988, 1992b; Officer 1984, 1992; Ross 1997, 2003). While being cohesive style regions – usually defined by bio-geological regions - internal variation (both synchronic and diachronic) is found in these.

The rock art of the Sydney Basin fits the definition for the Simple Figurative styles (Maynard 1976; Franklin 1984, 2004). Maynard's model predicts that this art is a relatively recent (Holocene) phenomenon which is supported by this region's archaeological context. Her original definition still provides a good general description of the region's art.

> the style is dominated by figurative motifs ... the majority of (these) ... conform(ing) to a pattern of crude naturalism. Whether the motif is engraved or painted, in outline or solid form, it usually consists of a very simple silhouette of a human or animal model. Most portrayals are strongly standardised. Human beings are depicted frontally, animals and birds in profile, snakes and lizards from above. Normally only the minimum visual requirements for recognition of the motif are fulfilled by the shape of the figure. (Maynard 1976:200-1)

Certain variations to her definition are necessitated by the current research. For instance, human figures are sometimes depicted in profile, while some animals (e.g. the echidna) are not always (Officer 1984, McDonald 1987). Franklin's work on macropods and men in the region indicates that Form, Technique and Motif are equally important in differentiating between the different regional Simple Figurative styles (Franklin 1984:89). Each of these assemblage characteristics is equally able to provide stylistic information which can characterise the region (see also McDonald 1993b). This is significant when evaluating quantitative as opposed to qualitative data (viz. Ashton 1983), and for the analyses undertaken for this research.

Previous work in the Sydney region – the growth of a data base

The art of the Sydney region has long had a fascination for European observers (e.g. Phillip 1789) and more recently rock art analysts. The first major publication of systematic rock art recordings from the Sydney region was by W.D. Campbell (1899). Campbell recorded (from the back of his horse) a vast number of (mainly) engraving sites encountered during his duties as Government Surveyor.

From the 1940's onwards there was an increase in the publication of engraving sites in the region. Fred McCarthy was the most prolific recorder and publisher (see bibliography) over a forty year period. Ian Sim was another prolific recorder with his published material being augmented by an enormous body of field notes and drawings, now archived by the AHIMS as the 'Sim Collection'. John Lough also recorded a large number of sites; many re-recordings of earlier recorders. These, however, were not made accessible to subsequent researchers and very few are published.

While early interpretations of the Sydney engravings were based loosely on borrowed ethnographic material from other regions (McCarthy 1939a, Elkin 1949) the first systematic analysis of this art was undertaken when Lesley Maynard (then McMah) undertook her Honours research in 1965. This was the first quantitative analysis of any body of art in Australia, the aim being '(to) produce, first a typology of the engravings, and second, a spatial distribution of traits, based on the typology' (McMah 1965:7). The results of the analysis (which involved the use of hand-sorted punch cards) indicated that:

1) there are definite patterns of distribution in both north-south and east-west planes; and,

2) the differences between one end of the range and the other may be ascribed to cultural causes - except those obviously resulting from the stimulus of different environments. (McMah 1965:75)

McMah's geographic divisions were too coarse to allow more than a glimpse of trends from north to south, but she did identify two distinct artistic units. One of these was located around the Upper Hawkesbury while the other was south of Botany Bay. She distinguished these areas largely by the presence of particular motifs. The first of these broad style areas has not survived the proliferation of data. The boundary for the second style area has been redefined as the Georges River (McDonald 1985a).

Since McMah's seminal work, research has been directed at localised assemblages and/or specific research questions. Konecny (1981) was concerned with both engraved and pigment art and investigated possible aggregation locales (viz. Conkey 1980) within the Sydney area. While identifying no specific aggregation locales, Konecny suggested that functionally similar sites could be identified in different areas in both media.

Laurajane Smith's work was concerned with identifying archaeological patterning across ethnographically reported tribal boundaries (Smith 1983:1). Her analysis was based in the Mangrove Creek/Macdonald River area, and involved both art components. Smith tested the linguistic boundary between the *Darkingung* and *Guringai* groups using Mangrove Creek (not Mooney Mooney Creek: Capell 1970:22; Map 1). Thus her analysis characterises variation within the *Darkingung* language area. Topography was considered to be a primary factor contributing to stylistic variation here.

Officer (1984) examined the art of the Campbelltown area on the Georges River. A sample of 57 shelter art sites (SPG 1983) was used to explore and describe the formal variability within a local body of art. Officer formulated a detailed hierarchical motif classification and argued for both functional and 'casual' interpretations for the considerable heterogeneity identified in this art body. An interrelatedness of motifs across the two art components was perceived, and Officer identified strong ties between the coast and hinterland, despite a linguistic boundary and other evidence for a cultural dichotomy here.

Franklin's (1984) work explored Maynard's (1976) definition of the Simple Figurative styles, and was thus involved with comparing the Sydney region with regional assemblages from Port Hedland, Cobar, south-east Cape York and the Grampians. This analysis confirmed previous conclusions made by McMah about the Sydney region.

Little art research has been achieved in conjunction with Cultural Heritage Management work, partly because the majority of urban development in the Sydney region occurs on the Cumberland Plain (see Figure 1.1) where there is no rock art. In several instances, however, a research oriented approach has been achieved (McDonald and Smith 1984, McDonald 1988a).

The Sydney Basin Art Style: a current definition

Schema

The word 'schema' is used to denote the manner of depiction (or abstraction) of an image from the object that is being depicted.

> A motif is a recurrent visual image which has a particular arrangement of components ... motifs are therefore the objectified expressions of schemata - the standardised pictorial forms which result from a consistent mental template (consistent with the cultural group of the artist). (Maynard 1977:396)

The art of the Sydney Basin has been described as a Simple *Figurative* style, because of the high level of recognition (for modern etic observers) between the art and a 'natural' assemblage (human figures, animals, birds, fish etc.). *Simple* Figurative also implies that the schema is not a complicated one: that a minimum amount of detail is provided.

The schemata used for the two art components in the Sydney Basin are very similar. This is manifested in the *Motif* range used, in the *Form* of these and - in particular - the general *Character* of the regional motif assemblage (as defined by Maynard 1977). The main difference between the two components is *Technique* but also *Size*. Size differences are mostly due to the differences in 'canvas' size (i.e. available rock surface). Extremely large motifs (life-size or larger) are occasionally found in shelter art assemblages (e.g. AHIMS #'s 45-2-48, 45-2-118[11]), while very small motifs (miniature or smaller than life-size) sometimes occur on open engraving sites (e.g. AHIMS #'s 45-2-224, 45-6-43).

There are also some differences in Form. Motif form for the pigment art comprises a variety of outline, infilled and combination forms. The vast majority (97%) of the engraved motifs are outline only. The only consistently infilled engravings are culture heroes which are generally decorated with series of pecked lines of dots. Intaglio forms (i.e. fully pecked infill) are extremely rare.

Chronology

Some Panaramitee style sites (cf. Maynard's 1976 definition) have been identified within the region. These assemblages contain predominantly tracks and circles produced in pecked intaglio, not pebraded outline form. Interestingly, these occur almost exclusively within shelter sites. It is possible that residual Panaramitee motifs are also located in open contexts, and that weathering and the association with figurative motifs is masking their presence (McDonald 1993b).

---

[11] AHIMS site identification numbers are used in the text in preference to the sample numbers used for sites in the multivariate analyses. This form of identification is potentially more meaningful to the reader (McDonald 1994: Appendices 5 and 6 identifies sites by both means).

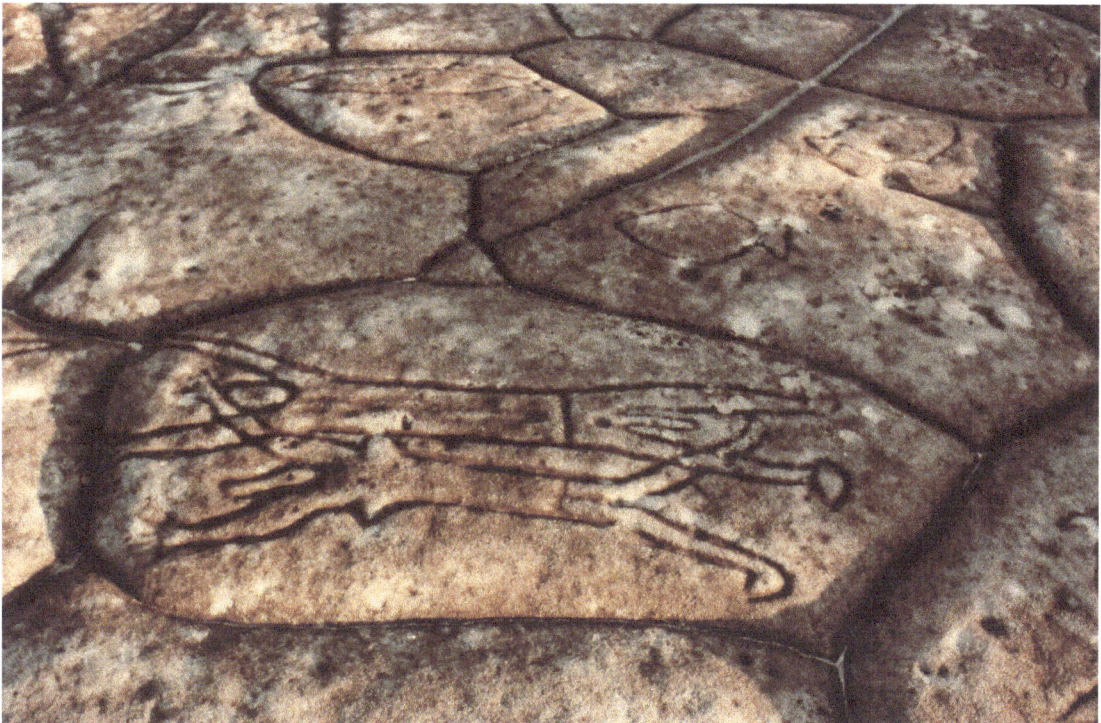

Figure 5.2: Engravings in Ku-ring-gai Chase National Park. Male and female figures and fish positioned in a tessellated pavement.

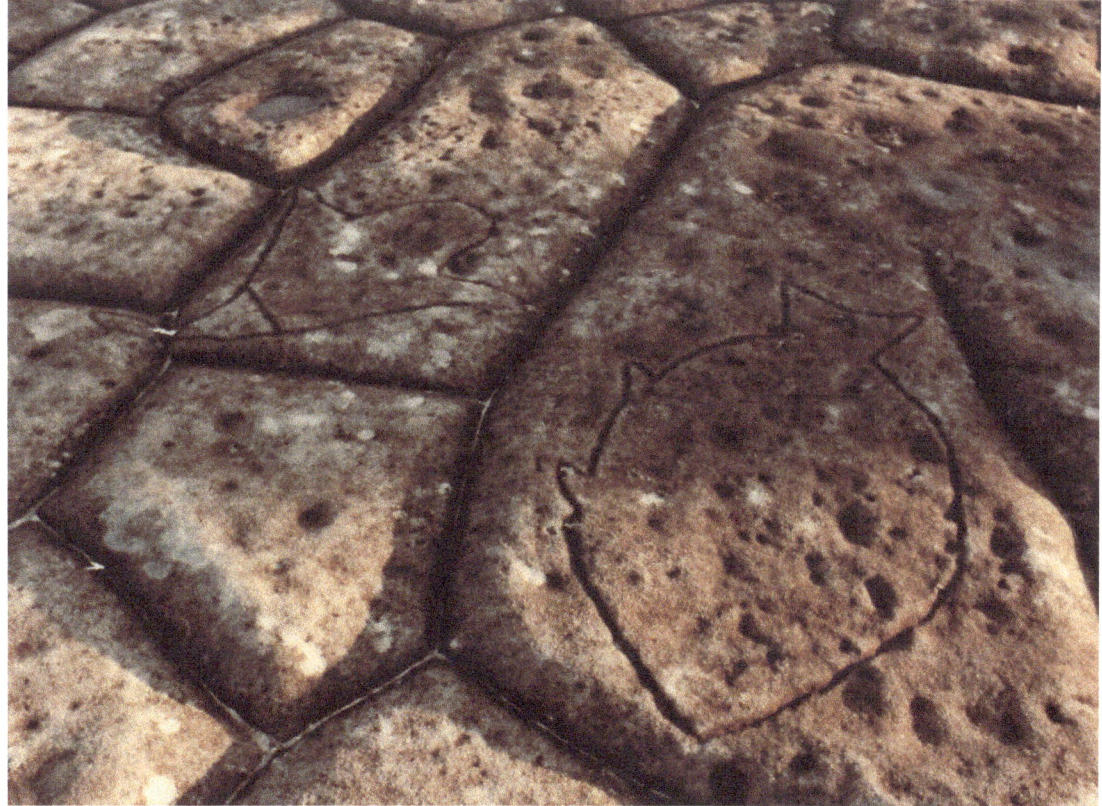

Figure 5.3: Engravings in Ku-ring-gai Chase. Fish and bird motifs positioned within and across natural tessellations.

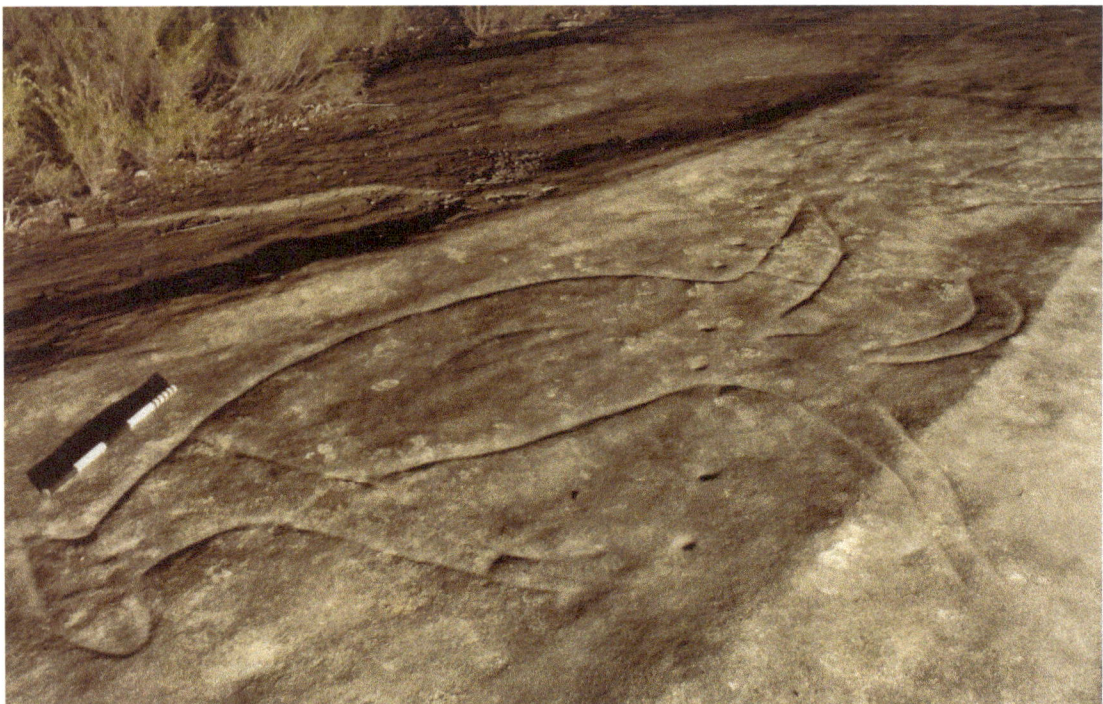

Figure 5.4: Large macropod engraving at Maroota, south of the Hawkesbury River. This macropod is being struck by two boomerangs. The row of pits crossing this motif traverses the entire assemblage.

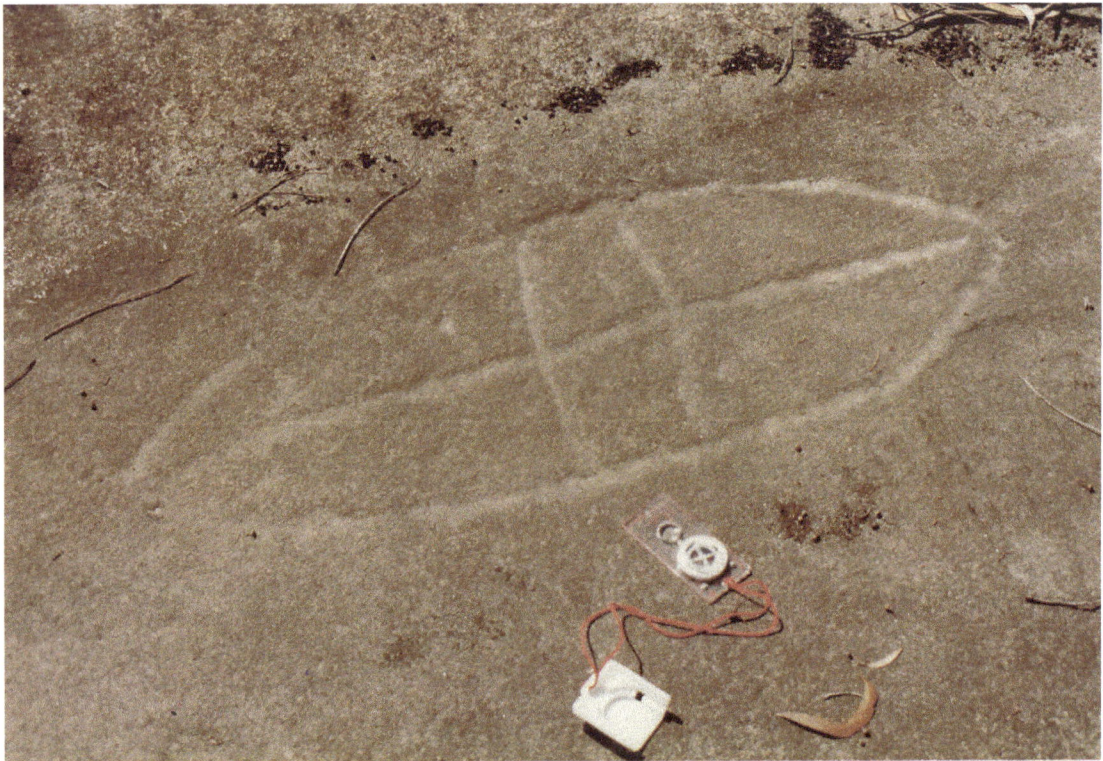

Figure 5.5: Engraved shield motif near Berowra Creek. The internal design on this shield matches early ethnohistoric descriptions of the St George Cross (Design 2B in analysis of compositional details).

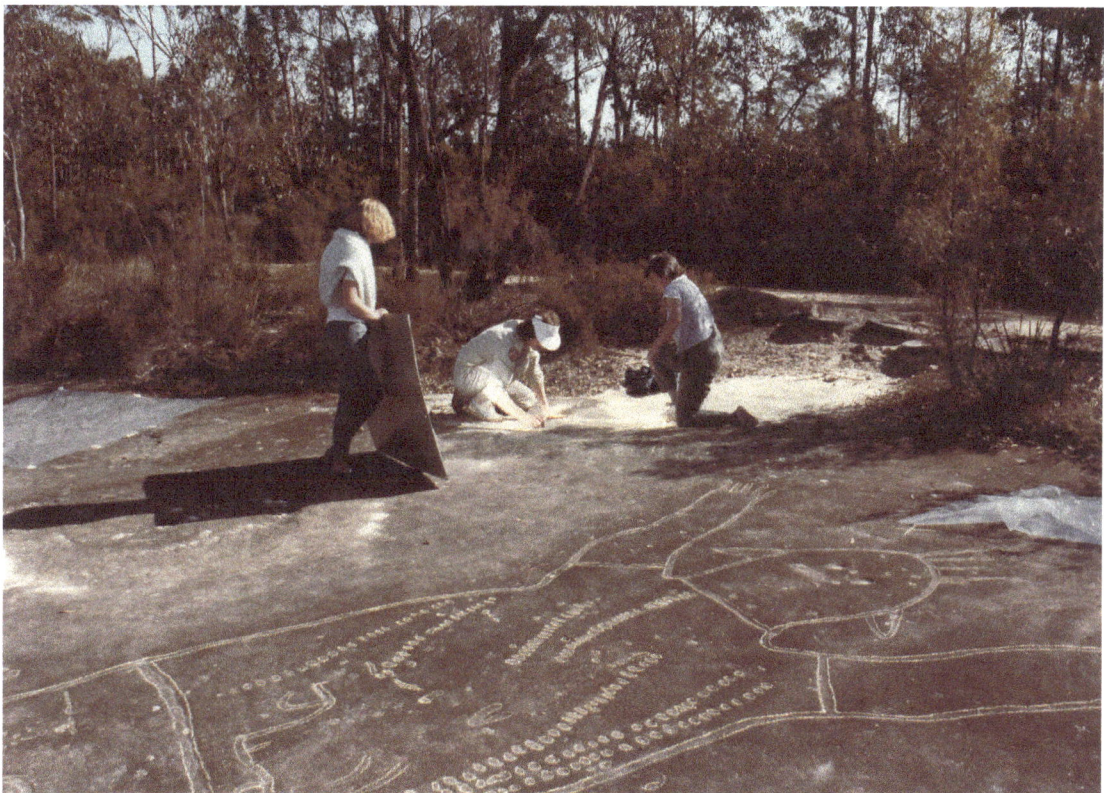

Figure 5.6: Devils Rock Maroota. Recording engravings in daylight using large acetate mirrors. In foreground is a Biaime culture hero. Compare the size of this motif with the humans in the background.

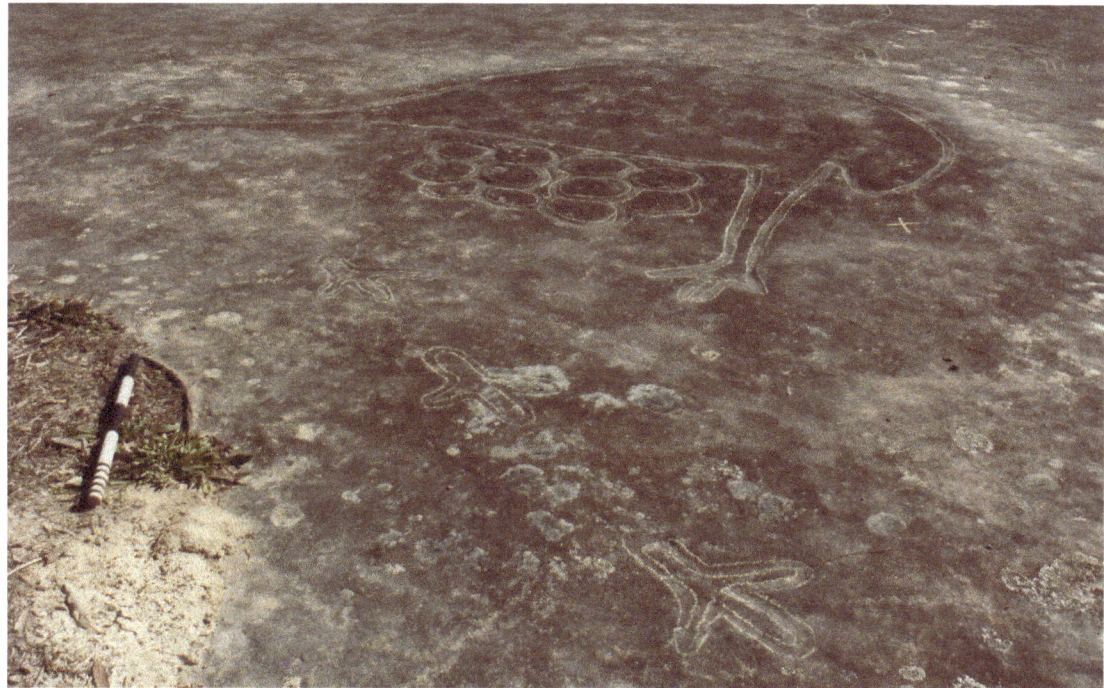

Figure 5.7: Large emu with clutch of eggs. The emu tracks are in outline form unlike the pecked intaglio version often found in the region's rockshelters. These motifs were chalked at night using oblique lighting. A s87 permit is required to undertake this type of recording.

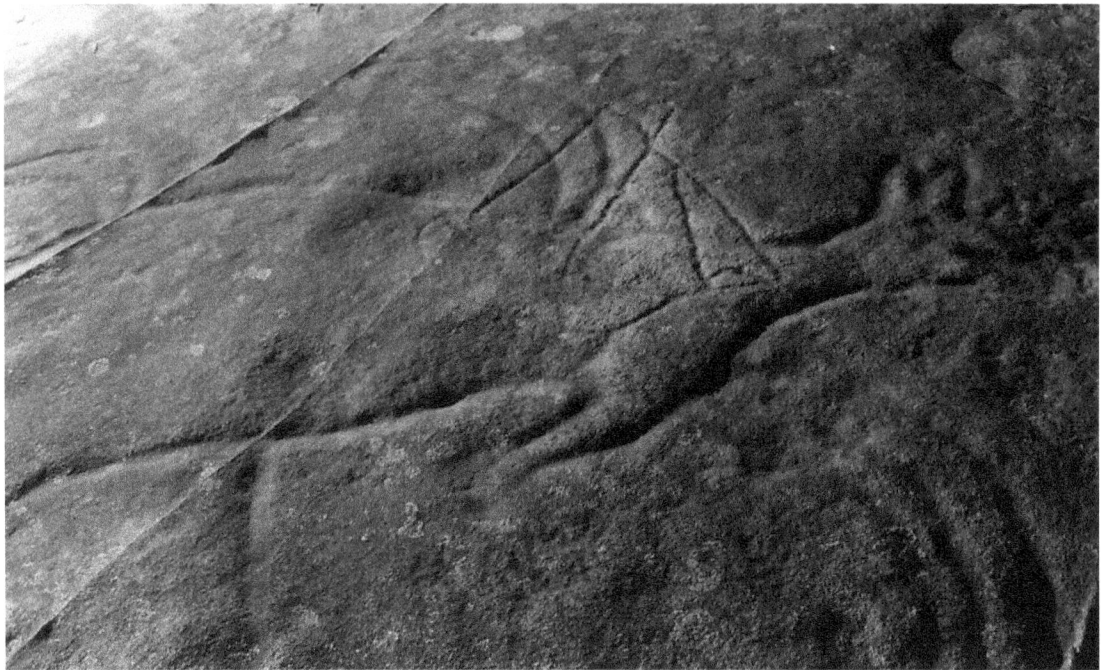

Figure 5.8: Macropod motif, superimposed by a post-contact sailing ship.

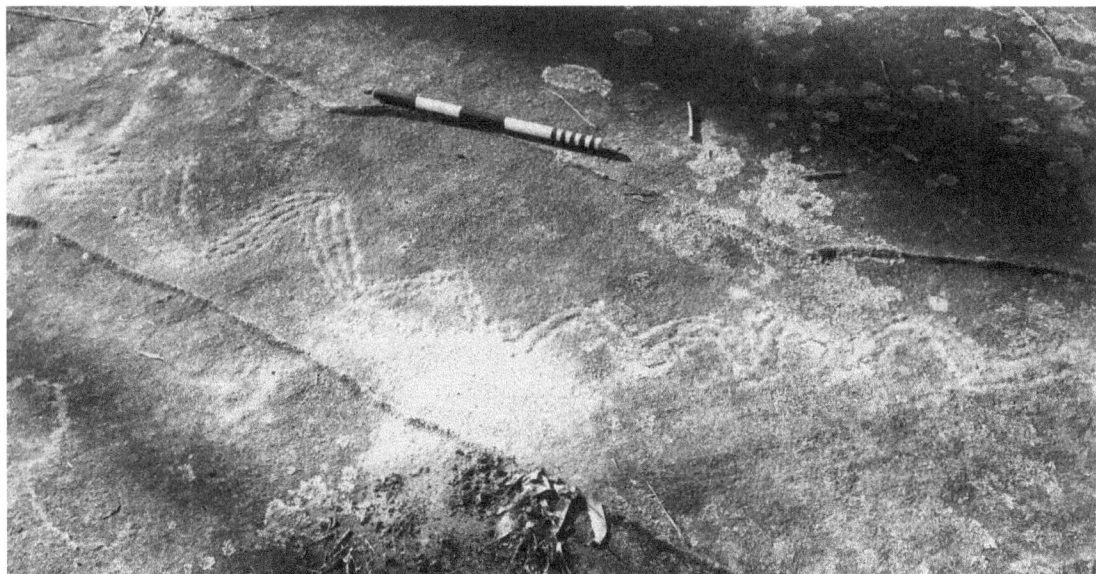

Figure 5.9: Engraved snake motif consisting of five parallel zig-zag lines. The individual peck marks are clearly visible.

The Samples

The samples for the two art contexts comprise 717 engraving sites and 546 shelter art sites. These figures represented 39.5% and 32.7% (respectively) of the known sites in the region in 1994. Based on these samples, the following regional characteristics of the art assemblages are summarised:

- motif and technique information;

- average assemblage size;

- topographic location;

- site associations;   and,

- the unique or unusual aspects of each component (e.g. engravings within shelters, the occurrence of open pigment art, vertical engravings etc.).

The Open Engraving sites

Assemblage Size

The average engraving assemblage contains 10.9 motifs. The largest site in the region (Burragurra; NPWS # 45-3-404) contains 174 motifs; and there are many sites (137: 19.1%) which contain only one motif.

The majority of sites (83.4%) have less than 16 motifs and about two-thirds (66.1%) have less than eight motifs (Figure 5.1). Sites with larger numbers of motifs are rare, and only 61 engraving sites (8.5%) have more than 30 motifs present. Four sites only (0.6%) have more than 100 motifs present.

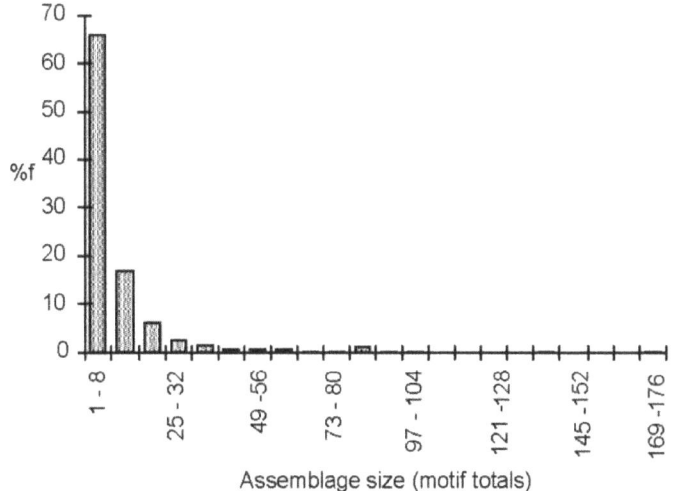

Figure 5.1: Engraving Sites. Percentage frequency of different assemblage sizes.

Site Nature and Topographic Location

The vast majority (97%) of open engraving sites are located on horizontal sandstone platforms. Only twenty-two open engraving sites have been identified on vertical boulders, most of these adjacent to waterways. These engravings on flat surfaces face the water. Several engravings on vertical faces also have been found on low clifflines adjacent to major creeks: others are located on the outside wall(s) of shelters. Given the interior dripline definition for shelters, this category excludes those which are located inside shelters.

One only open pigment site is known (AHIMS # 45-6-1411). This consists of a red pigment drawing on an open cliff face protected from the elements by a slight rocky projection. This motif is on Berowra Creek and is associated with vertical engravings.

Topographic position is summarised according to three broad categories: ridge tops, hill sides and valley bottoms. These locations were defined by Attenbrow (1987) and her definitions were used: the cut-off point for ridge tops and valley bottoms is 5m in elevation below and above (respectively) the ridgeline and valley bottom.

Distance from permanent drinking water was also recorded for each site. Distance to a second order stream was the general rule (McDonald 1996); with distance rounded off to the nearest 50m. Sites recorded by me during fieldwork had a greater precision. This type of data belies the presence of springs, rock holes and soaks which would have provided water, at least periodically, for knowledgeable Aboriginal inhabitants.

More than half of the engraving sites (401 sites: 55.9%) are located on ridgelines. Hill side locations are the next most common (296 sites: 41.2%) while valley bottom locations are relatively rare (20 sites: 2.8%). Almost half (45%) of those in valley bottom locations were vertical engraving sites, on boulders adjacent to major waterways (e.g. Berowra and Cowan Creeks). Slightly more than half (59%) of the vertical engravings, however, are located more than 5m in elevation above the water and therefore in the hillside zone.

The average distance to drinking water from any engraving site in the region is 650m and the maximum distance is approximately 3km. The minimum distance is 2m (on-site rock wells and/or creeklines).

Site Associations

The main site type association for open engraving sites is grinding grooves, although this is fairly rare. Only 96 engraved platforms (13.4%) also having grinding grooves. Water channels (i.e. pecked and abraded lines which direct water seepage around potholes: McDonald and Smith 1984) occur on a small proportion of engraving sites (22 - 3.1%). All sites with water channels also have grinding grooves present. Of the sites which contain engravings and grinding grooves, 23% also have water channels.

Only a very small proportion of engraving sites have shelter art associated with them. By definition, this association can only occur when the open platform and shelter occur within the one (usually massive) outcrop of sandstone. Less than 10 (1.3%) such occurrences are noted in the sample. This does not indicate that Aboriginal artists did not perceive of a contextual relationship between the two art forms. Rather, the location for the production of the two art forms was, by the nature of the two physical contexts, quite distinct. Engraved motifs amongst pigment art *within* shelter art sites indicate that the two components were not strictly separated. This is discussed further below.

Another rare association for engraving sites is the association with stone arrangements (4: 0.6%). It is notable, however, that 17 such recorded site associations do occur within the region (but the art assemblages of 13 of these have not been recorded in sufficient detail for inclusion in the current sample). While stone arrangements are unlikely to be found on an engraving site, it is likely (21%) for a stone arrangement to have engravings associated with it.

Very few site records indicate an association of archaeological deposit on open engraving sites (e.g. stone artefacts). This may be due to recorder indifference (or ignorance) to this type of evidence as many engraving sites do have sparse scatters of lithic artefacts in pockets of soils and/or vegetation (personal observation). Open exposures of sandstone generally do not encourage the build up of deposit, and it may well be that artefact manufacture rarely occurred in conjunction with the activity of engraving.

While open engravings have sparse associated evidence, the majority (>95%) of the *vertical* engraving sites are associated with extensive open midden deposits.

Motifs

A total of 7,804 motifs were analysed from the 717 engravings sites in the sample (see Table 5.1 and Table 5.2). The most common motif in the region is the human footprint ('mundoe' - 17%), followed by unidentifiable motifs[12] (15%) and fish (12%). Several other individual motifs figure reasonably strongly: bird tracks and macropods (7% each) and men (5%). The remainder, however, are present in relatively low percentages.

The clumped motif percentage frequencies (Table 5.2) clearly reveal the subject preferences of the region and indicate that Maynard's definition for the region needs refining. There is a focus on tracks, followed by a preference for marine animals, land animals, anthropomorphic representations and items of material culture. Birds and 'other' motifs (circles, complex-non-figurative motifs and contact motifs) are less common (see Figure 5.2-5.9).

Maynard's distinction between the Panaramitee and Simple Figurative styles is based not only on differing techniques and forms, but also on the motif ranges and preferences. Bird and animal tracks dominate the Panaramitee at 60% (Maynard 1976:193). Conversely the Simple Figurative was described as having predominantly figurative motifs (78% for Sydney - Maynard 1976:193) with animal and human tracks 'nowhere near as dominant as in the Panaramitee style' (Maynard 1976: 200-201). Maynard estimated that track motifs accounted for 5% of the Sydney engravings. The current data overturn this perception: while tracks in the Sydney Basin are not as predominant as in the Panaramitee, they are still the dominant motif type in the region. The possibility that some of the Sydney region's macropod and bird tracks on open site are residual Panaramitee style is discussed below.

The Georges River style boundary

The presence of a major style boundary was identified in the south of the region by McMah (1965) and confirmed by the first stage of the SRAP (McDonald 1985a). A number of motifs are completely absent from sites south of the Georges River. The number and density of engraving sites diminished south of the Georges River (open engraving sites are absent from the artistic repertoire south of the region: Officer 1992, 1993). While the geographic distribution of some motifs can be explained in terms of environment[13] several motifs are restricted to either side of the Georges River 'style' boundary (McDonald 1985a).

South of the Georges River:

- there are no profile anthropomorphs, culture heroes, emus and contact motifs;

- there is a much higher proportion of unidentified motifs;

- the proportions of tracks (particularly *mundoes*) are appreciably lower;

- anthropomorphic, marine and terrestrial depictions are more numerous and there is a reduced motif repertoire and lower number of tracks.

---

[12] Unidentifiable motifs are generally an indication of indistinct or incomplete engravings rather than motifs outside the motif classification.

[13] Whales are restricted to the coast and appear to be concentrated around Broken Bay: all are east of Berowra Creek. These continue down the coastline. There is another cluster of whales around the mouth of the Port Hacking, and the most southerly example is located just south of Bundeena.

Table 5.1: The Sydney Basin Engraving component: motif frequency and %f.

Variable *	Motif	Total	% frequency
1	Man	422	5.4
2	Woman	79	1.0
3	Anthropomorph	183	2.3
4	Profile Anthropomorph	78	0.9
5	Culture Hero	36	0.5
6	Macropod	543	7.0
7	Snake	56	0.7
8	Other Land Animal	312	4.0
9	Emu	76	0.9
10	Other Bird	166	2.1
11	Fish	905	11.6
12	Eel	182	2.3
13	Whale	101	1.3
14	Other Marine Animal	156	2.0
15	Shield	232	3.0
16	Boomerang	303	3.9
17	Axe	45	0.6
18	Other Material Object	218	2.8
19	Unidentified Open	474	6.1
20	Unidentified Closed	710	9.1
21	Hand	19	0.2
22	Human Foot	1,360	17.4
23	Roo Track	190	2.4
24	Bird Track	543	7.0
25	Circle	309	3.9
26	Complex-Non-Figurative	70	0.9
27	Contact Motif	36	0.5
28	Total	7,804	(99.8)

*Variable numbers are those referred to in subsequent analyses.

Table 5.2: The Sydney Basin Engraved Component: clumped motif frequency, %frequency of totals and identifiable motifs

Variable	Motif	Total	% f	% identifiable
1	Anthropomorphic	798	10.2	12.1
2	Terrestrial Animal	911	11.7	13.8
3	Birds	242	3.1	3.7
4	Marine Animals	1,344	17.2	20.3
5	Material Objects	798	10.2	12.1
6	Tracks	2,112	27.1	31.9
7	Other	415	5.3	6.3
8	Unidentified	1,184	15.2	-

As well as these assemblage differences, there are differences in the method of depiction - or the schema - of the art north and south of this boundary.

Macropods and other zoomorphs are the most obvious indicator of this: these motifs are depicted in profile with all four legs south of the Georges River, while to the north they are shown with only two legs [there are two known exceptions to this: a macropod located just south of Port Jackson with four legs; and another near the Lane Cove River; personal observation].

The whales south of the Georges River are less highly stylised than their northern counterparts, with few containing decorative infill or anatomical details (such as gills, eyes).

Another difference is the method of depicting echidnae. North of the Georges River these are depicted in profile; to the south they are depicted from beneath as 'pelts' (Officer 1984:49). Officer also makes the observation that there is similarly a mixed perspective ('simultaneous projection') demonstrated in the pigment art around Campbelltown (Officer 1984:49). The beaked anthropomorphs are a good example of this stylistic convention.

The Shelter Art sites

Assemblage Size

The average shelter art site assemblage contains 26.4 motifs. A total of 14,424 motifs were counted from the sample of 546 sites. The largest assemblage (Swinton's; AHIMS # 45-3-252) contains 857 motifs. A small a number of sites (31: 5.7%) contain only a single motif.

The majority of sites (478: 87.5%) have <30 motifs while almost half (285: 52.2%) have <10 motifs. Only 30 sites (5.5%) in the region have more than 100 motifs; eight sites only have more than 200 motifs (Figure 5.10).

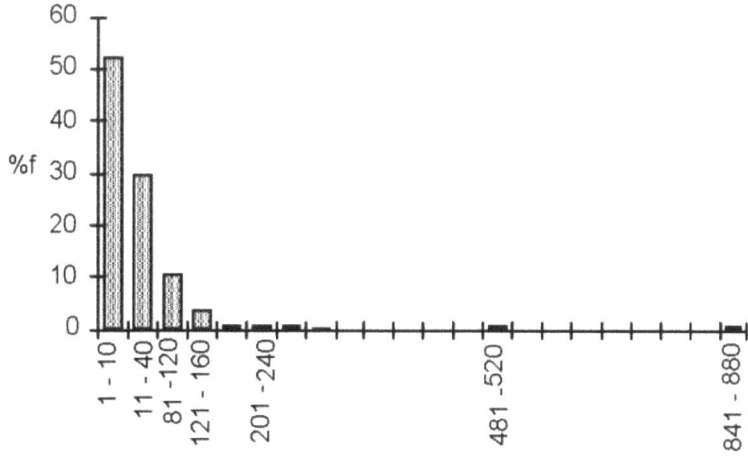

Figure 5.10: Shelter Art Sites: assemblage size information.

While showing the same general pattern as the engraved assemblage (i.e. a proliferation of very small sites with a moderate number of medium-sized sites and a few very large sites) the shelter art assemblages are slightly larger than their engraved counterparts. The average pigment assemblage is twice as large as the average engraved assemblage, while largest shelter art site contains almost five times as many motifs as the largest engraving site.

Site Nature and Topographic Location

All of the art in this component occurs within sandstone overhangs or shelters. No decorated deep caves occur in the region. Art within shelters is commonly located on the back wall, but often occurs on the ceiling and also on the inside lip of the dripline. The entire range of differently-sized shelters has been decorated, although this aspect has not been quantified. The smallest recorded art shelter site was 'Little end shelter' which measures 2.5m x 2.0m x 1.0m (see Figure 2.8). The largest recorded art shelter is at 'Sphinx Trig' in Ku-ring-gai Chase National Park (AHIMS #45-6-258) which measures 40m x 10m x 8m and contains 91 motifs.

The majority of shelter sites (70%) is located on the hillslopes, while the remainder are fairly evenly divided between the ridge top zone (13.7%) and the valley bottom zone (16.7%). The average distance to permanent drinking water for this site type is 262m. The greatest distance from water recorded was 2km while one site was recorded as having water permanently on-site

- with a creekline flowing over the shelter. Given most shelter sites are on hillslopes or in valley bottoms, this increased proximity to water compared to engravings is not surprising.

Site Associations

The major site type association for this art component is occupation deposit: both archaeological and midden deposits. Of the sample used, 138 sites (25.3%) had surface evidence for occupation deposit. Nineteen sites (13.7%) also contained grinding grooves.

Differences were observed in sites either side of the Georges River. Art sites south of this River have a much lower association with deposit (25/181 sites: 13.8%) than the northern sites (113/365 sites: 31.0%). South of the Georges River, seven of the 25 sites with deposit (28%) also contain grinding grooves; while north of the Georges River 12 of the 113 sites with deposit (10.6%) also contain grinding grooves.

Observer bias may be responsible for creating some of these differences and it is likely that the proportion of shelters with both art and deposit is in fact much higher. Indeed, many of the sites recorded during Stages II and III of the Rock Art Project were found to contain signs of occupation deposit which had not been recorded on the original recording. Many art recorders have been interested in recording only the art and not other less obvious forms of occupation evidence. Of the 214 shelter art sites in the current sample recorded by systematic and/or detailed survey (Attenbrow 1987, Gunn 1981, McDonald 1986a, 1987, 1988a, 1990b) 122 sites (57%) also have occupation deposit. Only 43 (20%) of these site definitely have no occupation deposit (i.e. they have sloping rock floors). Many of these sites (49; 23%) have floors assessed as having potential for archaeological deposit (PAD). Taphonomic factors contribute to this problem. Attenbrow (1987) tested shelters with PAD and revealed that almost 90% of these contain subsurface deposits.

Twenty-six shelter art sites (5%) also contain grinding grooves. Nine of these occur south of the Georges River, 17 to the north. The incidence of pigment art and grinding grooves without occupation deposit is rare.

Motifs

A large proportion (41%) of the 14,424 motifs counted for this art component is unidentifiable (Table 5.3). This reflects the nature of the art - complex, less formalised and often heavily superimposed. Often the poor preservation of the art results from instability of the sandstone surfaces within shelters (Spate and Jennings 1983; although see Watchman 1994).

The average number of identifiable motifs in the region is 15.4. Swinton's shelter, as well as having the greatest number of motifs recorded, also has the highest number of identifiable motifs - 653. There are numerous sites (51) which contain no identifiable motifs. The vast majority of the unidentified motifs are depictive[14] and dry pigment (i.e. drawn). Some are painted and some are also stencilled (see below).

Of the identifiable motifs, stencilled hands and hand variations predominate (49%). This reflects the dominance of stencilling as a technique, but also the fact that the depictive art is significantly less formalised. It also reflects a degree of unavoidable recorder bias: a partial hand stencil is still identifiable as part of a stencil of a hand, whereas the classification of depictive motifs require that motifs with insufficient diagnostic information must be classified as 'unidentified'. A quadruped without legs and/or tail could be either a kangaroo or a dingo: and is thus recorded as unidentifiable.

Of the identifiable depictive motifs, macropods are the dominant subject (9%) followed by anthropomorphs (7%) and other land animals (5.5%). All of the motif classes used in the engraved

---

[14] The term 'depictive' is used to describe non-stencilled motifs, this including both figurative and non-figurative forms.

assemblage are also present in this component, although percentage differences do occur (section 5.6, below).

Table 5.3: Shelter Art component: summary motif frequency information.

Variable	Motif	Frequency	%f	%f identif..
1	Man	239	1.7	2.8
2	Woman	107	0.7	1.3
3	Anthropomorph	570	4.0	6.7
4	Profile Anthropomorph	88	0.6	1.0
5	Culture Hero	18	0.1	0.2
6	Macropod	792	5.5	9.3
7	Snake	176	1.2	2.1
8	Other Land Animal	474	3.3	5.5
9	Emu	36	0.2	0.4
10	Other Bird	317	2.2	3.7
11	Fish	185	1.3	2.2
12	Eel	161	1.1	1.9
13	Whale	2	0.0	0.0
14	Other Marine Animal	32	0.2	0.4
15	Shield	57	0.4	0.7
16	Boomerang	184	1.3	2.2
17	Axe	81	0.6	1.0
18	Other Material Object	122	0.8	1.4
19	Unidentified Open	1,151	8.0	-
20	Unidentified Closed	4,770	33.1	-
21	Hand	3,601	25.0	42.3
22	Human Foot	66	0.5	0.8
23	Hand Variation	588	4.1	6.9
24	Roo Track	87	0.6	1.0
25	Bird Track	24	0.2	0.3
26	Circle	82	0.6	1.0
27	Complex-Non-Figurative	129	0.9	1.5
28	Contact Motif	45	0.3	0.5
29	Other	240	1.7	2.8
Total id		8,503	59.0	-
Total		14,424	(100.2)	(99.9)

The clumped motifs (Table 5.4) indicate that tracks (human hand, feet and animal and bird tracks) dominate (51%), followed by terrestrial animals (17%) and anthropomorphic depictions (12%). There are a relatively low number of birds, marine animals and material objects and a high number of 'other motifs' (6%). This latter category includes circles and complex-non-figurative motifs, as well as a high number of other repeatable motif classes (i.e. Simple-Non-Figuratives and many rare occurrences of special motif forms). Diversity within this motif class was more extensive than found in the engraved component.

Table 5.4: Shelter Art component: clumped motif frequency and % frequency.

Variable	Motif	Total	%f	% identif.
1	Anthropomorphic	1,022	7.1	12.0
2	Terrestrial Animal	1,442	10.0	17.0
3	Birds	353	2.4	4.2
4	Marine Animals	380	2.6	4.5
5	Material Objects	444	3.1	5.2
6	Tracks	4,366	30.3	51.3
7	Other	496	3.4	5.8
8	Unidentified	5,921	41.0	-

There are major differences in motif preferences between the sheltered and open art assemblages. These will be discussed in detail below.

Technique

The recording of technical information for unidentified motifs has a demonstrated a more complete understanding of the overall range of techniques employed in this art component (Table 5.5). If only identifiable motifs had been used for these analyses, stencilling would have dominated the results to a greater extent than is real. The majority of the art (66.1%) is depictive. Stencilling is common (32.6%) while engraving is rare (1.3%).

Of the 9,527 depictive pigmented motifs recorded, the vast majority (91%) are drawn. Most of the remainder (8%) have been painted, while a small number (1%) are drawn and painted. These motifs are predominantly (93%) monochrome. Bichrome (i.e. two colours) motifs are relatively rare (7%), while polychrome (i.e. three or more colours) is rarer (1%). Most of the polychrome motifs consist of three colours (usually red, black and white) although several of motifs in this category include four colours (yellow also).

The way the database was counted does not allow for the splitting colour information according to stencilled or depictive motifs. The following colour summary is thus based on a combination of the two techniques.

Despite the dominance of stencilling, black[15] is the predominant colour used in the region, accounting for 46.2% of the pigment art. White is the next most dominant (34.6%), followed by red (16.6%) and yellow (2.8%). These colour proportions vary significantly in localised areas of the Basin (see Chapter 9).

Stencilling is mainly restricted to white and red pigment. Yellow stencils occur in localised areas and, as indicated, black stencils are rare. The use of two colours within one stencil (bichrome) occurs, as has the mixing of pigments to achieve non-primary colours – e.g. the mixture of red and white pigment to make pink. Off-white (cream) stencils have also been recorded, but these are more likely to have resulted from the use of less pure pipeclay. Bichrome and black stencils had not been documented in the region prior to this work.

Of the 9,715 depictive motifs (engraved and pigment), most are in outline and infill form (38%), although outline-only forms are nearly as common (34%). Solid infill motifs are less common (28%). Linear (geometric or patterned) infill in depictive motifs rare (2.5%: see Figure 5.12).

The assemblages at 57 shelter sites include engraved motifs. While the proportion of engraved motifs within the entire shelter art assemblage is very low (c.1%), the percentage of sites at which this technique has been employed is fairly high – c.10% of the regional sample. Engraved motifs within shelter art sites are another feature which differentiates art production on either side of the Georges River. While there is not a complete absence of engravings from the shelter sites south of the Georges River, the proportions vary significantly. There are only two such sites (out of 181 sites: 1.1%) south of the Georges River; the other 55 sites (15.4%) are located north of Port Jackson and the Parramatta River.

Most of these engraving are pecked tracks or circles. However, three distinct types occur (McDonald 1991). These are (Figure 5.16, Figure 6.7, Figure 8.7):

1) fully pecked, intaglio motifs; usually of circles, or bird and/or kangaroo tracks;

2) miniature Sydney style engravings; these are the same form, character and technique as the open engraving site variety, but in miniature (i.e. < 20cm max. dimension);   and,

---

[15]Black pigment is very rarely used for stencilling. Black stencils have been recorded at Yengo 1 and are reported in one site in the Woronora catchment (Caryll Sefton, pers. comm.).

3) incised or scratched motifs, usually incorporated with pigment motifs. This type has the same motif character as figurative pigment motifs.

The use of the engraving technique within the shelter art assemblage will be discussed further.

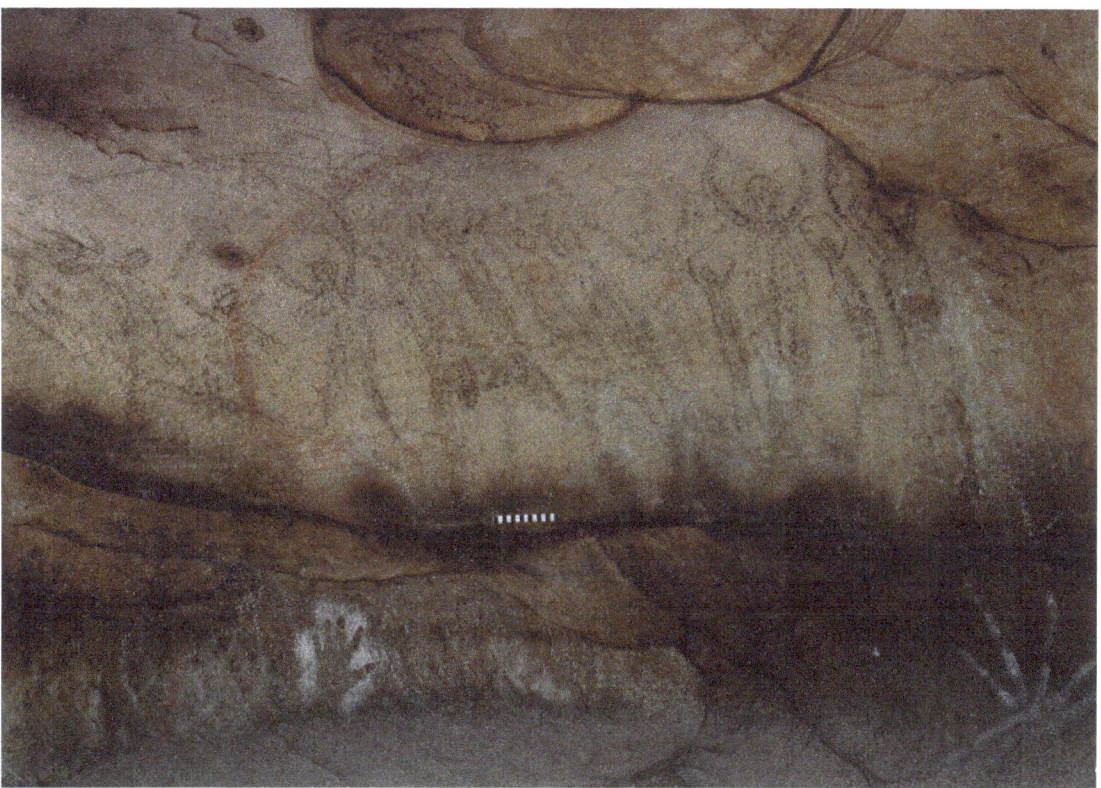

Figure 5.11: Frieze of charcoal women and men beneath a red outlined wombat. Hand stencils, white paintings and drawings and engraved macropod tracks were also in this assemblage near Warre Warren Creek.

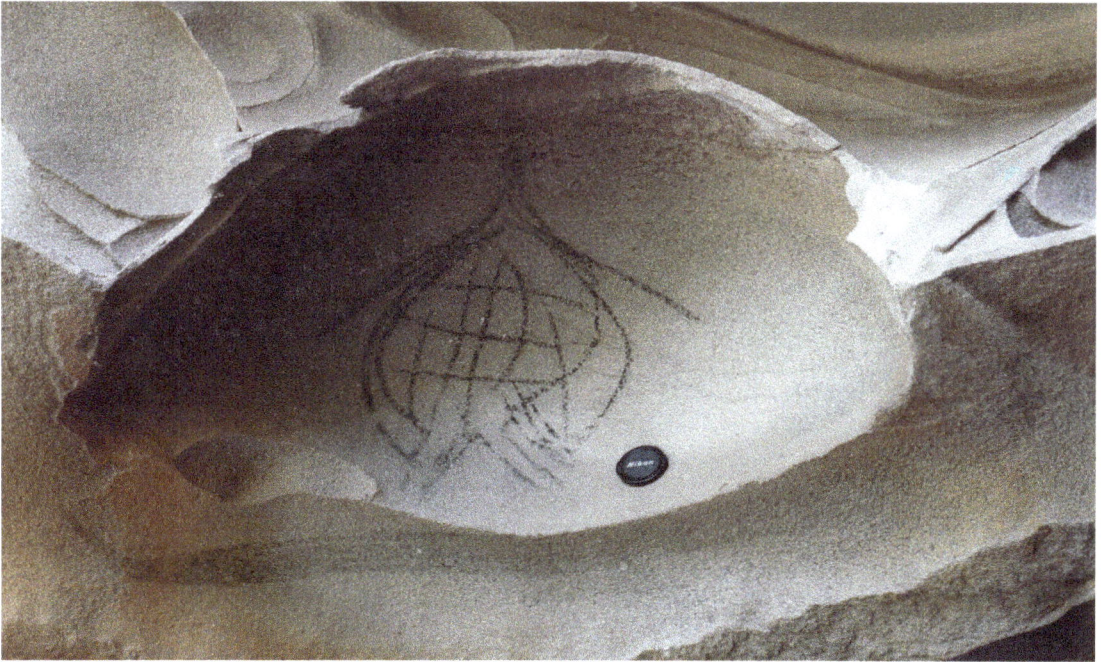

Figure 5.12: Black outlined and infilled turtle with linear infill positioned in a scalloped alcove. A faded, infilled snake motif runs along the surface below this alcove.

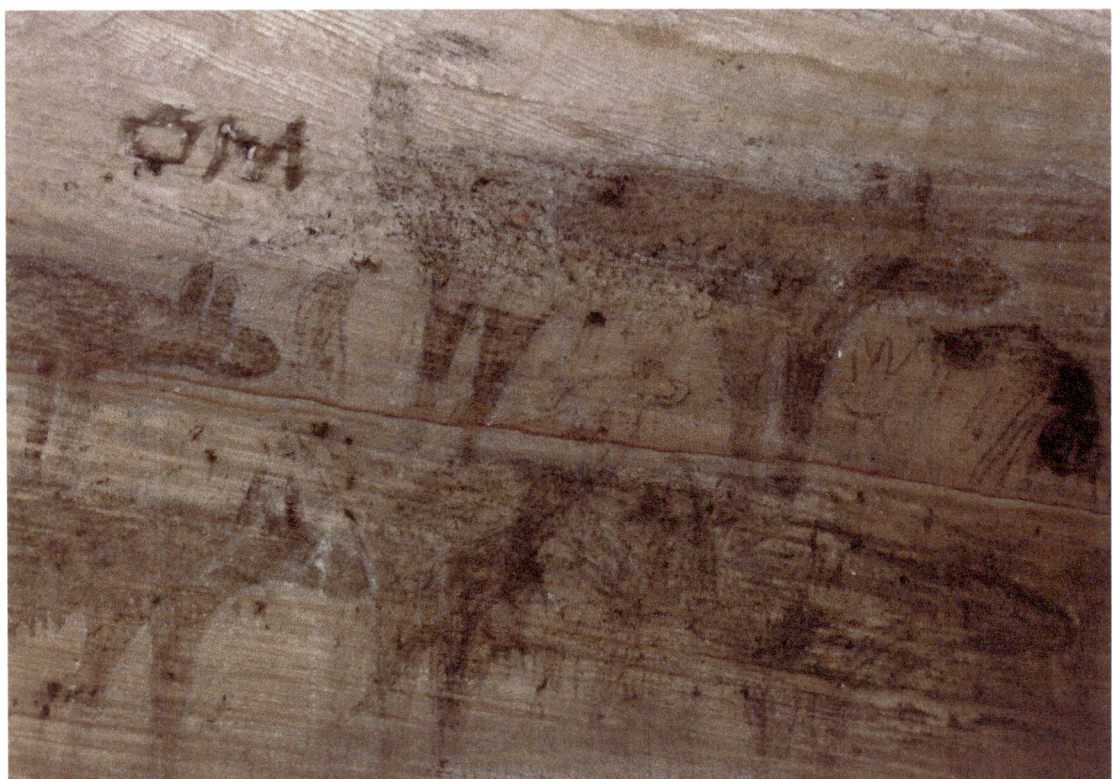

Figure 5.13: Black infilled and white outlined dingoes and macropods on Yatala Creek near the Colo River. These motifs are large (top dingo 1.25m long). Water seepage affects this art, making some colour run.

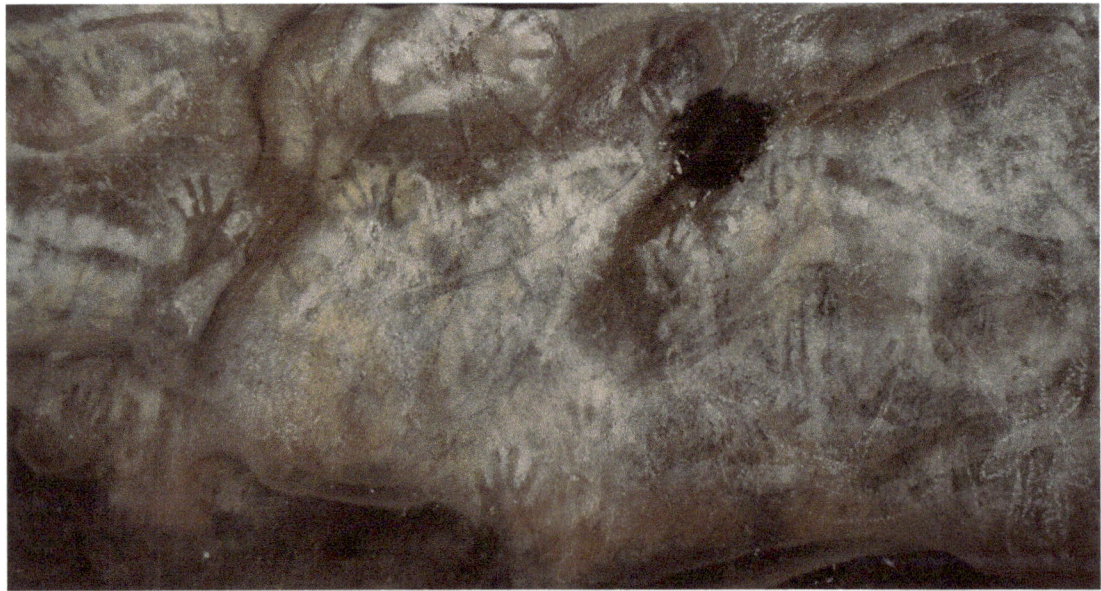

Figure 5.14: Swinton's shelter, with its heavily superimposed art. Stencils and drawings in red, black and white. The panel shown has a sequence that ends with a red horned anthropomorph, superimposed over white stencils.

Chapter 5: the rock art of the Sydney Basin

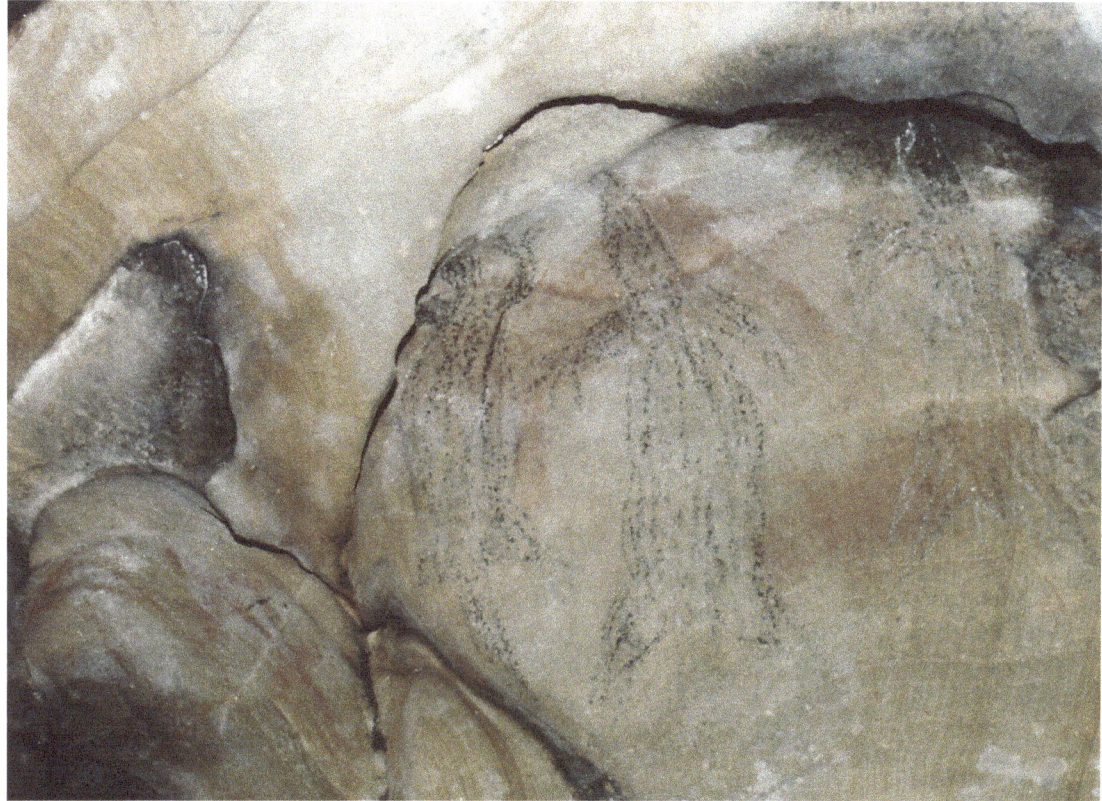

Figure 5.15: Black and white bichrome goannas (Phase 3) over red painted goannas (Phase 2) at the northern end of Swinton's shelter.

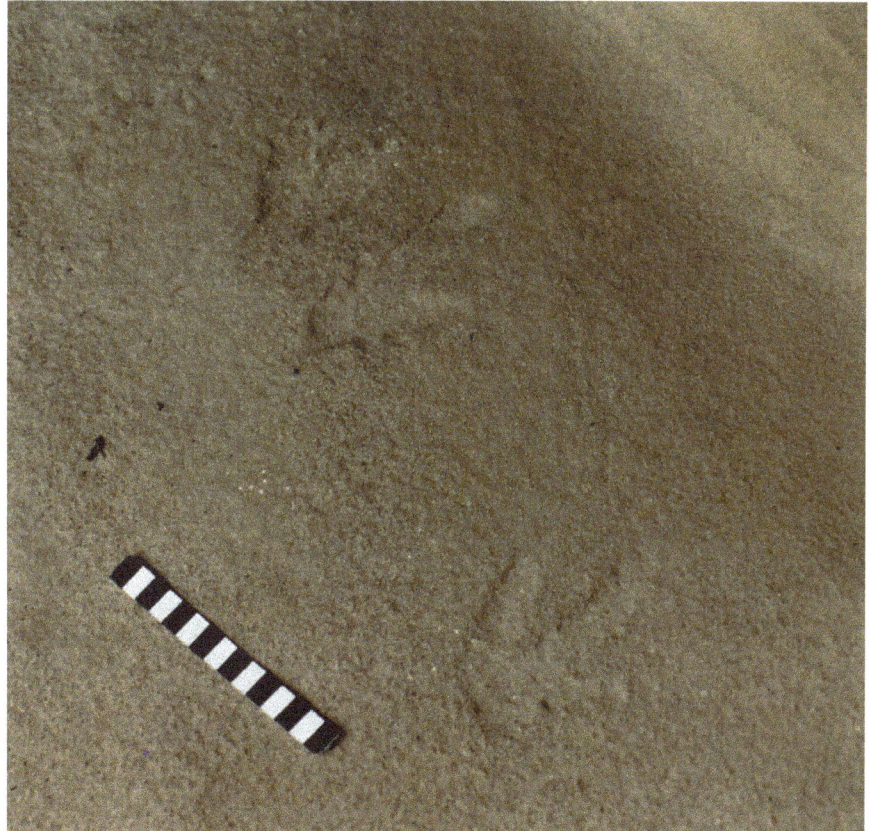

Figure 5.16: Three Birds Site: Engraved bird and macropod tracks on shelf at back of shelter just above floor level.

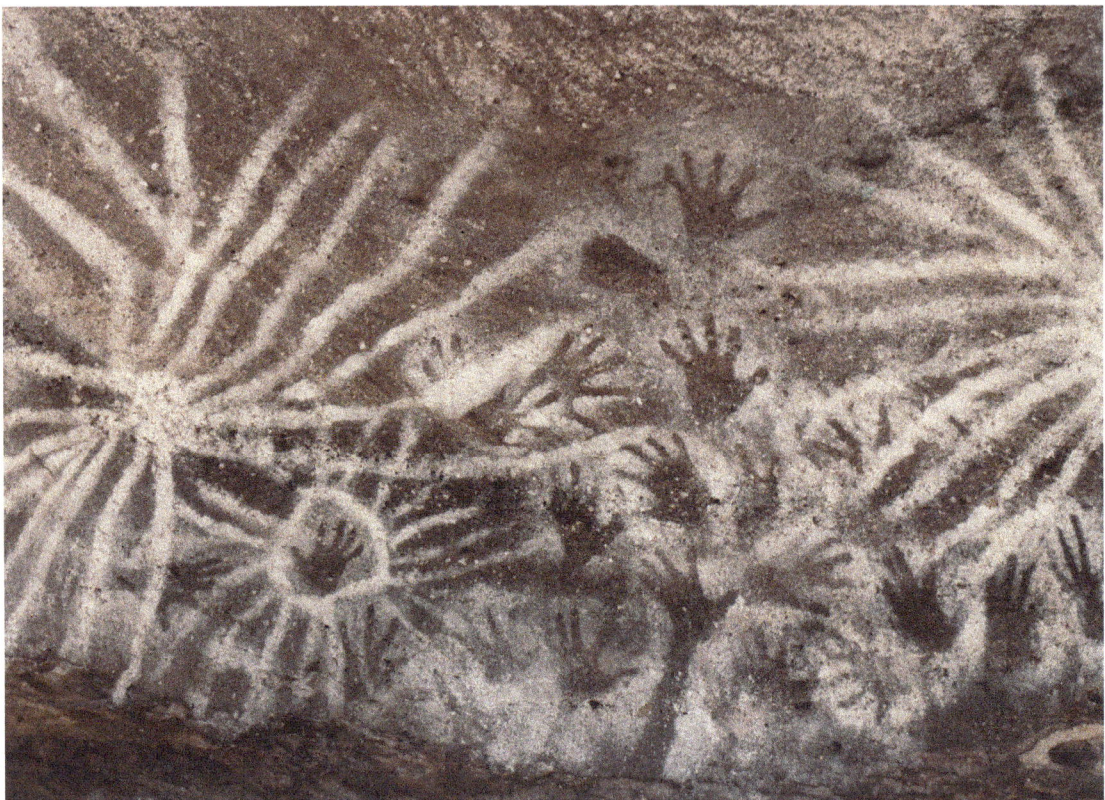

Figure 5.17: Cafe's Cave, Putty: near the northern limit of the study area. White painted complex-non-figuratives (CXNF), hand and axe stencils.

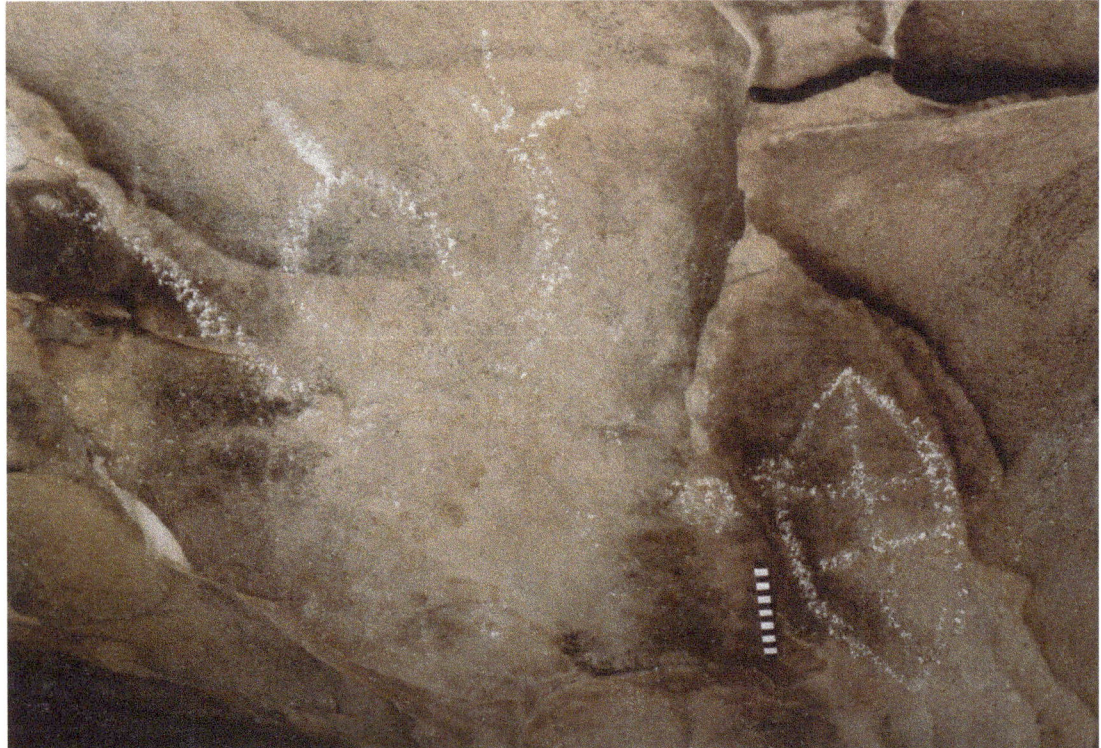

Figure 5.18: Tic Alley: white painted anthropomorph and shield with internal design (2B). These motifs are painted over black drawings.

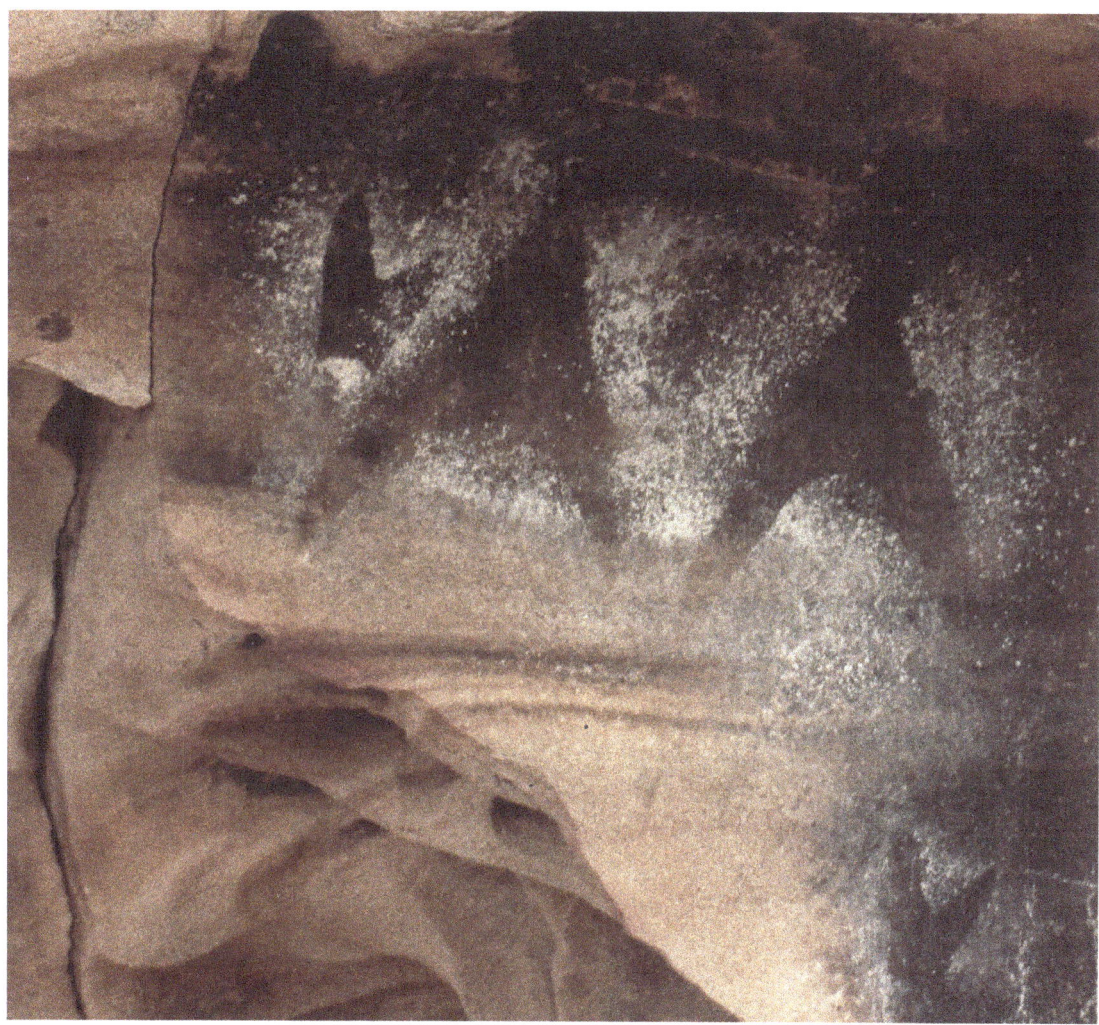

Figure 5.19: Gunyah Beach, near the confluence of Cowan Creek and the Hawkesbury River. Stencilled fish tails. This site also contained two small engraved whale motifs.

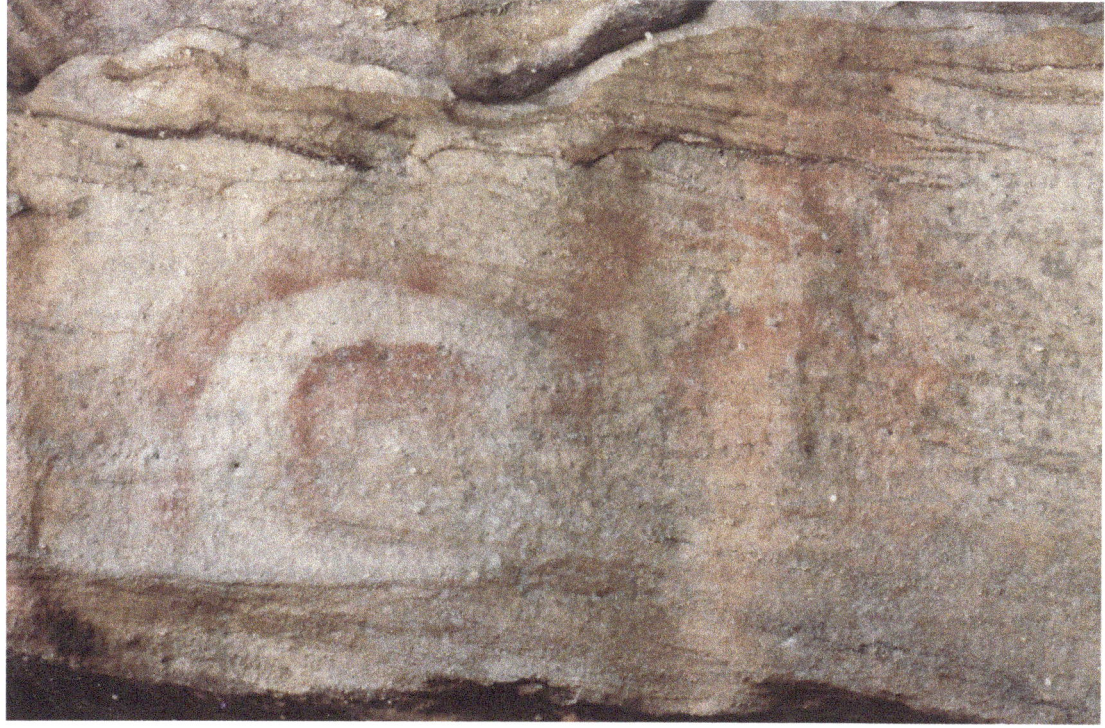

Figure 5.20: Boorai Creek, Wollemi National Park. Red stencilled kangaroo tails. This site also has stencils of boomerangs, hands and a kangaroo skin bag.

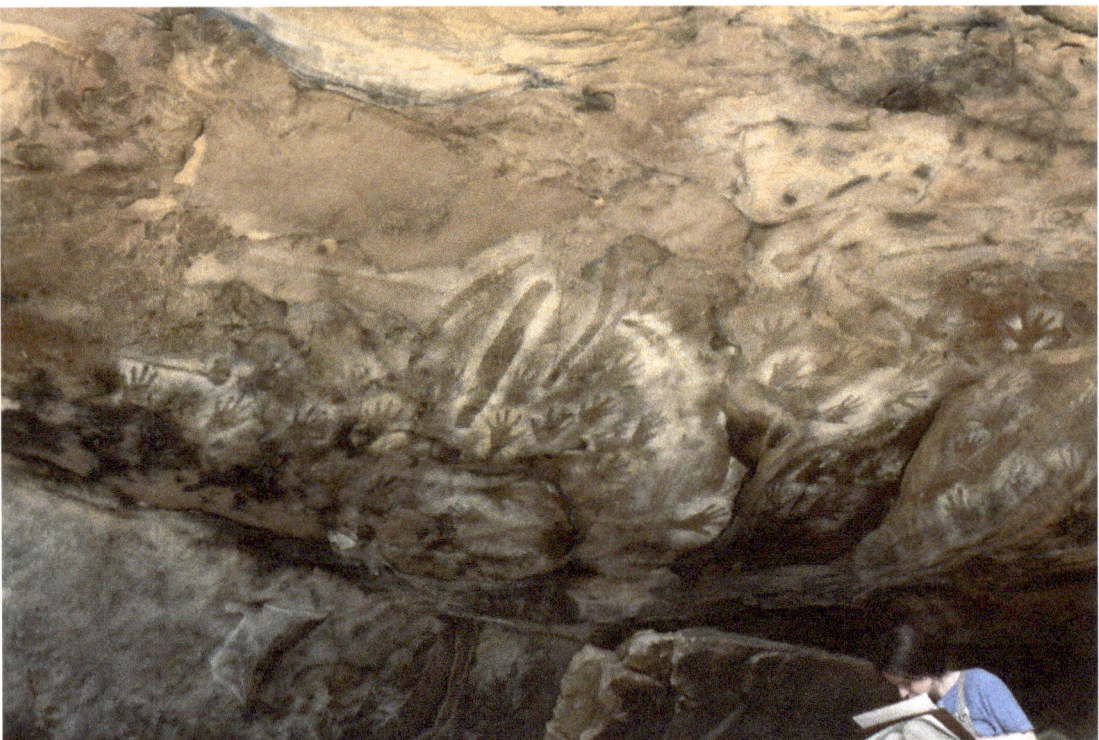

Figure 5.21: Colo Heights. Panel of stencilled material objects including boomerangs, parrying shield, axe, woomera and large hand stencils.

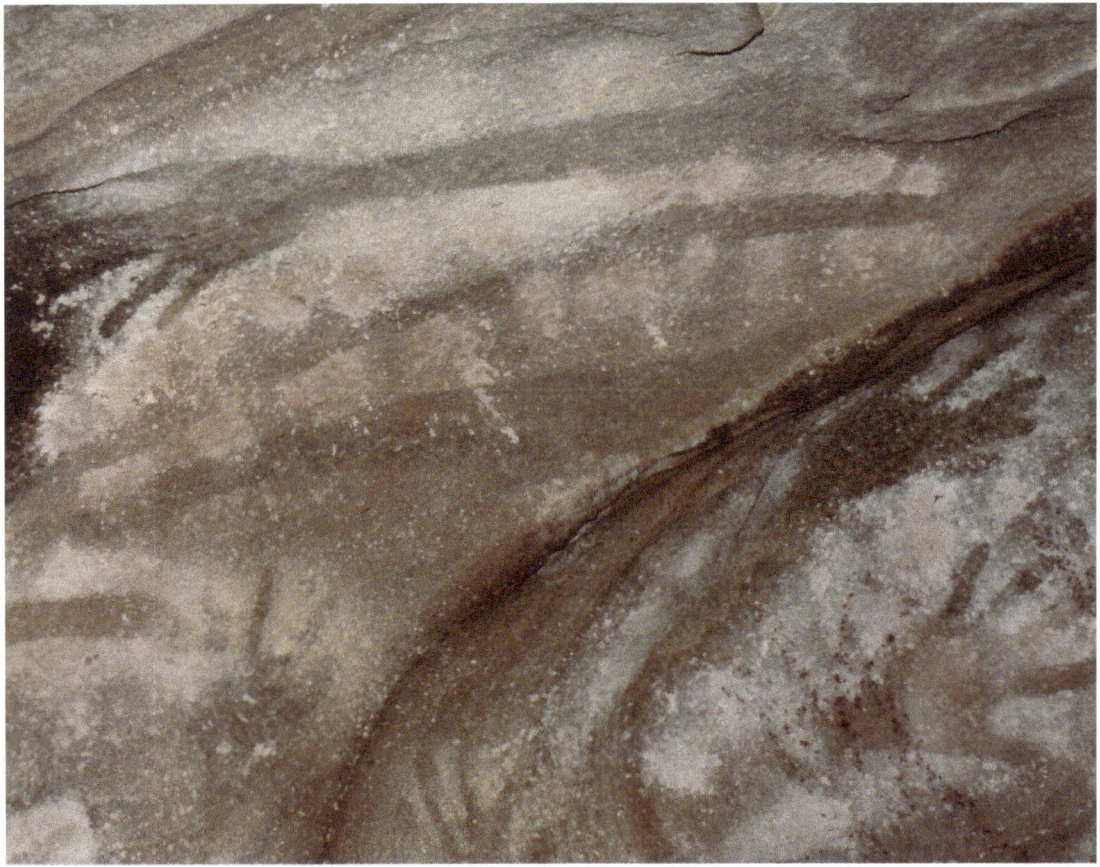

Figure 5.22: Swinton's shelter. Pink and bichrome stencils of boomerangs, and hand stencils. A red horned anthropomorph's hand is visible (bottom right) superimposed over stencils.

Table 5.5: Shelter Art component: summary technique frequency information.

Variable number	Technique description	Frequency	% (total) frequency*	% internal frequency
1	outline	3316	(67.4)	34.1
2	infill/solid	2747		28.3
3	outline and infill	3652		37.6
4	dry pigment	8637	(66.1)	90.6
5	wet pigment	771		8.1
6	wet and dry pigment	119		1.2
7	linear infill	244		2.5
8	stencil	4709	32.6	32.6
9	1 colour	8827	(66.1)	92.7
10	2 colours	647		6.8
11	3 colours	53		0.6
12	black pigment	6921	98.7	46.2
13	white pigment	5195		34.6
14	red pigment	2487		16.6
15	yellow pigment	430		2.8
16	engraving	188	1.3	-
17	Total motifs	14,424		
18	Total recognisable motifs	8,503	59.0	

*Percentages in brackets represent the proportion of the data base for which this information was counted i.e. the engraved motifs did not have colour information; variables 1 - 3 include pigmented and engraved motifs (but not stencils); variables 4 - 6 and 9 - 11 refer only to pigmented depictive motifs (not stencils or engraved motifs). The variable number column indicates the numbering system used in subsequent quantitative analyses.

Comparison of the two art contexts: Sheltered and Open Art Sites

The Sydney Basin with its two synchronous art forms is unique. In no other area of Australia does this occur to the same extent. In most regions one medium has been practised to the exclusion of the other (e.g. Arnhem Land, Kimberley, Pilbara, Laura), or the two are diachronically distinct (western NSW, central Queensland).

The two overriding assumptions about the Sydney region art components are that they are relatively recent and that they are roughly contemporaneous. Given the assumption of contemporaneity, the two art components allow quite specific questions of comparison. These questions can be at a very simple level.

How do the two components compare in regard to the motifs used? and,

How do the distributions in one component, compare with those of the other component?

At a higher level, stylistic patterning may be attributed to spheres of mediation in the society which produced the art. The meaning of the art cannot be explored, as the artists are no longer present. But by establishing certain parameters, the analysis of the two components enable a detailed picture to emerge of how the art may have functioned across the region. General comparisons are explored first.

Assemblage Size

Sheltered art assemblages are generally 2.5 times larger than open engraving assemblages. The largest pigment site contains five times as many motifs as the largest engraving site. This could be due to the fact that the average engraving's production time would be significantly greater than that for any pigment motif (Clegg 1981). The differing motif totals may reflect a commensurate amount of artistic activity.

Interestingly, both components demonstrate similar proportions of larger to small assemblages: there is a relative infrequency of very large site assemblages, but a proliferation of small, perhaps single episode sites. This suggests that there are particular art foci around the region, and that these may represent a different type of activity to the interspersed sites with less intensive art production (e.g. casual sites vs. aggregation sites).

Topographic location

A longstanding preconception about Sydney engravings has been that these were located on ridgelines while pigment art shelters were located in valleys. This division is not substantiated by the current work. Certainly, only a small proportion of the shelter art sites are located on ridgelines, but relatively equal proportions of engraving sites occur on hillslopes and on ridgelines.

A comparison of engraving sites on ridgelines with those on hillslopes and in valley bottoms has been undertaken (McDonald 1998b) to determine the relative stylistic homogeneity of these. A sample from Ku-ring-gai Chase National Park (within a single language group) was used for this comparison: topographic diversity here is pronounced and there is a high number of engraving sites generally. As well as open engraving sites on horizontal platforms, engravings on vertical surfaces close to the water's edge also occur. This analysis confirmed stylistic differences which can be explained in terms of topographic location and social context.

The average distance to drinking water from shelter sites is considerably less than the average distance from the open engraving sites. A greater association between art and occupation evidence in shelters is one possible explanation for this, especially since the contemporaneity of art and occupation can often be demonstrated (Chapter 6). The fact that more shelter sites occur in valley bottom and hillside locations (and are topographically closer to creeklines) could also explain this finding.

Motif

The motif range and preferences of the two art components were compared. This had not been done by previous research in the region (Franklin 1984, McMah 1965, Maynard 1976, Officer 1984, Smith 1983) and, as indicated above, is not possible in any other Australian art region.

This has been done at a simple level (Table 5.1 and Table 5.3; Figure 5.23 and Figure 5.24) and on the basis of variance described by multivariate analyses (Chapters 11 and 12).

There are obvious similarities and differences between the two components' motif assemblages. In both components, tracks dominate the identifiable motifs (Table 5.2 and Table 5.4). With the shelter art sites, this is due to the overall predominance of hand stencils; in the engraving component it results from the predominance of human tracks ('mundoes') and bird and macropod tracks. Given stencils are likely individual or personalised markers (Moore 1977), and that hand stencils place the artist very firmly amongst the art, the presence of human tracks in great numbers amongst the engravings is an intriguing similarity. Is it a coincidence? The insertion of people into both art assemblages does suggest a semiotic implication, although engraved mundoes are not in the same representational class as stencilled hands: one is conventional and the other is replication.

Another similarity between the two components is found in the proportions of anthropomorphs, birds and terrestrial animals. Major difference between the two components can be seen in the frequency of marine depictions and material objects (high in engraving sites; low in shelter art sites). The greater number of 'other' motifs in the shelter art assemblage is partly indicative of the more 'stylistically unfettered nature' of this medium (Officer 1984:72).

The proportion of unidentifiable motifs within the shelter art component is significantly greater than that found in engraved sites (41%:15.2%). This is partly due to the nature of production and to the more fragile nature of the shelter art generally. Pigment art has a greater susceptibility to natural deterioration due to the instability of sandstone surfaces e.g. surface exfoliation, pigment flaking etc. except perhaps where silica coatings have helped to stabilised these (Hughes 1978, Watchman 1994). The more ad hoc nature of pigment art production (i.e. freehand drawing using unprepared surfaces/pigments) may also have contributed to this high proportion of unidentifiable motifs. Superimpositionning of motifs (and the obscuring of earlier art) is more common in pigment sites then on engraving sites.

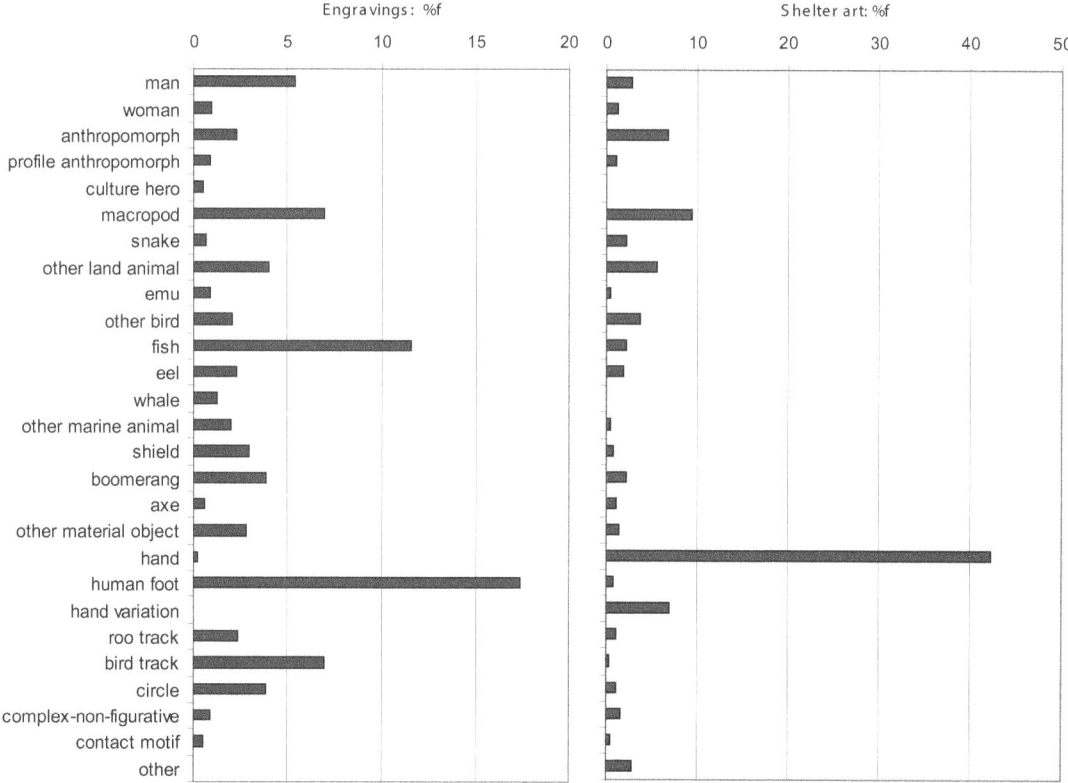

Figure 5.23: Motif preferences for the two art contexts. Motif classification (excluding unidentified motifs).

Composition

It cannot be assumed that any rockshelter or engraving assemblage is a single artistic event; indeed many pigment assemblages suggest several artistic phases. Occasionally, shelter and open engraving sites indicate that the assemblage was probably produced in a single artistic episode. Compositions include recognisable 'scenes' (hunting, fishing, corroborees etc.) or repetitive designs (using combinations of motifs, often stencils).

Most of the vertical engraving sites consist of complex compositions involving the sharing of internal lines/features for decorative effect, and a positioning of the design relative to the shape of the available surface. While composition is often recognisable in open engraving sites, by association and positioning of motif type (e.g. the Ku-ring-gai Fish Shoal site: Campbell 1899: 62; or the overlapping whales at West Head: McCarthy 1954: Figure 9) it is never as conclusively so as in the vertical sites (Figure 11.19).

Compositions within the shelter art sites take the form of obvious motif organisations [e.g. Native Animals (McDonald *et al.* 1990: Figure 5.2), Dingo and Horned Anthropomorph (Macintosh 1965: Plate 1)], as well as the positioning of hand and other stencils to make patterns or repetitive designs [e.g. Devils Hole, Cafe's Cave (Figure 5.17), Swinton's]. These types of compositions are good indications for the simultaneous production of many sites' assemblages.

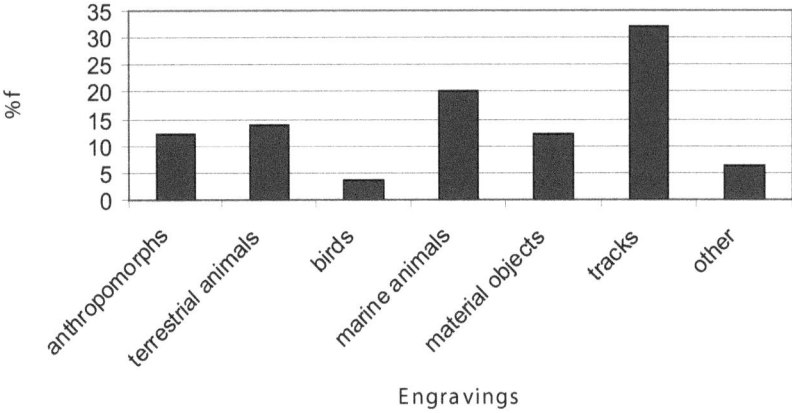

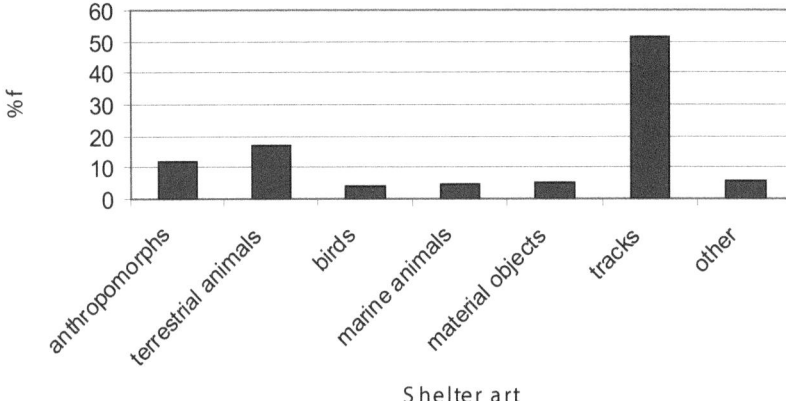

Figure 5.24: Subject preference for the two art contexts. Clumped motif classes (excluding unidentified motifs).

6

EXCAVATIONS AT YENGO 1 AND YENGO 2

The following three chapters detail the excavation results from the four art shelters excavated for this research. This chapter details the results of excavations at two sites in the north-west of the Sydney Region near Mount Yengo (Figure 6.1). The main shelter (Yengo 1; NPWS #37-5-1) has art (stencils, paintings and engravings), occupation deposit and grinding grooves. The engraved vertical panel at the front of the shelter has a complex of weathered pecked circles. Prior to the excavation it was apparent that these were truncated by the deposit. The second shelter (Yengo 2: NPWS #37-5-2) is located 10m north of the main site. This shelter has art (stencils, drawings and paintings) and occupation deposit.

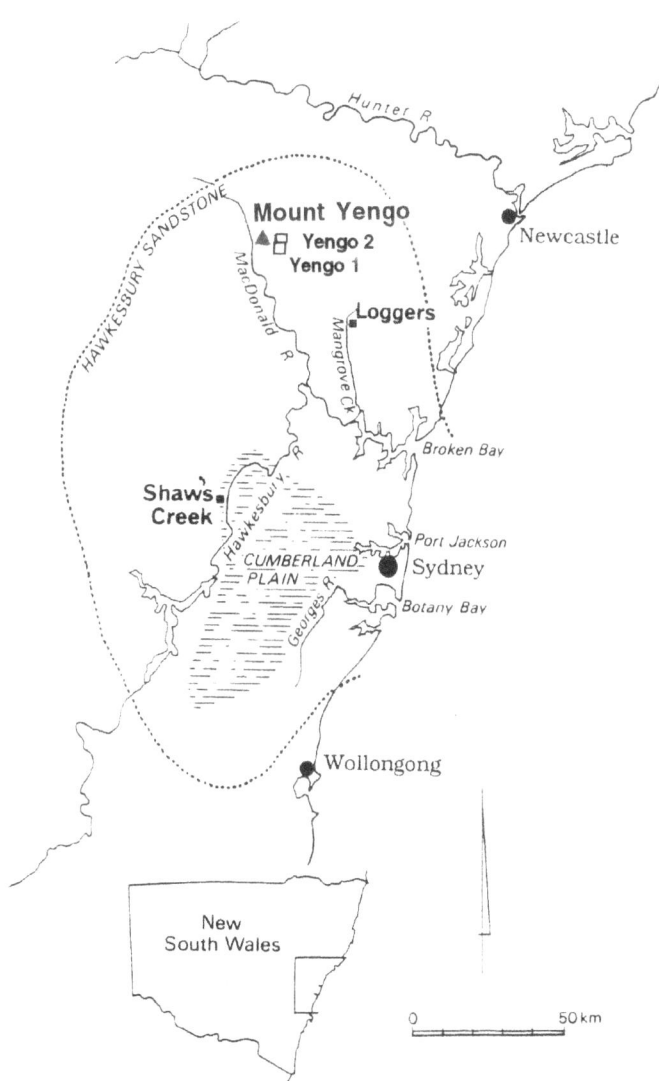

Figure 6.1: Locality Map showing site context.

The excavation aims for Yengo 1 were threefold;

1) to determine whether the engravings continued beneath the deposit;

2) to obtain evidence which might establish the age of the engravings; and,

3) to investigate the contemporaneity of the archaeological components at the site.

The parietal art was recorded in October 1986 (McDonald 1987) and two field seasons were spent excavating the site. The first of these was in September-October 1987, the second in September 1988. During the first season approximately 1.6 cubic metres of deposit were excavated, yielding 10,100 artefacts. Towards the end of this first season extremely heavy rain caused flooding in the trenches, and made the drawing of sections impossible. The second season reopened the site to complete drawing the sections, and to investigate the nature of a large boulder buried in Square 1. Six charcoal samples were submitted between seasons for radiocarbon dating. Three more samples were subsequently submitted, to refine the sequence.

The excavation questions for Yengo 2 had a different orientation. These were based on the fact that the art assemblage in this shelter was different from that in the main shelter. Considering the proximity of the two sites, establishing the contemporaneity of occupation for the two sites was thus of some interest. The research questions for Yengo 2 were:

1) was there archaeological deposit at the site; and, if so,

2) to establish the nature and age of the deposit; and,

3) to establish the relationship between this site and Yengo 1.
This site was excavated in September 1988.

Environmental Context

The two sites are located on the east facing slope 250m from Big Yengo Creek[16] and 20m in elevation above it. At this point the creek is 8km south of its headwaters in the Hunter Range, and 30km north of its junction with the McDonald River. Mt Yengo (elevation 668m) is located approximately 4km west-south-west of the sites. The Hunter Range forms the divide between the Hunter and Hawkesbury Rivers catchments.

Big Yengo Creek is reputed to contain water for almost all of the year. While currently the creek has a sandy bottom, prior to the 1950 floods, the creek bottom was pebbly (Ken May, ex-property manager, pers. comm.). The gently sloping flats along both sides of the creek have been cleared for grazing. Many of the side gullies have also been cleared, and many have had dams constructed within them. Only a few very large trees remain within these grassy flats. The steeper slopes contain undisturbed dry sclerophyll with a medium understorey. *Angophora floribunda* (Rough-barked apple) dominates the upper storey, with *Casuarina* spp. and *Acacia* spp in abundance. Native Cherry (*Exocarpus cipressiformis*) was observed in front of Yengo 1, and seeds from this species were found throughout the deposit.

In this part of the Sydney Basin the Hawkesbury Sandstone formation overlies the Narrabeen Sandstone and the Hunter Range is towards the most northerly extent of this bedrock. The hillslopes at *Big Yango* contain many sandstone overhangs and boulders. Seven other shelters with art have been identified in the vicinity, and systematic survey would undoubtedly locate more.

[16]The property on which these sites are located is '*Big Yango*' while on maps the orthography for the creek is Big Yengo and the mountain is spelt Mount Yengo. The sites have been named after the mountain and are therefore spelt 'Yengo'. Yengo is reputedly an Aboriginal word meaning mountain (Sim 1966a: 38). The sites are thus called Yengo 1 and Yengo 2. This was also done to avoid confusion with Moore's (1981) Yango Creek site (YC/1).

Mount Yengo is the highest peak in this hinterland country. It has a Tertiary basalt (Alancime Dolerite) cap: the remains of a large sill (AGSHV 1981). Mount Wareng (elevation 594m) is a similarly shaped geological feature to the north.

The Sites

Yengo 1 (AHIMS #37-5-1)

This site is a medium-sized shelter measuring 14m x 6m x 1.9m at the dripline. It has an easterly perspective and is a very pleasant morning habitation (Figure 6.2 and Figure 6.3). The deposit is dry and the shelter is fairly well protected from the prevailing winds. There is an extensive pigment assemblage on the ceiling and back wall of the shelter, the interior vertical panel of the roof fall boulder is engraved and the sloping back shelf of the shelter is replete with grinding grooves (Figure 6.2, Figure 6.5).

When the site was first visited in 1986, no artefactual material was observed on the surface of the deposit, although some fragmented faunal material was observed near the entrance to a wombat burrow at the northern end of the shelter. From the interior size of the burrows it was clear that the deposit was >50cm deep and that the potential for occupation material was high. While some deposit derives from slopewash at the northern end of the shelter (and over the dripline), the deposit is mostly fine-grained grey ashy material of primarily cultural derivation. A large roof-fall boulder at the northern end of the shelter acts as a successful barrier for the deposit. This large boulder contains the engraved circles, bird and macropod tracks (Figure 6.4). There is a slight lip in the sandstone outcrop at the southern end of the sandstone which further acts to foil gravity and the downslope movement of deposit.

The Art

Of the 36 engravings on the vertical boulder (Figure 6.4, Figure 6.7), all are pecked and most are circles (90%). Two of these circles have a pecked central dot. The macropod and bird tracks are intaglio (i.e. pecked solid) while the circles consist of thick pecked forms. They have not been subsequently abraded, as with most circles on open engraving sites in the region. These motifs are classified as a regional variant of the Panaramitee (following Maynard 1979, Rosenfeld *et al.* 1981). Most of the art in the Sydney Basin is of the Simple Figurative style. It was thus assumed (based on Maynard's classification) that this art should be older than the majority of the sites found in the Sydney Basin.

The pigment art at Yengo 1 consists of stencils, paintings and drawings in white, red, yellow, pink and black pigment (Figure 6.8 to Figure 6.11). The assemblage is large (just over 500 motifs) the second largest recorded in the Sydney Basin (McDonald 1987, 1990a). The predominant motif is the hand stencil and there are numerous hand stencil variations (hand and arm, hand and wrist, finger manipulations). The art was recorded in detail with the assistance of Laurajane Smith and Warren Bluff (see Table 6.1).

The predominant colour used in the pigment art at the site is white (73%), followed by yellow (10.1%) and black and red (4.9% each). Stencilling is the predominant technique at the site (82.9%) followed by engraving (7.1%), drawing (5.8%) and painting (3.6%). Motifs with a mixture of wet and dry pigment are present but rare (0.6%).

Of the 50 depictive pigment motifs, most (58%) are infilled only; 34% are outlined and 8% are outlined and infilled. All of the identifiable motifs are either black or white. Most of the depictive motifs (78%) are either no longer identifiable or are generally unclassifiable. No depictive motifs were executed in yellow pigment: this colour is restricted to stencilling.

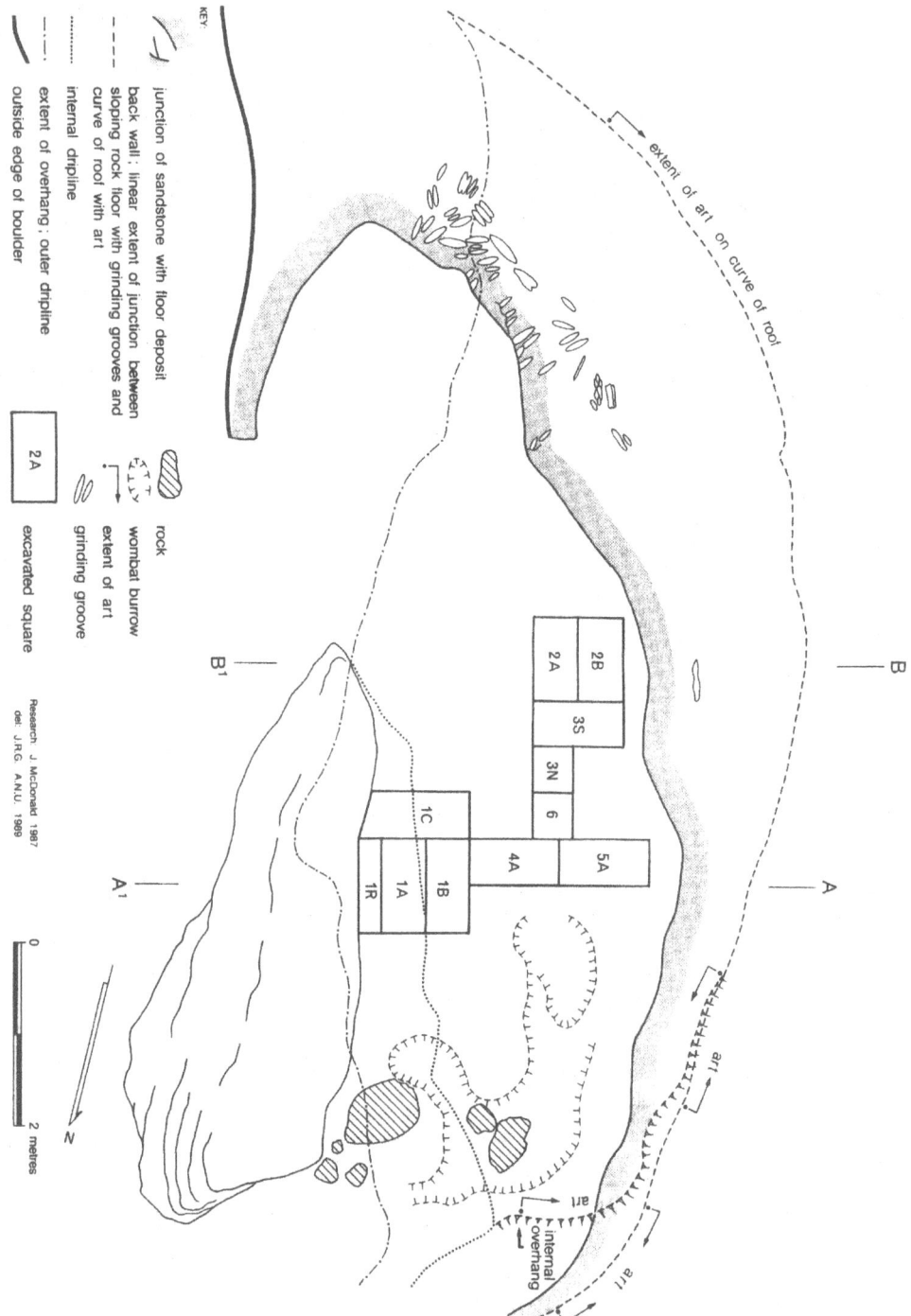

Figure 6.2: Site Plan, Yengo 1.

Figurative motifs include two anthropomorphs, an emu and an eel. Complex non-figurative motifs (CXNF) are slightly more common, as are bird tracks. The CXNFs are in white paint or drawn charcoal, and are predominantly geometric (i.e. zigzags, parallel lines with an emu track on the end of one: Figure 6.11). Several hand stencils also contain the black outline drawings of hands - either the result of tracing around the hand after the stencil had been completed, or by tracing around the inside edge of the stencilled hand's outline. This same technique has also been applied to two axe stencils and an unidentified stencilled implement. These motifs have the potential to be dated by AMS techniques (McDonald *et al.* 1990; McDonald 2000c).

Hands are the predominant stencilled subject but axes, boomerangs and other material objects (clubs, straight sticks and several unidentified objects) also feature. Yellow is the second most common colour used in stencilling, followed by red and black. There are eight black stencilled hands. It is not clear whether this is charcoal or manganese.

Chapter 6: Excavations at Yengo 1 and Yengo 2

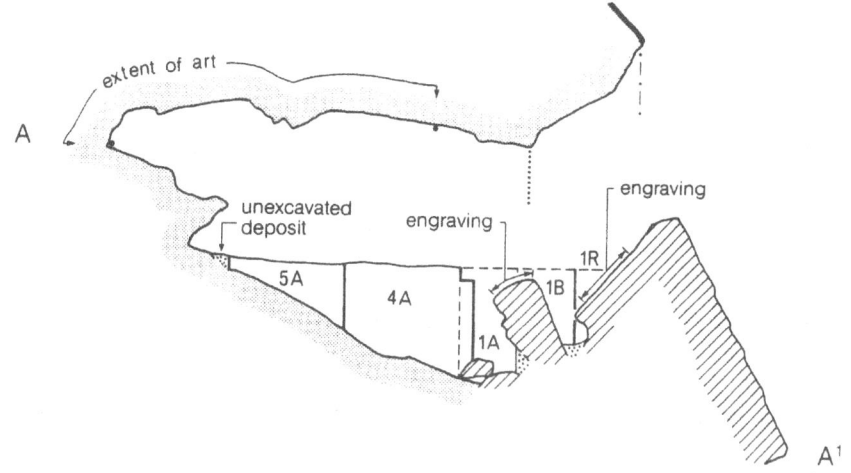

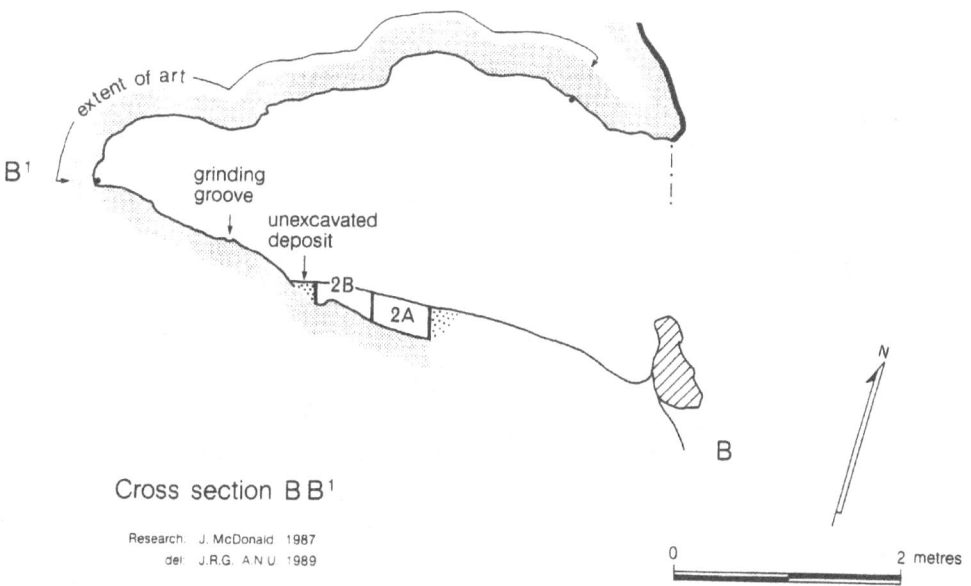

Figure 6.3: Cross-sections, Yengo 1.

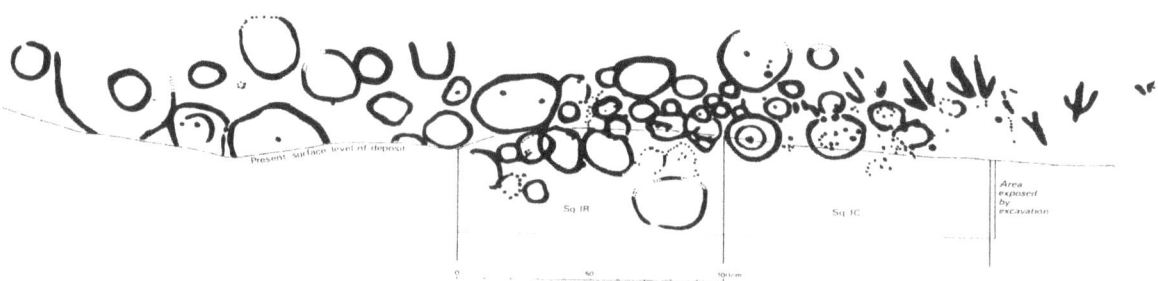

Figure 6.4: Engraved panel at the front of Yengo 1, showing the engravings and their (excavated) depth below the surface level.

Table 6.1: Yengo 1. Art Assemblage. Motif and technique information.

Motif	Colour				Technique							Total
	Red	White	Black	Yellow	Dry	Wet/*	Stencil	Eng'v	Out.	Infill	O/I	
Anthrop.		1	1		2				1	1		2
Emu			1		1					1		1
Eel		1			1				1			1
Boomerang		2					2					2
Axe		8	2		2		8		1	1		10
O. mat obj.		8	1		1		8				1	9
Hand	14	277	9	49	2		347		2			349
Hand var.	2	49		2			53					53
Bird track		2				2		1		3		3
Roo track								5		5		5
Circle								30	28		2	30
CXNF		2	2		2	2			1	3		4
Unid lines	1	4			2	3			5			5
Unid solid+	8	17	9		16	11/3*			6	21	3	30
Total	25	371	25	51	29	18/3*	418	36	45	35	6	504

+ Four of these are bichrome (i.e. black and white)　　　　　　　　　　　　　　　* wet and dry

Several features of the site's parietal art are interesting or unique. This is the only known site in the region where a paint wash has been used either to cover existing art or to prepare the surface for subsequent art. Pink, tan and white paint have been used in four locations covering areas ranging between 1.5m x 0.4m to 0.8m x 0.8m. At the northern end of the site, hand stencils are superimposed over the wash. Several stencils (hands and axes) were also observed to be partially covered by wash.

Black hand stencils are rare. A few have been recorded in the south of the region (Caryll Sefton, pers. comm. 1988). The use of two colours in a single stencil (bichrome technique) is also rare: otherwise recorded only at Swinton's, near Mangrove Creek (personal observation).

Prints (positive stencils) are also rare in the region: four hand prints (one yellow and three tan) occur in Yengo 1. The use of non-primary colours, or at least of mixed pigments, in this case pink (presumably red + white mixed) and orange (yellow + red mixed), is also uncommon. A tan colour is used quite commonly at this site. This may be a mixture of yellow and red, or could be a very fine clay mud.

The Hand Stencils

The size range of the hand stencils at the site was recorded to indicate the likely age range and gender of the population participating in the art's production. It was hoped that this would give some insight as to what sort of site this may represent. Two measurements were consistently made for all hands: width and height. Width was measured between the tips of the thumb and the little finger; height was measured between the tip of the middle finger and the heel of the hand. The latter measurement is considered the most reliable indicator of the size of the hand being stencilled, since the former can vary depending on the amount of finger splay.

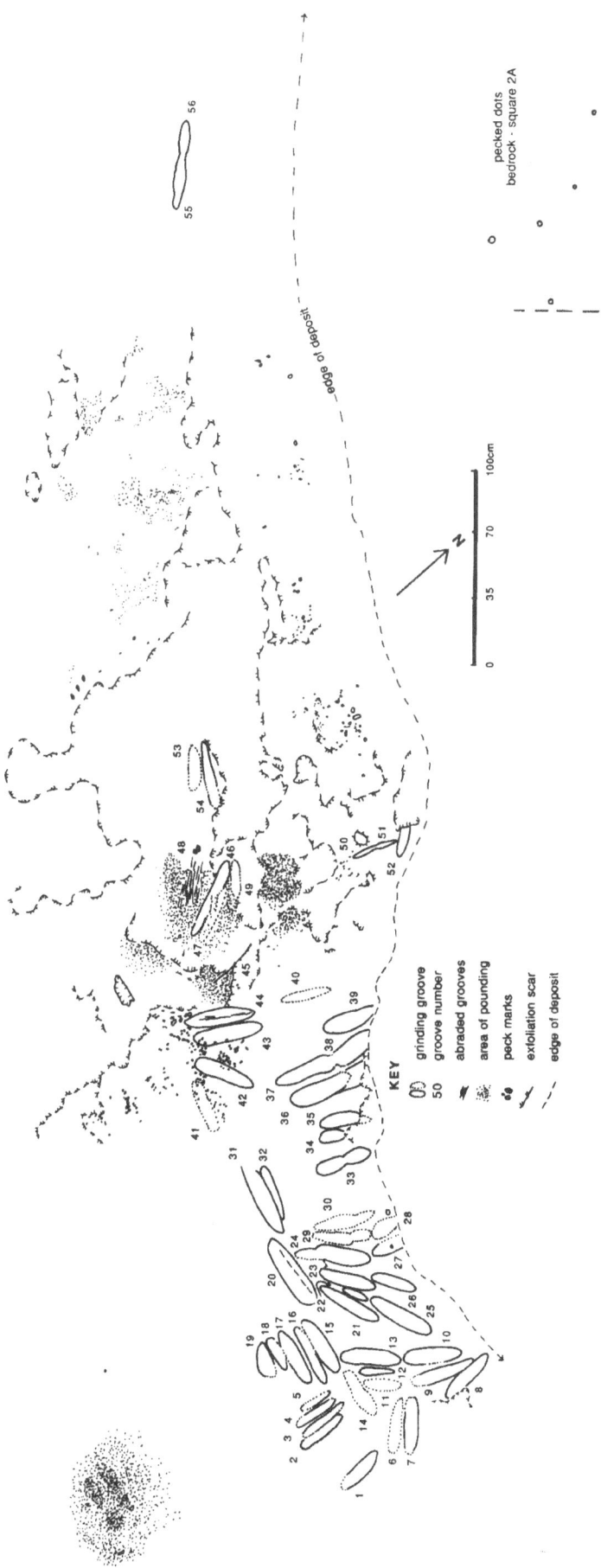

Figure 6.5: Yengo 1: grinding grooves - pecked and pounded areas.

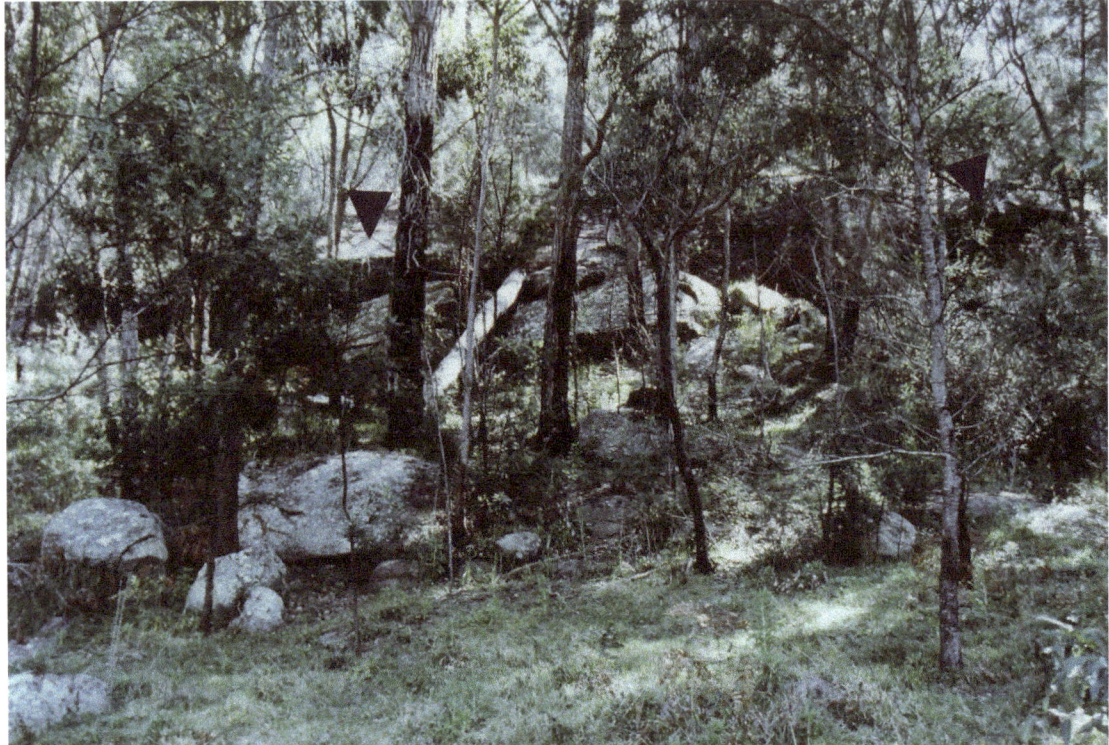

Figure 6.6: View upslope towards sites Yengo 1 and Yengo 2 (#'s 37-5-1 and 37-5-2). Sites are arrowed - Yengo 1 is on the left.

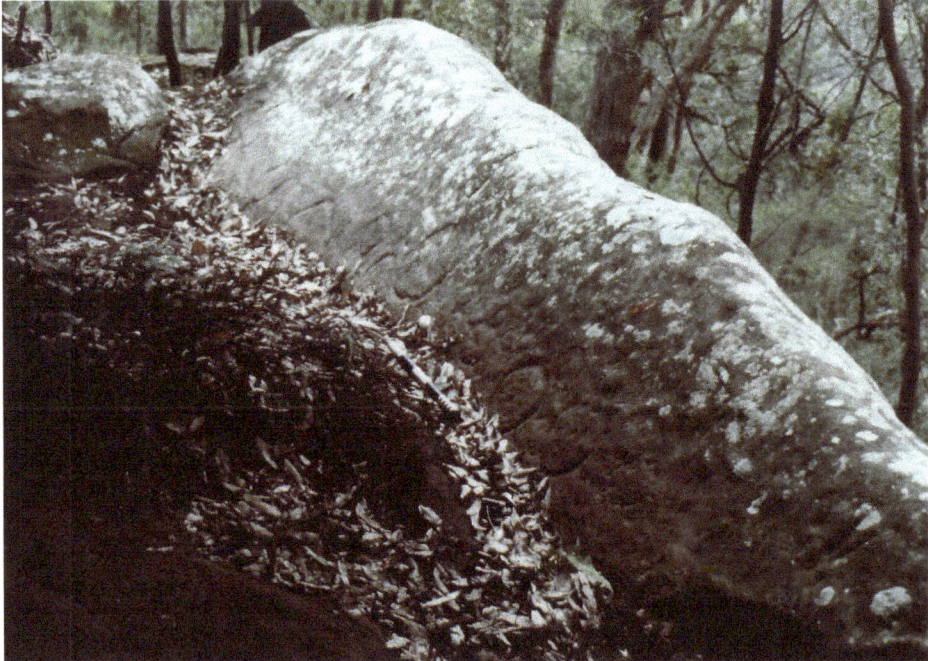

Figure 6.7: Yengo 1. The engraved panel at front of shelter prior to excavation. View from the south-west.

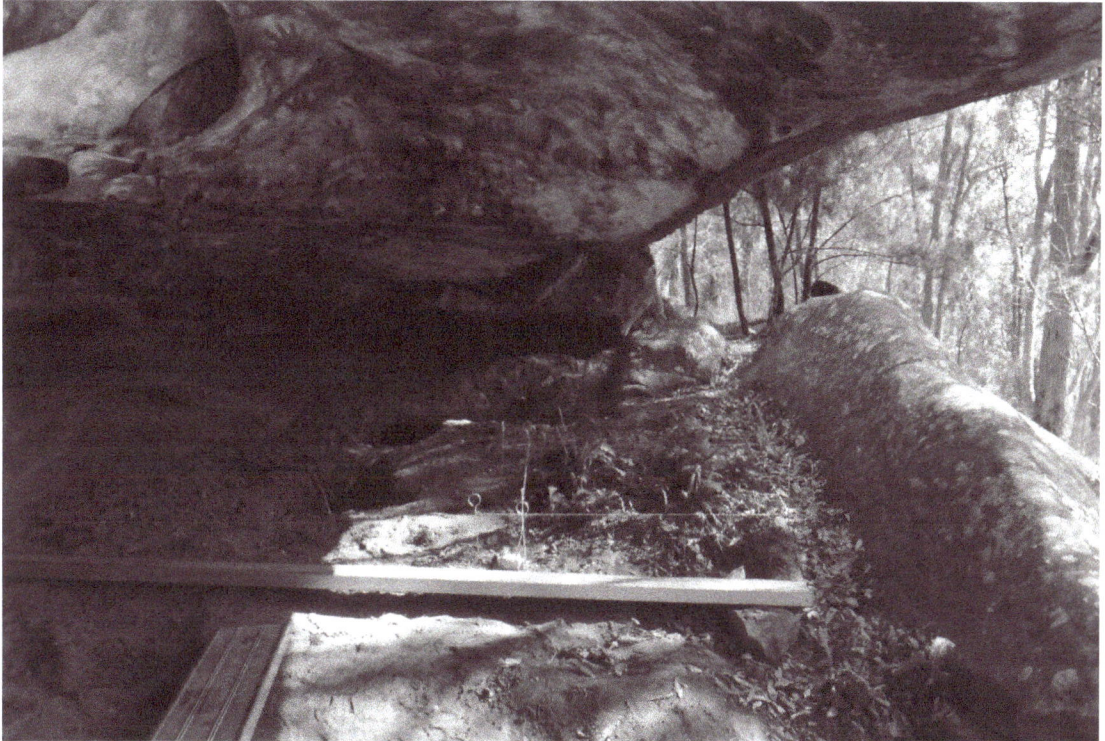

Figure 6.8: Yengo 1. Interior view of shelter showing relationship between engraved boulders stencilled art on ceiling and deposit and set up of the initial square (1A) adjacent to engraved boulder. Planks were used because of friable deposit.

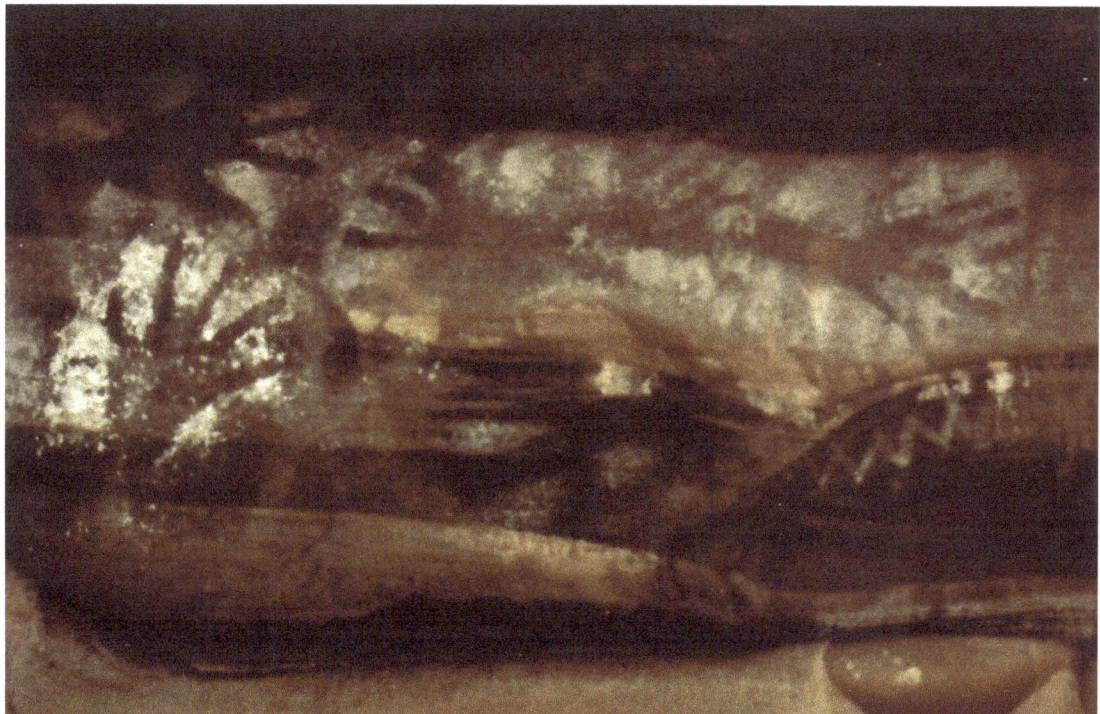

Figure 6.9: Yengo 1. Detail of stencils and white painted motifs on rear wall.

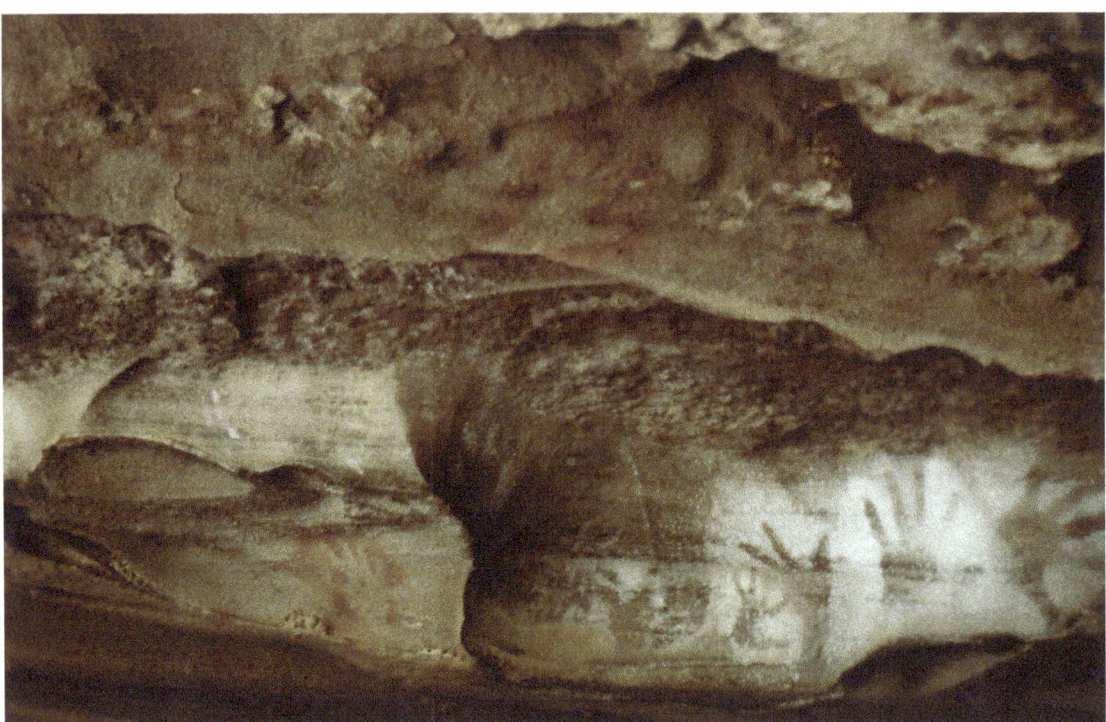

Figure 6.10: Yengo 1. White and red hand stencils on rear wall of shelter.

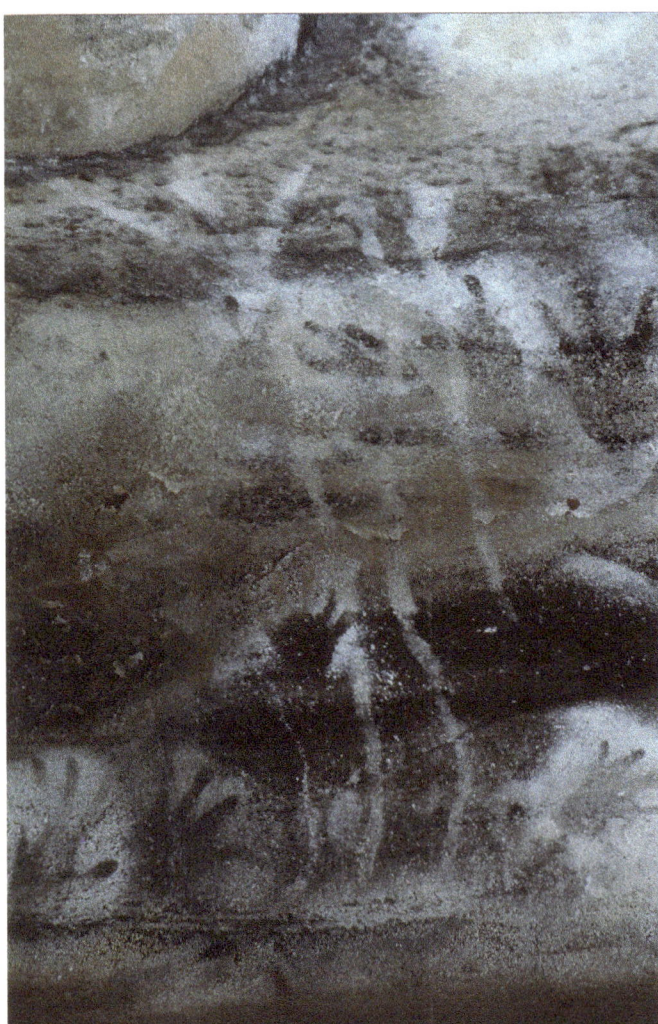

Figure 6.11: Yengo 1. White stencils and painted complex-non-figurative design on back wall of shelter.

Most of the hand stencils have the fingers splayed normally (i.e. not stretched). However a number of stencil variations are also present. Most of the stencil variations involve the stencilling of wrist and/or arm, but there are numerous examples of finger manipulations ('mutilations') and other variations of finger position.

Left or right?

Left and right handedness of stencils was also recorded: assuming that hands are placed palm-down on the rock surface for stencilling (Table 6.2). Most of the hand stencils are left hands. With the white stencils and stencil variations, left hands occur at a ratio of approximately 2:1 over right hands. With all other colours the ratio is closer to equal.

Size

Of the 392 stencils recorded at the site, 237 could be measured. Size, colour and side (left or right) were measured to assess whether particular colours might have been used by certain age groups. If so, then certain colours or sides may have been restricted to certain size ranges (i.e. age groups).

Table 6.2: Yengo 1. Left and right handedness of the coloured stencils.

	Hands			Hand variations			
Colour	Left	Right	?	Left	Right	?	Total
White*	105 (29)	55 (31)	49	25 (1)	15 (1)	5	316
Yellow≈	8 (15)	10 (9)	11	1	1		55
Red¥	3 (1)	3 (2)	4				13
Black	3 (1)	3	1				8
Total	119 (46)	71 (42)	65	26 (1)	16 (1)	5	392

* includes cream   ≈ includes tan and orange   ¥ includes pink
(x) = stencils for which side can be determined but full measurements not possible.

Measurements from living Aboriginal populations from central and northern Australia[17] were compared (Abbey 1975). The relevant measurements are hand length – the distance from end of radius to tip of middle finger - comparable to the length measurement here on the hand stencils. Abbey's results show:

Men	range = 14.0 - 21.8 cm	mean = 18.2cm
Women	range = 14.7 - 20.2 cm	mean = 17.2cm

There is considerable overlap between the two gender groups and the lower end for the range for males is lower than that recorded for females. Differentiating gender in stencils therefore, on the basis of size alone, is not possible. No doubt there is also considerable overlap between the hand sizes of adolescents (particularly boys) and women, as well as the fact that some men are small and some women large. Also to be considered is the fact that the contemporary sample derives from central Australia. While this was the only data available, it is possible that the Sydney (coastal) Aborigines were more robust than this sample[18] and that their hand sizes may have been marginally larger. The stencil analysis however, demonstrates a similar set of size ranges to those recorded by Abbey.

The presence of babies' hands amongst the stencils is a better indicator of the presence of women at a site than the presence of medium-sized hand stencils. Similarly, it is probably only the very large stencils (>21cm) which could be definitely identified as being male - except perhaps

[17] Abbey's population sample included men and women (aged 21-60+) from Yuendumu, Haast's Bluff, Maningrida, Yatala, Beswick and Kalumbaru.
[18] Analysis of long bones has indicated a greater robusticity among coastal and Murray River skeletal remains than those from the arid zone (Dr Denise Donlon, pers. comm., 1992).

where there was direct association between the hand stencils and male hunting equipment (e.g. an axe). While men's and women's hand sizes, and most likely adolescent's hand sizes also involve considerable overlap, the size of children's and babies' hands are more reliably identified. On the basis of a study of the hands of 35 children under four[19], the following measurements are used as the basis for identifying the stencils of infants and children at the site (children <12 years grouped in a size range between infants and small adults):

Babies/infants under 4   <11 cm

Children 4-12 years   11 - 14 cm

An analysis of size indicated the following (Table 6.3 and Figure 6.12: based on the 267 stencils which were fully measurable).

The largest hand stencils at the site are 22cm long. There are five only of these largest stencils; four are white left, one is a white right. The descriptive statistics for the different stencils (Figure 6.12) indicate considerable overlap between the bulk of the stencils in terms of median size and 50% of the range.

Table 6.3: Yengo 1. Size ranges of hand stencils (group n≥3).

Hand	Minimum	Mean	Maximum	St. Dev.
White Left	13.5	18.8	22	1.8
White Right	9.5	18.8	21	2.9
White Var. Left	15	17.5	21	1.8
White Var. Right	15	17.7	21	2.1
Yellow Left	13	16.9	20.5	2.6
Yellow Right	11	16.7	19	3.0
Red Left	17	18	19	2.1
Red Right	16	17	21	2.7
Black Left	14	17.3	21	3.5
Black Right	15.5	16.8	18	1.3

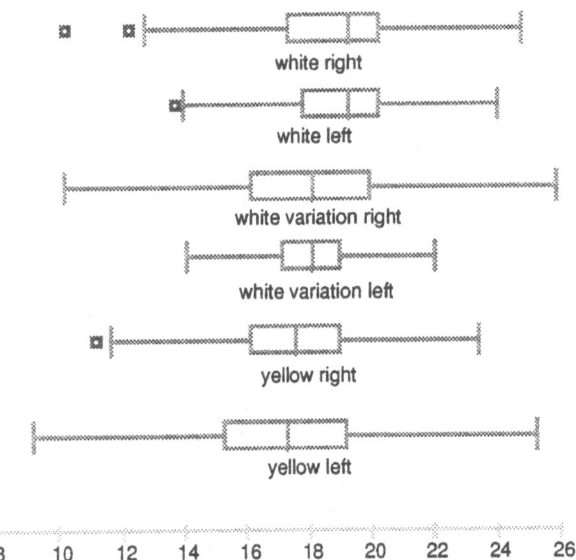

Figure 6.12: Yengo 1. Stencils. The range, median and variance (measurements in cm). [Box indicates the 25th and 75th percentile. Whiskers extend to values which represent 1.5 times the median to the corresponding edge of the box. Values outside this are outliers].

Several stencil types, however, show a tighter clustering in terms of their overall size range, particularly white left hands (plain and variations). These both have the smallest standard deviations (Table 6.3: with the exception of black right stencils where n=3) as well as the largest mean size. This may indicate that the subject hand sizes for these were perhaps restricted to men (this mean size is similar to that found by Abbey), or at least to adults.

Only the two white left hand stencil groups have a mean size larger than Abbey's mean male hand

---

[19] Research undertaken by the author at the Laurel Tree House Child Care Facility, University of Sydney, 1992. The subjects made hand prints which were labelled and thence measured.

Table 6.4: Yengo 1. The grinding grooves (measurements in centimetres).

No.	Length	Breadth	Depth	No.	Length	Breadth	Depth
1	25	8	1.8	2	33	5	0.4
3	25.5	3.7	0.5	4	23	4	0.3
5	17	3	0.2	6	27	5.5	0.3
7	26	7	0.4	8	30	7	1.0
9	33	6.5	1.0	10	32	9	1.8
11	18	5	0.5	12	18	3	0.2
13	30	7	2.5	14	33	7	2.5
15	30	10	2.1	16	28	5.5	0.3
17	26	7	1.0	18	20	4	0.2
19	20	6	0.2	20	40	7	2.0
21	34	8	1.3	22	18	3	0.2
23	30	6	1.4	24	40	10	2.1
25	35	8.5	1.2	26	15	7	1.5
27	15	7	1.1	28	10	8	1.0
29	30	6	0.3	30	30	5.5	0.3
31	35	7	1.2	32	33	4.5	0.3
33	30	5.5	0.5	34	12	7	0.3
35	22	8	0.4	36	37	9	1.6
37	33	7	0.9	38	20	8	1.1
39	26	7	0.4	40	27	4	0.2
41	23	3.5	0.2	42	28	5.5	1.0
43	30	7	1.0	44	31	7.5	0.6
44a	(abraded area within #44)				20	4.5	0.1
45	27	5	0.4	46	33	3.5	0.2
47	33	4	0.3	48	25	10	0.2
49	16	5	0.4	50	24	2	0.3
51	12	2	0.3	52	16	7	0.3
53	20	3	0.1	54	33	5	0.5
55	18	4.5	0.5	56	19	4	0.5

size. Two children's sized stencils [13.5cm and 14cm long] were also recorded amongst the white left hands. The yellow stencils and the right handed black and red stencils all have a mean smaller than Abbey's mean women's hand sizes. Children's and babies' sized hands were found stencilled in white and yellow, but not in black and red. No hand stencil variations were made using children's hands. A total of 10 very small stencils were recorded at the site. Three only were identified as babies'; the remaining seven were identified as children-sized.

While the occurrence is relatively rare (4.2%), children and babies obviously took part in the production of stencil art at the site. While it cannot be demonstrated conclusively, it is argued that women also likely to have taken part in the production of this art form; the yellow stencils and the right handed black and red stencils have a mean size well below the mean male hand size. Given the overlap found by Abbey in men's and women's hand sizes, distinguishing between men's and women's hand stencils based on size alone is a fruitless endeavour. From the presence of children's and babies' stencils amongst the art it would seem likely that women were not only present at the site but also participated in the production of this art form (McDonald 1992b).

The Grinding Grooves

The grinding grooves and areas of pecking and abrasion were traced on polythene and measured in the field. A total of 55 grinding grooves, two areas of abrasion and one abraded groove were recorded, as were five discrete areas of battering/pounding and numerous pecked marks (Table 6.4, Figure 6.5, Figure 6.13).

The average length of the 55 grinding grooves was 26cm (s.d. = 7.5). The maximum length was 40cm and the minimum length was 10cm. The average width of the grooves was 6.1cm (s.d.

= 1.9). The narrowest groove was 2cm wide and the widest was 10cm. The longest groove was not the widest groove. The grooves were on the whole fairly shallow, with the average being 0.8cm deep (s.d. = 0.7). The deepest groove was 2.5cm deep, and the shallowest was 0.1 cm deep.

As well as sharpening axes, there is evidence that other activities were undertaken on the gently sloping back wall of the shelter. Three areas of abrasion (#s 44a, 45 and 48: Figure 6.5) include one discrete incised groove and two areas where these incisions occur more as collections of grooves. These indicate the sharpening of either wooden spears or perhaps women's digging sticks (McDonald 1992b).

Also present were five discrete areas of pounding, i.e. bruising or battering of the surface. Most of these occur on the relatively flat shelf, not the sloping surfaces. Where these occur in superimposition with grinding grooves, the areas of pounding predate the grinding grooves. Several of these areas have been affected by exfoliation of the case hardened surface in this area - as have several grinding grooves (i.e. groove #s 35, 52: Figure 6.5). There is also evidence that some surface exfoliation predates both the grinding and pounding activity.

There are also numerous discrete pecked dots as well as clusters of dots (as opposed to the pounded areas) scattered across the back wall area. On the bedrock surface beneath Squares 2 and 3, five more pecked dots were also discovered (Figure 6.5). This suggests that this type of activity may predate the site's most intensive occupation period, and could be related to the production of the pecked art on the shelter's boulder. Pecked dots are a common component of the Panaramitee (Clegg 1987, Edwards 1971, McDonald 1983).

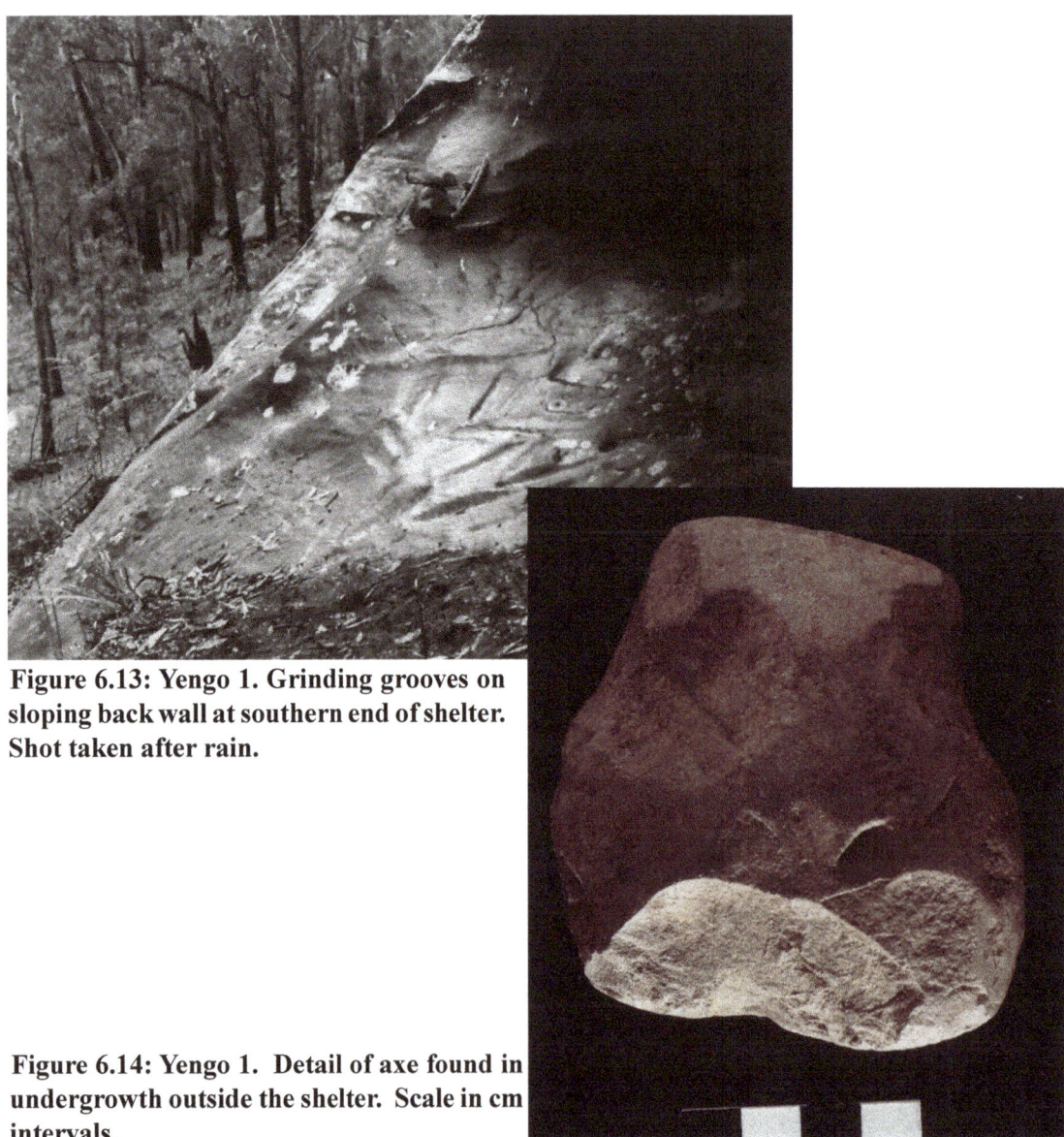

Figure 6.13: Yengo 1. Grinding grooves on sloping back wall at southern end of shelter. Shot taken after rain.

Figure 6.14: Yengo 1. Detail of axe found in undergrowth outside the shelter. Scale in cm intervals.

One ground edged axe head was found in the undergrowth at the front of the shelter. This measured 8.0cm x 7.5cm x 3.8cm, and was made of fine grained basic material (Figure 6.14). One ground edge was 5cm long while the length of opposite edge would have been in the order of 6.5cm. There are a few remnant stains/adhesions suggestive of hafting resin and the head is waisted. Both ground edges show evidence of usewear (chopping and battering), and there is a sheen near the shorter cutting edge in addition to the grinding striations. Fragments of ground edged material were also excavated from the deposit.

Yengo 2 (AHIMS #37-5-2)

This site is located in a medium sized shelter measuring 7.0m x 4.0m x 1.35m (Figure 6.15 and Figure 6.16). It faces east and is 10m north of Yengo 1 (Figure 6.6) at the same contour. The art assemblage in this site is very different to that found in the main shelter (Table 6.5). Here drawings and stencils (40.4% each) are co-dominant while paintings represent 17% of the assemblage (Figure 6.17, Figure 6.18). Mixed dry and wet pigment motifs occur but are rare (2%). White pigment dominates (58.6%), but not to the same extent as found in Yengo 1. Black is the next most commonly used colour (31.1%) followed by red (13.1%) and yellow (5%).

Table 6.5: Yengo 2. Art Assemblage. Motif and technique information.

Motif	Colour				Technique							
	Red	White	Black	Yellow	Dry	Wet/*	Stencil	Eng'v	Out.	Infill	O/I	Total
Anthropomorph	3	1	11		9	6			2	11	2	15
Macropod+	3	7	4		9	1/1			7		4	11
Snake			1	1	1	1			1	1		2
Mammals^+		4	2		4				2		2	4
Reptiles≈+		3	2		2	-/1			1		2	3
Eel		2			2				2			2
Axe		1					1					1
Club		1			1					1		1
Hands		26		3			29					29
Hand variation		8					8					8
Bird tracks+		1	1		1						1	1
CXNF		1	2		2	1			1		2	3
Unid. lines	2	1	1		2	2			4			4
Unid. solid	5	2	7	1	7	6	2		8	4	1	15
Total	13	58	31	5	40	17/2	40	-	28	17	14	99

+ Seven of these are bichrome (i.e. black and white)   ^ 2 possums, 1 koala, 1 flying fox
* wet and dry   ≈ 3 goannas

The 40 stencilled motifs comprise mostly hands (72.5%) or hand variations (20%). There is one stencilled axe. All but three of the stencils are white: the others are yellow. Two areas of blown red pigment were identified, although these are not identifiable stencils.

Depictive motifs account for c.60% of the assemblage, and most are identifiable (72%) and figurative. The subject range includes mainly anthropomorphs and macropods (25% and 19% respectively) with a variety of other land animals also depicted (goannas, snakes, possums, a koala and a flying fox). Bird tracks and complex-non-figurative motifs are also present amongst the assemblage. Outlined depictive motifs were the most common (48%) followed by infilled (29%) and then outlined and infilled (24%). Colour usage in the depictive assemblage is quite different (cf. the stencils): black predominates (47%) followed by white (35%), red (15%) and yellow (3%).

The following superimpositions were observed:

Over >

white stencils	dry red outline
dry red solid	dry black solid
dry pink outline	dry black solid
wet white outline	dry black solid
dry white outline /infill	dry black solid
dry and/or wet red outline	dry black solid
dry black outline /infill	dry black solid
wet white outline	dry black outline /infill
black and white o/i	dry red solid

The proposed sequence based on these superimposition relationships is as follows:

Earliest	dry black solid
	dry red or pink solid and/or outline
	dry white outline and infill
	dry black outline and infill
	wet white outline
	black and white bichrome (some wet/dry combination)
Most recent	white hand stencils

There is a suggestion of motif preference changes with these changes in technique. Most of the earliest art (dry black solid) consists of small anthropomorphs (<25cm in size) while the middle phase consists of the widest range of subjects. These are (mainly) kangaroos, anthropomorphs (all >30cm; most >50cm in size) and the full range of figurative motifs present at the site. Black and white bichromes are restricted to macropods, goannas, possums and bird tracks.

Stencils

Side and size for the hand stencils were recorded, although the sample size here is considerably smaller. There are similarities and differences between this and the Yengo 1 assemblage. Left hand stencils again dominate, representing 65% of stencils for which side was identifiable and 87.5% of the stencil variations (Table 6.6). A relatively small proportion (30%) of the stencils could not be identified for size and/or side. This reflects less superimpositionning rather than the condition of the art.

Table 6.6: Yengo 2. Handedness (left or right) of the coloured stencils.

	Stencils			Hand variations			
	Left	Right	?	Left	Right	?	Total
White	8 (4)	8	6	7	1		34
Yellow	2 (1)						3
Total	10 (5)	8	6	7	1		37

(#) = stencils for which side can be determined but full measurements not possible.

Half of the hand variations here consist of the 'hand + arm' variety. The remaining four consist of two pairs of stencils positioned wrist-to-wrist (Figure 6.19). One of these pairs also has the fourth and fifth fingers positioned close together.

All of the stencils at this site are adult sized. The mean sizes for the two of the three main varieties are all larger than their counterparts in Yengo 1, and the standard deviations are generally smaller (Table 6.7). The smallest hand was 15cm in length. The largest stencils (n=3) in this site are 21cm.

Stencilling in this site is a more restricted phenomenon both in terms of the overall site assemblage and compared with the extensive usage of this technique in the shelter next door. From the stencils' measurements it would appear that the population producing these stencils was also more restricted. All are adult sized, most of these from the middle range. One stencilled axe amongst the assemblage is the only possible gender association. Superimposition analysis indicates that stencil production occurred late in this art sequence. Black + white bichromes and white paint also appear late in this shelter's art sequence.

Table 6.7: Yengo 2. Size ranges of hand stencils (where n >3).

Colour/side	Minimum	Mean	Maximum	St. Dev.
White left	16	17.8	21	1.6
White right	15	18.25	21	2.1
White var. left	16	18	21	1.8

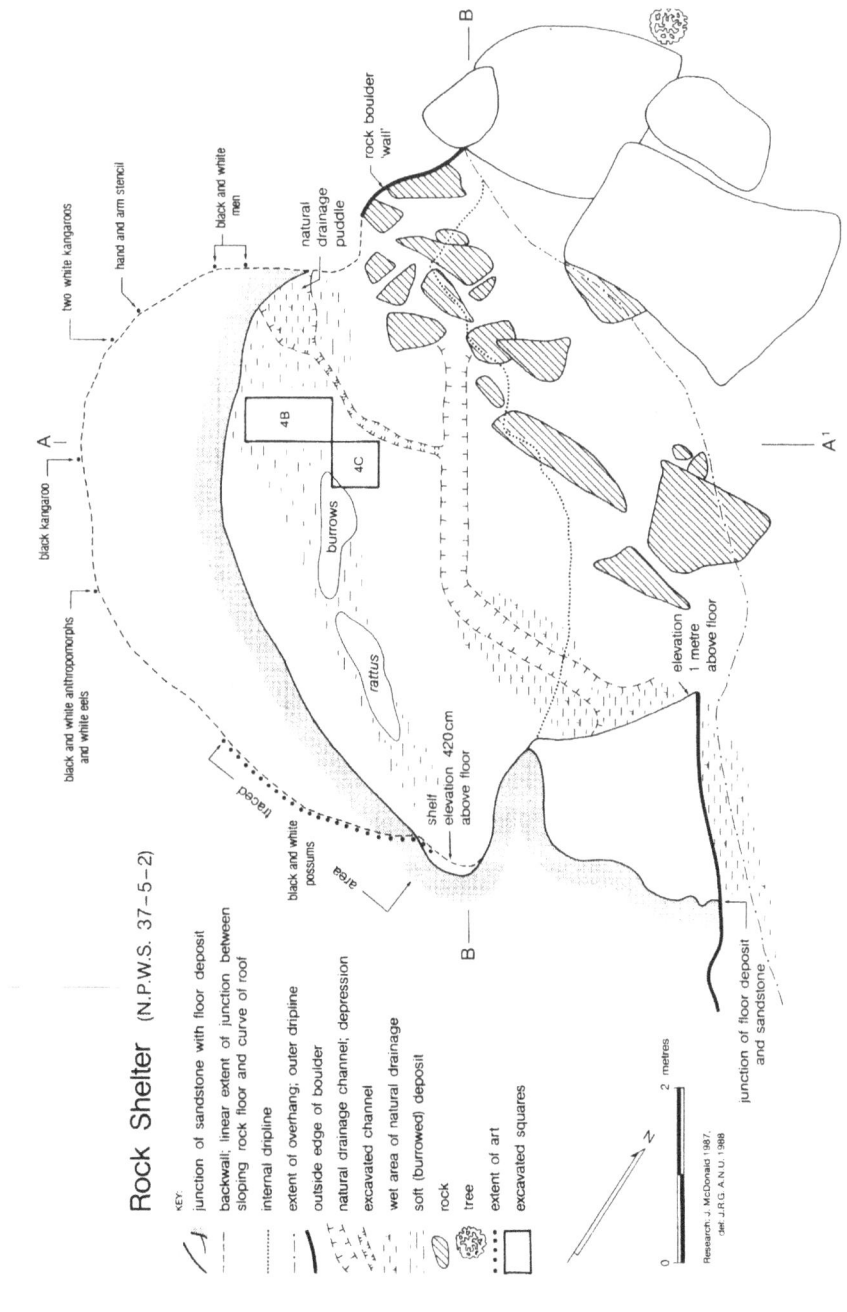

Figure 6.15: Yengo 2. Site Plan.

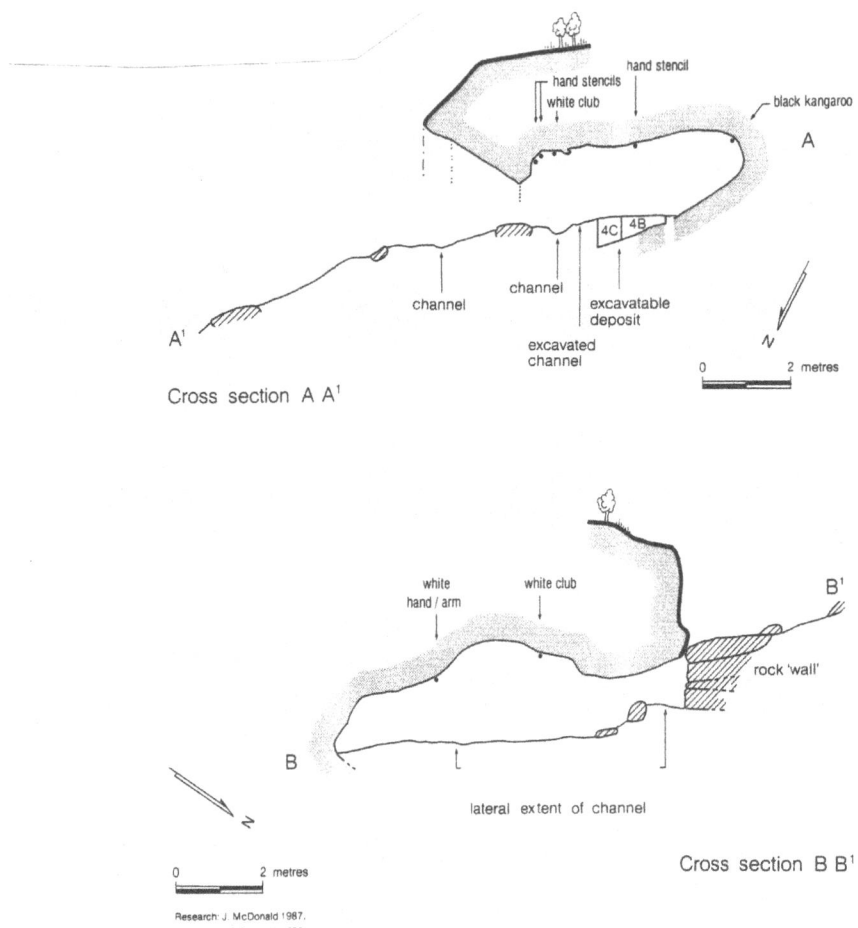

Figure 6.16: Yengo 2. Cross-sections.

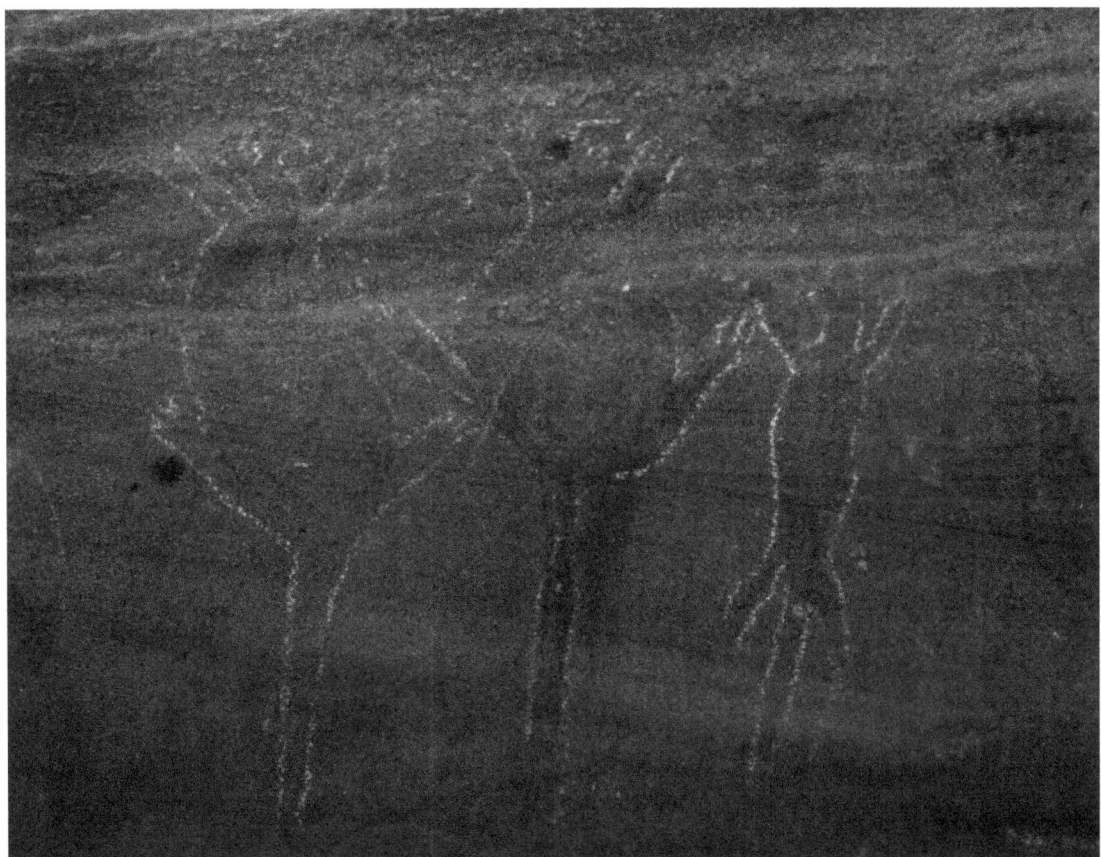

Figure 6.17: Yengo 2. Black and white possums and goannas. There are superimposed over red solid motifs.

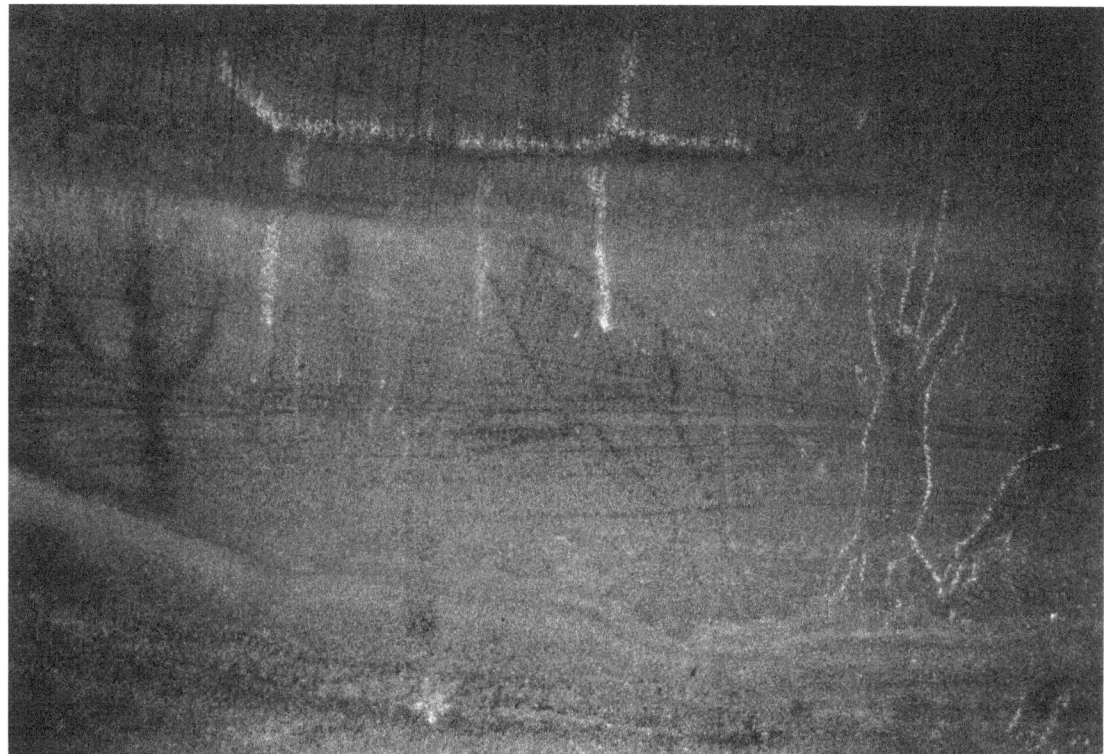

Figure 6.18: Yengo 2. Black and white goanna, black drawn and white painted non-figurative motifs and red solid figure on southern side of shelter.

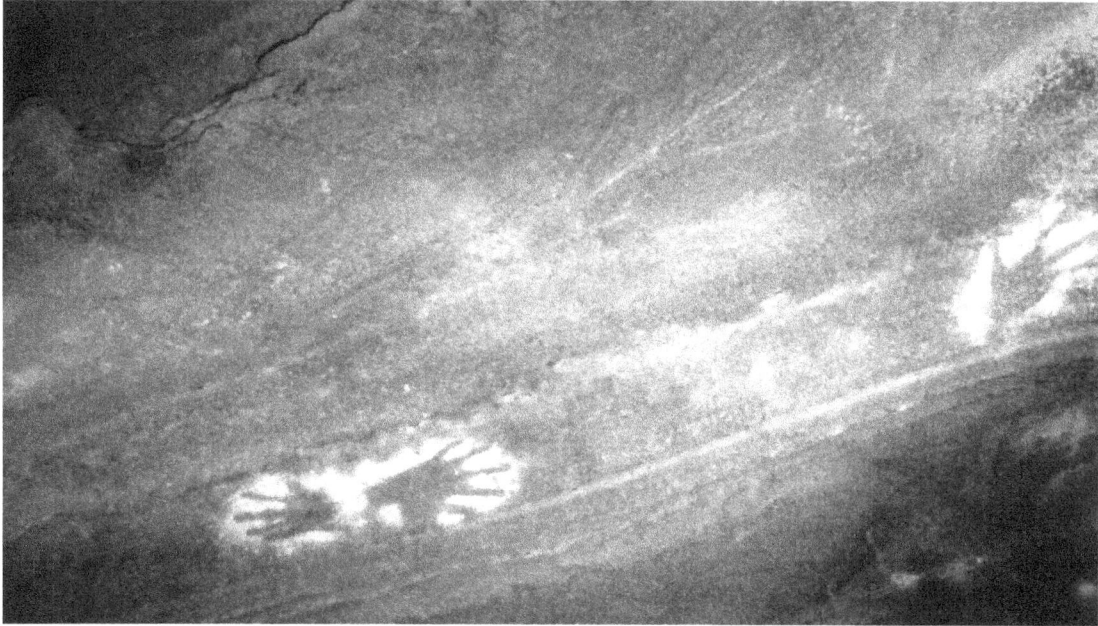

Figure 6.19: Yengo 2. White hand stencil variations - wrist-wrist combination on ceiling of shelter, superimposed over red outline macropod.

Table 6.8: Yengo 1. Excavated Pit Dimensions.

Square	Length (m)	Breadth (m)
1A	1.0	0.5
1B	1.0	0.5
1C	1.2	0.5
1R	1.0	0.26
2A	1.0	0.5
2B	1.0	0.5
3S	1.0	0.5
3N	0.5	0.5
4A	1.0	0.5
5A	1.0	0.5
6	0.58	0.5

Excavation Procedures

Yengo 1

In 1987 a total of 11 pits were excavated (Figure 6.2), removing a total of 2.2 tonnes of deposit. In 1988, square 1 was reopened as was part of trench 4A/5A. The base of the fallen boulder in square 1 was not exposed during the first season's excavation and it was not clear whether it post-dated the engraved boulder (or whether this was a large manuport or part of the original roof-fall episode). A total of 483kg of deposit were excavated from square 1C (1.2m x 0.5m) in 1988.

The majority of the pits measured 1.0m x 0.5m (Table 6.8). The unorthodox placement (and occasionally size) of the pits resulted from the morphology of the shelter and from the need to avoid several areas of obvious and severe disturbance caused by extensive wombat burrowing. Initially the pits were positioned to investigate two separate areas of deposit: adjacent to the engraved panel (square 1) and in the flat central floor area (Squares 2 and 3). Once excavation commenced, however, it became apparent that site formation processes were complex. Because of this and the very friable nature of the deposit, the two main investigation areas were joined by two trenches. One of these (Squares 4A and 5A) connected the engraved panel (square 1) with the back wall. The other trench (Squares 3 and 6) connected the central area of the deposit (square 2) with the first trench. All squares (except 1A) were excavated to bedrock. A total of 123 spits were excavated (Figure 6.20 to Figure 6.23).

Once excavation commenced it became apparent that burrowing activity was not restricted to wombats and that surface disturbance was not a clear indicator of subsurface burrowing activity. Almost all squares contained bush rat (*Rattus fuscipes*) burrows - prehistoric, recently collapsed and extant varieties. During our excavation a trap set with peanut butter was used (by Jon Saunders, Zoologist) to catch a current resident that kept re-excavating its burrow into our trenches every night. Thankfully none of the wombat burrows were currently in use!

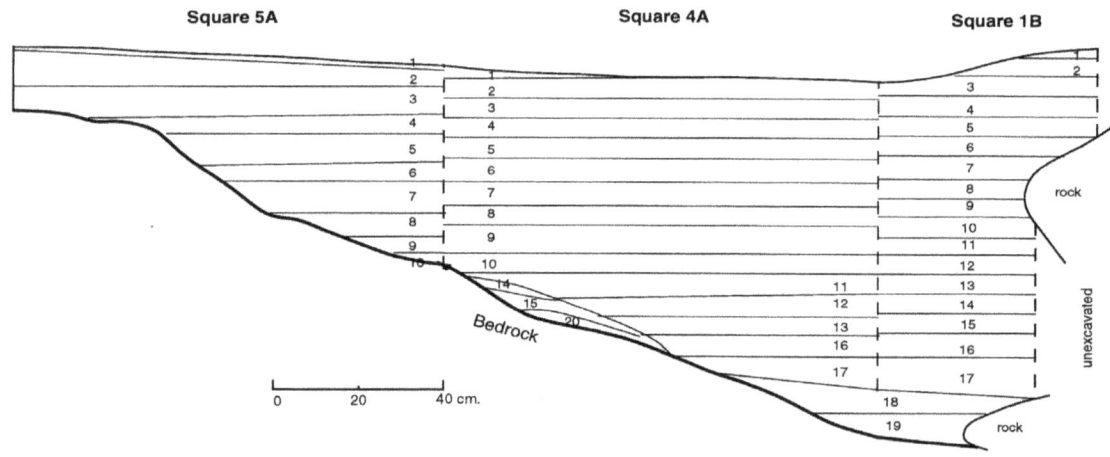

Figure 6.20: Yengo 1. Squares 1B, 4A and 5A, northern baulk showing excavated units.

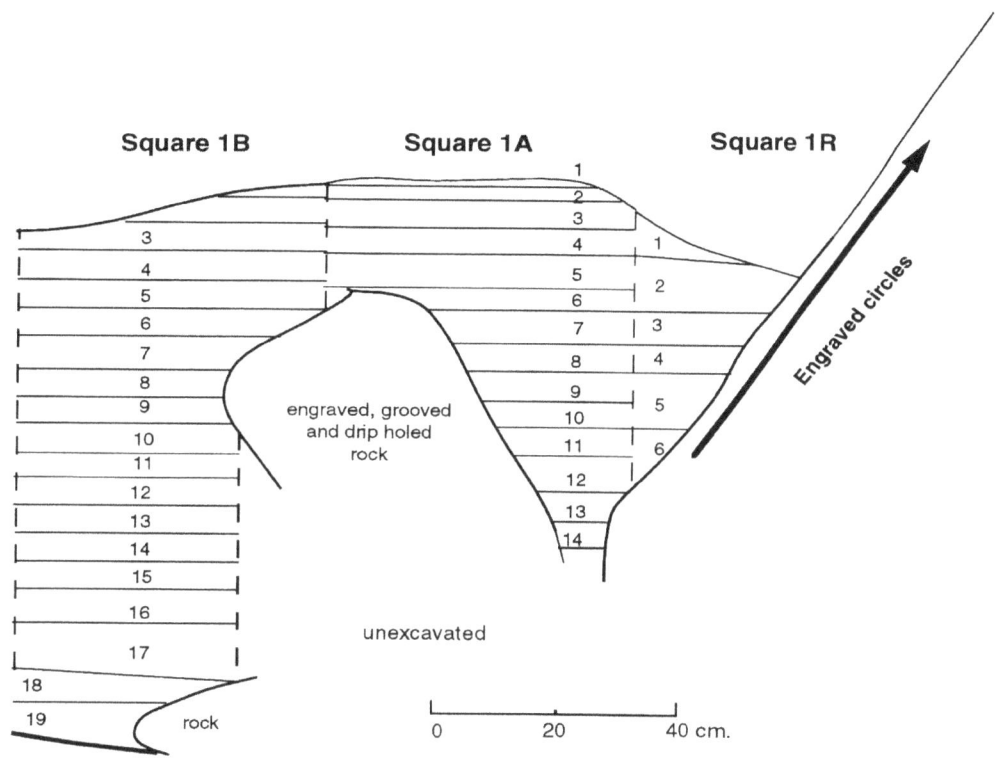

Figure 6.21: Yengo 1. Squares 1A, 1B and 1R, northern baulk showing excavated units.

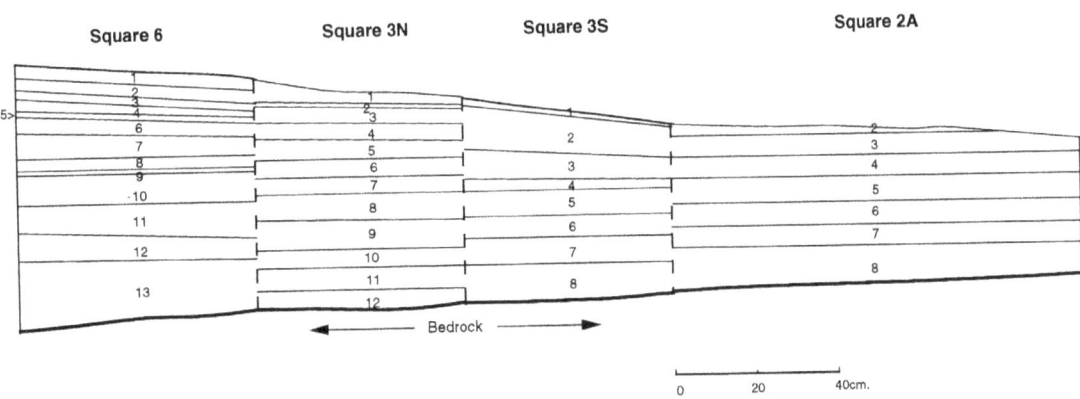

Figure 6.22: Yengo 1. Squares 2A, 3S, 3N and 6, eastern baulk showing excavated units.

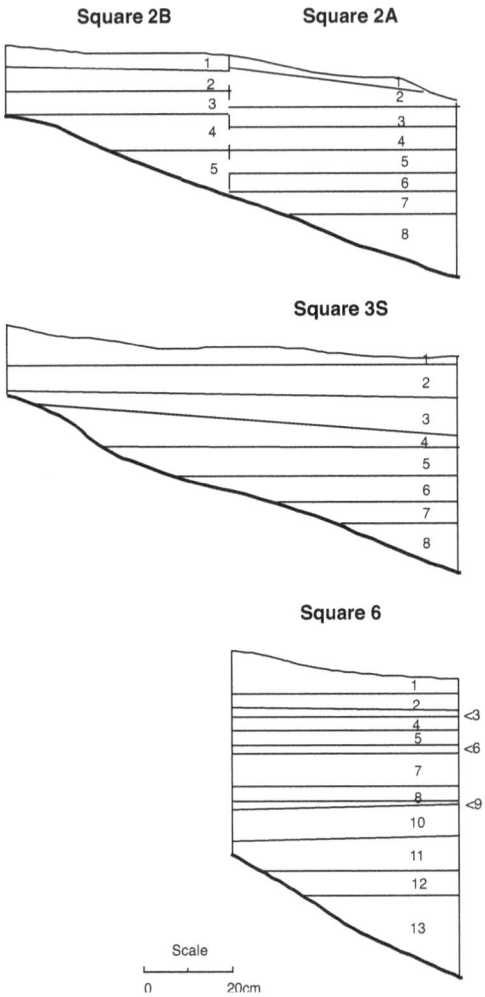

Figure 6.23: Yengo 1. Squares 2A, 2B, 3S and 6E, northern baulks showing excavated units.

Yengo 2

A total of nine person days were spent on this excavation in September 1988. Two test pits were excavated (Figure 6.15) removing a total of 494kg of deposit. Sixteen spits were excavated in these two test squares (Figure 6.25).

Bush rats had also burrowed in this site and heavy rain throughout the excavation period demonstrated that the floor can be inundated by water from the northern end of the shelter. Two pits were excavated to provide a good sample of artefactual material. While the deposit contained archaeological material, the density of this was extremely low. Saturation of the deposits meant no stratigraphic divisions were observable here. Decomposing sandstone was encountered immediately above bedrock (Figure 6.24).

Results - Yengo 1

Stratigraphy

The stratigraphy at the site is complex, as a result of the shelter's morphology and different depositional processes. Bedrock in the shelter consists of a sloping rock floor. The large engraved boulder at the front of the shelter retains deposit only at the northern end, and in this area there

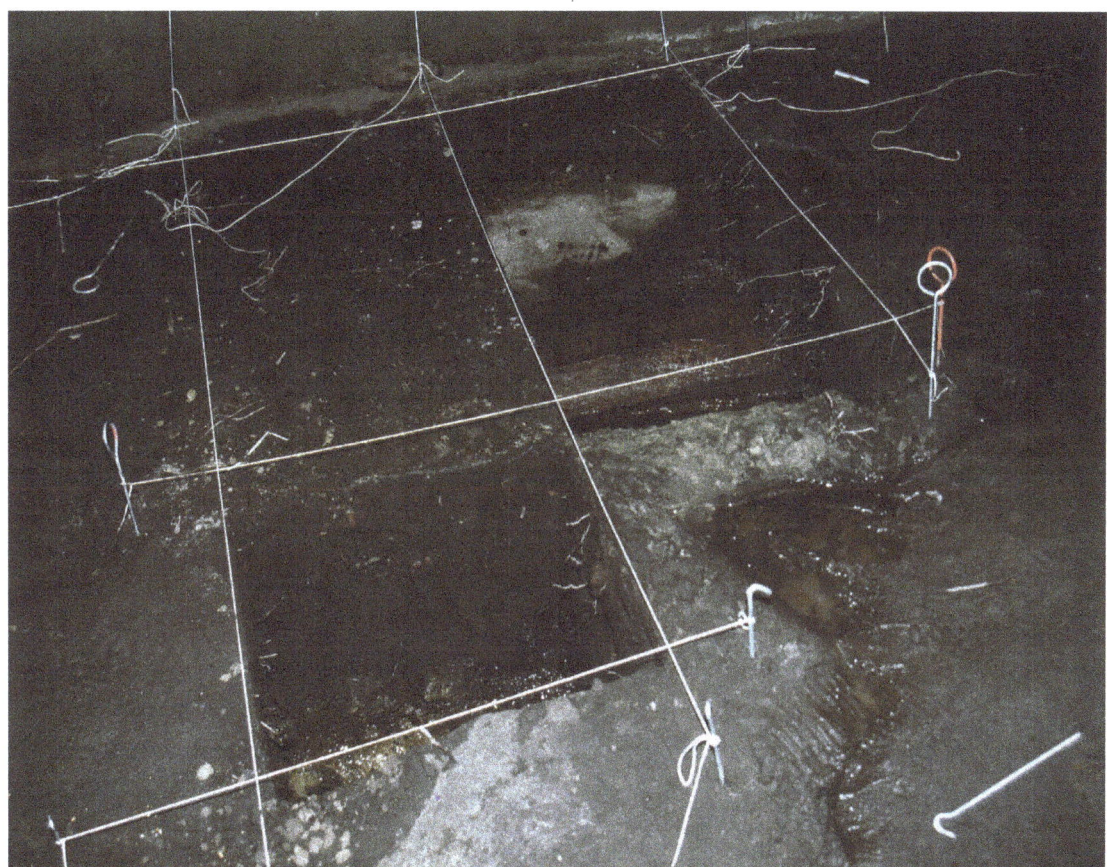

Figure 6.24: Yengo 2. Completed excavation. Square 4C in the foreground; 4B in background. Note the lack of obvious stratigraphy.

is also a considerable influx of deposit through dripline activity. Several of the layers of deposit identified are only present in excavated squares at the front of the site. The steeply sloping bedrock means that at the back of the shelter, the most recent layers are adjacent to bedrock. Deposition has been accelerated in the front squares, because of incoming sediment. The sand fraction in the front squares is considerably higher than those in the central area of the site, where the deposit is fine grained and ashy. Through time, the area of flat, deposit covered floor increased laterally, bringing the ceiling to within reach.

The squares at the site are viewed as two groups:

1. Square 1A, 1B, 1R and 1C which contain the deepest and earliest deposit and have been affected by soil deposition from outside the dripline; and,

2. Squares 2, 3, and 5A which are relatively shallow and mostly recent for which sedimentation is primarily cultural.

Squares 4A and 6 show combinations of these formation processes.

Stratigraphic Layers

Section drawings were done for all excavated squares except Pits 1A, 1R and 1B. Torrential rain (10cm in 24 hours) fell towards the end of the first season before the sections in this area of the site had been drawn. This rain severely affected the surfaces of the balks at the front of the site. The main reason for the return season was to reopen this trench to interpret its stratigraphy. Unfortunately torrential rain again fell and this problem was not perfectly resolved. It would

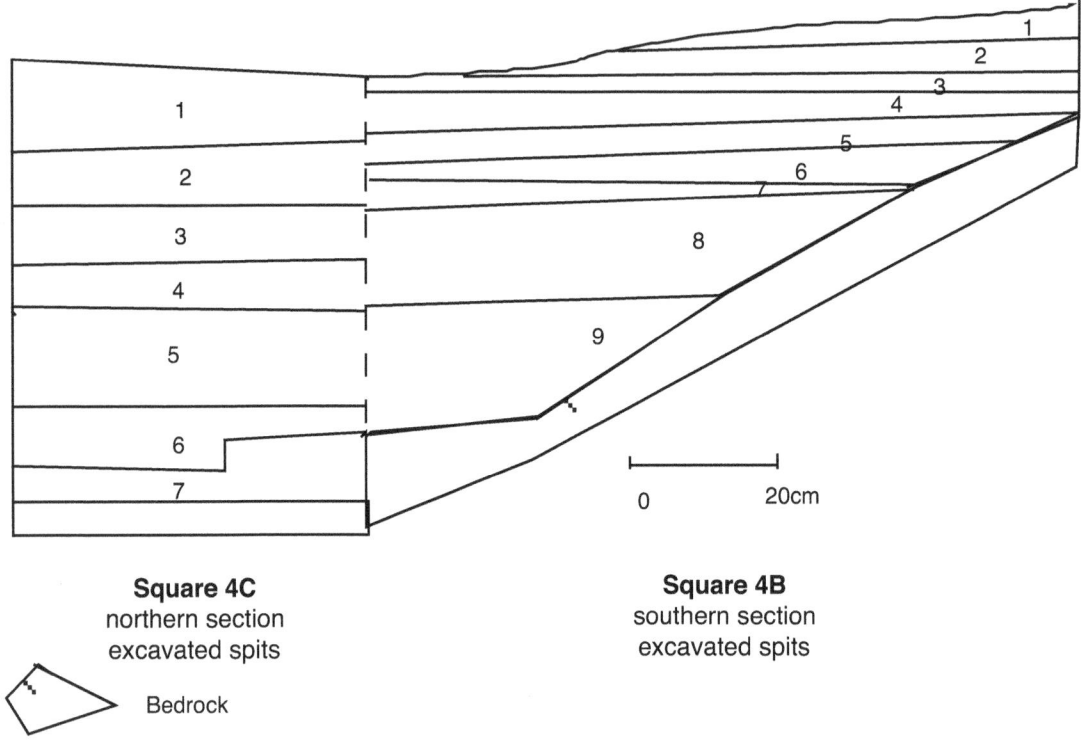
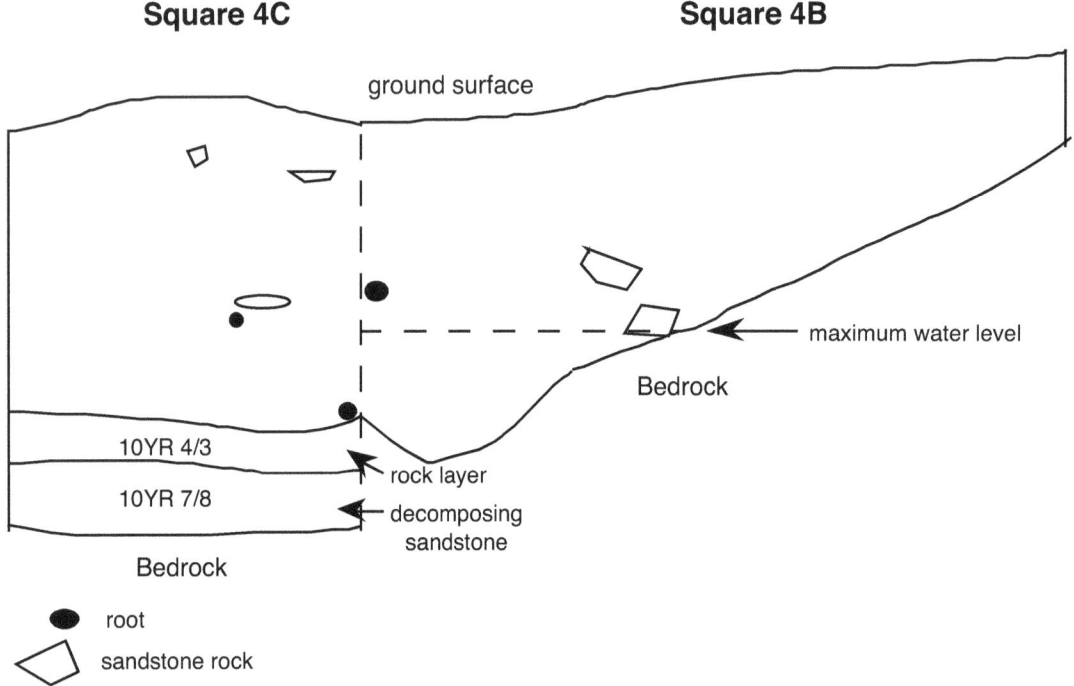

Figure 6.25: Yengo 2. Squares 4B and 4C. Excavated spits & stratigraphic section.

appear that this type inundation has probably continuously affected the soil matrix at the front of the site: the complex stratigraphic detail observable inside the shelter is non-existent outside the protection of the shelter.

Seven stratigraphic layers (as distinguishable from analytical units) were identified at the site (Figure 6.26- Figure 6.28);

Layer I   Loose fine deposit, high leaf litter content. Disturbed. [2.5YR 5/2; pH = 7]

Layer II   Fine, ashy compact grey deposit, rich in charcoal, artefacts and faunal remains. [2.5YR 5/1; pH = 8.5]

              In Squares 2A, 2B, 3S, 3N, 6 between Units II and III is a red/grey ashy lens, with high charcoal content [5YR4/4; pH 8.5 – 9]

Layer III   Fine ashy compact dark grey deposit, rich in charcoal, artefacts and faunal remains [5YR 5/2; pH = 8.5]

Layer IV   Orange mixed sandy deposit with high ash and charcoal content

Layer V   Red/brown/buff, compact sandy deposit, low charcoal [pH = 8.5]

Layer VI   Yellow/buff fine, compact deposit, low charcoal [10YR 6/3; pH = 8.5–9]

Layer VII   Very compact buff, with high roof-fall component [10YR 6/3]

NB.   Individual hearths, extant and prehistoric burrows and ash lenses have been identified on the section drawings and are described in the detailed discussion for each square.

The deposit in the central area of the shelter was extremely fine, dry and alkaline. The densest layers were rich in artefactual material and contained large quantities of highly fragmented faunal remains (over 600g of bone were recovered from eight of the squares).

Given the stratigraphic complexities of the site, the high degree of disturbance in some squares and the large size of the excavated stone tool assemblage, a sample of the excavated material was analysed in detail (see below). Stratigraphic components in all excavated squares are given (Table 6.9 and Table 6.10). Detailed stratigraphic notes are presented for the four analysis squares (1B, 4A, 3S and 6). The correlation of excavated spits with analysis units/stratigraphic layers for the analysis squares is given (Table 6.10).

Square 1B

Unit I   Loose fine deposit, high leaf litter content, large pieces of charcoal observed. Many small ferns: well established root system. Very uneven surface.

Unit II   Fine, compact brown deposit, rich in charcoal and with increasing artefacts. Top of boulder exposed[20]: many small rocks (<3cm diameter) concentrated at northern end.

Unit III   Increasingly compact dark grey deposit, rich in charcoal and artefacts. Many pieces of red coloured roof fall observed. A rat burrow intrudes into the north-western face (a 10cm baulk was employed along 50cm of wall to prevent collapse).

[20]This boulder, the top of which was uncovered in Unit II, was excavated beneath only in square 1C (see Figure 6.28). This appears to be contemporaneous roof fall with the large engraved boulder at the front of the shelter.

Unit IV	More red/brown with decreased charcoal throughout this layer. Artefacts present in relatively large numbers, deposit more compact than above, and moisture content increasing. High proportion of sandstone rubble/ roof-fall.
Unit V	Lighter brown, damp compact sandy deposit, low charcoal and artefacts.
Unit VI	Lighter coloured compact deposit, low charcoal.
Unit VII	Combined square 1B/4A see square 4A notes.
Square 4A	(Figure 6.26)
Unit I	Loose light grey mottled deposit with leaf litter, twigs, ferns. Very disturbed.
Unit II	More compact grey/brown deposit, rich in charcoal and artefacts; some bone. The burrow identified in sq 1B, also affects this square. Boundary of burrow easily defined; this surrounded by looser, mottled orange deposit: otherwise deposit quite firm.
Unit III	Fine ashy compact dark grey deposit, rich in charcoal and artefacts but no faunal remains. Base of burrow above protrudes 4cm into this layer.
Unit IV	Orange mixed sandy deposit with high ash and charcoal content. Many artefacts observed while digging.
Unit V	Red/brown yellow/buff, compact sandy deposit, areas of charcoal staining; floor area diminishes dramatically due to sloping bedrock.
Unit VI	Hardened yellow compact deposit, low charcoal. A few large artefacts encountered, but density much lower.
Unit VII	Very compact buff, with high roof-fall component: initially mistaken as bedrock. Dug with Geo-pick.
Square 3S	(Figure 6.29)
Unit I	Loose fine deposit, high leaf litter content. Disturbed. Two prehistoric burrows enter square from north, one transects square. Animal scats insect cases and other evidence of recent deposition.
Unit II	Fine, ashy compact grey deposit, rich in charcoal, artefacts and faunal remains. Deposit compact until scraped with trowel, then becomes fine powder.
	Red/grey ashy lens, with high charcoal content: thicker at eastern end. Deposit more friable than layer immediately below.
Unit III	Fine ashy compact dark grey deposit, rich in charcoal, artefacts and faunal remains. Artefacts decline towards base of layer. Another prehistoric burrow runs along eastern baulk, and extends down through this and the layer below (all material from identified burrows was excavated and bagged separately to the remaining, undisturbed portions of the square).

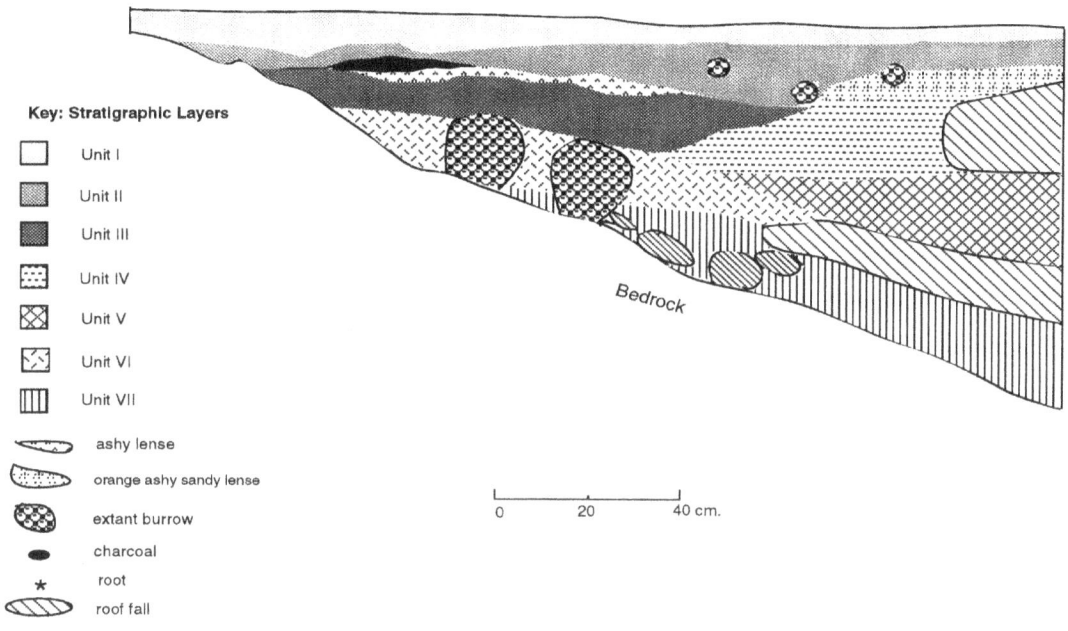

Figure 6.26: Stratigraphic section. Northern baulk of Squares 4A and 5A.

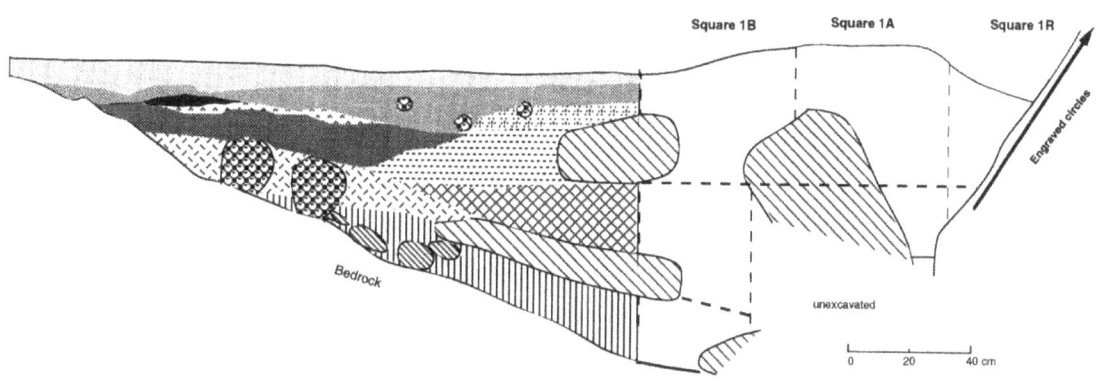

Figure 6.27: Trench aligned with Square 1, showing relationship of engravings to site's internal stratigraphy (Figure 6.26).

Unit IV	None
Unit V	None
Unit VI	Hard yellow deposit, low charcoal.
Unit VII	Extremely hard compact yellow buff, with base of prehistoric burrow adjacent to junction of this and unit VI.

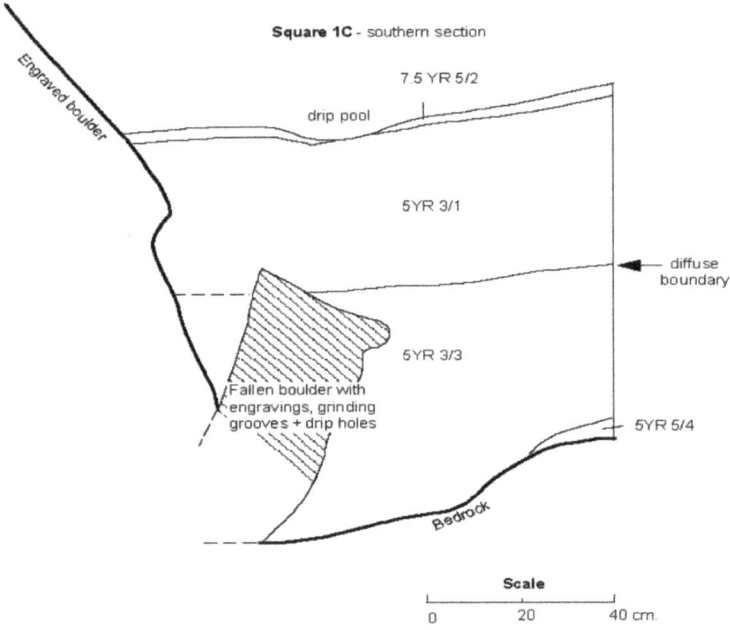

Figure 6.28: Yengo 1. Southern baulk, square 1C, showing location of buried boulder relative to engraved boulder (schematic boundaries).

Square 6   (Figure 6.29)

Unit I   Loose fine deposit, high leaf litter content. Disturbed.

Unit II   Fine, ashy compact grey deposit, rich in charcoal, artefacts and faunal remains. Red/grey ashy lens with high charcoal content. Located across northern side of the square. Yellow lens identified beneath this in centre and northern baulk of the square. Across most of the square it is a pinker and less ashy than identified elsewhere.

Unit III   Fine ashy compact dark grey deposit, rich in charcoal, artefacts and faunal remains. Collapsed burrow (prehistoric) towards base of this layer (10cm diameter) from centre north to south east corner.

Unit IV   Orange mixed deposit with less ash and charcoal content than found in layer above.

Unit V   None

Unit VI   Buff fine, compact deposit, low charcoal. Appeared sterile while digging.

Unit VII   Very compact buff, for approx 12cm immediately above bedrock. Two roots (1cm diameter) traversed square adjacent to bedrock.

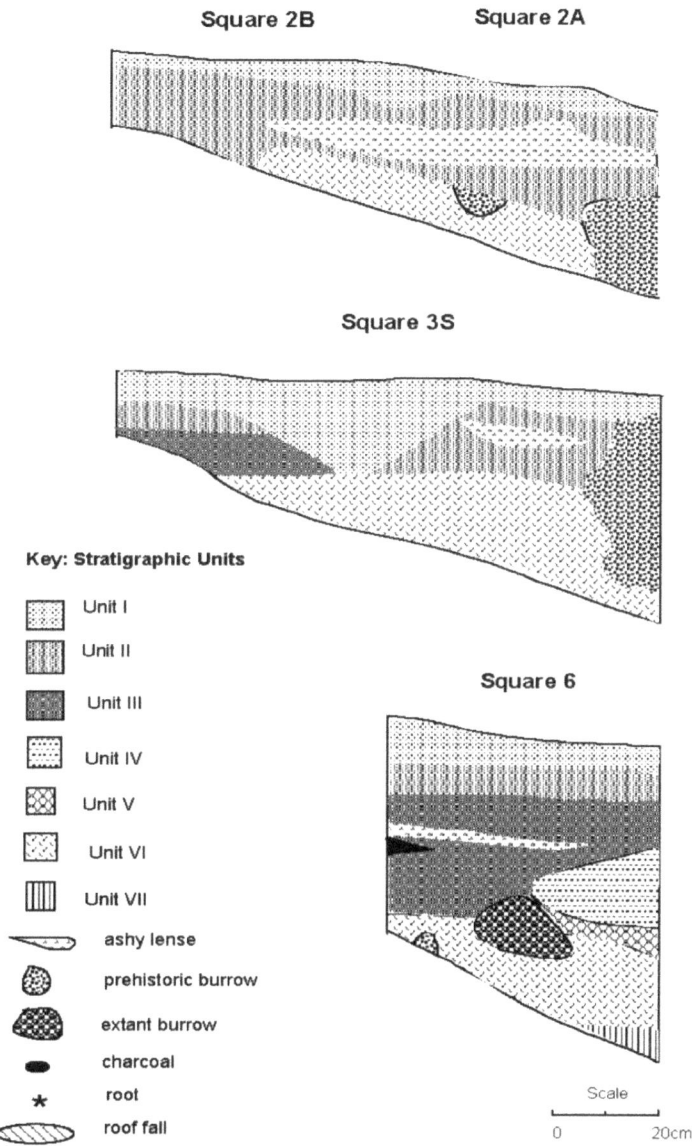

Figure 6.29: Yengo 1 Squares 2A, 2B, 3S and 6. Stratigraphic sections northern baulks.

Table 6.9: Yengo 1. Excavated Depth, Spits and Stratigraphic Layers. All excavated squares.

Square	Depth below surface (cm)	No. of Spits	Stratigraphic Layers
1A	66	14	I, II, V
1B	97	18	I, II, III, IV, V, VI, VII
1C	65	10	I, II, V, VI
1R	51.5	6	I, II
2A	32	10	I, II, VI
2B	31	6	I, II, VI
3S	46	8	I, II, III, VI, VII
3N	49	12	I, II, III, VI, VII
4A	93.5	20	I, II, III, IV, V, VI, VII
5A	58	10	I, II, III, VI, VII
6	53	13	I, II, III, IV, VI, VII

Table 6.10: Yengo 1. Correlation of spits with analytical units and stratigraphic layers: Analysis squares (1B, 4A, 3S and 6) only.

Analytical Units	Excavated Spits				Stratigraphic Layer
	Square				
	1B	4A	3S	6	
1	1,2,3	1,2	1	1,2	I
2	4	3	2	3,4	II
3	5	4	3	5,6	III
4	6	5	4	7	III
5	7	6	5	8,9	III
6	8	7	-	10	IV
7	9	8			IV
8	10,11	9			IV
9	12	10			V
0	13	11			V
1	14	12			V
2	15	13			V
3	16,17	16	6	11	VI
4	18	17	7	12	VI
5		14,15,20	8	13	VII
6	(4A/1B)18,19				VII

Dates

Nine charcoal samples were submitted to the ANU Radiocarbon Lab for age determinations (Figure 6.30; Table 6.11 and Table 6.12). These established that initial occupation of the shelter dates to the pre-Bondaian period, commencing around 6,000 years ago. It would appear that occupation of the shelter at this time was sporadic and continued for c.1,000 years. There may have been a hiatus between this and subsequent occupation of the site, although artefact accumulation rates suggest this not to be the case. The next phase of occupation ended around 3,000 years ago.

Table 6.11: Yengo 1. Radiocarbon determinations.

Square/spit/ sample number	Lab-ID	Depth below surface (cm)	Stratigraphic Layer	Age
2B/4/2	ANU-6058	16.5	II	540 ±180 BP
1R/3/2	ANU-6057	17	II	260 ± 120 BP
3S/4/4	ANU-6217	10	III	101 ± 1.1%M
4A/6/1	ANU-6216	30.5	III	1,530 ± 110 BP
4A/10/X	ANU-6054	52	IV	1,950 ± 400 BP
1A/10/2	ANU-6215	55	V	2,750 ± 220BP
4A/10/2	ANU-6056	52.5	V	2,840 ± 240 BP
4A/10/1	ANU-6055	54.5	VI	4,590 ± 300 BP
4A/1B/19/1	ANU-6059	93.5	VII	5,980 ± 290 BP

The most intensive occupation of the shelter commenced sometime after 2,000 years ago. Occupation continued from this time well into the last millennium. There is no evidence of post contact use, either in the dates, the art or the excavated artefacts.

Two of the dates received (ANU-6057, ANU-6217) are considered to be unreliable indicators of age. The former, being located as it was beneath the dripline may well have been contaminated by more recent charcoal percolating down through the deposit. The latter sample may have been affected by bush rat burrowing in the spit below (3S/5): the sample when collected was noted to be underneath an *in situ* piece of sandstone associated with a hearth. Contamination of this sample with younger material therefore must be attributed to below, rather than from above.

Table 6.12: Yengo 1. Features dated by the charcoal samples. Identification number is that used on stratigraphic sections (Figure 6.30).

Id. No.	Sample Id.	Feature
1.	ANU-6054	Charcoal rich hearth, associated with artefact concentration.
2.	ANU-6055	Interface between units IV and VI, western end of square, base spit.
3.	ANU-6056	Charcoal from eastern end of square, colour dichotomy, later identified as unit V.
4.	ANU-6057	Adjacent to engravings, associated with large *in situ* artefacts.
5.	ANU-6058	Directly above hardened layer; artefact bone and ash rich spit.
6.	ANU-6059	Basal occupation, above small piece roof fall, adjacent to bedrock.
7.	ANU-6215	Artefact rich layer adjacent to engraved /grooved boulder.
8.	ANU-6216	Artefact rich layer in burnt orange area, western end square.
9.	ANU-6217	Artefact rich layer beneath sandstone fragment above hardened layer: affected by bush rat burrowing?

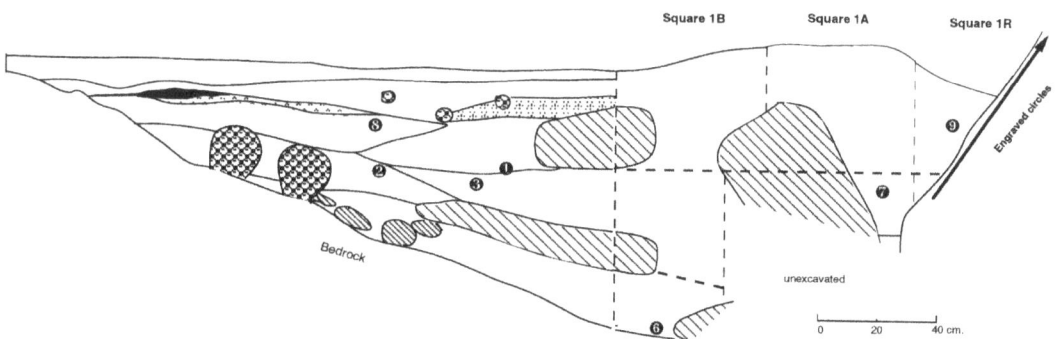

Figure 6.30: Schematic northern section (Trench 1) showing engraved boulder and locations of charcoal samples (for ID number see Figure 6.12).

The Excavated Lithic Assemblage

A total of 10,060 stone artefacts were excavated from the 10 counted excavation pits[21]. Artefacts were found in all but seven of the 117 counted spits. The assemblages from all counted squares were classified and quantified. General patterns in horizontal and vertical distribution were analysed for all squares, but because of the large assemblage retrieved, the complexity of the stratigraphy and since many of the squares were severely affected by burrowing, it was decided to analyse only a sample of the excavated squares in detail. Squares 1B, 3S, 4A and 6 were selected for analysis for the following reasons:

- the stratigraphy in these squares was well controlled and clear;

- these squares contain high numbers of artefacts, and provide a good sized samples for analysis;

- both the central living floor area and the deeper deposit are well represented by these squares.

Square 4A was severely affected by extant burrowing. Excavation in this square proceeded extremely careful to isolate the material from the burrows. This square provided the most complete sequence of the site's strata as well as five of the radiocarbon samples.

The four analysis squares represent 56% of the volume of deposit excavated (0.9 cubic metres), and c.52% of the artefact assemblage retrieved (Table 6.13). It was considered that the

[21] The material from Square 1C has been neither sorted nor counted.

aims of the excavation would be achieved by analysing this sample and that many complications arising from stratigraphic anomalies and disturbance from burrowing would be avoided. Data from the sample squares was amalgamated into two pairs of analysis squares (called Squares 1B-4 and 3S-6) for the analysis of vertical (diachronic) patterning.

Artefact Density

Artefact densities were calculated using excavated bucket (deposit) weights. The weights of sandstone fragments and gravel (weighed in the field) were subtracted from these totals. This ensured comparability between squares with varying amounts of roof-fall. A separate calculation was made to enable comparison of roof-fall components throughout the deposit (see McDonald 1994: Appendix 1: Table A1.A:1). Cubic metre calculations were based on spit volume (length x breadth x depth) and weighed excavated volume. The calculated ratio was 0.00073 cubic metres of deposit/kilogram. Density calculations were based on this ratio. Densities are also expressed as artefacts/kg. The two methods of describing artefact density were calculated to enable comparison with other excavation data from this northern part of the Sydney Basin (Attenbrow 1987, Moore 1981).

Assemblage Characteristics

The majority (98.4%) of the lithic material at the site is unmodified debitage. Almost 70% of the assemblage is also <1cm in size, and is classified as microdebitage. Only 161 artefacts with macroscopic evidence of usewear and/or retouch were present amongst the assemblage (1.6%). Most of these (47.2%) were modified amorphous flakes and flake pieces. A total of 87 cores (0.8%) were present amongst the assemblage.

Table 6.13: Yengo 1. Average artefact densities (density$_1$ = artefacts/m^3; density$_2$ = artefacts/kg).

Square	Artefacts	Kilograms of deposit	Cubic Metres	Density ^{1}	Density ^{2}
1R	21	87.75	0.06	350	0.2
1A	127	287.75	0.21	605	0.4
1B	423	397.75	0.29	1,459	1.1
2A	2,161	161.75	0.12	18,008	13.4
2B	1,113	123.5	0.09	12,367	9.0
3S	2,289	210.75	0.15	15,260	10.9
3N	982	123.75	0.09	10,911	7.9
4A	1,028	424.25	0.31	3316	2.4
5A	822	184.5	0.13	6,323	4.5
6	1,094	201.3	0.15	7,293	5.4
TOTAL	10,060	2,203.05	1.6	6,287	4.6

Artefacts with Retouch/Usewear

The majority of the artefacts with macroscopic evidence of use or retouch are flakes (22.4%) or flaked pieces (24.8%). No detailed analyses (e.g. residue or functional) have been undertaken of these artefacts. A range of edge damage characteristics was identified during the artefact analysis. This included fine, scalar flaking; chunky stepped flaking; notching; crushing; and bifacial flaking. Concave and straight edges, and 'noses' and notches were also identified; as were the presence of edge angles (particularly, steep and fine). A range of activities are suggested by this range of assemblage characteristics (e.g. butchering, wood working, knapping and so on). It would appear that the site's lithic assemblage represents that of a generalised site and not a specialised site. This is discussed further below.

Cores

Of 87 cores retrieved, 37 were in the analysed squares (Table 6.14). These represent only 0.8% of the assemblage. Most of the cores (69%) are quartz. All other raw material found at the site are represented by this artefact type, albeit in small numbers: Silicified tuff[20] and chert (8: 9.3%), FGS (6: 7%), quartzite (4: 4.6%), FGB (5: 5.7%), volcanic (2: 2.2%) and silcrete and other (1: 1.1% each).

As would be expected (given the dominance of quartz) most (53%) of the cores are bipolar. However, not all the quartz cores were made with this technique. Other raw materials were used with this technique, including one silicified tuff core. Multiplatformed cores were next most common (35%) while single platformed cores were less common (13%).

Table 6.14: Yengo 1. Cores.

Square	Bipolar	M - platform	S - platform	Total cores	%f
1R	1	2		3	3.4
1A	6	1	1	8	9.3
1B-4	7	8	4	19	21.8
2A	6	5		11	12.7
2B	6	1	1	8	9.3
3S-6	9	7	2	18	20.1
3N	5	4	2	11	12.7
5A	6	2	1	9	10.3
TOTAL	46	30	11	87	99.6

Backed Artefacts

Fifty-eight backed artefacts were counted in the assemblage, representing 36% of the modified assemblage. Asymmetrical (Bondi) points were in the majority (91%); the remainder were symmetrical (geometric) backed artefacts.

A large proportion (21: 36%) of the backed artefacts was broken, and many (19: 33%) had evidence of chord usewear. Four had both evidence of use and breakage. One of the backed artefacts with R/U on the chord also had macroscopic evidence of residual hafting material. Utilised and/or broken backed artefacts occurred in all squares with backed artefacts. Broken and used backed artefacts appear to be concentrated in the central area of the site, and most were found in Squares 2A and 3N (Table 6.15).

Different causes are likely for these different states of modification to this implement type; i.e. breakage often happens during manufacture and/or use, depending on the nature of the break (Baker 1992); and usewear on the chord indicates that the blade has been used, i.e. for cutting, prior to discard (McDonald *et al.* 1994; Robertson 2005). Broken and/or used points may also have been replaced from a haft in a composite weapon, e.g. as barbs in a spear (McDonald *et al.* 2007). Differential distributions for these breakage characteristics may indicate either a manufacturing or a gearing up area.

Table 6.15: Yengo 1. Distribution of retouched and broken backed blades across the site.

	1B	1R	2A	2B	3S	3N	4A	5A	6	Total
R/U on chord			4		2	5	2	2		15
broken	1		2	3	5	1	2	1	2	17
broken and R/U			2	1				1		4
Tot. broken R/U	1	-	8	4	7	6	4	4	2	36
Total BB	1	0	11	7	13	8	6	7	5	58
% BB Broken /R/U	100	-	73	57	54	75	67	54	40	69

[22]This material was known as indurated mudstone at the time of this analysis; lithological analysis has demonstrated that it is indeed tuff (Kamminga 1997).

In square 3N, all but one (83%) of the broken backed blades has R/U on the chord, while the remaining piece is the medial-distal piece of the backed artefact. This assemblage may indicate gearing up activity. In square 3S, however, the majority of the backed artefacts are fragments with little evidence for use. This area of the site may represent primary manufacture.

Ground Fragments

Some (17%) of the modified material in the assemblage provides evidence that ground edged implements, were used at the site. This supplements the considerable evidence provided by the 55 grinding grooves on the sloping back shelf of the shelter and the eight stencilled axes on the ceiling. These fragments with grinding evidence indicate the re-forming or breaking up of ground edged implements. Almost all of these were made on fine grained basic (FGB) material; with one exception was a coarser grained basic (in spit 6/8). All grinding is located on the dorsal surface of the flakes. None of these artefacts has subsequent retouch or usewear. Several have evidence of hafting material adhering to the old surface. One complete and very battered axe was located at the front of the site in the undergrowth (see section 3 above and Figure 6.14).

Other Implement Types

The other formal implement types identified was the scraper. A total of eight of these were located, one of which (from 3S/2) has evidence of residual hafting and is very worn down on its distal end. Three of these tools are thumbnail scrapers (all quartz; one from spit 4A/6; two from spit 6/8).

Hafting Residue

Six artefacts in the assemblage have macroscopic evidence of residue suggestive of gum or resin hafting. This residue was identified on a range of artefact types: the surface ground edged axe, two backed artefacts, one scraper and two amorphous retouched artefacts. This material has not been analysed in detail but is certainly worthy of further investigation. Residue which may be blood was also identified on three artefacts. Macroscopic plant residue was identified on one fragment of ground edged material.

The Analysis Squares

The two analysis Squares (1B-4 and 3S-6) contained 1,451 and 3,383 artefacts respectively (Table 6.16). Of these 547 and 1,017 were >1cm (Table 6.17).

Assemblage Characteristics

As with the total assemblage, the majority of the material in these analysed squares consists of unmodified debitage. Most of the 4,834 artefacts are unmodified debitage and most are <1cm in size.

Raw Material

The dominant raw material in each sample was quartz (56% and 55%) followed by silicified tuff (17% and 15%) and quartzite (9% and 11%). The other raw materials classified amongst the assemblage - silcrete, fine grained basic (FGB), fine grained siliceous (FGS), volcanic and 'other' made up the remainder of the assemblage (18% and 19% in the two squares). The proportions of raw materials in the two squares are relatively equal and they remain consistent regardless of size (Table 6.18 ; Figure 6.31 and Figure 6.32).

Size

The majority (67.5%) of the non-modified artefactual material found at the site is <1cm long. The next largest category (1-3cm) contains the next most artefacts (29.3%), and very few artefacts were found that were more than 3cm long (see Table 6.19).

Table 6.16: Yengo 1. Artefact Totals per Spit both Analysis Squares.

Spit	1B	%	4A	%	3S	%	6	%
1	-	0	13	1.3	166	7.2	55	5.0
2	1	0.2	35	3.4	188	8.2	57	5.2
3	6	1.4	23	2.2	833	36.4	113	10.3
4	5	1.2	74	7.2	836	36.5	341	31.2
5	20	4.7	202	19.6	212	9.3	348	31.8
6	20	4.7	209	20.3	49	2.1	87	8.0
7	44	10.4	194	18.9	5	0.2	70	6.4
8	52	12.2	89	8.7	0	0	17	1.6
9	126	29.6	56	5.4			6	0.5
10	46	10.8	22	2.1	-	-	-	-
11	9	2.1	34	3.3				
12	27	6.4	18	1.8				
13	7	1.6	16	1.6				
14	6	1.4	0	0				
15	15	3.5	4	0.4				
16	13	3.1	9	0.9				
17	10	2.4	15	1.5				
18	16	3.8	9	0.9				
19	-	-	1	0.1				
20	-	-	5	0.5				
	423	99.9	1028	100.1	2289	100.0	1094	100.0

Table 6.17: Artefact Totals per Spit; Squares 1B, 4A, 3S, 6. Artefacts >1cm.

Spit	1B	%	4A	%	3S	%	6	%
1	-	-	8	2.4	67	9.7	27	8.2
2	1	0.5	8	2.4	40	5.8	16	4.9
3	6	2.9	12	3.6	210	30.5	29	8.8
4	3	1.4	35	10.4	299	43.4	99	30.2
5	15	7.1	65	19.3	63	9.1	112	34.1
6	14	6.7	74	22.0	10	1.4	26	7.9
7	24	11.4	50	14.8	0	0	17	5.2
8	30	14.3	23	6.9	0	0	0	0
9	42	20.0	12	3.6	-	-	-	-
10	22	10.5	10	3.0	-	-	-	-
11	5	2.4	9	2.7	-	-	-	-
12	9	4.2	12	3.6	-	-	-	-
13	5	2.4	5	1.5	-	-	-	-
14	6	2.9	0	0	-	-	-	-
15	2	0.9	4	1.2	-	-	-	-
16	11	5.2	4	1.2	-	-	-	-
17	5	2.4	5	1.5	-	-	-	-
18	10	4.8	0	0	-	-	-	-
19	-	-	0	0	-	-	-	-
20	-	-	1	0.3	-	-	-	-
	210	100	337	100.4	689	99.9	328	99.9

Table 6.18: Yengo 1. Raw Material Percentages in Squares 1B-4 and 3S-6.

SQ	Qz	%	Silc	%	FGS	%	ST	%	FGB	%	Q'zte	%	Volc	%	Other	%
1B	269	63.6	19	4.5	13	3.1	60	14.2	11	2.6	35	8.3	6	1.4	10	2.4
4A	543	52.8	64	6.2	43	4.2	185	18.0	53	5.2	99	9.6	27	2.6	14	1.4
3S	1242	54.3	69	3.0	200	8.7	338	14.8	149	6.5	279	12.2	11	0.5	1	0.0
6	623	56.9	50	4.6	60	5.5	157	14.4	71	6.5	92	8.4	38	3.5	3	0.2
Tot	2677	55.4	202	4.2	316	6.5	740	15.3	284	5.9	505	10.4	82	1.7	28	0.6

Qz = Quartz   Silc.= Silcrete   FGS = Fine Grained Siliceous   ST = Silicified tuff
Q'zt = Quartzite   Volc = Volcanic   FGB=Fine Grained Basic

Table 6.19: Yengo 1. Size ranges, non-modified artefacts, Squares 1B-4 and 3S-6.

Raw material	<1cm	1-3cm	3-5cm	>5cm	Cores	BP	R/U	Total
Quartz	1,922	710	11	0	25	3	6	2,677
Silcrete	109	83	2	0	0	5	3	202
S. Tuff	534	180	6	0	1	7	12	740
FGB	138	125	6	0	2	0	13	284
Quartzite	315	164	6	1	2	8	9	505
FGS	195	107	6	0	5	2	2	317
Volcanic	43	36	1	0	1	0	1	82
Other	9	13	4	0	1	0	0	27
Debitage Total	3,265	1,418	42	1	37	25	46	4,834
%f	67.5	29.3	0.9	0.0	0.7	0.5	1.0	100.0
Total incl r/u	3,274	1,493	64	3				4,834
%f	67.7	30.9	1.3	0.1				100.0

Very different size ranges are found amongst the cores and modified material (cf. Table 6.19 and Table 6.20). Most modified artefacts and cores (69%) are in the 1-3cm category and the next most common size range is 3-5cm (20.4%). Only nine of these artefacts (8%) are in the <1cm size range and two artefacts are larger than 5cm. Three of the modified artefacts in the <1cm range are distal tip fragments of backed artefacts. The largest artefact at the site is a large quartzite pebble chopper (9.5cm maximum dimension) found in spit 1B4/11.

The fact that such a large proportion of the assemblage is <1cm indicates that knapping was common at the site. The shattering properties of quartz and predominance of the bipolar knapping techniques are also contributing factors. Much of the increase in artefact deposition rates over time is likely to be due to the higher proportions of bipolar quartz in the upper levels (Hiscock 1986). This possibility, and distance to raw material resources (another potential cause for decreased artefact size) is investigated further below.

Table 6.20: Size ranges: modified artefact and core assemblages; Squares 1B-4 and 3S-6.

	<1cm	1-3cm	3-5cm	>5cm	Total
Cores					
Bipolar		14	2		16
Multi-Platform		7	7	1	15
Single Platform		3	3		6
Backed artefacts	4	19	2		25
R/U Flakes		16	2		18
R/U FP	4	10	2	1	17
Ground fragments	1	6	4		11
Total	9	75	22	2	108
%f	8.3	69.4	20.4	1.9	100.0

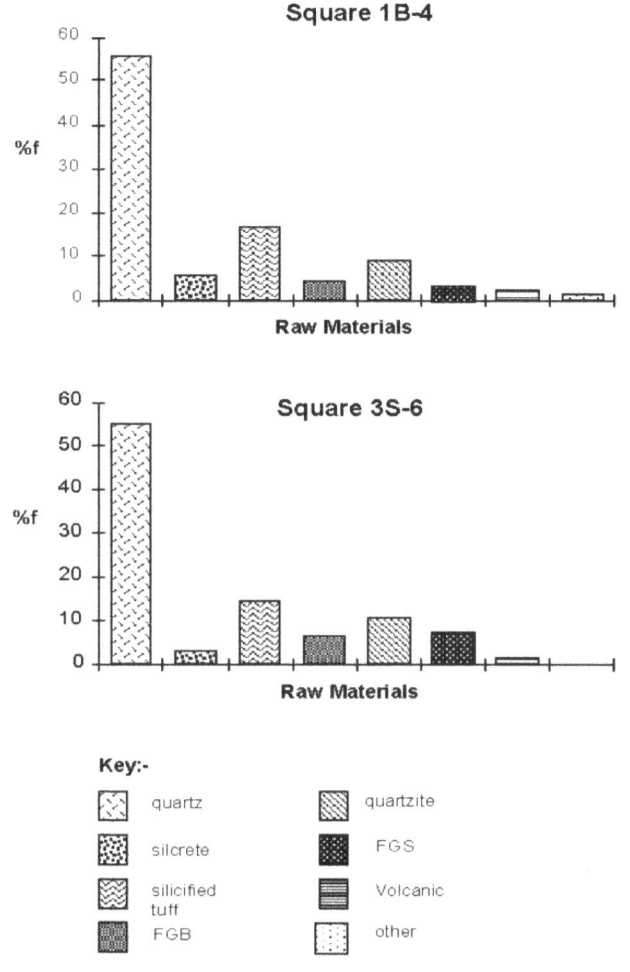

Figure 6.31: Yengo 1. Proportions of raw materials in the two analysis squares. All artefacts.

Artefact Types

The two analysis squares contain 25 backed artefacts, 46 artefacts with retouch/usewear (including ground fragments) and 35 cores. These artefacts indicate different raw material preference. Some of these differences were marked in contrast to the general assemblage raw material proportions (see above).

The raw materials used for cores show the greatest similarity to the general raw material proportions: quartz dominates (cf. Figure 6.31 and Figure 6.33). The dominance of quartz in the assemblage generally, and in the cores, reflects the dominance of bipolar knapping during the later occupation of the site. In contrast, no silcrete cores were found in the assemblage and the proportions of the less common materials do not match those demonstrated by the general assemblage. This indicates that the import and knapping of some materials (such as silcrete) at the site must have occurred only rarely. Given the high numbers of silcrete backed artefacts in the assemblage, this finding is significant.

Most (43%) of the cores in the two analysis squares are bipolar, followed by multiplatformed (40.5%) and single platformed cores (16%). Half of the single platformed cores are blade cores. Some of the multiplatformed cores were identified as being multi directional - and are possibly the type identified by Baker (1992) in the Hunter River central lowlands as representing an alternating knapping technique - one of a number of strategies used for blade production (and see also Hiscock 1986, 1993).

Backed artefacts reveal a significantly different and more restricted range of raw material preference than any other artefact class (cf. Figure 6.31 and Figure 6.33). Quartzite, silicified tuff and silcrete predominate, followed by quartz and then FGS. Volcanic materials, FGB and 'other' raw materials have not been used for this artefact type.

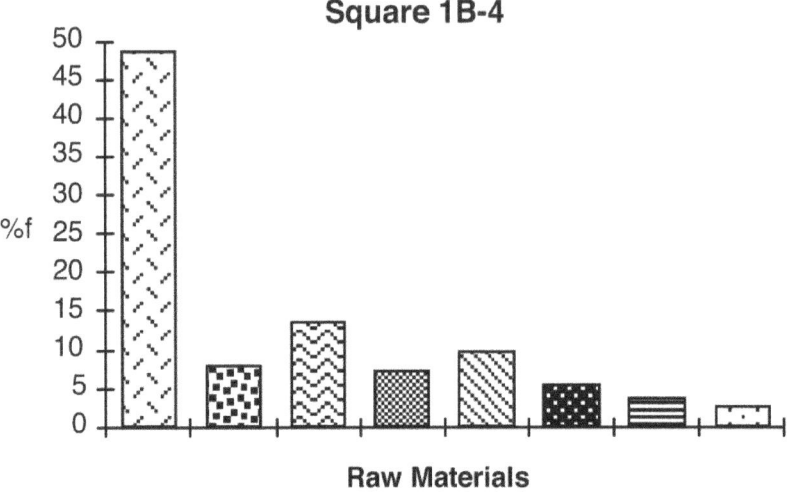

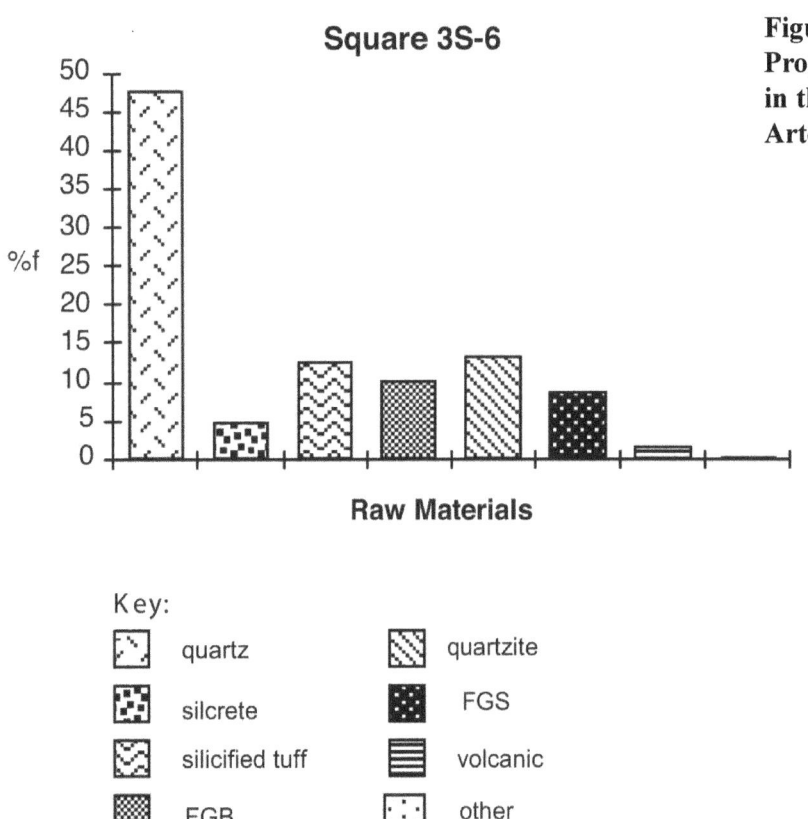

Figure 6.32: Yengo 1. Proportions of raw materials in the two analysis squares. Artefacts >1cm.

Three only of the backed blades were symmetrical, the remainder were asymmetrical. No eloueras were present. Two of the silicified tuff backed artefacts were very long (3S/4: 3.5cm x 1.0cm x 0.4cm; 6/8: 4.7cm x 0.7cm x 0.6cm: a similar size to those found in a silicified tuff knapping floor in the Rouse Hill area: JMcD CHM 2005a), most the whole backed pieces fall within the 1-3cm size range. The one whole backed artefact smaller than this was a triangular symmetrical piece (measuring 0.9cm x 0.8cm x 0.3cm) from 3S/5. This too was made of silicified tuff.

Artefacts with R/U also show a significant difference in raw material usage (cf. Figure 6.31 and Figure 6.33). FGB predominates (thanks to the 10 fragments with ground facets), followed closely by silicified tuff and quartzite. Quartz, silcrete, FGS and volcanic materials follow. 'Other' raw materials have not been utilised. These findings suggest several things. That:

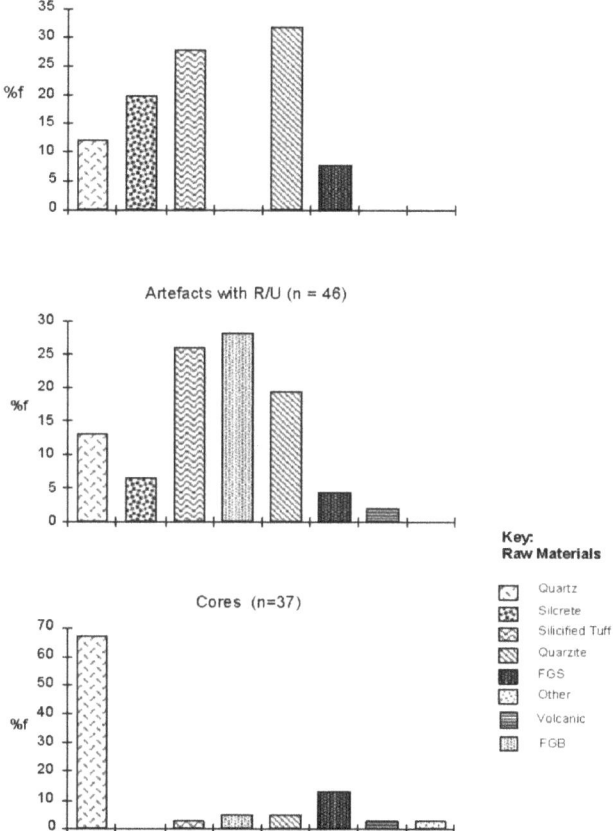

Figure 6.33: Yengo 1. Raw material proportions for backed blades, artefacts with R/U and cores.

- different raw materials were used for different types of tools;

- while quartz dominated the lithic assemblage, only a relatively small proportion of artefacts made of this material were used for subsequent tasks (at least on site). This suggests that the general assemblage proportions then do not necessarily reflect the intent of lithic manufacture and use at the site, but perhaps the flaking characteristics of certain raw materials (Hiscock 1986). Another possible interpretation is that local raw materials were used for certain, non-specific tasks while exotic raw materials were used for selected tasks (as represented by formalised tool types);

- some backed blades (particularly the silcrete and quartzite ones) were not made on the site, but were brought there, perhaps for replacement in a 'gearing up' situation, or perhaps as trade items;

- if cores of silcrete were brought to the site for knapping of that material, these were removed subsequently, perhaps for use elsewhere; and,

- silicified tuff was a preferred raw material for generalised tools: more than 25% of the artefacts with r/u were made of silicified tuff while this raw material only represents 15% of the assemblage >1cm.

Vertical Distribution of Material

The discussion of vertical patterning at the site is restricted to the data from the two paired analysis squares. The artefactual material at the site was found to be unevenly distributed throughout the vertical sequence.

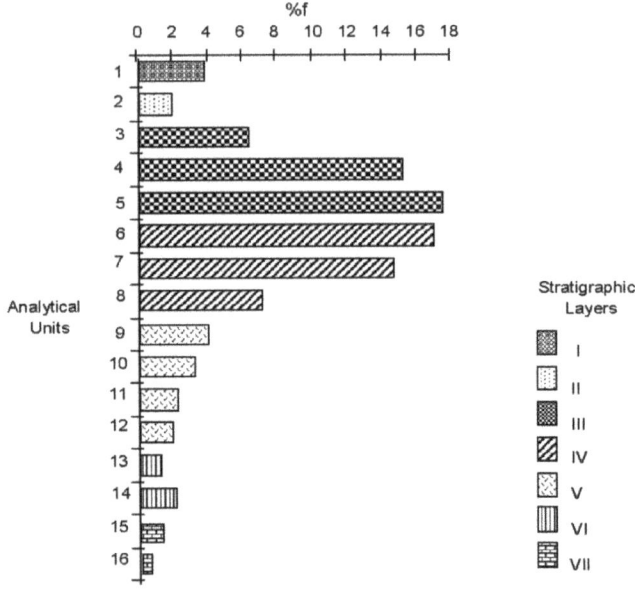

Figure 6.34: Yengo 1 Square 1B-4. Artefact Distribution (%f) by analytical unit.

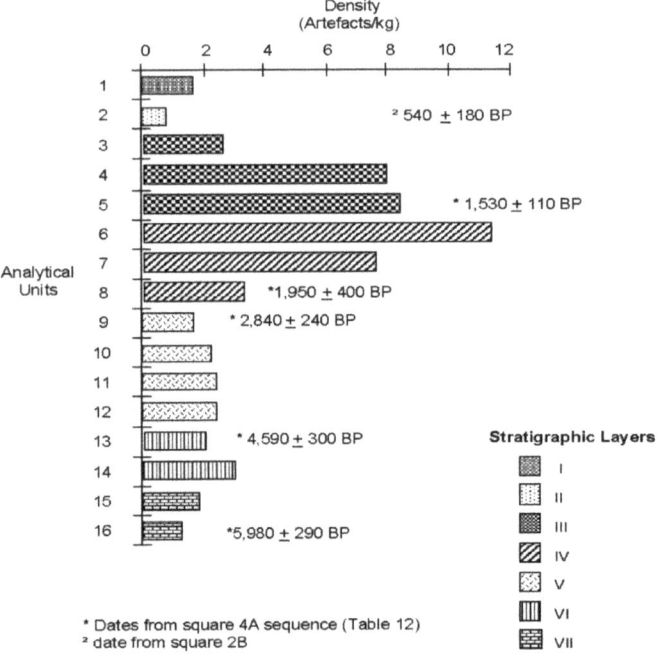

Figure 6.35: Yengo 1. Square 1B/4. Artefact density (per kg) through time.

Artefact Density and Percentage frequency

Units 4-7 (square 1B-4) and units 3 and 4 (square 3S-6) contain both the largest number and greatest density of artefactual material. Layers III and IV represent the most intensive occupation period in square 1B-4 (Figure 6.34 and Figure 6.35). In the shallower square, Layer III alone contains the bulk of the artefactual material. Layer IV is very shallow and only partially present of square 3S/6. Over 92% of the artefact assemblage occurs in the top four Layers.

The distribution plots for artefact density and percentage frequency suggest that Layers III and IV may be one cultural period: the increase and decrease throughout the units demonstrate a classic battleship curve.

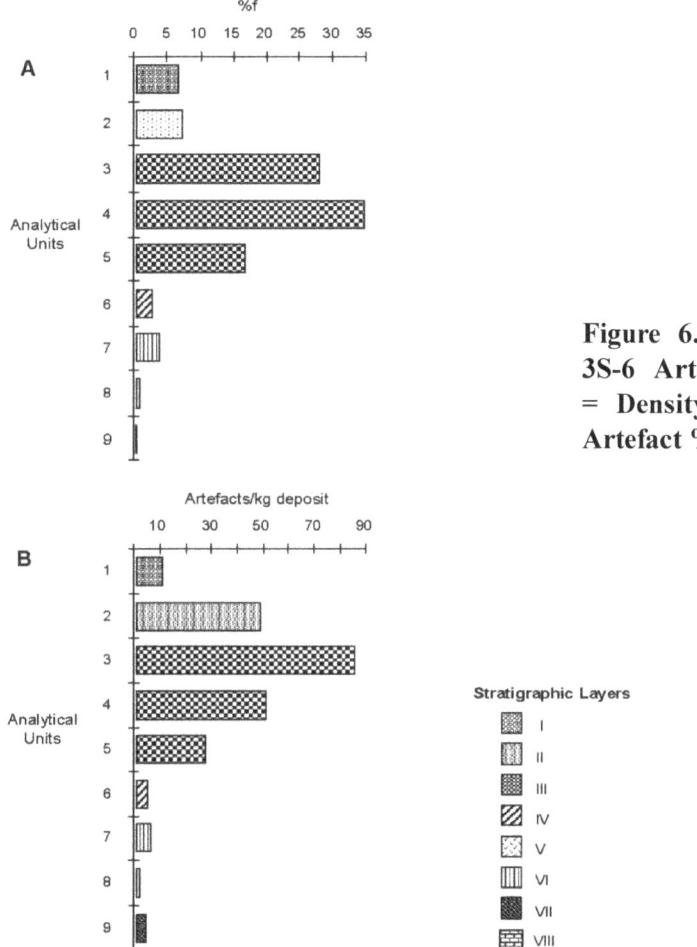

Figure 6.36: Yengo 1. Square 3S-6 Artefact Distribution. A = Density (artefacts/kg); B = Artefact %f.

Artefact deposition rates

An age-depth curve was calculated (Figure 6.37), although there were some concerns about the utility of this due the complexity of the deposit, the sloping bedrock of the shelter and variable deposition factors (cultural, natural). It cannot be assumed that the rate of deposition throughout the sequence or across the site was constant, since the area being deposited over laterally increased with time. As most of the dates were submitted from square 4A, the analytical units from this square were used to make this calculation. The reliable upper date from square 2B was used to provide perspective on this more recent deposit. Artefact totals are from the combined 1B/4A analysis square (Table 6.21).

The artefact accumulation rate demonstrated (Figure 6.38) is similar to that achieved using density and percentage frequency measures (cf. Figure 6.34 and Figure 6.35). The most intensive periods of artefact accumulation occurs in Layers III and IV. Rates are extremely low in the upper and lower layers. Artefact accumulation rates suggest a hiatus in site usage between the earliest and the main occupation. It is tempting to accept the age-depth interpretation, as it relates to this hiatus, given the similarity in the patterning between the age/depth results and the density measures throughout the sequence. Given concerns about the validity of this method at this particular site, this potential hiatus is explored further using other components of the assemblage. The dates for the sequence support a hiatus between the early and middle Bondaian phases of occupation.

Table 6.21: Yengo 1. Age-depth calculations. Artefact totals and bone weights per 100 years (refer Figure 6.37, Table 6.9).

Analytical Unit	No. of Artefacts	No. of Years	Artefacts/100 yrs	Bone (g)/100 yrs
1	55	300	18	10.8
2	28	150	19	56.1
3	93	200	47	59.35
4	221	220	100	20.8
5	254	190	134	6.3
6	246	230	107	1
7	213	210	101	
8	105	250	42	
9	58	2,100	3	
10	47	1,150	4	
11	33	210	16	
12	29	210	14	
13	19	200	10	0.83
14	31	350	9	
16	10	400	2.5	

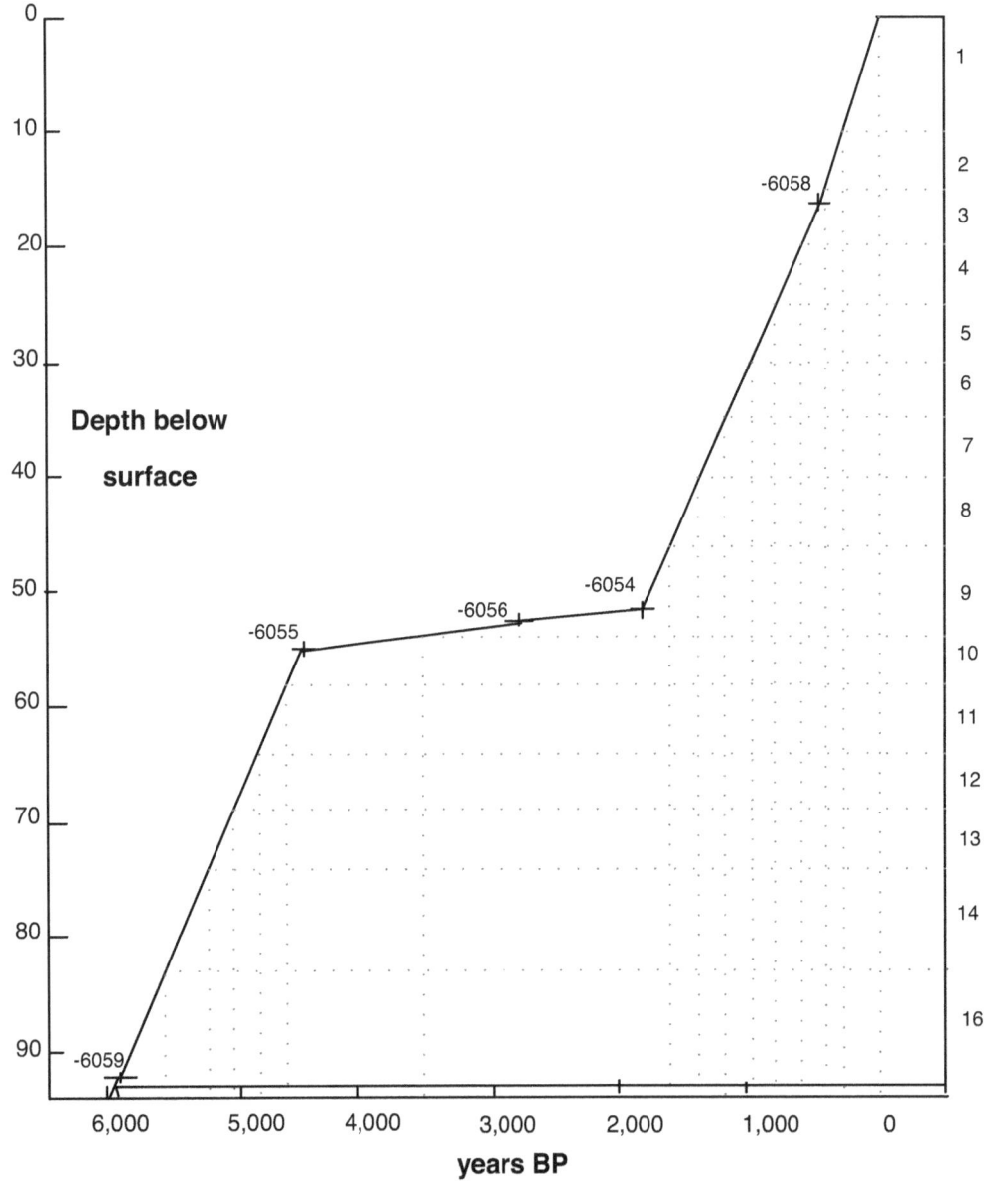

Figure 6.37: Yengo 1. Age depth curve for square 4A. Upper date from square 2B. Analytical units shown on vertical axis.

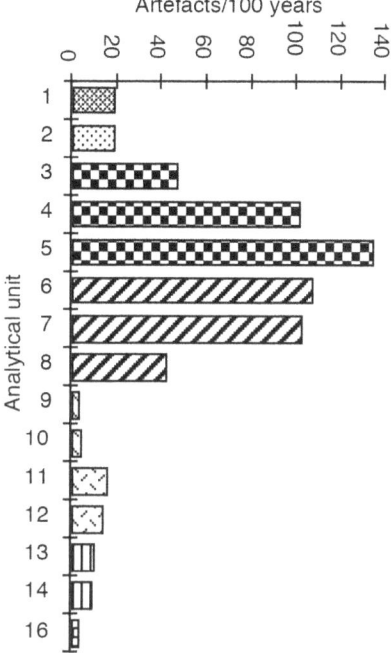

Figure 6.38: Yengo 1. Artefacts accumulation per 100 years: square 1B-4.

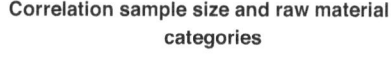

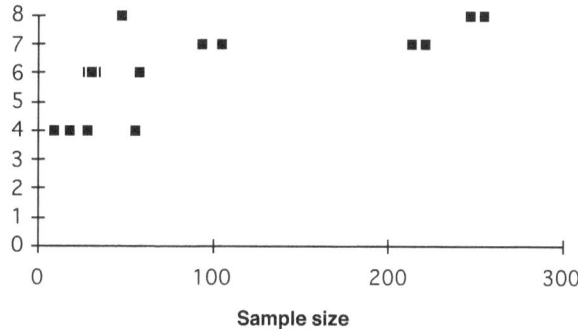

a. Square 1B-4.

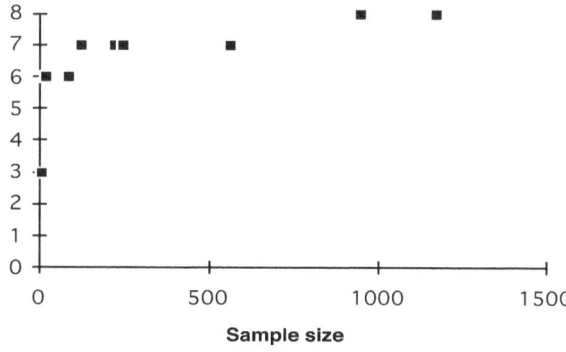

b. Square 3S-6

Figure 6.39: Correlation between sample size and raw materials per analytical unit. All artefacts.

Raw Materials

There is a trend over time in preferences for particular raw materials. In order to test whether varying sample sizes were responsible for patterning, a simple correlation matrix for raw material and sample size was produced (Figure 6.39). This demonstrates no positive correlation between sample size and raw material variety per analytical unit. Thus the general trends are interpreted as having validity, i.e. they are not induced by variable sample sizes (James 1993).

In square 1B-4 the diachronic trend is characterised by an initial focus on silicified tuff, quartzite and 'other' raw materials, followed by a marked increase in the use of fine-grained siliceous materials (Figure 6.40 and Figure 6.41). In the upper layers (analytical units 1-9; stratigraphic layers I-IV) there is a proliferation in the range of raw materials being used, with quartz predominant. In square 3S-6 a similar, if compacted, version of this same pattern is seen. FGS is the dominant earlier material, replaced by quartz in the middle and uppermost spits.

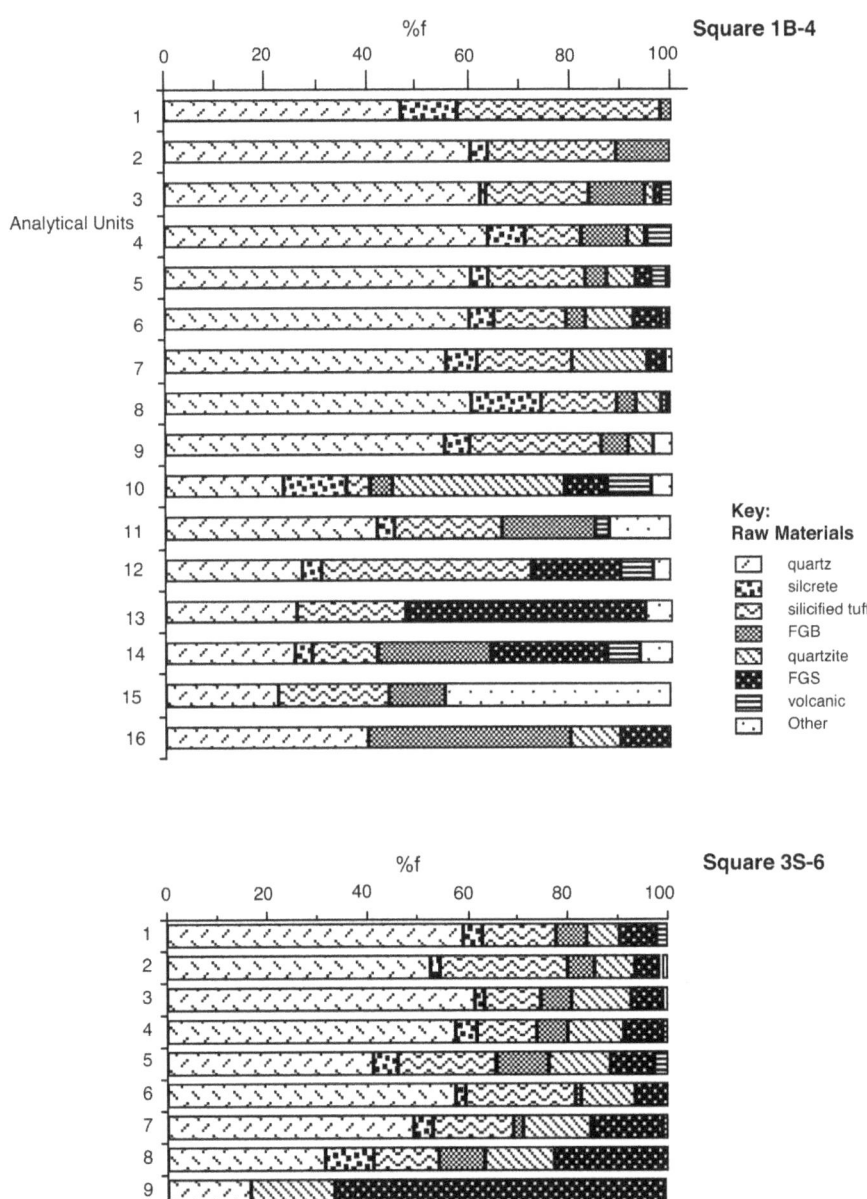

Figure 6.40: Yengo 1. Vertical trends in raw material proportions. Both analysis squares. All artefacts.

Changing raw material preferences over time (Figure 6.42) confirms that in the upper four Layers, quartz predominates while in Layers V-VII the stone artefact assemblage is primarily non-quartz.

Size

The possibility of change over time in the general size of the assemblage was investigated. The proportion of debitage >1cm for the four main raw materials (quartz: 28%; silicified tuff: 28%; quartzite: 38%; and FGS: 38.5%) was analysed.

No clear pattern of decreasing size with time was demonstrated. This was unexpected given the trend demonstrated in other sites of similar age [e.g. Upside-Down Man, Cherrybrook (CB1) - McDonald 1985b], the generally recognised trend from Capertian to Bondaian assemblages, and the increased proportion of bipolar quartz in the middle and upper units. Small sample sizes in many of the lower units could be implicated in this lack of patterning although the simple correlation test (Figure 6.39) tends to discount this. The fluctuating pattern in the upper units in particular cannot be thus explained (i.e. given the consistently large samples in these units).

The sizes of cores and artefacts with R/U were thus analysed to further explore assemblage size characteristic through time (Table 6.22, Table 6.23; Figure 6.44). All size categories for these artefact classes were included in this analysis.

The sample sizes in most layers (except III) are extremely small. Sampling is thus likely to affect this analysis. While tools <1cm only occur in the top layers, and the only tool >5cm occurs in the lower layers, there is no substantive trend of decreasing size with time. Similarly while the largest cores are at the bottom of the sequence, proportionally, cores in the 1-3cm range predominate throughout.

Tools and cores are predominantly and consistently in the 1-3cm size range throughout the Yengo 1 sequence.

Table 6.22: Yengo 1. Size ranges of cores in the analysis squares (per stratigraphic layer).

Layer	<1cm	1-3cm	3-5cm	>5cm
I	-	1	-	-
II	-	1	5	-
III	-	18	7	1
IV	-	2	1	-
V	-	2	1	-
VI	-	-	2	1

Artefact types

The combined analysis squares had 73 implements or artefacts with R/U. This sample represents more than 44% of the modified assemblage from the site.

Table 6.23: Yengo 1. Size ranges of artefacts with R/U in each stratigraphic layer. Includes ground fragments and backed blades.

Layer	<1cm	1-3cm	3-5cm	>5cm
I	1	2	1	-
II	-	5	1	-
III	5	34	5	-
IV	-	9	3	-
V	-	-	1	1
VI	-	2	-	-
VII	-	1	-	-

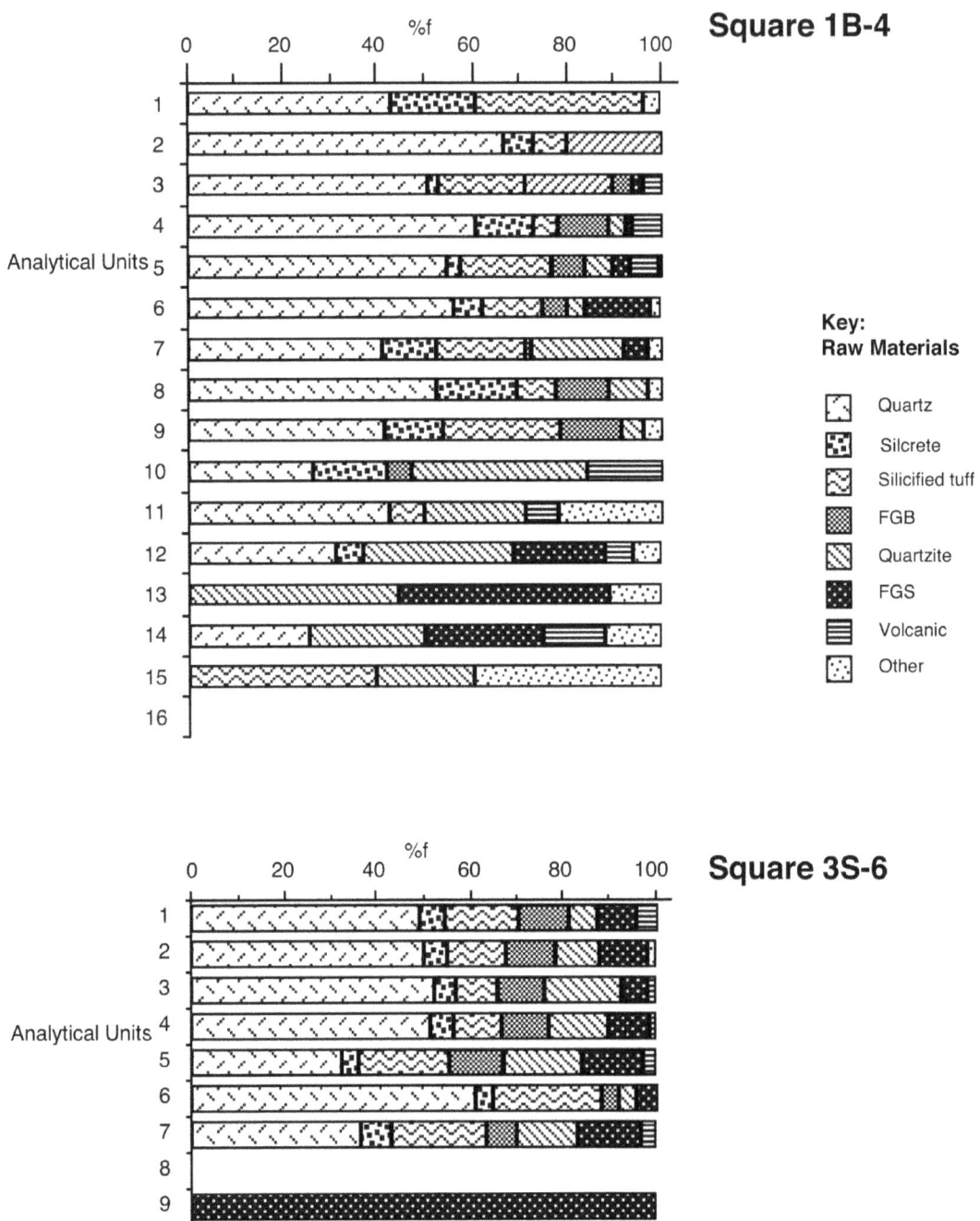

Figure 6.41: Vertical distribution of raw materials. Artefacts >1cm.

Amorphously retouched flakes and flaked pieces

Most of the artefacts with retouch/usewear are located in Layers III (square 3S-6) and/or IV (Square 1b-4: Figure 42). This distribution matches that of overall artefact density (cf. Figure 6.34). Tools are found in varying frequencies in all Layers.

Changing raw material preferences for implements were analysed, removing artefacts with ground facets (but including otherwise modified FGB tools). This indicated that during the earliest phases at the site (specifically Layers VI and VII), silicified tuff only was used for tools (Figure 6.46). This changed in Layer V to exclusive use of quartzite, while in the most recent layers there was a proliferation in the range of raw materials used (Figure 6.45). The absence of R/U material above Unit 5 in square 1B/4 suggests that during the most recent phase of occupation stone tool use may have occurred in a more restricted area and that the central area may have been the focus for stone tool production (i.e. in square 3S-6). The spatial distribution of flaking debitage indicates a similar spatial distribution to artefacts with R/U (cf. Figure 6.53 and Figure 6.54).

A more spatially restricted focus for stone tool use and manufacture generally could be explained by the decreasing head room in the northern area of the site with the build up of the more recent deposits. Currently, the central area of the site is more amenable to general living than further north, where it is necessary to bend over to move around. Head room, however, would not have been restricted to persons sitting, which presumably includes people knapping.

Backed Artefacts

Backed artefacts were found throughout the central part of the sequence in Layers V - III (Figure 6.47). This distribution is consistent with the age determinations received for the site (Table 6.11). Layer V is Early Bondaian while Layer III is (a late) Middle Bondaian (Attenbrow 1987). The absence of backed implements and dates in the two top units are consistent with this being a Late Bondaian assemblage (cf. Hiscock 1986).

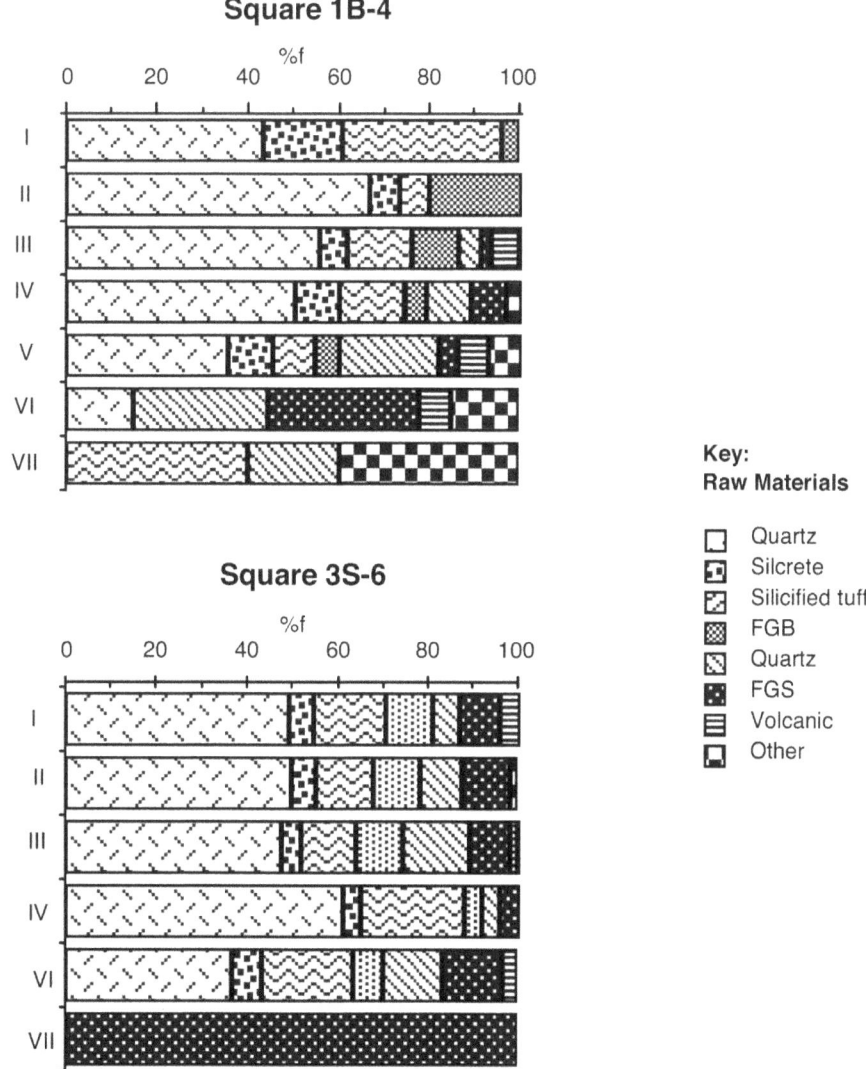

Figure 6.42: Yengo 1. Vertical distribution of raw material per stratigraphic layer. Both squares. Artefacts >1cm.

There are indications of changing preferences in raw material use with this artefact type over time (Figure 6.47), although the small sample sizes make definite statements impossible. Silcrete and indurated mudstone, on the other hand, are used for backed blade production throughout the sequence. Quartzite appears later in the sequence (in Layer III). Quartz and FGS appear only in the middle of the sequence for this artefact type, during the period of most intensive artefact deposition and predominance of quartz usage.

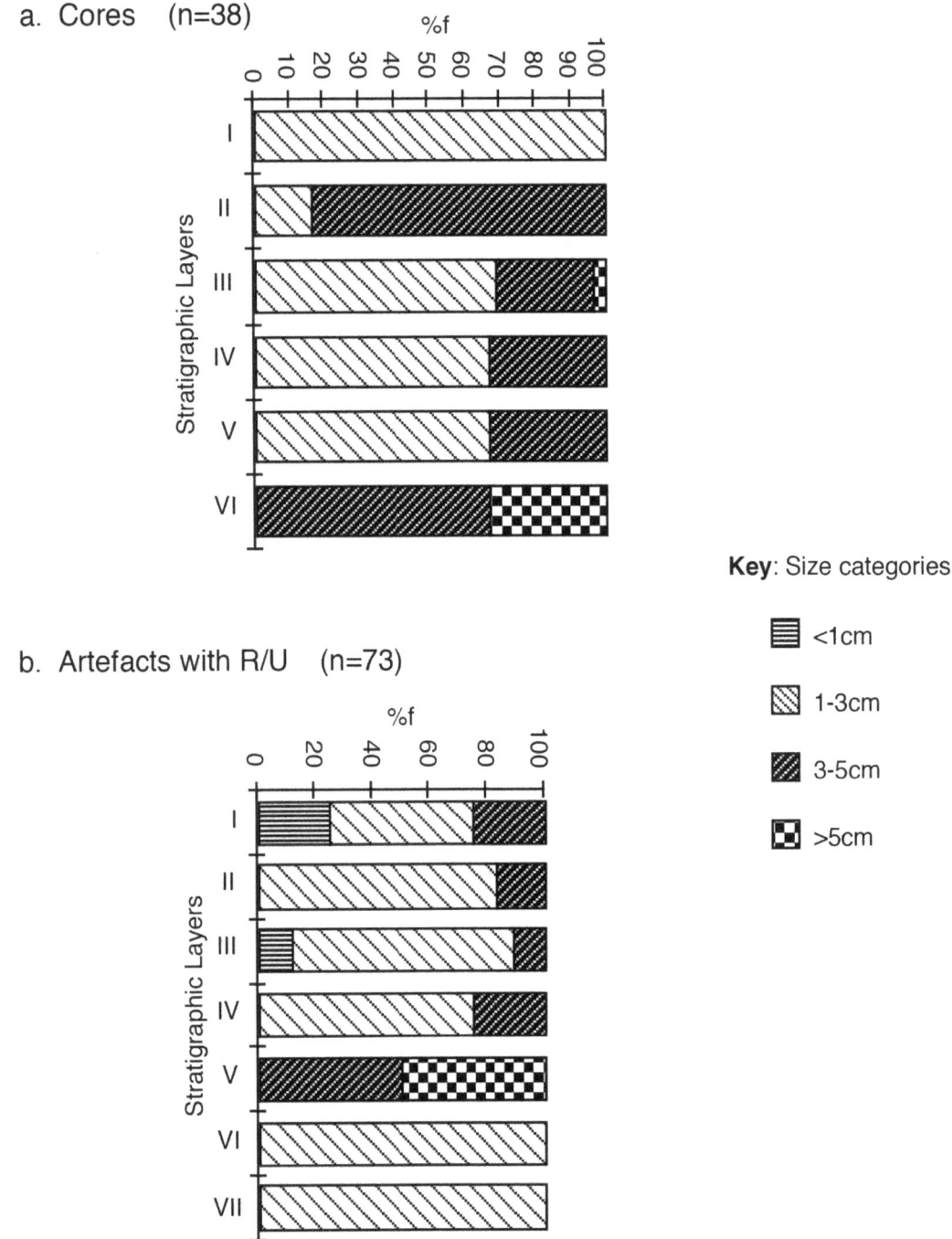

Figure 6.43: Yengo 1. Proportions of size categories over time per stratigraphic layer. Cores and artefacts with R/U.

Ground Fragments

Ground fragments are found only in the top three Layers (Figure 6.47). The use of ground edged implements at the site is dated to within the last 1,500 years. These data support Attenbrow's (1987) suggestion that there is an increase in the deposition of ground material in the late Bondaian (as represented here by Layers I and II).

Cores

Cores are found throughout the sequence, but most occur in the upper four Layers. This patterning generally matches the assemblage distribution, although there are proportionally more cores in Layers V and VI than would have been expected on the basis of overall assemblage proportions (Figure 6.49; cf. Figure 6.34). This could suggest an earlier knapping focus in the more northerly part of the site, prior to diminishing headroom. The presence of cores throughout the sequence demonstrates that artefact knapping took place at this site throughout the period of its occupation.

An analysis of core type was undertaken to investigate possible changes in reduction strategies with time (Table 6.24). Three types of cores were distinguished between in this analysis: Bipolar, Multi-platformed and Single platformed. This analysis demonstrated a clear trend from an early use of multi-platformed cores to a later focus on bipolar cores (Figure 6.48). The single platformed cores were occasionally found to be blade cores. Approximately half the multi-platformed cores were multi directional ones, a type identified by Baker (1992) as one of a series of microblade reduction strategies in the Hunter central lowlands. These core types accord well with other evidence for technological trends. There are no bipolar cores in the earliest levels (i.e. pre Layer V). The very small samples sizes in most units/layers are acknowledged.

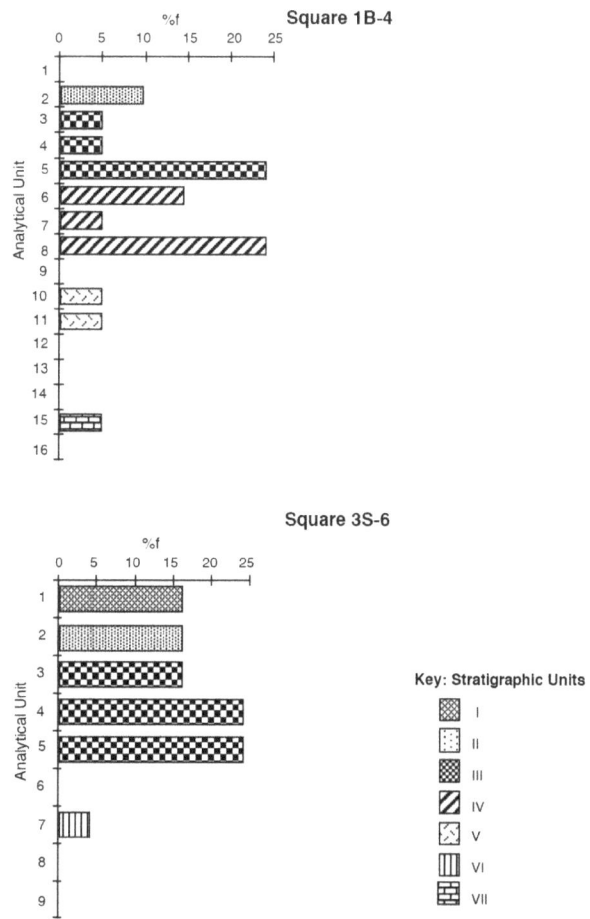

Figure 6.44: Yengo 1. Vertical distribution of artefacts with R/U.

Table 6.24: Yengo 1. Core platform characteristics per stratigraphic layer.

Layer	Bipolar	%f	Multi	%f	Single	%f	Total
I	2	100					2
II	1	100					1
III	13	50	10	38.5	3	11.5	26
IV	1	25	3	75			4
V			2	66.7	1	33.1	3
VI			1	100			1
VII							0

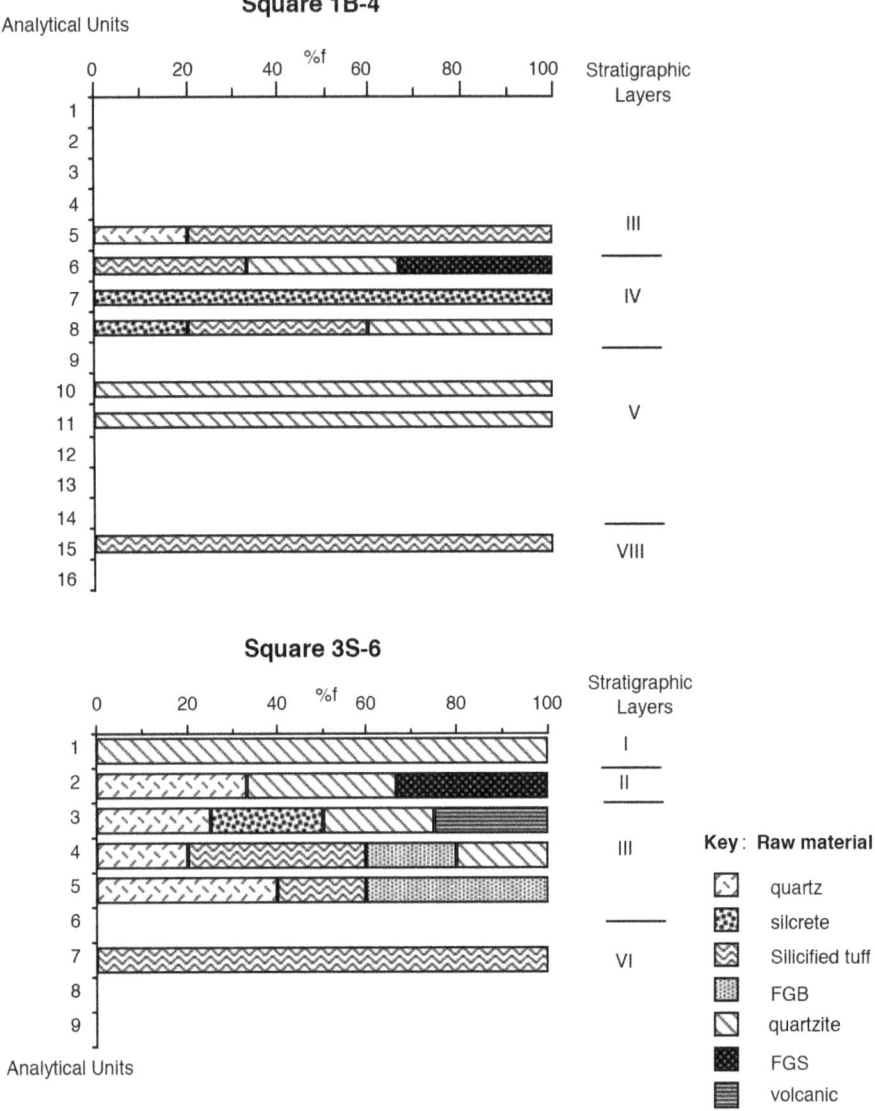

Figure 6.45: Yengo 1. Changing raw material preferences over time. Artefacts with retouch/ usewear. Both squares. Ground fragments excluded.

Summary

Changes over time in artefact densities, artefact accumulation rates, the types of artefacts and raw materials used indicate that the site was occupied over several different culture periods. The artefacts and the dates received are consistent with the site's early sporadic use during the (late) pre Bondaian period with an increased site usage through the Early Bondaian. The most intensive period of site usage was during the middle Bondaian, but occupation continued into the late Bondaian. There is no evidence of contact occupation either in the deposit or in the art assemblage.

This site usage pattern accords well with the catchment patterns identified in Upper Mangrove Creek (Attenbrow 2004). It is not matched, however, by any other individual site in the local area (Attenbrow 2004, Hiscock 1986, MacIntosh 1965, Moore 1970). The significance of this, in terms of the mosaic of site and/or localised patterns of usage and general site function will be discussed further.

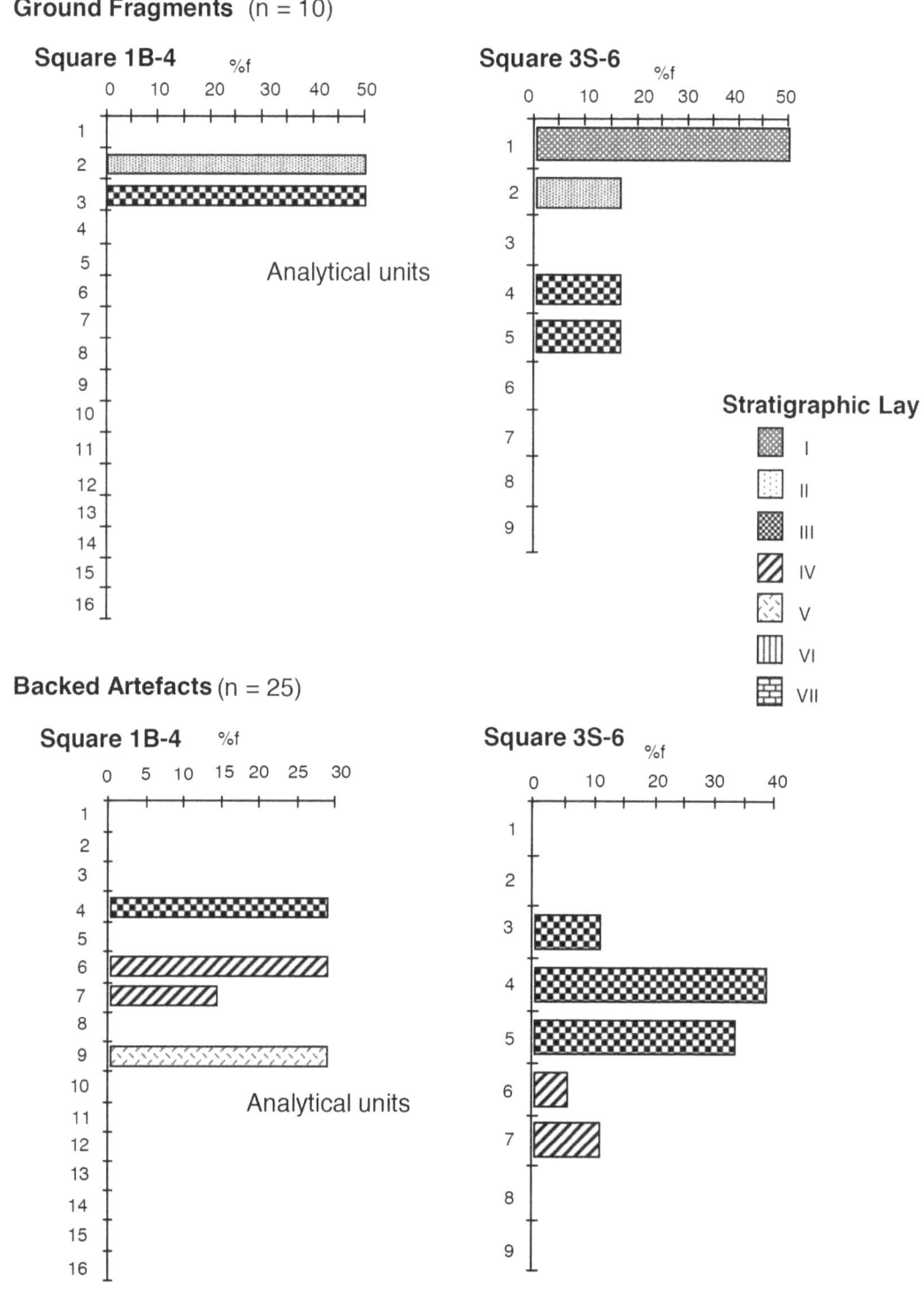

Figure 6.46: Yengo 1. Vertical distribution of backed blades and ground fragments in both analysis squares. Stratigraphic layers indicated.

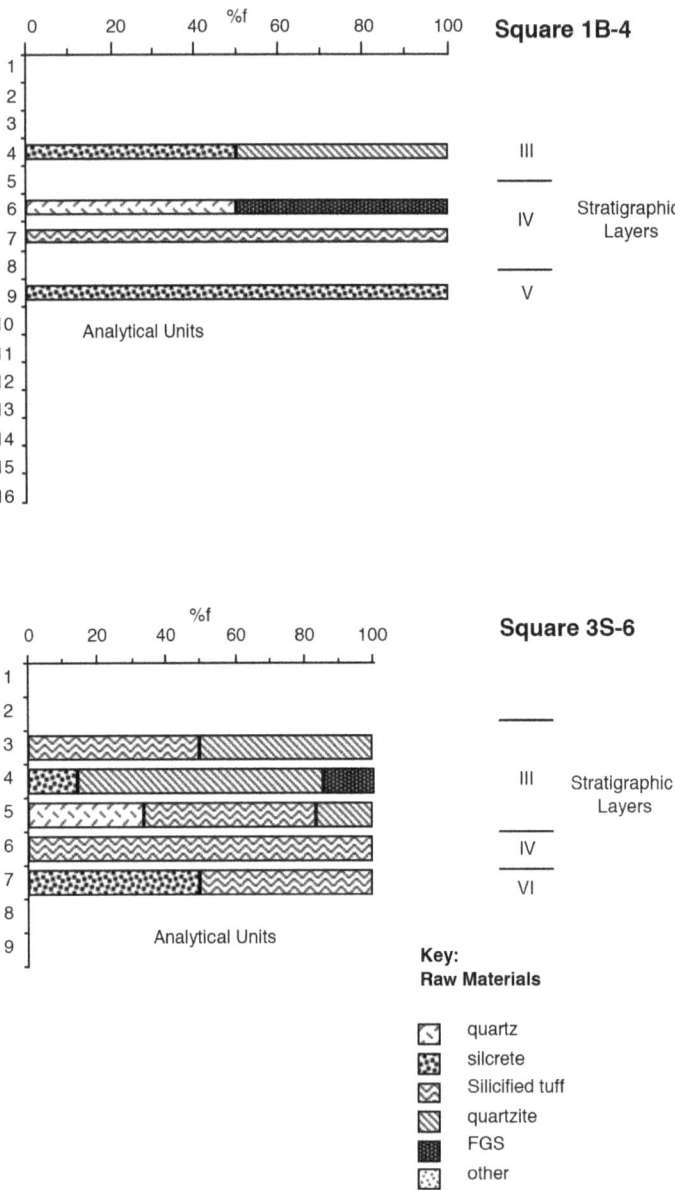

Figure 6.47: Yengo 1. Raw Materials use in backed artefacts throughout the vertical sequence.

Faunal Remains

Culturally deposited faunal remains were found only in the upper three stratigraphic layers (Figure 6.50). Very small amounts of bone were found in the lower levels, but without exception these were found in burrow affected deposit. The vertical distribution of the faunal remains is not matched by the lithic material (cf. Figure 6.34, Figure 6.35). The faunal material peaks slightly later in the sequence, at a time when there is a slight decline in artefact densities. Bone accumulation rates (grams per 100 years; cf. Table 6.21, Figure 6.38) demonstrate this patterning clearly. Preservation conditions (i.e. pH, moisture content) throughout the levels in question are equal. The faunal assemblage was analysed by Dominic Steele (1994).

The bones displayed relatively good preservation characteristics: although it was uniformly and highly fragmented. Less than 1% of the bone fragments by number were identifiable to species level (Table 6.25).

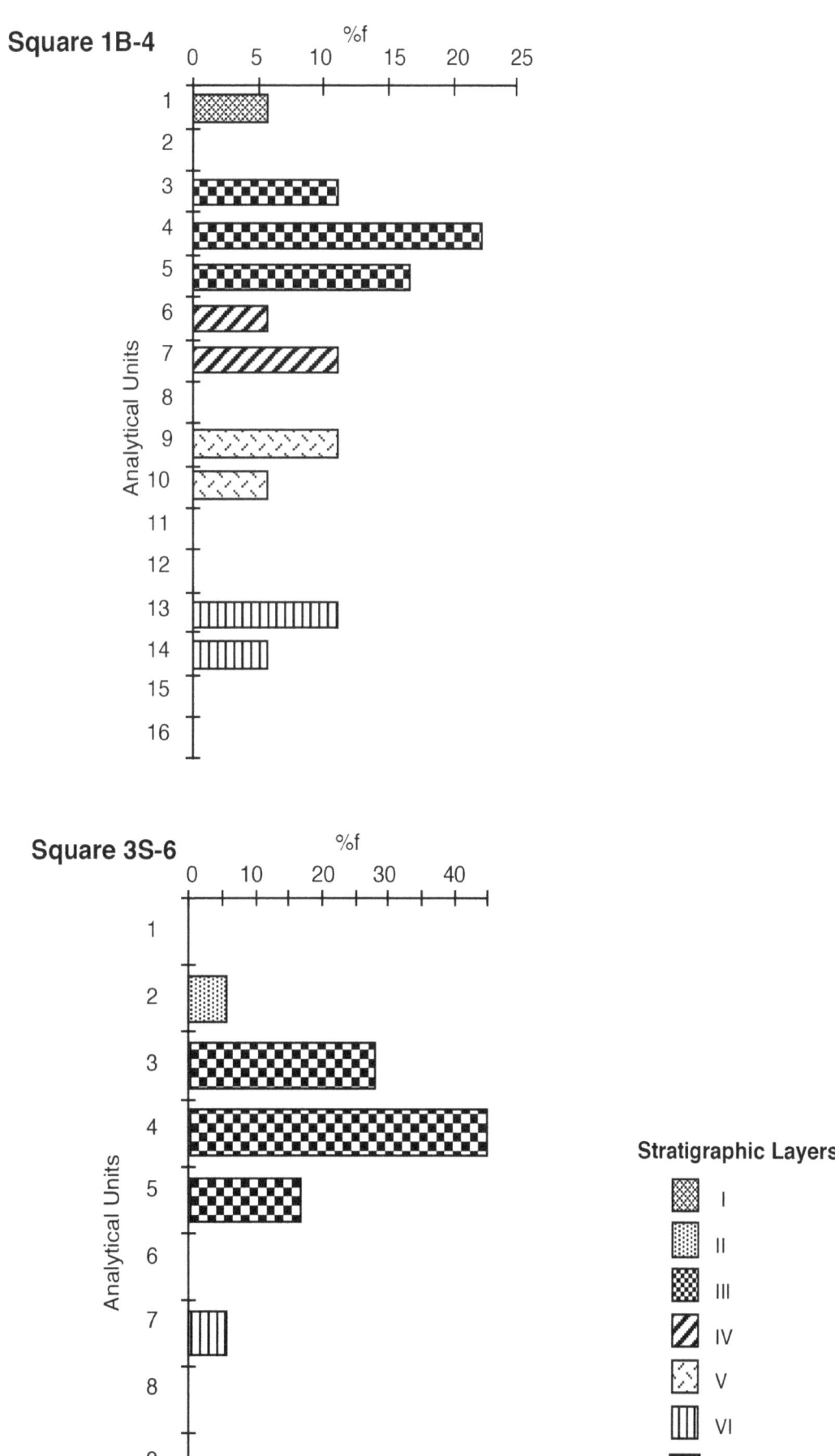

Figure 6.48: Yengo 1. Vertical distribution of cores per analytical unit. Both analysis squares.

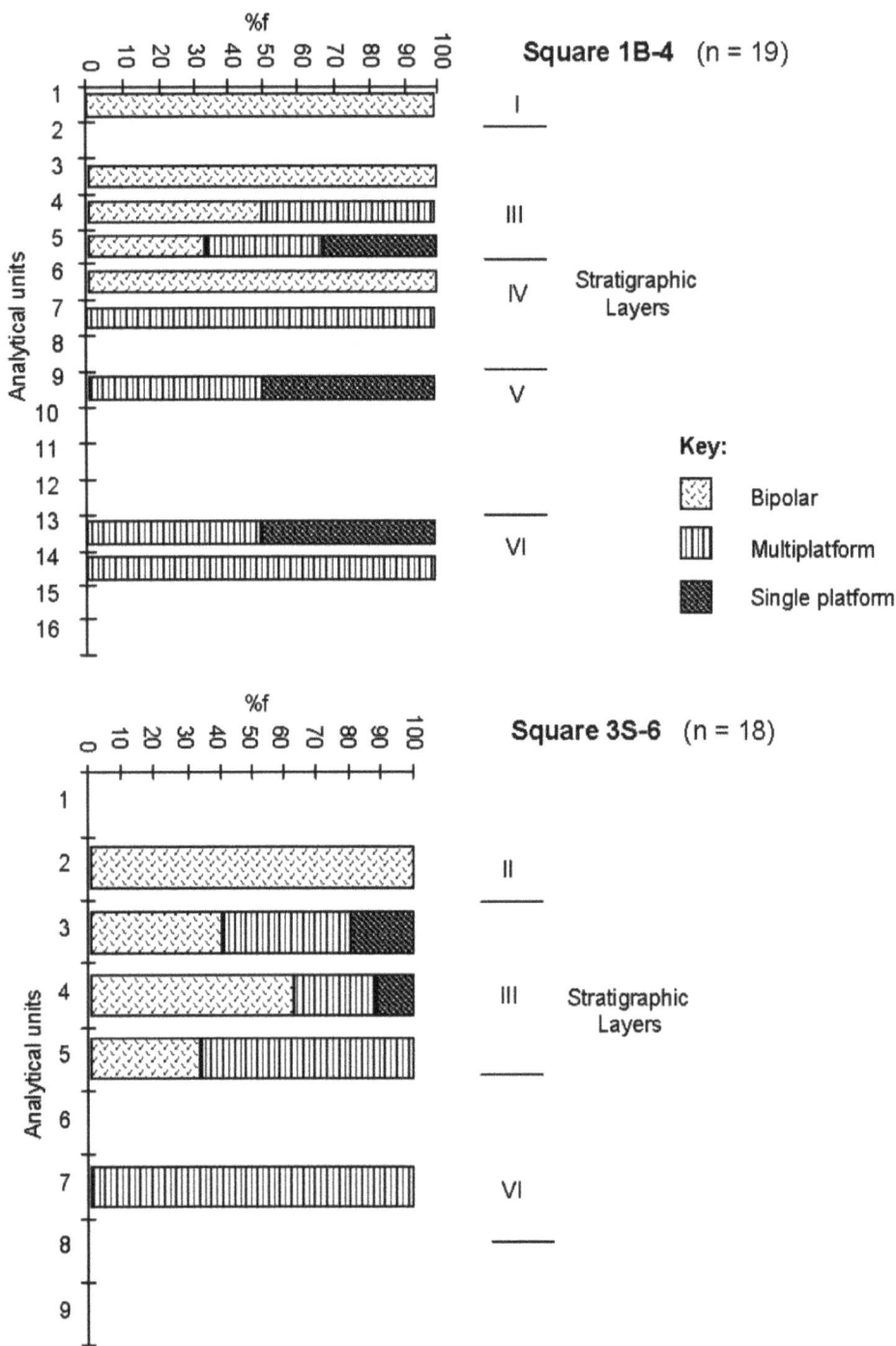

Figure 6.49: Yengo 1. Changing platform characteristics of cores through time. Both analysis squares.

This pattern, similar to that found at Gatton by Morwood (1986), indicates that declining lithic artefact deposition doesn't necessarily indicate decline in site usage so much as a change in site usage (see below). Steele (1994) demonstrated a similar pattern in overall bone distribution with the bone occurring only in the five uppermost analytical units.

The identified component of the Yengo 1 collection was 1,051 bone fragments. This represents c.15.7% of the total assemblage. Bones which are likely to derive from animals of comparable size to wallabies and pademelons dominate the sample (63.2%). Elements of large animals (i.e. mostly kangaroo) represent 16.2% of the collection. Bones of small mammals represent 20.6% of the collection.

Exploring temporal change in the faunal composition at Yengo 1 is difficult, largely because a high degree of fragmentation has resulted in a small proportion of identifiable material. Despite the limited depth of deposit from which faunal remains derive, the relatively limited proportion of the assemblage for which identification was possible, and the rapid decline in sample size with depth, changes can be seen in taxonomic composition through time (Steele 1994).

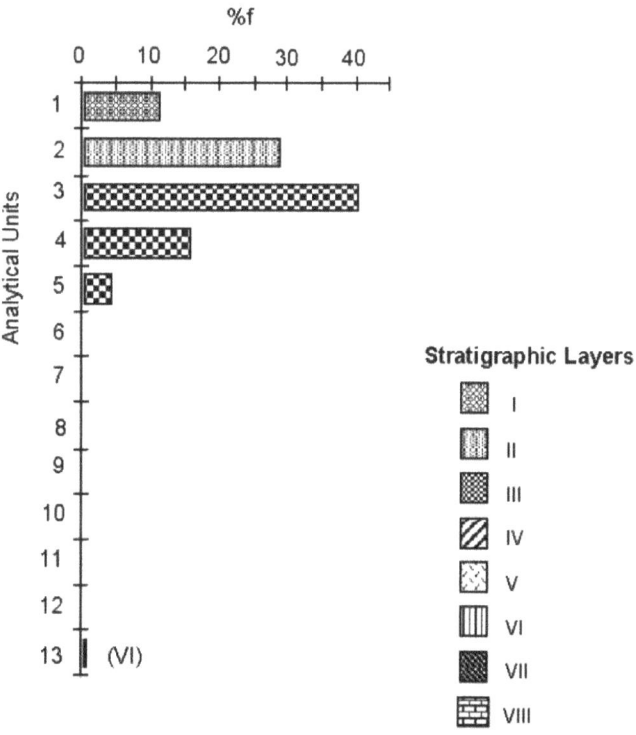

Figure 6.50: Yengo 1. Vertical distribution of bone. Based on bone weights %f. Squares 3S, 3N, 4A and 6 (data from Steele 1994: Table A1.4).

Table 6.25: Taxonomic composition of the Yengo1 faunal assemblage.

Taxa	Weight (grams)	%f	No.	%f
Mammals	287.5	55.5	1,127	16.8
Reptiles	4.98	0.9	100	1.5
Birds	0.52	<0.1	4	<0.1
Fish	0.02	<0.1	2	<0.1
Unidentified	233.9	44.3	5,466	81.6

Stratigraphic Layers I and II in Squares 3N, 3S, 4A and 6 produced 237 bone fragments for which various levels of taxonomic assignment have been made. This represents almost 41% of the identified sample (Table 6.26). Around a third of these elements derive from small animals. Varieties identified include possum, potoroo and bettong, bandicoot, glider, wallaby, dingo and grey kangaroo. Small numbers of bird and fish bone also derive from these levels. A little over 57% of the bones derive from medium sized wallabies and other unidentified similarly sized mammals. The remaining bones (8.9%) from these layers reflect large animals such as the eastern grey kangaroo. Reptile bones in this layer included two types of lizard and three snake varieties.

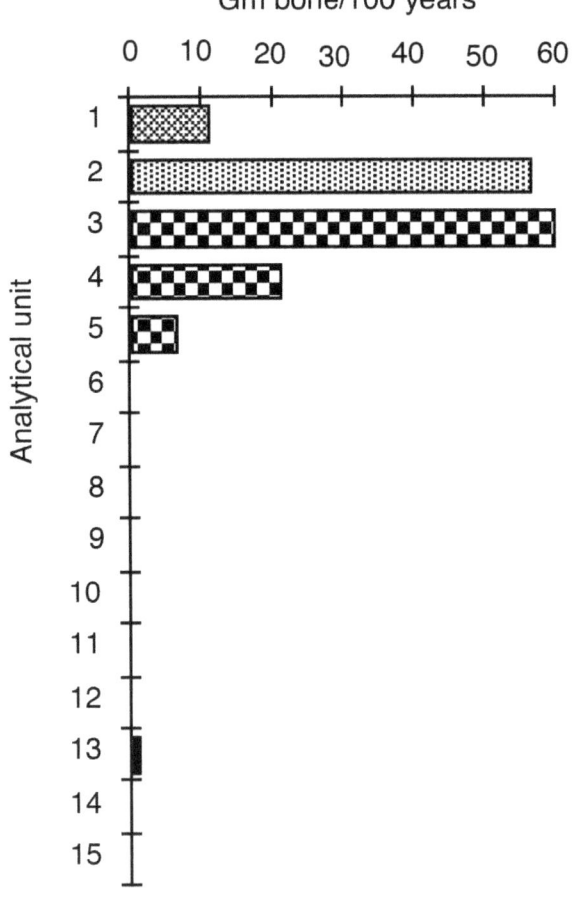

Figure 6.51: Yengo 1. Bone accumulation rates. Based on bone weights %f. Squares 3S, 3N, 4A and 6 (data from Steele 1994: Table A1.4; see Table A1.22).

Table 6.26: Breakdown of animal varieties identified at Yengo 1. Categories of Small, Medium and Large by NISP.

SMALL ANIMALS	
Eastern bettong	2
Rufous bettong	1
Long-nosed potoroo	3
Brush-tail possum	7
Long-nosed bandicoot	1
Short-nosed bandicoot	1
Lesser gliding possum	1
Southern Bush rat	1
Native Bush rat	2
Unidentified Mammals	198
Total	**217**
MEDIUM ANIMALS	
Swamp Wallaby	6
Red-necked wallaby	1
Parma wallaby	1
Brush-tailed rock wallaby	1
Dingo	1
Unidentified small macropod	20
Unidentified medium macropod	39
Unidentified mammal	595
Total	**664**
LARGE ANIMALS	
Eastern grey kangaroo	4
Unidentified large macropod	6
Unidentified mammal	160
Total	**170**

A different taxonomic composition is suggested by the faunal remains in stratigraphic layer III. This sample comprises 385 bone fragments. Of these, only 18.4% derive from small animals (a single bandicoot element and a number of unidentified mammals). A range of animals represented in the later deposits (i.e. possum, potoroo and bettong) are entirely absent from these earlier deposits. Also poorly represented in this mid-phase are reptile bones.

The earlier deposits are dominated by the presence of bones of medium-sized animals (swamp and rock wallaby, and a number of other unidentified macropods). This component of the assemblage comprises almost 73% of the identified bones. The proportion of large mammal and macropod bones for this period of occupation is equivalent to that for the later phase. The faunal sample relating to the shelter's earliest period of usage from these squares is too small for comparison with the subsequent site occupancy.

There is a clear contrast in the composition of the assemblage from the upper and lower deposits. The bones of medium to large macropods and unidentified mammals are a proportionally more dominant taxonomic component in the lower and earlier excavation contexts compared with the upper layer.

Pigment

Pigment was found throughout the sequence in seven squares but predominantly in the upper three layers. Red fragment were found in the basal layers (Table 6.27). Most of the pigment and pipeclay recovered consisted of very small fragments or nodules. None of the material recovered showed use striations, and the very small size of all pieces recovered would account for this. Layers II and III contained the majority of the pigment. As well as a greater quantity, the greatest variety of colours is also present at this level.

Table 6.27: Yengo 1. Distribution of pipeclay and pigment across the site and throughout the sequence.

Layer	1B	2A	2B	3S	3N	4A	6
I		1R		1B			
II	1R	2R/2W	1R				
III	1R/1 Y/10			5R/5 W/10	3R/3W	2W	2R/1W
IV						1R	
V						1R	
VI	1R	4R					1R

R=red   W=white Y=yellow   C=cream   B=black

Activity areas: spatial patterning

The squares in the central, driest part of the site contained the highest number of lithic artefacts. Squares 2 (A and B) and 3 (S and N) covered <40% of the area and <28% of the volume excavated, but contained more than 60% of the artefactual material retrieved. Most artefacts (2,329: 21.3%) were found in square 3S: the lowest number (21: 0.2%) were found in square 1R (Table 6.13; Figure 6.52 and Figure 6.53).

As well as the majority of artefact discard occurring in the central area of the site, the more recent periods of deposition were the most prolific in terms of artefact discard. Over 92% of the assemblage is in Layers I-IV. The squares in the central area are predominantly made up of these layers.

Artefact Density

Square 2A had the highest density of all excavated squares at the site, while square 1R had the lowest density (Table 6.13; Figure 6.53 and Figure 6.54). The maximum artefact density recorded was in spit 2A/7 with a projected density of 64,640 artefact/m3 (50.5 artefacts/kg). The central area of the site generally had much higher densities of artefactual material, while the squares closer to the dripline had the lowest densities. This pattern is the converse of that found in many excavated sites - where artefact density is highest in the dripline area (e.g. Mill Creek 11 (Koettig 1985), Cherrybrook (McDonald 1985b), Sandy Hollow, Yango Creek and Macdonald River 1 (Moore 1970, 1981), Ken's Cave, Native Wells 1 and 2 (Morwood 1984), and etc.).

The pattern at Yengo 1 may be due largely to the shelter's sloping bedrock morphology and natural deposition processes: there is considerable influx of (non-cultural) deposit over the dripline area. However, the generally higher frequencies of artefacts and the nature of the material being deposited in the central area of the site also suggests that artefact discard occurred here at a higher rate. The most intensive period of shelter usage, is represented mainly by the deposit which built up progressively at the rear of the shelter.

As over 92% of the artefactual assemblage occurred in the top four Layers, the spatial patterning described here pertains mainly to the more recent phases of occupation at the site.

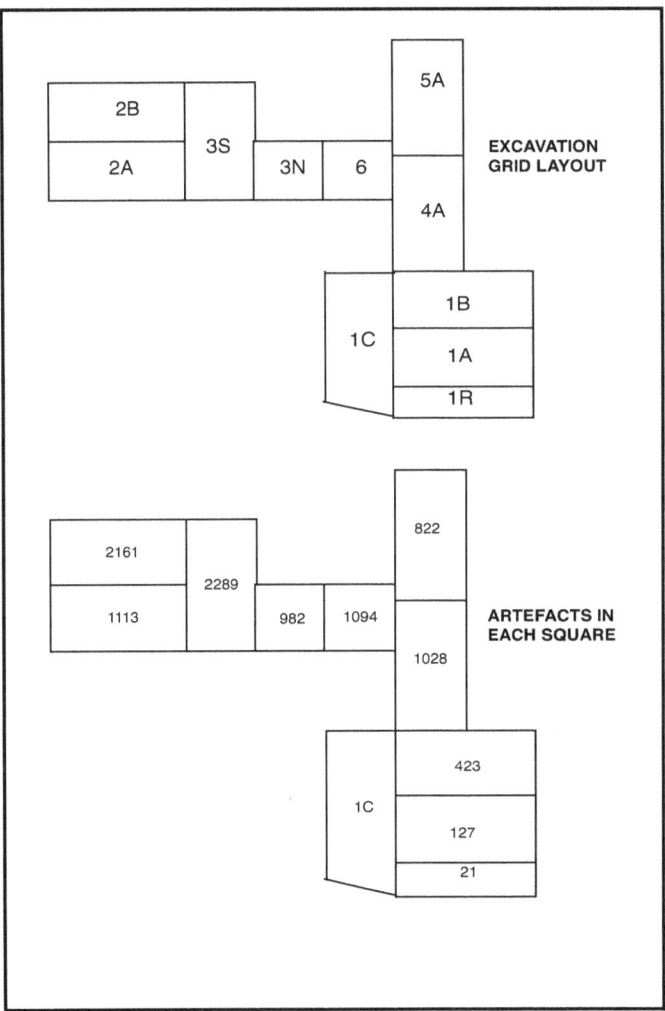

Figure 6.52: Yengo 1. The excavation grid layout and number of artefacts retrieved from each pit.

Implements and other artefact types

The distribution of implements (particularly backed pieces and fragments with evidence of grinding), artefacts with retouch/usewear and micro-debitage all indicate that a variety of knapping, artefact manufacture and/or retooling took place in specific locations across the site (Table 6.28; Figure 6.54 and Figure 6.55). Combined with the economic remains this provides insight into the complex living floors, which were intercepted by these excavations.

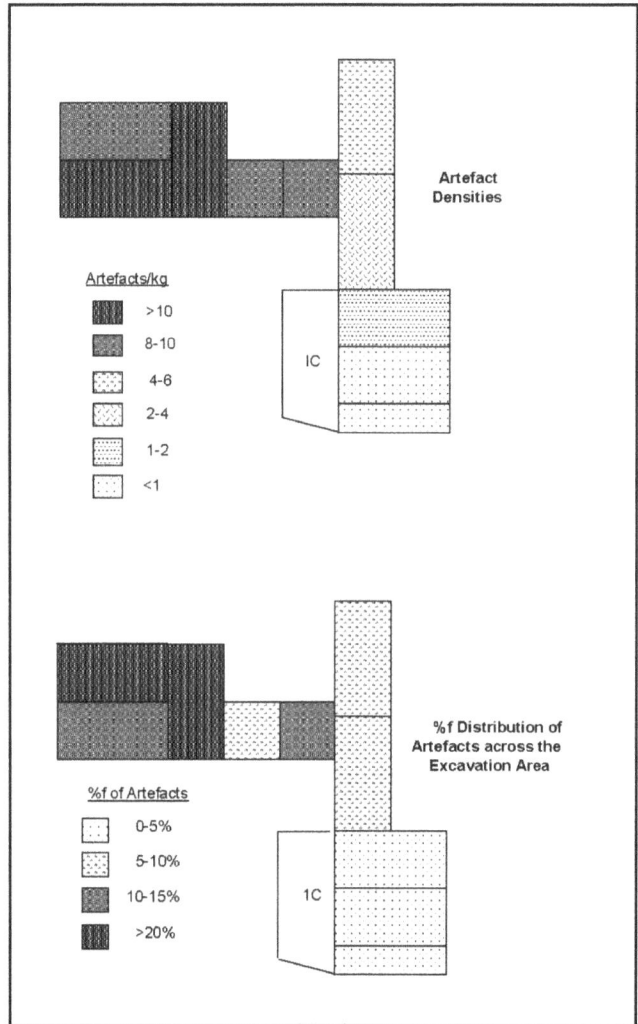

Figure 6.53: Yengo 1. Artefact densities and percentage frequency distributions of lithic material across the site.

The distribution of micro-debitage suggests that knapping occurred in most areas of the site (except perhaps beneath and outside the dripline: Squares 1A, 1B, 1R). The type of debitage suggests that knapping activity was concentrated in the central area of the site (particularly Squares 2 and 3).

The spatial distribution of backed artefacts suggests that the manufacture and/or discard/ retooling of these also took place in the central area of the site, and that there was a secondary such activity area, slightly further north (Squares 4A and 5A).

Fragments of ground artefacts (i.e. flakes or flaked pieces with evidence of previous grinding, from breaking up ground edged implements) were discard occurs across most of the back of the site but concentrated in square 2A. Similarly, the distribution of artefacts with evidence of (mostly) usewear and (some) retouch was again concentrated around square 2: material in 4A again suggests another activity area located slightly further north (perhaps beyond the excavation?).

Other Economic Remains

Shell

Fragments of freshwater mussel shell (*Hyridella* sp.) were recovered from four of the excavated squares (Squares 2A, 3N, 3S and 6). This was found in the top two Layers.

Pigment

This was found throughout the sequence in seven excavated squares. None was found in the squares beneath or outside the dripline and this is likely to be due to preservation factors (particularly increased moisture). Pigment was predominantly found in the uppermost three layers. Most of the pigment consisted of very small fragments or nodules, and none of the recovered materials had striations or facets indicating use. Most of the ochre was dark red, although several orange pieces were also discovered. Pipeclay varied in colour between white and cream. The spatial and vertical distribution of the material was discussed (Table 6.27).

Seeds

Seeds were found through the upper part of the sequence. Most of these were Native cherry (*Exocarpus cipressiformis*), although the seeds of Geebung (*Persoonia spp.* - 1) and Sarsaparilla (*Hardenbergia* - 1) were also identified. Four other types are still unidentified. At the northern end of the shelter there is an extant Native cherry, which may well be the source of this seed type. Most of the seeds were found in the central area of the site (square 2A spits 1, 2, 4 and 8; 2B/3). Several seeds were also found in spit 4A/2.

Faunal material

Six squares (2A, 2B, 3N, 3S, 4A and 6) out of the eleven excavated at Yengo 1 produced approximately 6,698 fragments of bones with a combined weight of 526.9g (Steele 1994). Most of this material was located in the central driest part of the site (Figure 6.56). Square 2A had both the highest density and largest number of fragments.

Despite the fact that only a small proportion of skeletal elements were identifiable to species or family level, a relatively wide variety of mammal species are represented within the identified sample. A variety of macropods are present: bones of wallabies (e.g. swamp, parma, red necked and brush tailed) are dominant. A range of smaller mammals such as dingo, bandicoots, possums, potoroos and bettongs were identified. Assessment of the size, thickness and general morphology of the unidentified mammal bones suggests the majority of fragments derive from animals of wallaby size through to bandicoots and bettongs. Approximately 58% of the fragments are likely to derive from 'medium sized' animals, whilst a little over 15% are of a size suggestive of large macropods.

Reptile bones identified include at least two varieties of snake and three kinds of lizards. Only four bird elements were identified. The collection also included two small fish vertebrae. Considering the highly fragmented state of the Yengo assemblage (including bones of more robust mammals), it is possible that the unidentified bone included highly fragmented bird and reptile bone.

The vast majority of the mammal bones identified consisted of fragmented long bones. The highly fragmented nature of the assemblage is exemplified by the fact that few of these exceed 2-3cm in maximum dimension.

Just over 20% of the Yengo 1 bone fragments (16.5% by weight) had evidence for burning. Charred and calcined bones were found in equal proportions horizontally and with depth for all squares. Most of the bones in the assemblage were a brown colour: likely to be the product of staining from minerals in the deposit.

Steele found no strong trends in the frequency or proportional occurrence of burnt and calcined bone with deposit depth. He argues that the preservation factors in Layers I - IV appear both constant and equivalent. The Yengo 1 assemblage also displays a high and uniform degree of fragmentation. Few bones are complete, and most consist of small incomplete fragments. Steele concluded that (1994: 21):

Table 6.28: Yengo 1. Distribution of implements, R/U and micro-debitage (artefacts <1cm) across the site.

	SQ1B	SQ1R	SQ2A	SQ2B	SQ3S	SQ3N	SQ4A	SQ5A	SQ6	TOTAL
R/U F	2		8	5	9	3	4	1	4	36
R/U FP	4	2	10	11	5	1	6		1	40
BB	1		11	7	13	8	6	7	5	58
Ground	1		6	4	3	2	4	4	3	27
TOT R/U	8	2	35	27	30	14	20	12	13	161
%R/U F	5.6	0	22.2	13.9	25	8.3	11.1	2.8	11.1	(22.4)
R/U FP	10	5	25	27.5	12.5	2.5	15	0	2.5	(24.8)
BB.	1.7	0	18.9	12.1	22.4	13.8	10.4	12.1	8.6	(36.0)
Ground	3.7	0	22.2	14.8	11.1	7.4	14.8	14.8	11.1	(16.8)
% Tot R/U	5.0	1.2	21.7	16.8	18.6	8.7	12.4	7.5	8.1	-
Artefacts<1	213	9	1597	827	1600	701	691	563	766	6967*
Total	423	21	2161	1113	2289	982	1028	822	1094	9933*
%<1cm	50.4	42.9	73.9	74.3	69.9	71.4	67.2	68.5	70.0	69.8
%R/U	1.9	9.5	1.6	2.4	1.3	1.4	1.9	1.5	1.2	1.6

*Totals do not include figures for square 1A - in which no R/U was located.

1. The Yengo 1 assemblage is largely cultural in origin. Scavengers appear to have contributed little to the accumulation, fragmentation and spatial configuration of the assemblage. A very small proportion of the bones may reflect natural deaths (e.g. small reptiles and mammals);

2. The bones appear to have been quickly incorporated into the occupation matrix (a low frequency of bones had evidence for weathering);

3. The bulk of the fragmentation appears to have occurred while the bones were dry and/or old and following their incorporation into the sediments. The uniform and extensive fragmentation is likely to be the result of chemical deterioration accelerated by trampling;

4. The high fragmentation rate has reduced the level and detail to which the assemblage may be taxonomically characterised. Less than 0.1% of the bones were identifiable to species level;

5. Differential preservation is likely to have reduced the quantity of bone originally deposited at the front of the shelter. Cultural practices (e.g. site maintenance), combined with a variety of non-cultural processes, are likely to have removed and/or reduced a proportion of larger bones from the assemblage;

6. The majority of the bones display relatively good preservation characteristics. The decrease in the quantity of bone in the lower stratigraphic layers may be a product of decay, but would also appear to be the result of a low discard frequency of faunal remains during the earlier occupation of the shelter.

Results - Yengo 2

Analysis of this site's material was not attempted until 1992. When it came to undertaking this analysis, it was found that in my four year's absence from the ANU, the material from the Mt Yengo sites had been moved several times (due to ongoing renovations around the Department) and that the artefacts and charcoal from Yengo 2 were missing. It has therefore not been possible to make a detailed study of the artefacts, nor to submit any charcoal from this site for analysis. A subsequent search may find this lost material. In the meantime, the nature of the deposit is described on the basis of field notes.

Stratigraphy

The only stratigraphy observed during excavation of pits 4B and 4C was the change at the base of the deposit from dark grey brown to yellow (i.e. decomposing sandstone). This was only observed in the deeper (and wetter) of the two pits (Figure 6.25).

The field notes indicate that a charcoal rich layer, 'possibly a lens?' was encountered in spit 4C/4. This layer also contained two 'nice cores'. This was immediately above an increase in the quantity of gravel.

Heavy rain during the field season meant that the pits sustained considerable water inundation. This indicated that primary sedimentation here is due to a natural influx of deposit. During excavation, channels were made at surface (Figure 6.15) in an attempt to divert water around the test pits (Figure 6.24). This was only partially successful due to a natural sink effect in the vicinity of the pits. The area identified prior to excavation as being somewhat softer than the remainder of the floor, and which had evidence of rat burrowing, is obviously affected by the periodic inundation.

The excavators noted (with increasing vigour!) that the site was not a good habitation, being very dark and (increasingly) wet: the deposit was described as - being 'dark, rooty, wet and ... altogether quite unpleasant to excavate' (*field notes*). It was so dark in this shelter during most of the day (early morning being the exception) that caving lamps were used to aid excavation. Sieve notes indicate a low density of artefactual material was retrieved. This is an apposite reflection of the poor habitation conditions provided by the site.

Cultural Material

Field notes indicate the presence of charcoal and nature of artefacts found during the dry sieving undertaken on site (Table 6.29). This list is not exhaustive as microdebitage was recovered during wet sieving of small sieve residues back at the Lab. It is also only an indication of charcoal presence as this was not always mentioned. It is, however, a fairly good indication of the distribution (and paucity) of artefactual material larger than 1cm. No faunal remains or seeds were recovered.

Table 6.29: Yengo 2. Location of artefacts and charcoal throughout the two test pits.

	Square 4B		Square 4C	
Spit	Artefacts	Charcoal	Artefacts	Charcoal
1	-	XX		
2	-	XX		
3	-	XX		
4	-	XX	2 cores	XX
5	-	X		X
6	2Q	X	1 blue	X
7	1Q	X	-	-
8	-	-	-	-
9	-	-	-	-

Q = quartz   blue = porcellenite

Discussion and Interpretation

The excavation programme and subsequent analyses of Yengo1 and Yengo 2 achieved most of the designated aims.

Yengo 1

The major findings of this research were, that:

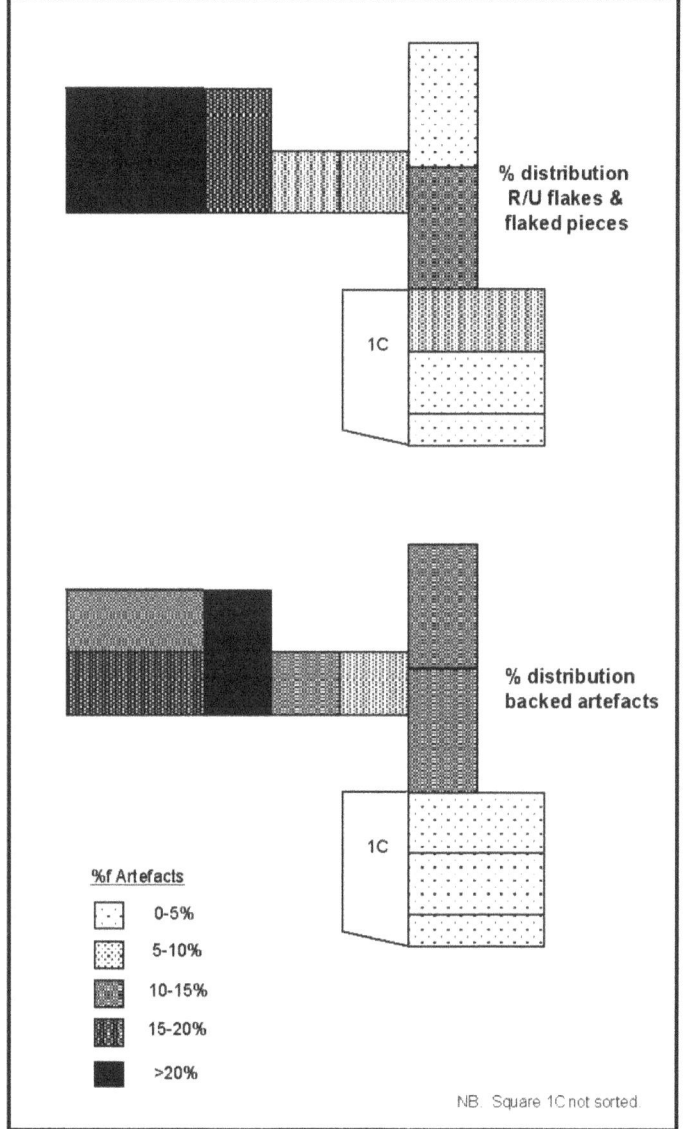

Figure 6.54: Yengo 1. Spatial distribution of backed artefacts and flakes and flaked pieces with retouch/usewear.

- the engravings continue beneath the deposit;

- the occupation deposit varies (in nature and intensity) over time;

- based on stratigraphic evidence the relative age of both the engraved panel and the pigment art can be argued;   and,

- based on the excavated assemblage, the relative age of the grinding grooves and some of the pigment art can be inferred.

The engravings were found to continue some 35cm below the current floor level. The most intensive period of usage at the site resulted in the initial covering of the engravings. Based on stratigraphic evidence, this indicates that the engravings predate the main occupation of the site.

The initial occupation of the shelter is dated to 5,980 ± 290 BP (ANU-6059). A second phase of occupation ended around 2,840 ± 240 BP (ANU-6056) after which the most intensive occupation of the shelter commenced at c.1,950 ± 400 BP (ANU-6054). Occupation of the site continued until after 540 ± 180 BP (ANU-6058).

As the deposit at the site built up, the ceiling came within reach of the artists using pigment at the site. The stencil art appears to correlate with the main or latter occupation of the site. The

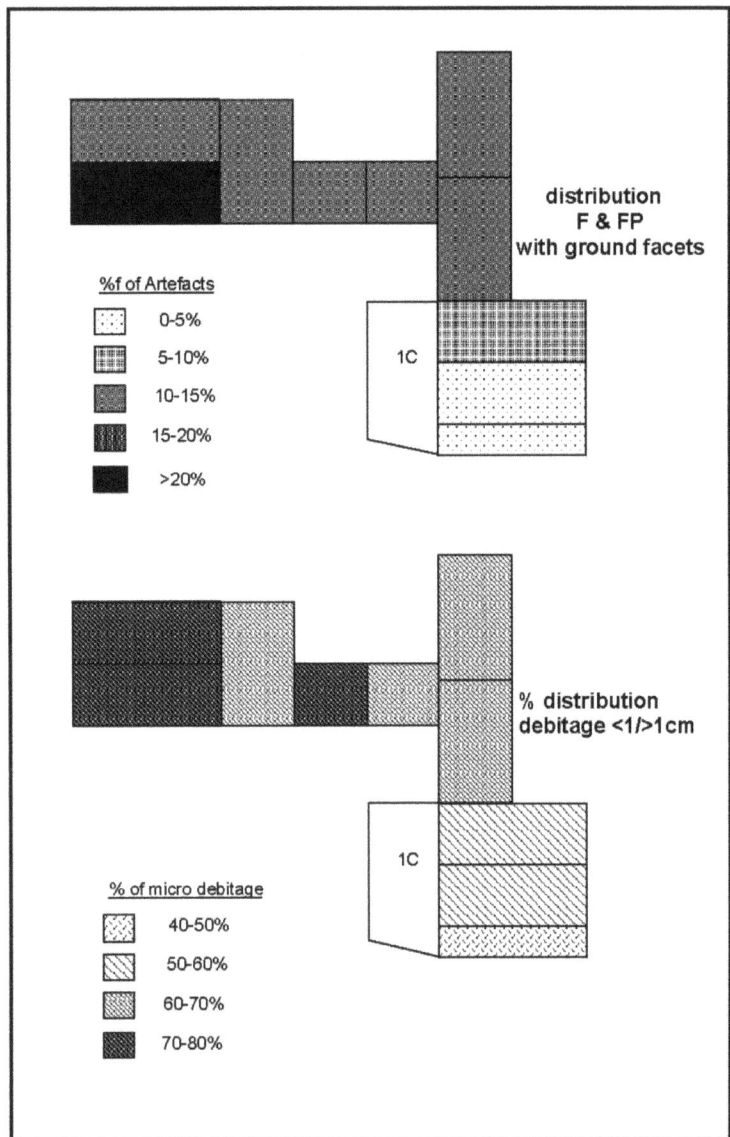

Figure 6.55: Yengo 1. Spatial distribution of ground pieces and micro-debitage (<1cm).

grinding grooves at the site are located around the sloping rock ledge, particularly at its southern end. Ground edged artefacts were found amongst the excavated assemblage in the top three units of the deposit, making the dated use of ground edged material at the site younger than 1,500 years. It is assumed that the use of the grinding grooves at the site is similarly placed within this time frame.

Artefact accumulation rates

An age-depth curve was calculated. There were some concerns about the utility of this due to the complexity of the deposit generally, the sloping bedrock base of the shelter and the variable cultural and natural deposition factors. As most of the dates were submitted from the sequence in square 4A, this square was used to make this calculation. The artefact accumulation rates correlated well with the patterns demonstrated by fluctuating artefact density over time.

The Lithic Assemblage

The excavated artefact assemblage provides information on the changing nature of occupation evidence at the site.

During the earliest period of occupation (Layers VI-VII), the artefact discard rate was low. There were no backed implements, no ground edged material, and quartz was used only as a minor

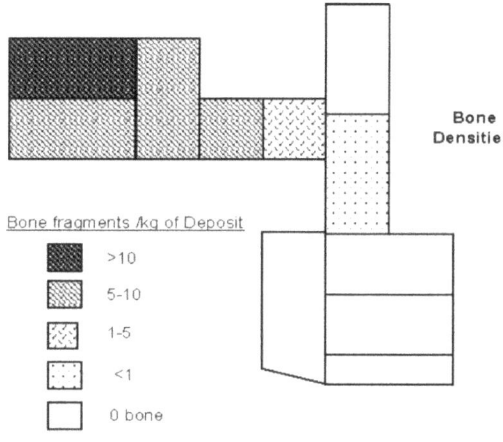
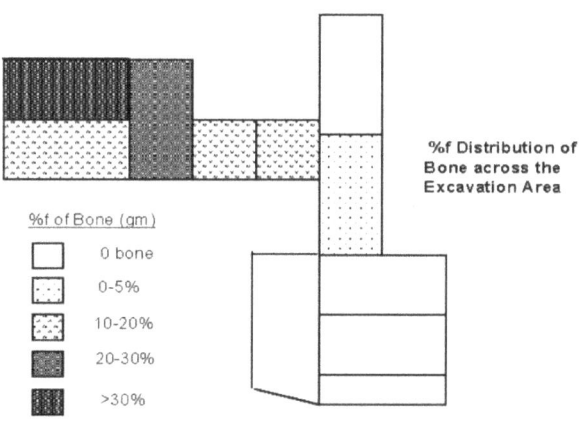

Figure 6.56: Yengo 1. Spatial distribution of bone fragments and %f distribution of bone densities (based on weights; Table A1.14).

raw material. Fine grained basic, silicified tuff and fine grained siliceous material dominated. No core tools or scrapers were found in the lowest layers, but sample sizes from these levels were very small and retouched artefacts extremely rare. Where modified artefacts were found, these were exclusively made of silicified tuff. One chopper tool was located low in unit V (spit 4A/12). The assemblage characteristics and dates support the interpretation of this occupation as pre Bondaian (JMcD CHM 2005a, b) or Mangrove Creek Phase 1 (following Attenbrow's 1987 definitions).

In the next phase of occupation (Layer V), artefact discard rates were again fairly low. Silicified tuff dominated the raw materials being used, although quartz became increasingly important. Backed artefacts were present only in the top of this unit. During this phase, all modified artefacts were made of quartzite. No ground edged material was present. The very low accumulation rates at the end of this phase suggest a hiatus in occupation between this phase and the next. This is supported by the dates received.

At the beginning of the second millennium BP there was a marked increase in artefact discard rates. The peak period of site usage appears to have started after 2,000 BP and to have peaked by 1,500 years ago. This continued for some time thereafter (Units III-IV). During this phase, backed artefacts were made and ground edged implements were introduced. Quartz was the dominant raw material, but there was a wide range of other raw materials being used. Modified artefacts, particularly backed artefacts, were found on a wide range of raw material types.

The terminal phase of site usage (Units I-II) saw a decline in artefact accumulation rates. There was also a decrease in the range of raw materials being used, and backed artefacts dropped out of the assemblage. Ground edged implement fragments increased. This phase is dated to around 500 years ago.

These four occupation phases correspond with the four typological phases identified for Mangrove Creek (Attenbrow 1987: Table 4:7). The Yengo 1 dates are in general accord with

Attenbrow's (1987:189) dates with the exception that the Middle Bondaian phase continues after 1,600 years BP at Yengo 1. The Yengo 1 dates suggest the following time frame for Attenbrow's phases (rounded to 50 years: cf. Table 6.11):

Phase 1	c. 6,000 - 4,600 years BP
Phase 2	c. 4,600 years to c. 2,850 years BP
Phase 3	c. <1,950 years to >540 years BP
Phase 4	c. <540 years BP

The Faunal Remains

More than 20 mammal species were identified from amongst the 527gm of bone at Yengo 1. Preservation conditions for faunal material were good but fragmentations rates high. A variety of macropods are present, with large kangaroos (Eastern grey), wallabies (e.g. swamp, parma, red necked and brush tailed), dingo and a range of smaller mammals such as bandicoots, possums, potoroos and bettongs. Several varieties of snake and three kinds of lizards were also identified. There was meagre evidence for bird and fish remains.

Despite a relatively good weight of bone being retrieved a relatively small proportion of the assemblage comprised identifiable fragments (cf. other shelters in the region with faunal remains: Loggers and Mussel; Aplin 1981, Angophora Reserve; McDonald 1990, Wood 1989).

Very little bone was deposited prior to c.1,500 years ago and peak deposition appears to have been between this time and c.500 years ago. This vertical patterning does not match that of the lithic material: the faunal material peak post-dating the most intensive period of artefact accumulation. A similar pattern has been identified elsewhere (e.g. Morwood 1986, Attenbrow 1987).

Over the period of deposition, there has been a change in the species represented. The bone from Layer III is dominated by elements of medium and large animals: c. 90% of the bone from these lower units is from animals of kangaroo and wallaby size. The paucity of small macropod and mammal bones in these lower levels is not a product of differential survival: there is a clear contrast in the composition of the upper and lower assemblages.

In stratigraphic layers I and II, 33% of the taxa are small animals. Varieties identified include possum, potoroo and bettong. Almost 60% of the bones here are from medium sized wallabies and an unidentified range of similarly sized mammals. Almost all of the reptile bones derive from these Late Bondaian layers. Less than 9% of the bones from these layers are from large animals.

This change over time in focus from larger animals to a proliferation of smaller species and a range of habitats - is similar to that observed by McBryde (1976) in the New England region of New South Wales and by Morwood in south-east Queensland (1986). Aplin's (1981) analysis of Logger's shelter (in Mangrove Creek) produced a similar pattern (although not at Mussel shelter). The postulated correlations of faunal change with technological change (namely the loss of backed blades) and shifts in procurement strategies has been argued as evidence that 'patterns of change in ... resource structure, technology, economy and symbolic behaviour were functionally related' (Morwood 1986:118). The Yengo 1 data supports this argument (see below).

The Engravings

The engravings located on a vertical interior surface of the boulder outside the dripline of the shelter were found to extend some 35cm below the current ground surface. The deposit adjacent to the engraved boulder is affected by water percolation, and the depositional processes at the front of the shelter are different to those operating in the shelter's interior.

On the basis of stratigraphy in the trench perpendicular to the boulder and from an age determination received from close to the boulder (ANU-6215) the following conclusions are drawn.

1. The engravings were produced prior the main occupation of the shelter. Deposit dating to >2,800 years ago covered the base of the engravings;

2. On the basis of artefact accumulation rates, there is a possible hiatus in site usage between the second and third phases of site usage. This would appear to be the only break in usage, representing slightly <1,000 years. There appears to have been continual low density occupation between the earliest (basal) occupation and subsequent occupation. The only date for Layer V is some 2,000 years younger than for Layer VI, but this date is from the top of unit V. Artefact accumulation rates reinforce no hiatus at this point;

3. The engraved boulder and the smaller buried boulder (in Squares 1A, 1B and 1C) are contemporaneous in age and part of the same roof fall episode. The smaller boulder would have been in place throughout the occupation of the site, and thus have been in this position when the engravings were being produced;

4. It is unlikely that the engraved circles were produced after the time when there was any accumulation of deposit between the engraved boulder and the smaller boulder. The distance between the base of the lowest circle on the larger boulder and the junction of the two boulders is <40cm. Other researchers (Flood and Horsfall 1986, Morwood 1992, Rosenfeld *et al*. 1981) have defined 35cm as the requisite height for a contemporaneous floor level and vertical engravings;

5. The deposit which resulted in the initial covering of the engravings predates the hiatus in site usage (between the Early and Middle Bondaian). While there is no evidence directly linking the engravings to the Pre-Bondaian occupation of the site, the engravings definitely predate the Middle Bondaian occupation of the site;

6. It seems likely that the engravings also predate the Early Bondaian phase of site occupation, since it is the layer of deposit associated with this assemblage which also diminished the floor height below the circles such that their production would have been difficult. On this basis, and the assumption that some occupation evidence would be contemporaneous with the production of the engraved circles, it is argued that the engravings are Pre-Bondaian in age;

7. Pecked dots were discovered on the sloping bedrock floor beneath square 2A. These are covered by deposit from Layer VI, and therefore predate some of the earliest occupation of the site.

It is concluded that the engraved circles, dots and kangaroo tracks were probably produced during the earliest occupation of the shelter, between 5-6,000 years ago. They may have been produced prior to occupation of the rockshelter although it seems likely that the act of producing this art is likely to have been part of a suite of behaviours and that the people producing the art had to live somewhere. The minimum age for their production is greater than c.3,000 years - and this estimate is conservative - given that this date is associated with deposit which resulted in the engravings being covered.

Stylistically, the engraved panel and pecked dots have more in common with the Panaramitee style, than they do with the Sydney Basin engraving style. The Panaramitee has a more restricted figurative range and concentration on pecked geometric designs (such as circles) and kangaroo and bird tracks (Clegg 1987; Maynard 1979; Franklin 1991, 2004; McDonald 1983, 1993a). Most Sydney region engravings are figurative (animals, birds, human figures etc.). Although tracks are not uncommon, human tracks predominate in this style, and circles are rare. Further, the main technique employed in the Sydney Basin style is pecked and abraded outline: the Panaramitee technique consists of pecked intaglio (solid) motifs.

The Panaramitee was initially perceived as being of great antiquity (although see McCarthy 1979, 1988) based on the identification of extinct megafaunal tracks and animals (Basedow 1914, Mountford and Edwards 1963). Edwards (1971) identified that this art tradition appeared to be remarkably consistent over enormous distances in the arid zone and that most of these art sites were deeply patinated. Maynard (1979) suggested an age bracket of between 7-10,000 years on this art form.

There have been numerous more recent attempts to date the Panaramitee and its regional variants. Much of this effort has been directed at sites where these engravings are found in rock shelters with associated occupation deposits (e.g. Flood and Horsfall 1986, Morwood 1992, 2002; Rosenfeld *et al.* 1981). These sites have provided dates ranging between $1,570 \pm 60$ BP (Beta-3777) at Green Ant; $13,200 \pm 170$ BP (ANU-1441) at Early Man and >14,000 BP in Sandy Creek 1 (all in north Queensland). Efforts in the arid zone with open sites and a range of dating techniques have provided similarly wide ranging results (indicating problems of sampling as much as the likely age bracket for this style). Dates of between 1,400 and 35,000 years were obtained using cation-ratio dating (Nobbs and Dorn 1988) this indicating problems with the sampling technique (see Watchman 1992). A date of c.10,000 years at Sturt's Meadows was obtained using AMS on charcoal from carbonate crusting (Dragovich 1986).

There are obvious dissimilarities between sites in the coastal or montane regions of Australia where Panaramitee-like engravings occur and the arid zone Panaramitee tradition proper (Rosenfeld *et al.* 1981:88-9; Rosenfeld 1991; cf. Morwood 1984). Many of the regional variants contain relatively variable motif assemblages, and are not classic Panaramitee. A recent analysis by Franklin (2004) has indicated that there is:

> variation within the Panaramitee style both in the arid zone and outside it. The Panaramitee style cuts across environments ... there are similarities between engraving sites across the continent, but ... individual regional manifestations also occur... the Panaramitee is marked by regional variation (Franklin 2004: 135).

The Yengo 1 engraved panel is a regional variant of the Panaramitee. It is clearly of a different style to the majority of the engraved art found in the Sydney Basin. The pre-Bondaian age for this engraved panel supports the general contention that this art style predates the proliferation of Simple Figurative styles around Australia during the Holocene. A small number of shelter sites scattered around the Sydney Region contain residual Panaramitee engravings, indicating an earlier, lower density artistic tradition predating the main late Holocene occupation and artistic period of the region. This earlier lower density art phase matches other forms of occupation evidence - also lower density - confirming a continuing tradition over time for the decorating of shelter occupation sites of the region.

<u>The Pigment Art</u>

It is argued that the pigment art, primarily white stencils, was produced during the most intensive period of the site's occupation. This conclusion is based of the following:

1. The ceiling is beyond easy reach of anybody standing on the sloping floor of the shelter, prior to the accumulation of 30cm (min) of deposit. While the art on the rear wall and

back ledge of the shelter could be reached easily at any time, this part of the assemblage is contemporaneous with the remainder of the pigment art;

2. Most of the excavated pipeclay and ochre is found in Layer III;

3. White pipeclay is found only in the top three Layers of the deposit. Constant acidity with depth argues for constant survival rates throughout the deposit.

Small quantities of red pigment were found in the lowest units of the deposit. Ochre may have been used here for purposes other than the production of parietal art (e.g. body painting) from the earliest times of the site's usage. The Yengo 2 art sequence has an early use of red pigment. Elsewhere in the Sydney Basin the early use of red pigment in shelter art sites has been identified - as it has elsewhere across the continent (Cook *et al.* 1990; Loy *et al.* 1990; Roberts *et al.* 1997). Given the absence of excavated pigments in the Yengo 2 deposits, the presence of a mainly red and black depictive assemblage and the presence of red pigments in the earlier layers at Yengo 1, it is possible that the depictive assemblage in Yengo 2 is older than the stencilled assemblage in Yengo 1. All white stencils in the Yengo 2 site post-date the remainder of the assemblage. The fact that the contemporaneity (or otherwise) of these two sites has not been demonstrated, means, for the time being, that this remains speculation.

Stencils

The stencilled assemblage can be used to infer a number of things about the population which took part in its production. The range of hand sizes, for instance, indicates a mixed population - i.e. men, women and children. The stencilled material objects are primarily identified as men's tools (e.g. axes, clubs and boomerangs). Four straight sticks, however, were also stencilled and, while these lack any diagnostic features, these may have been women's digging sticks.

The range of stencil variations may also have provided other information about the art's audience. Mostly hand and arm, and hand and wrist stencils were recorded, with relatively few (14) finger 'mutilation' stencils. Two of these consisted of all fingers bent into a fist (one with no thumb) while two comprised hands wrapped around the natural fluting on the ceiling. One pair consisted of two hands, wrist-to-wrist with three fingers only showing. Two had no thumb and four had two or more fingers bent over to the first joint. One had the little finger dislocated while the last had the third and fourth fingers positioned together. None of these variants actually record finger mutilation. Rather, all were produced by manipulating the position of the fingers.

Several of these finger positions were identified by Wright (1985) as illustrating sign language hand positions [following Morwood (1979), Walsh (1979)]. For instance, the fist is used to denote 'the men coming up are friendly' (Wright 1985: Table A1.2; citing Spencer 1928). Other alternative interpretations include the depiction of clan totems or levels of initiation. The fist, again (in central Queensland), is recorded as representing the large eagle-hawk; all fingers together, is recorded as representing a fish (Wright 1985: Table A1.2; citing Roth 1897). Most of the stencil examples examined by Wright were associated with mortuary sites. More than 17%, however, were not. While central Queensland hand signals are possibly of little relevance to Sydney it is possible that certain hand positions did represent totems. Mathews (1897c) provides a list of the animal and bird species which were of totemic import to the Darkingung people. Unfortunately no ethnographic work was undertaken in this region on the localised use of sign language.

Forge proposed that stencils are not art - since they are mechanically executed.

> they are the equivalent to signing the visitor's book, or names and signs scrawled all over any permanent surface ... They mark the presence at some time of an individual, they are not

mediated through any symbolic symbol, they are not part of a culture. ... It may be part of a culture to make such marks, either at times of ritual or on first reaching a certain point, etc. ... However, any such cultural requirement refers only to the fact of stencilling not to the form (Forge 1991: 40).

While this argument requires consideration in broad scaled stylistic analyses – when trying to 'reconstruct an art system' (Forge 1991: 44), it begs the question when addressing general questions about parietal art production. A stencil, in itself, may not be an art form (aesthetically speaking) - but this form often contributes significantly to pigment art assemblages in this and other regions in Australia (e.g. Ross 2003). In other parts of the world, the technique is also used to create highly aesthetic figurative motifs (e.g. the Upper Palaeolithic black dotted horse in Trois Frére: Bahn and Vertut 1988).

The stencilling of hands and other objects certainly records a different range of information to depictive paintings and drawings. But this surely is also an example of a cultural group's choice in the use of pigment, media and technique to record images on a rock surface. The proportions, the colour usage, and subjects used in this art form vary synchronically across the region, indicating that different groups used stencils in varying ways. In some parts of Sydney (e.g. around Campbelltown), there are very few hand stencils in most shelter art sites; in other areas (e.g. this northern portion) they predominate in most assemblages and are a highly developed art form (McDonald 1987). Surely this indicates stylistic information when viewing an art system as a whole?

The Grinding Grooves

The grinding grooves, like the pigment art, are thought to coincide with the most intensive, later usage of the shelter. This conclusion is based of the distribution of ground edged fragments throughout the excavated sequence.

The pigment art also contains eight stencils and/or drawings of hafted axes. This further indicates that these implements were in use during the period of the shelter's pigment art production. Two of the white stencils have been 'coloured in' with charcoal lines. These motifs have the potential to be dated using the AMS technique (McDonald 2002c; McDonald *et al.* 1990; Rowe 2001).

What Type of Site?

Before excavation commenced it was obvious that the site was a major focus for art production and that the shelter had also been used extensively in the sharpening of (presumably) axes. The back ledge and sloping shelf also contains evidence of battering and rubbing, suggesting that this surface could have been used for knapping (i.e. as an anvil), seed preparation (i.e. in mortar and pestle fashion). As well as axes or hatchets, the sharpening of thinner, pointier objects has also occurred along the back ledge. These thinner objects may well have been (men's) wooden spears, or they could have been women's digging sticks (McDonald 1991b).

Extensive stencilling of hands and other objects and, in a more limited fashion, painting and drawing has occurred here. This site has the second largest assemblage that has so far been counted in the region: only Swinton's shelter (in the Mangrove Creek catchment, with 847 motifs) is larger. Yengo 1 represents a major artistic focus. Swinton's has a relatively smaller proportion of stencils than Yengo 1: 65% as opposed to 83%. The fact that Yengo 1 has such a strong focus on stencilling and so few depictive motifs is felt to indicate something about the nature of the site.

While no effort can be made to interpret the art, possible functions can be suggested on the basis of what stencils may represent. Moore (1977) suggested seven possible functions for stencils on the basis of ethnohistoric research and 'a number of published Aboriginal explanations' (1977:318). These were (as):

i) individual signatures of artists or to record a visit;

ii) a memorial to be mourned over after death at a mortuary site;

iii) messages to the spirit ancestors;

iv) a secular message to other Aborigines;

v) a record of an historical event (telling a story);

vi) a story telling device to record the myths at a sacred site;

vii) a means of using the power of a sorcery site (Moore 1977: 322).

This list indicates that both secular and ritual explanations for hand stencils, and no doubt many examples could evoke several explanations, depending on context (the level(s) of initiation of the producer, the informant and the observer). Some of the potential explanations for the Yengo assemblage can be eliminated on the basis of association.

There are, for instance, no associations between stencils and depictive motifs. Moore's explanations 5-7 all involve the placement of stencils over or around major figurative components in an assemblage. Explanation 2 can also be removed from consideration given the absence of human skeletal remains at the site; stencils as memorial or mortuary devices other areas in Australia have all involved burials occurring in the same site (e.g. central and northern Queensland highlands: Morwood 1979, 1992).

Analysis of the stencilled hands indicates that most of the stencils here were of left hands, and that there was a range of hand sizes, from babies and children up to large adults. Given the overlap in the hand sizes of gendered groups in living populations (see above) detailed gender information about the Mt Yengo participants cannot be discerned. However, the presence of babies' hands among the assemblage is argued as being a good indication of the presence of women at the site: while the mixed size ranges also suggest that women also participated in the stencil production. It would seem likely that the use of stencils at this site is secular, and it would seem reasonable to assume that one possible explanation of the stencils was to record the number (and identity) of people who camped there - presumably on a number of occasions.

Forge's (1991) argument supports stencils as signatures, or statements of presence and/or involvement. Based on this, and the extensive domestic occupation evidence obtained from the deposit, the art in the Yengo 1 site is interpreted as being secular.

The stone tool assemblage and faunal remains indicate that the site, during its period of most intensive occupation, was probably a base camp. There is evidence for a range of activities being carried out at the site (e.g. backed blade production and gearing up [replacement] strategies, the consumption of food), the production and/or resharpening of ground edged implements, as well as a range of general purpose stone tool use and production. While no detailed usewear analysis was undertaken, a macroscopic inspection of utilised tools indicates a range of edge angles - suggesting a variety of activities: wood working, butchering and activities involving 'softer' processes (cf. Gorman 1992). Intact hearths across the site indicate that there were probably several camping foci within the shelter: discrete areas of higher density knapping and patchy areas of higher density faunal material lend support.

Meehan defines base camps as representing a long term focus for occupation for the territorial group - 'men of several land owning units whose territories formed a contiguous area, together with their wives and unmarried children' (Meehan 1982:13). Such sites are generally located a variety of distances from primary resources and may be occupied over many months and (in tropical Australia) sometimes years. Food debris disposal at base camps is patterned in a complicated yet predictable way on the periphery of each hearth complex (Meehan 1982:114).

Such sites include evidence for more intensive occupation, including site structure (hearths, living areas, disposal areas, etc.) as well as evidence for all members of the group.

The Yengo 1 archaeological remains, during the period of most intensive occupation, certainly demonstrate such requisite types of evidence. This and the range and density of hand stencils at the site, are interpreted as indicating that this site presented a focus in the landscape for the people living in the northern reaches of the Darkingung territory during the last two millennia BP.

Interpretation of the earlier uses of the site is based on less tangible evidence. Occupation remains are sparse and relatively low density. The engraving of circles and bird and roo tracks has been interpreted in other regions as indicating an art system which is more ambiguous and generally for 'restricted' rather than 'public' contexts (Morwood 1988: 33). Such art is thought to indicate more open social networks and lower population densities (Gamble 1982, Morwood 1988, Smith 1989; although see McDonald 2005a).

It would appear likely that the site had a very different function during its earliest use as an art shelter. Whether this function was ritual or secular cannot be said. However,

> corporate territorial expression through the indelible marking of place with a stylistic graphic system may have been a powerful means of asserting corporate rights and relationships. (Rosenfeld 1993: 77)

There is every reason to assume that this site was a focus for groups living around Mount Yengo throughout prehistory, for as long as this area has been populated. It is also possible to argue that this focus became less ritualised over time - at the peak of its usage the site was used and decorated by the entire group as a base camp. The decline in usage over the last millennium (indicated by declining artefact densities) may indicate a continuing change in focus. The faunal evidence indicates a continuing intensive use of the shelter in its terminal phases for the consumption of food, if not the production of stone tools. This would tend to suggest a change in the nature of habitation in the site rather than a decrease in population density (i.e. Morwood 1986) and perhaps a move by an increasingly large population into open locations for camping (as suggested by the ethnohistoric literature at contact). Obviously an excavation programme targeting the broad open lower hillslopes below the site could further explore this latter possibility.

Yengo 2

The excavation aims for Yengo 2 had a different orientation because of the fact that the art assemblage in this shelter was so different from that in the main shelter. Given the proximity of the two sites, establishing the contemporaneity of their occupation and the likely age of their art assemblages was of interest. An analysis of the art in this shelter indicates at least three artistic episodes, the last of which seems to coincide with the prolific art production in Yengo 1.

The small amount of material retrieved from this second excavated site is lost. It is possible to state, however, that there was sparse occupation deposit in the shelter, but that the majority of the sedimentation here is through natural processes.

There was an absence of pigment/ochres in the deposits in Yengo 2 and correlating the art with the deposit in the shelter may well have been inconclusive. The sparseness of occupation deposit can be explained by the unpleasantness of the shelter for camping at any time other than on the hottest day. It is dark, dank and the deposit and sloping walls are extremely damp.

The tantalising question is - why was art produced in this shelter? The sandstone surfaces in this site are smooth textured and provide an excellent medium for drawing. This, however, is a minor consideration given that the art is just about invisible at all times of the day - bar early morning on a sunny day - or perhaps by firelight. The evidence for domestic usage here was sparse and this pigment art may represent a different function to that found in the larger shelter –

with its abundant evidence for domestic use. There is evidence in Yengo 2 for several episodes of art production, perhaps spanning a considerable time period.

When this research was undertaken, AMS dating techniques were in a nascent stage of development (cf. Chaffee *et al.* 1993, 1994; Geib and Fairley 1992; McDonald *et al.* 1990; Watchman and Lessard 1993). The Yengo 2 art assemblage was inspected in 1988 with a view to collecting samples for dating. This inspection indicated that there were very few black motifs with visible charcoal fragments on them. The lack of visible charcoal is a possible indication of age (i.e. the charcoal may have chemically bonded with the rock or had longer to be worn away) but is more likely to be due to the water percolating through the wall (flushing away adhering fragments). The surface conditions in this site indicated other potential difficulties: there is macroscopic lichen growth on several parts of the wall (again, due to the damp conditions) which would be potential sources of modern carbon contamination. Salts and other accretions were also identified over several of the motifs. These accretions may be silica or oxalates - which would perhaps present an alternative dating opportunity (Watchman 1993b; Watchman and Lessard 1993).

Given the identified potential contamination problems as well as (the then) lengthy delays being experienced in AMS technology (the success of the method created extensive waiting periods for sample counting: John Head, pers. comm., 1994), it was decided not to pursue this line of inquiry for this research. Future work may well be able to elucidate the age of the Yengo 2 art assemblage and assist in interpreting the variability observed between this and the main shelter site.

The Yengo Sites in a Regional Context

Moore (1981) investigated contact between the Hunter and Hawkesbury River valleys. Based on excavations at site YC/1 on Yango Creek, a tributary of Wollombi Brook (flowing north to the Hunter), and site MR/1 on the lower Macdonald River, a tributary of the Hawkesbury, Moore made the following conclusions:

> Contacts between the Wanaruah [in the Hunter] and the Darkingung in the south, along the Boree Track, seems to have been mainly for joint ceremonials and for trade; this is reflected in the rock art between the Hunter and the Hawkesbury and also in the presence of Hunter valley cherts and quartzites in the tools and wastes excavated in MR/1 ... The apparent intensification of occupation in sites at both ends of the Boree Track about 2,000 BP has already been mentioned. ... contact between the Wonaruah and the Darkingung may only have been established towards the end of the third millennium. (Moore 1981:423)

Moore also concluded that the presence of Hunter chert (now called silicified tuff) from around 5,000 BP, was an indication of the Darkingung travelling north for this raw material, rather than evidence for earlier trade between the two groups. This is now discounted by more recent work on the Cumberland Plain which indicates that the early use of this material occurred across the Sydney region and that the likely source of this is the Nepean-Hawkesbury gravels (JMcD CHM 2005a and b).

Moore also draws a distinction between the art of the Darkingung and that of the Wanaruah.

> [Darkingung art is characterised by] comparatively realistic representations of people and animals, ... in charcoal or ochre, groups or lines of small human figures dancing and carrying out other activities, and various types of stencils. ... immediately to the north ... there is an extensive use of stencils of hands, sometimes including the whole arm, of weapons and tools such as boomerangs, spear throwers, axes, etc. ... Also common are series of straight white lines ('tally marks') and radiate figures in white ('sun symbols'). Representations of animals are extremely rare. (Moore 1981:396).

Work on the Rock Art Project established that the distribution of Wanaruah elements (viz. the white painted radiate figures) and the concentration on white hand stencils is more widespread than suggested by Moore. White painted 'sun' motifs have been recorded as far south as the north bank of the Hawkesbury river and as far east as Mangrove Creek (e.g. Swinton's shelter). There is considerable evidence for a clinal sharing of stylistic characteristics from the two adjoining groups (as defined by Moore), with a strong Wanaruan artistic influence extending down the Macdonald River and surrounding ridgelines to the east and west.

By Moore's definition, the art in Yengo 1 is Wanaruan style, while the Yengo 2 shelter contains distinctive Darkingung characteristics. Moore's model would suggest the Yengo 2 art predates the main (pigment) art activity of the Yengo 1 site: perhaps being a relic of the days when the Darkingung foraged northwards in search of raw materials from the Hunter. The discovery of a porcellenite artefact in the Yengo 2 deposits (Table 6.29) certainly demonstrates raw materials being obtained from further north; e.g. Rich 1992).

The Yengo 1 shelter was used prior to 3,000 BP, predating either of Moore's Hunter Valley shelter sites (e.g. Sandy Hollow, YC/1) and the majority of the dated sites from the Hunter region generally. More recent work has discovered shelter deposits as old as 4,740± 70 BP (Wk-1191) on the Goulburn River (Haglund 1995) as well as a late Pleistocene open site in the Hunter uplands (Koettig 1987). Given the late Pleistocene site at Glennies Creek, site settlement patterns identified in the Mangrove Creek area, and Pleistocene occupation of the Cumberland Plain, we now know that there was widespread, if sporadic, use of both the Hunter and the Hawkesbury regions prior to the Holocene. Yengo 1 provides the most comprehensive evidence for the range of activities (e.g. art and lithics) which are likely to have occurred in the pre-Bondaian phase in this intermediate position between to two major biogeographic regions.

The lithic material at the Yengo 1 site fits well into Attenbrow's four typological phases (these are discussed in more detail in the Upside Down Man excavation report). Given the predominance of quartz in the deposits, the Mount Yengo archaeology appears to be more closely aligned with site occupation patterns identified in the Sydney region than those identified in the Hunter (Moore 1970, 1981; Hiscock 1986). Elements of the archaeology indicate a sharing of raw materials with the Hunter Valley (the single porcellenite artefact is the best evidence for this: silicified tuff and silcrete derive from both the Hunter and Cumberland Plain).

The art evidence with its predominance of white pigment and hand stencils indicates close cultural contacts with groups further north. This apparent contradiction in terms of artistic influence and the stone tool phases and emphases may well indicate that there was equal mobility from the north and the south.

This site complex fits well the definition of an aggregation locale (viz. Conkey 1980), whereby neighbouring language groups unite with the incumbent territorial group for a variety of reason. Ethnohistoric reports put Mount Yengo on a branch of the traditional access route along the Boree Track, between the Hunter and Macdonald Rivers. The site may represent an important meeting place between the two groups, the art being a record of large scale meetings: the combined evidence of art and stone suggests the presence of 'regional personnel' (Conkey 1980). The Yengo shelters in tandem also demonstrate considerable stylistic diversity, both relative to each other and compared with the sites in the immediate and broader scale (McDonald 1987, 1993b). This diversity fits another of Conkey's defined requirement for an aggregation locale (and see McDonald and Veth 2006)

The Yengo sites (particularly the main shelter) were a focus for groups living around Mount Yengo, for as long as this area has been populated. The early use of the shelter included the production of an iconic art system and relatively sparse occupation deposit. This focus might have become less ritualised over time. Subsequent use of the shelter changed to a proliferation of occupation activity - domestic, technological and artistic. This is consistent with a model of tightening social and territorial organisation (e.g. Rosenfeld 1993). The changes evident within the Holocene support a model of ongoing social change throughout this period and particularly a model whereby art assists in the facilitation of increased social contact.

7

EXCAVATION AT THE GREAT MACKEREL ROCKSHELTER

This chapter details the excavation and analysis of the Great Mackerel rockshelter in Ku-ring-gai Chase National Park. A total of 24 person days was spent on the excavation in November 1987. The art was recorded in 1986 for Stage II of the Sydney Basin Rock Art Project (McDonald 1987). Two person days were spent recording the art.

The aims of the excavation were to determine:

1) the nature of the occupation evidence at the site, including its contents and age;

2) whether there were two phases of occupation at the site (as suggested by the art assemblage);

3) whether the art was contemporaneous with the occupation deposit.

Environmental Context

Great Mackerel Beach is on the eastern side of Pittwater, towards the north of West Head in the National Park (Figure 7.1). The Great Mackerel rockshelter is located on the ridgeline south of the beach, 80 metres in elevation above sea level.

The shelter is surrounded by undisturbed dry sclerophyll forest. Dominant tree species include *Angophora costata* (Sydney smooth-gum apple), various Eucalypts and *Casuarina* spp. (She-oaks), and *Banksia serrata* (Old-man Banksia). There is a moderately dense understorey of grasses, *Xanthorrhoea* sp. and bracken fern.

The bedrock along the ridge top is of the Hawkesbury Sandstone Formation, while Narrabeen sandstone and shales are exposed further down the slopes. The nearest permanent water source is a creek feeding into the southern end of Great Mackerel Beach. While this is only 150m distant, it is 80m in elevation below the site.

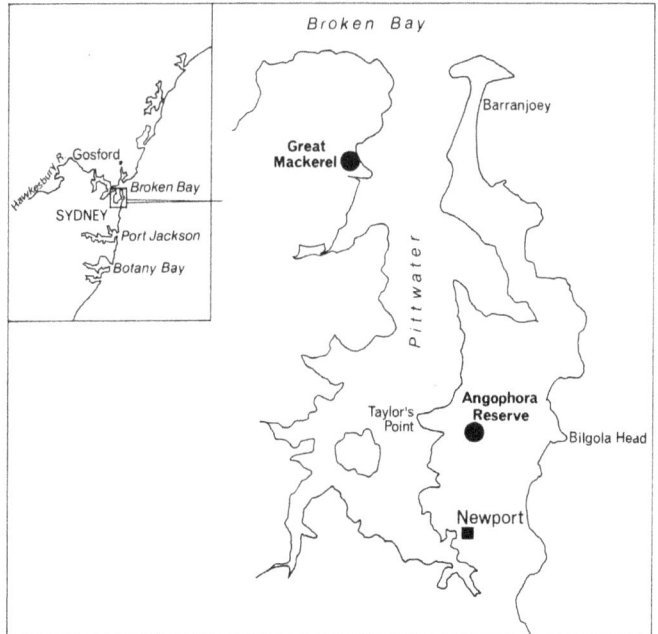

Figure 7.1: Locality Map showing Great Mackerel rockshelter and its context.

The Site

The Great Mackerel rockshelter measures 10m x 4.5m x 2.5m (Figure 7.6 and Figure 7.7) and has a northerly aspect. The archaeological components of the site include art (stencils, drawings and paintings) and occupation deposit (shell, stone artefacts, bone and pigment).

The Art

An assemblage of 114 motifs was recorded at the site. Most of these are stencils. There are two artistic phases at the site. The earlier art phase consists of red hand stencils only. The later phase includes white stencilling (hands and other objects), white drawing and painting, and charcoal drawing. The art is in good condition, while fairly faint, and there is limited evidence of exfoliation.

Motif and technical information were recorded at the site (Table 7.1). White is the dominant colour used (66%), followed by red (19%) and black (16%). Stencilling as a technique predominates (82%): drawn motifs represent 16% of the assemblage and painted motifs 3%. Of the twenty non-stencilled (i.e. depictive) motifs the majority (65%) are infilled, with smaller numbers outline only (15%) or outlined and infilled (20%). Three of the depictive motifs have linear infill.

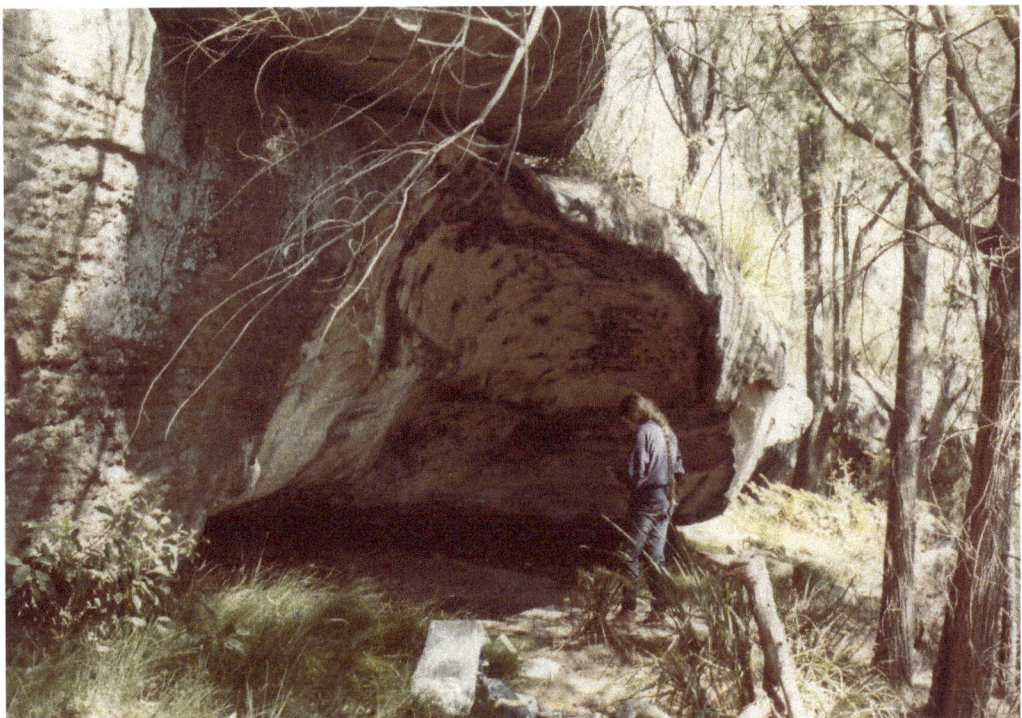

Figure 7.2: The Great Mackerel rockshelter. View from the north-east.

Table 7.1: Art Assemblage: Motif and Technique Information.

Motif	Colour			Technique						Total
	Red	White	Black	Drawn	Painted	Stencil	Outline	Infill	O/I	
Boomerang		2				2				2
Shield			3	3					3	3
Digging stk		4				4				4
Hands	21	58				79				79
Feet		3				3				3
Hand var.		3				3				3
Bird tracks*		2	6	6	2			6	1	7
Leaf		2				2				2
Twig		1				1				1
Unid lines			3	3			3			3
Unid solid	1			6	6	1			7	7
Total	**22**	**75**	**18**	**18**	**3**	**94**	**3**	**13**	**4**	**114**

*One of these is bichrome (i.e. black and white)

Stencils

The hand stencils were recorded for size, colour and handedness (i.e. left or right). All of the red hand stencils were superimposed beneath the white ones, and were considerably more weathered. Exfoliation has affected the red stencils but pre-dates the later art phase. No surface exfoliation has occurred since the later art phase.

Most (52%) of the twenty-one red hand stencils are left hands, 19% are right hands while 29% are indeterminate. Only 12 of the red hands were complete enough to be fully measured. The average span of these was 14.6cm, while the average length (measured from heel of hand to tip of middle finger) was 16cm. Three of these stencils were children's (<14cm length), six were medium sized (15-18cm length) while the remainder were large (>19cm long).

A majority (71%) of the 58 white hand stencils are also left hands, while 12% are right hands. Only 17% are indeterminate. The higher proportion of measurable white stencils indicates the better preservation of this sample, compared with the red ones. The average span size of the white hand stencils was 14cm while the average length was 16cm. Thirteen of the white stencils (32%) were children's, 19 (46%) were average and 9 (22%) were large. All three of the white foot stencils are fully measurable. These are also children's (one) and babies' (two) sized (16 and 12cm long respectively).

These stencil size ranges indicate the presence of a mixed family group. There was no exclusivity of particular members of the social group at this site. Several white stencil compositions also suggest the presence of a mixed social group - both in terms of the hand sizes used and the association of these with different items of stencilled material culture. Four of the stencil compositions include a baby's hand (10 x 10cm, 11 x 10cm, 11 x 11cm, 9 x 10cm) in direct association with a medium sized hand (a mother's?). Another indicator of women is a composition of stencilled digging sticks. Men are also indicated as participants in this art assemblage. At the western ends of the shelter there is a stencil composition of a large symmetrical boomerang (70cm x 5.5cm max.) held in place by two very large (23cm long) hands. These are clearly men's hands (see discussion in the Mt Yengo excavation report regarding gender differences in hand sizes).

Fifty-two shelter art sites have been recorded in detail in Ku-Ring-Gai Chase (McDonald 1987:15-44). The dominant technique here is stencilling (52%) with the hand being the most common motif (45%). Pigment usage was found to be fairly even (red - 33%; white - 33%; black - 32%). Most of the art shelters in this area contained relatively small assemblages (94% contain <49 motifs). Relatively little superimpositionning or compositional complexity was found. Fish (25%) and kangaroos (21%) dominate the (identifiable) depictive motifs, followed by men (6.8%), boomerangs (6.1%), birds (5.4%) and anthropomorphs, bird tracks and 'others' equally represented (4.7% each).

The Great Mackerel site is unusual in terms of its assemblage size; it's higher than average hand stencil component and the dominance of white pigment. In terms of motif content, the range is relatively restricted, and there is an emphasis on several unusual or uncommon motifs – e.g. bird tracks and stencilled material objects. There are fewer than average unrecognisable motifs. Given the general complexity of the site, this is unusual. Motif clarity - especially in light of the assemblage's general complexity - suggests that the art of the most recently produced phase may be relatively recent in age.

The Deposit

When the art at the site was recorded in 1986, the deposit was described as 'midden - surface scatter of shells (*Anadara, Pyrazus*); potential archaeological deposit; dark grey, greasy; between <3 cm to >20 cm deep.' Around 20 mud whelks (*Pyrazus ebenesis*) were observed lying on the surface beneath the low ceiling at the rear of the shelter. No surface artefacts were seen. The deposit appeared undisturbed and was covered by a carpet of *Casuarina* needles.

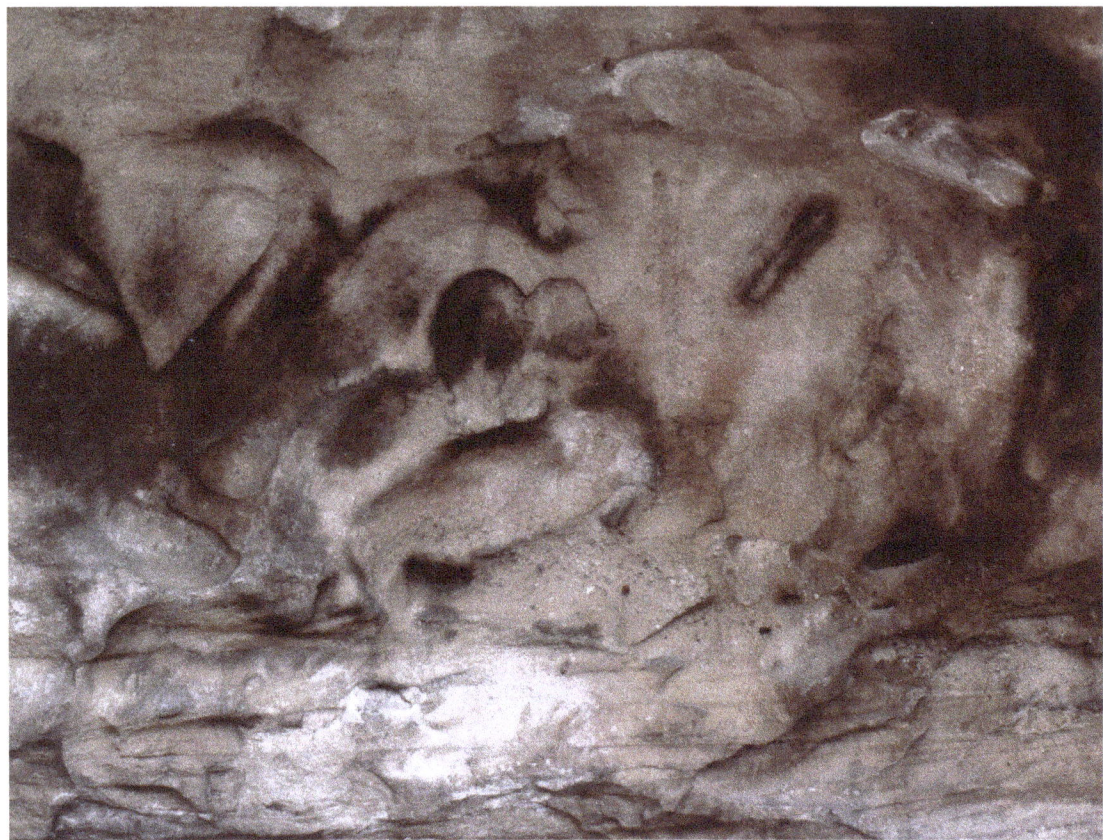

Figure 7.3: The Great Mackerel rockshelter. Art assemblage. Faint white hand stencil in varying sizes.

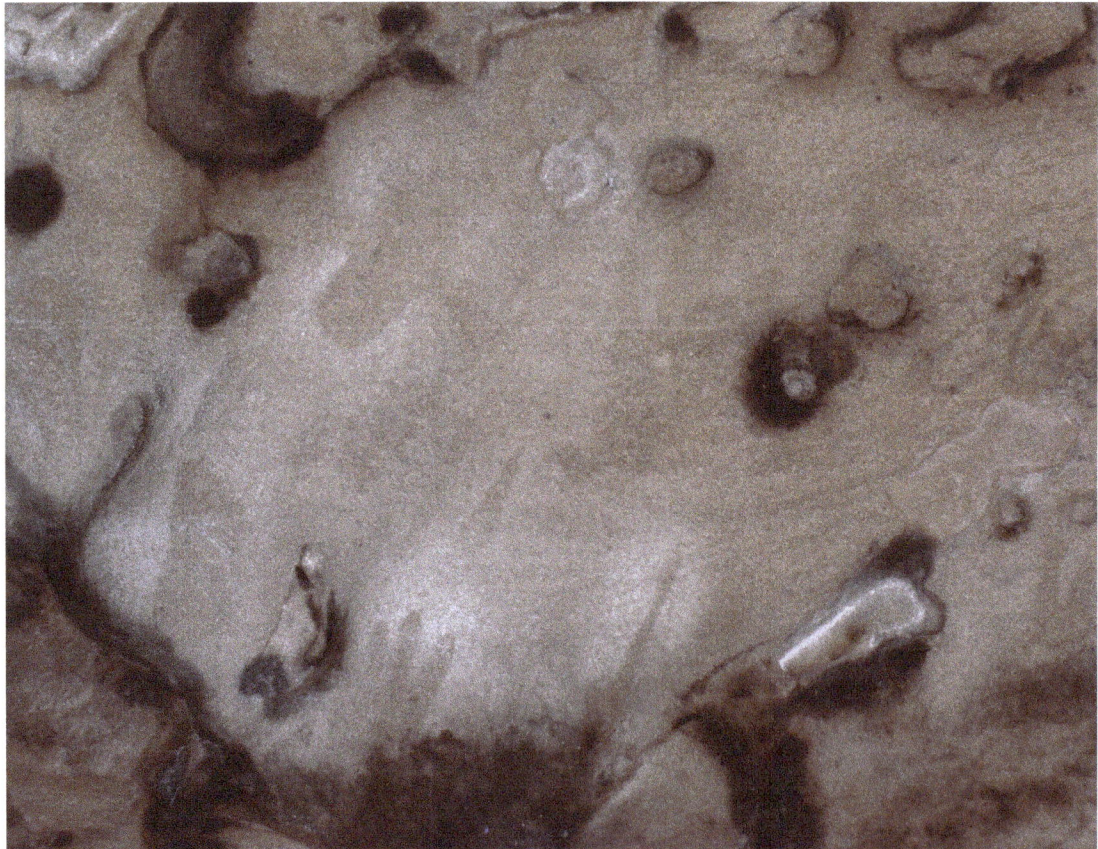

Figure 7.4: The Great Mackerel rockshelter. Stencil composition with feet, hands and digging sticks.

Chapter 7: Excavations at the Great Mackerel Rockshelter

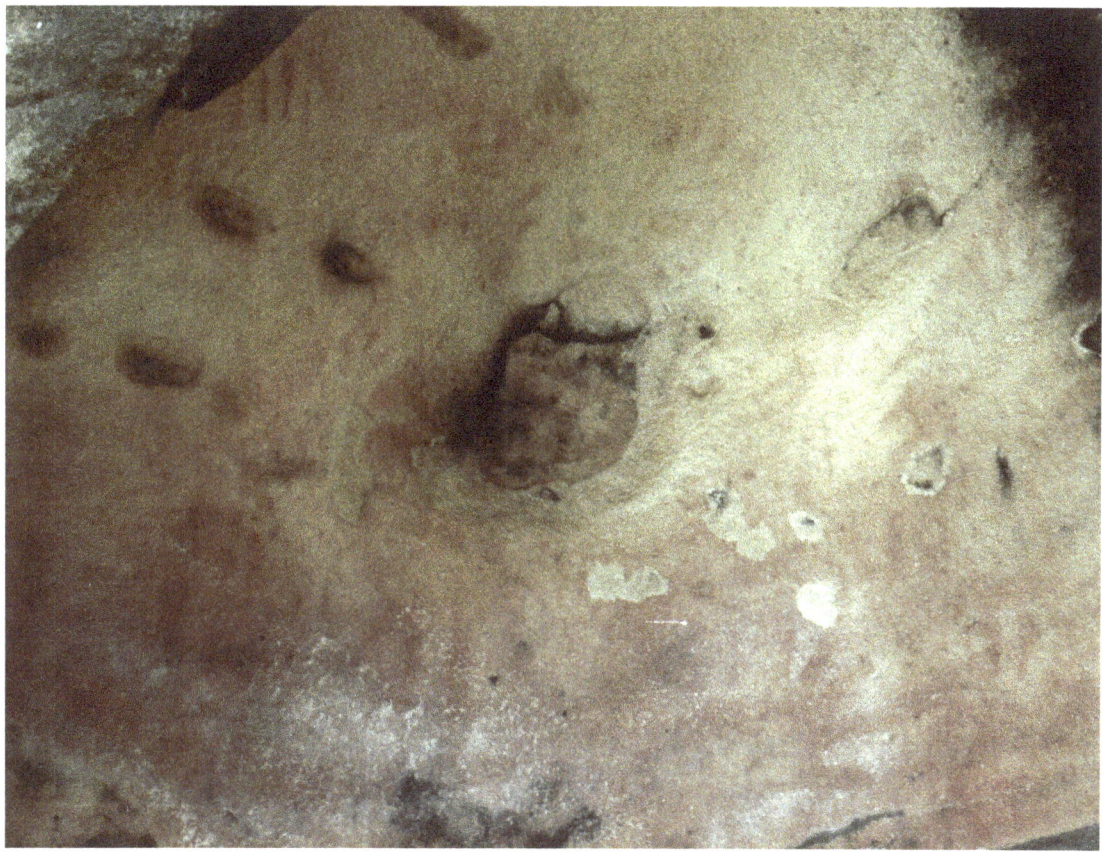

Figure 7.5: The Great Mackerel rockshelter, Panel A. Older exfoliated red hand stencils are superimposed by white stencils.

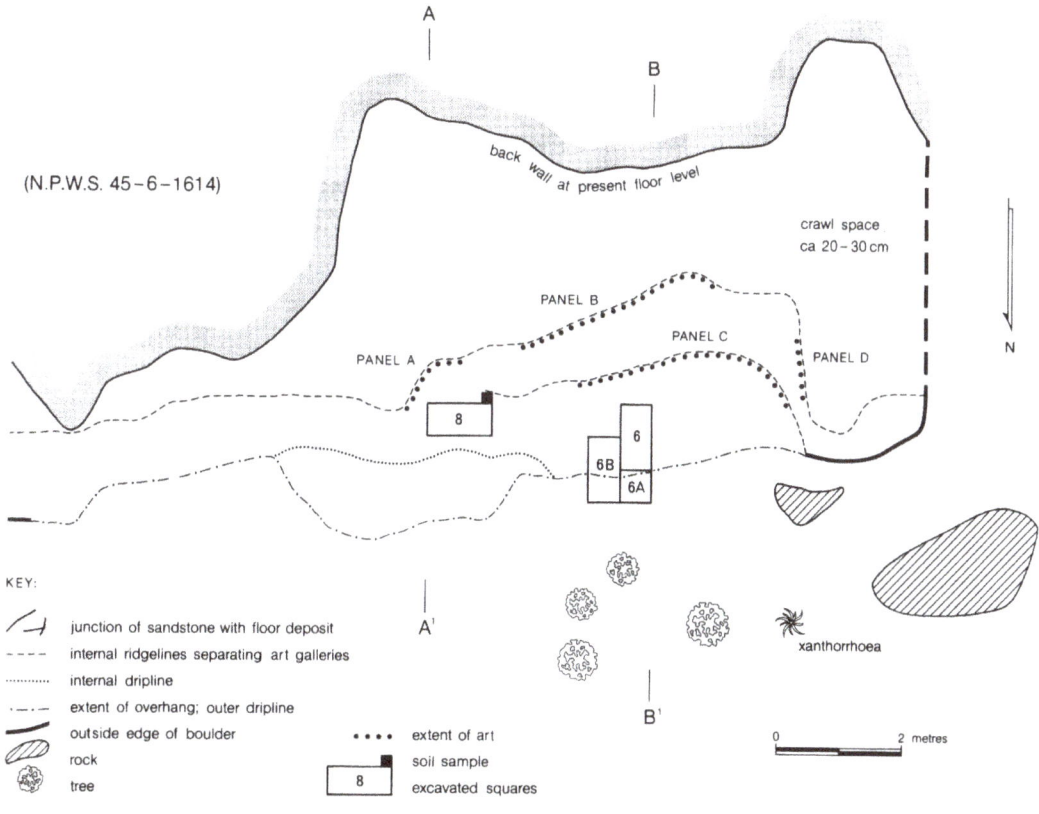

Figure 7.6: The Great Mackerel rockshelter. Site Plan.

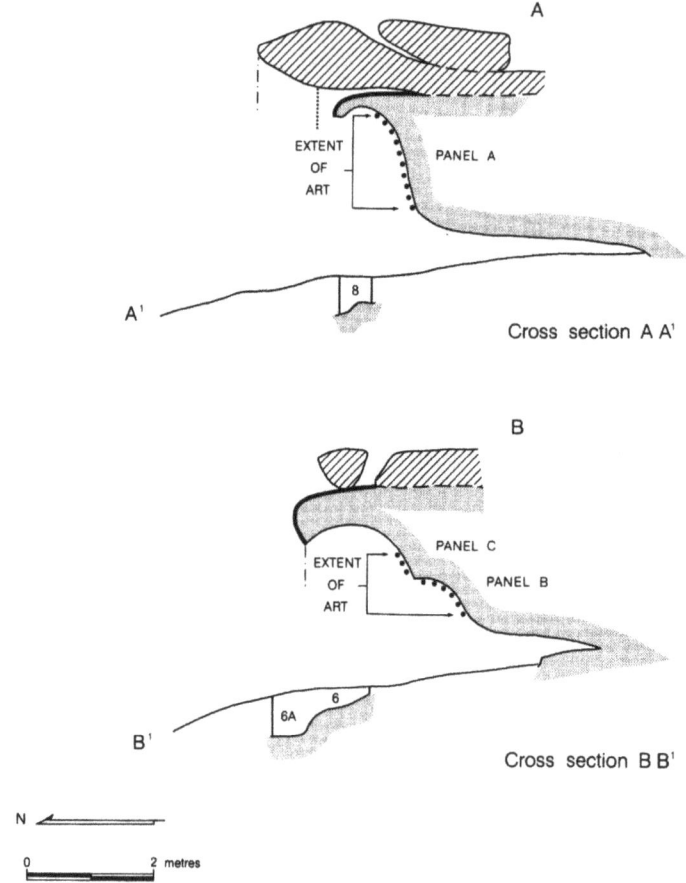

Figure 7.7: The Great Mackerel rockshelter. Cross-sections.

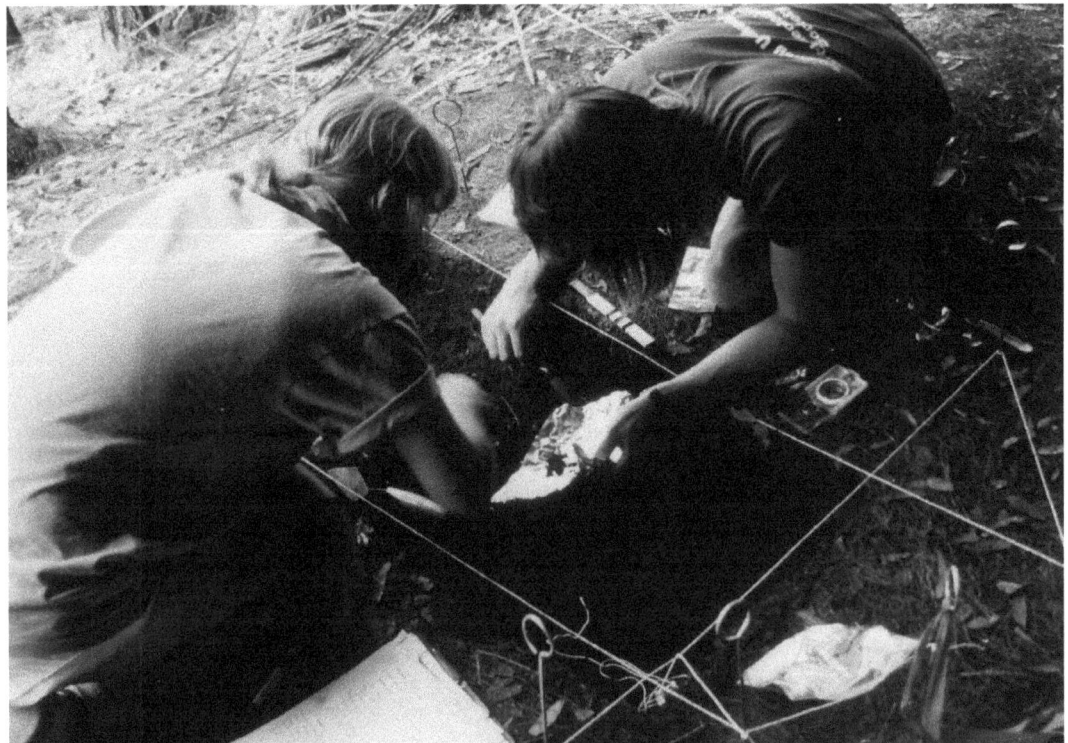

Figure 7.8: Excavating in square 6. Terri Bonhomme and Neville Baker provenancing shell clusters and collecting charcoal samples.

Field methods

The excavation at this site was a test excavation in accordance with a Preliminary Research Permit issued by the National Parks and Wildlife Service, NSW. The floor area was gridded out and the squares aligned to a horizontal datum (Figure 7.6). Excavation squares were selected based on proximity to decorated wall panels and the presence of a relatively flat, protected floor area (Figure 7.7). Initially two 50 cm x 100 cm trenches were excavated (Squares 6 and 8). Square 6 was extended with a 50cm x 50cm pit (6A) and another 50cm x 100cm trench (6B). A total of 1.75 sq metres was investigated.

Excavation in Squares 6 and 8 commenced initially in 2cm and then 5cm spits (Figure 7.11). In the basal layers of the squares the spits became deeper. This was done to allow for the contraction in area which resulted from the presence of roof-fall (sq 6) and the unevenness of bedrock (sq 8).

In Squares 6A and 6B, excavation was by 5cm and 10cm spits, or by stratigraphic units. These squares were excavated with the aim of increasing the sample for analysis. Excavated spit depths are shown in Figure 7.11. All deposit was dry sieved through nested 5mm and 2.5mm sieves on site. Unsorted residue from the 2.5mm sieves was bagged and retained for later sorting and analysis in the lab.

Results

Cultural Material

A total of 1,032.75 kg of deposit was excavated from the site. From this, slightly less than 6 kg of cultural material (Table 7.2) and 111.5 kg of roof-fall were obtained. There is a relatively even distribution of cultural material across the four squares, with the highest density of material being in square 6 and the lowest in square 8 (Table 7.3).

Table 7.2: Proportions of shell, bone, artefacts, charcoal and pigment in cultural material (all squares).

Cultural component	Total (g)	%f
Shell	4,369.5	74.2
Stone artefacts	886.3	15.0
Charcoal	620.8	10.4
Bone	21.2	0.3
Pigment	0.5	0.0
Total	5,898.3	99.9

Table 7.3: Distribution of Cultural Material by Square. Weights in grams.

Cultural component	6	6A	6B	8
Shell	1,322.2	1,193.1	952.7	901.4
Artefacts	153.1	178.2	423.4	131.6
Charcoal	137.9	134.2	303.7	45.0
Bone	8.6	0.8	9.5	2.3
Pigment	0.4	-	0.1	-
Total	1,622.2	1,506.3	1,689.4	1,080.3
Wt excavated deposit (kg)	132.0	251.9	330.9	251.9
Density = kg cult material/kg deposit	1.2	0.6	0.5	0.4

By weight, the predominant archaeological component of the deposit was shell, followed by artefacts, charcoal, bone and pigment. Shell and bone are present only in the upper levels (with very few lithic artefacts). Stone artefacts predominated in the lower levels (with a total absence of shell and bone). The nature of this dichotomy is discussed below.

Stratigraphy

Five stratigraphic layers were identified at the site (Figure 7.12) based on soil colour and texture. The presence/absence of midden deposit was a large factor in determining the nature of the deposit.

The Layers identified were as follows:

1. Surface leaf litter, *Casuarina* needles and loose grey, silty sand. Some shell. 10YR 3/3. pH = 8.5.

2. Black, greasy loam with high shell and charcoal content: the midden layer proper. 10YR 2/1. pH = 8.5.

3. Grey/brown loose sandy loam with high roof-fall content. Some fragmentary shell decreasing with depth. 10YR 3/1. pH = 7.

4. Mottled grey/brown loamy sand. No shell: very little charcoal. 10YR 5/2 - 5/3. pH = 5.

5. Loose yellow/brown sandy loam. Large quartz pebbles; extant roots. Decomposing roof-fall. Increasing clay fraction. Colour grading from 10YR 6/4 - 6/6 to 10YR 4/2 - 5/3. pH = 5.5 - 6.

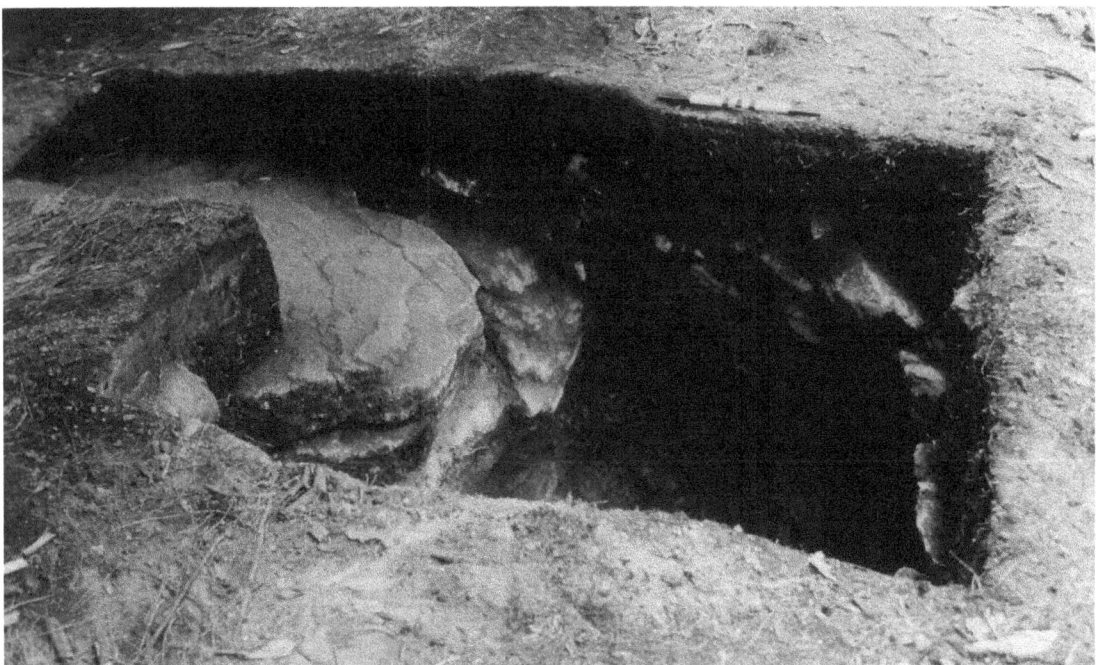

Figure 7.9: The Great Mackerel rockshelter. Squares 6, 6A and 6B at the completion of excavation. View to the western baulk from the east-north-east.

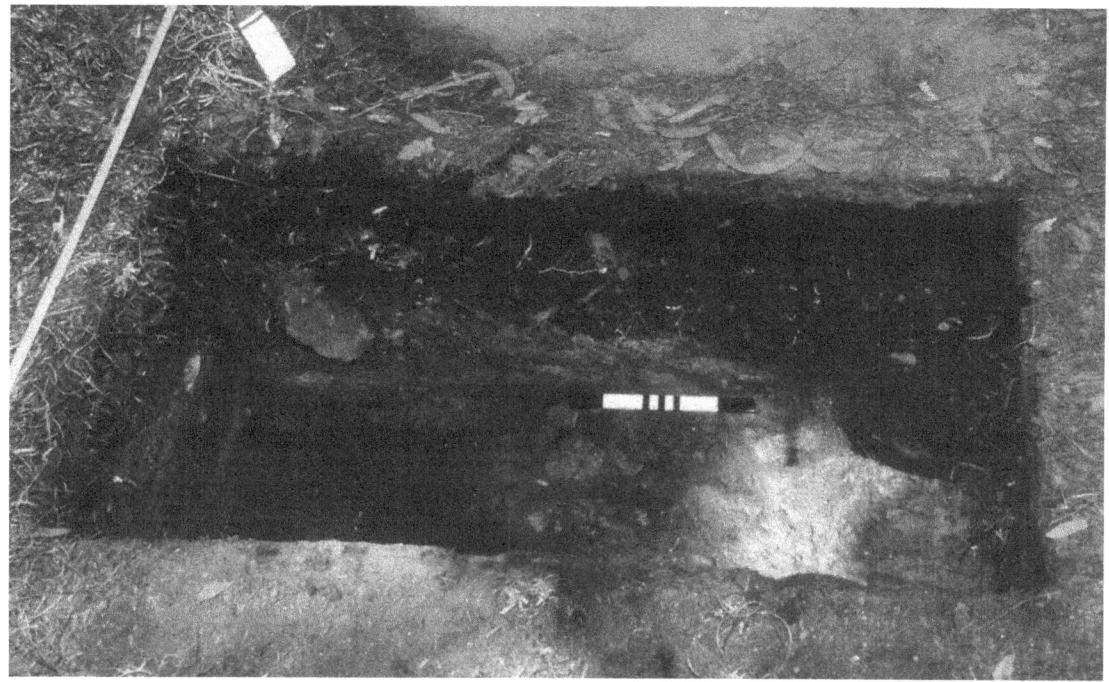

Figure 7.10: The Great Mackerel rockshelter. Square 8 excavated to bedrock. View of southern baulk.

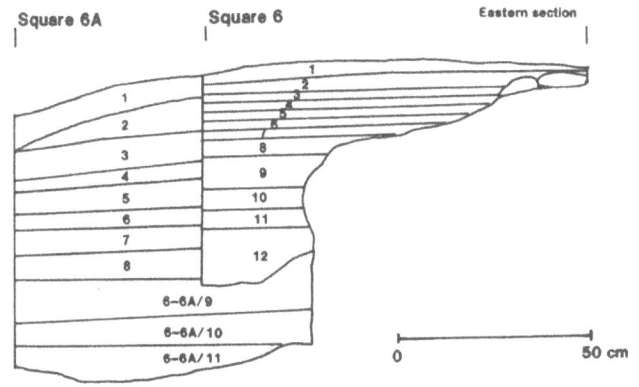

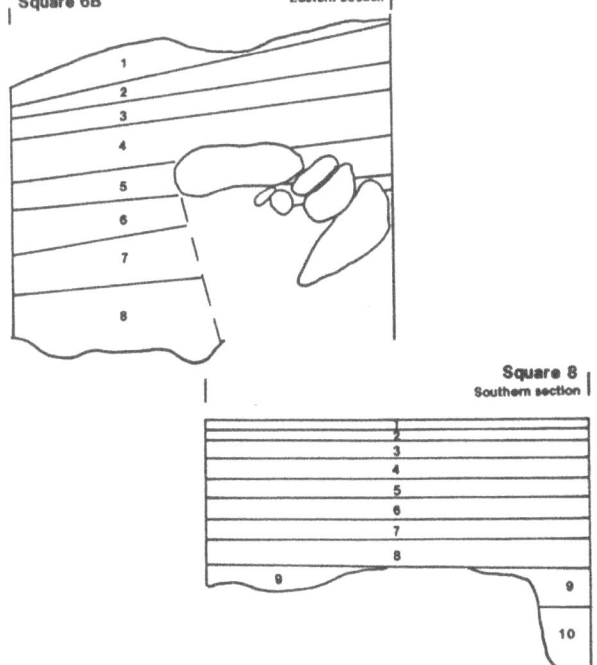

Figure 7.11: Excavated spits, Squares 6 and 8.

An intact hearth was identified in Squares 6 and 6B (10YR 4/2 - 5/2; pH = 8.5). In square 6B a slight variation was observed at the interface between Layers 3 and 4:

> 4a. Lighter grey interface. Same texture as 4 but more homogeneous in colour - 10YR 4/1. pH = 4.5 - 5.

In square 8 a variation to Layer 5 was identified:

> 5a. Light grey/white sandy deposit. pH = 6.

These Layers are depicted in the section drawings (Figure 7.12).

There are two major cultural strata at the site - the midden layer (Layers 1 - 3) and the underlying stone artefact layer (Layers 4 and 5). These can be subdivided on the basis of colour and textural differences resulting from site and soil formation processes. The increase in the acidity of the deposit with depth may account for the decrease in shell content in the lower layers – and certainly the alkalinity of the midden deposit results from the high number of shells and ashy deposit. There are major differences between the two layers in the number, size and type of stone artefacts which cannot be explained by taphonomy or differential preservation. The differences between the two cultural layers are discussed. Due to the relatively small sample sizes, the square 6 spits (6, 6A and 6B) were amalgamated for chronological analysis. Several of the square 8 spits were also amalgamated for comparability with those of Square 6 (Figure 7.11, Table 7.4).

Table 7.4: Correlation of excavated spits (Squares 6, 6A, 6B and 8), analytical units and stratigraphic layers.

Analytical Unit	Square				Stratigraphic Layer
	6	6A	6B	8	
I	1,2	1	1	1,2	1
II	3,4,5	2	2	3	2
III	6,7,8	3	3	4	2
IV	9	4,5	4	5,6	3
V	10, 11	6	5	7	3
VI	12	7,8	6	8	4
VII	*	9	7	9	4
VIII	*	10,11	8	10	5

* Squares 6, 6A excavated together from and below 6-6A/9

Dates

Five charcoal samples were submitted for dating to the ANU Radiocarbon Lab. These have been counted and the following age determinations obtained (Table 7.5; Figure 7.12).

Four samples (ANU- 6370-6373) were collected from square 6: the fifth was from square 6B. All counts but that made on ANU-6373 were from provenanced charcoal samples.

It is considered that ANU-6373 is anomalous. In light of the undisturbed stratigraphy it is considered either, that:

- there has been contamination by recent ground water seepage beneath the large slab of roof-fall in square 6; or;

- the fact that this sample was collected during sieving has resulted in this being contaminated.

The anomalous date is excluded from discussions as it is not considered a realistic indication of age.

Sample ANU-6615 addresses the earliest occupation at the site. Only very small quantities of charcoal and other cultural material occurred below this sample. This date is interpreted as indicating the initial, archaeologically visible, occupation of the shelter. The deposit below this level may represent a very sparse initial occupation of the shelter. As the deposit at this depth is largely decomposing bedrock is could also be the result of soil formation processes and the downwards treadage of sparse early material.

Table 7.5: Radiocarbon dates, depth below surface and association with stratigraphic layers.

No.	Lab No.	Field Number	Depth below surface	Stratigraphic Layer	Age Determination
1.	ANU-6370	6/3/2	6 cm	2	220 + 120 BP
2.	ANU-6371	6/6/1	12 cm	2	480 + 90 BP
3.	ANU-6372	6/9/1	22 cm	3	560 + 160 BP
4.	ANU-6373	6/12	44 cm	4	90 + 2.1%M
5.	ANU-6615	6B/6/1	51 cm	4	3,670 + 150 BP

An age-depth diagram indicates major differences in deposition rates between the midden layer and earlier occupation layer (Figure 7.13). The deposition rate for the basal 20cm has been calculated by extrapolating backwards. This appears to have accumulated over c. 2,000 years. The next c.30cm appears to have accrued over more than 2,500 years. The most likely interpretation is of a sporadic occupation extending over an extremely long time period. This is supported by the relatively low artefact densities (increasing later) throughout.

The age determinations for the site indicate that the midden layer is recent. Shell deposition commenced around 560±160 BP. Midden accumulated at an extremely rapid rate compared with the lower layers. More than half of the deposition during this midden period occurred before 480±90 BP. The site appears to have been most intensively used around 500 years ago. It was, however, probably in sporadic use until European contact. The use of fishhooks is dated to 220±120 BP which accords well with ethnographic accounts and current archaeological data. There are no well substantiated dates for this artefact type predating 840±160 BP (Wattamolla WL/-). Several sandstone fish-hook files have been found in layers dating to much older than this (1,930 ± 80 BP: Curracurrang 2CU5/-; 1,970 ± 80 BP: Currarong 1) but their interpretation is problematic in the absence of better supporting functional evidence.

Artefact accumulation rates (Figure 7.14) also show that the midden period represented the most intensive occupation at the shelter. Analytical units II and III had the highest rates of cultural material accumulation; units III and IV had the highest stone artefact accumulation rates.

Table 7.6: Age depth calculations. Number of artefacts accumulating per 100 year (refer Figure 7.14).

Analytical Unit	Number of Artefacts	Wt Cultural Material (g)	Number of Years	Artefact/100years	Deposit(g)/ 100 years
I	6	362.75	180	3.3	201.5
II	16	1411.55	150	10.7	941.0
III	59	1651.85	180	32.8	917.7
IV	142	1421.3	450	31.6	315.8
V	114	254.4	1,000	11.4	25.4
VI	101	245.8	1,230	8.2	20.0
VII	60	242.9	1,050	5.7	23.1
VIII	23	36.7	1,000	2.3	3.7

Shell

The 29 shellfish species found at the site represent the range of species readily available on the diverse western shores of Pittwater. This shoreline provides a variety of littoral and sub-littoral conditions as a result of the proximity of the estuarine, calm waters of Pittwater and the open sea pounding in through the mouth of Broken Bay. Pittwater contains a number of sandy beaches, interspersed with rocky platformed headlands. The species identified at the site, and the habitats from which they derive, are listed (Table 7.7).

Five shell fish-hooks were recovered during the excavation. These have been made from the small turban shell (*Turbo undulata*) and several pre-forms or blanks were located as well as the complete and broken ones (Figure 7.18).

Proportions of Shellfish Species

Both weights and minimum numbers of shells were counted in order to assess the relative importance of different species. Combined excavation spits/analytical units (Table 7.4) from Squares 6, 6A and 6B are described in these analyses as Square 6 (Table 7.8).

The Sydney Rock Oyster (*Crassostrea*), the limpet (*Cellana*), the hairy mussel (*Trichomya*) and the black periwinkle (*Nerita*) dominate in the comparison of minimum numbers (Table 7.8). These species all derive from rock platforms although some of the *Crassostrea* appear to have been attached to mangroves. In square 6 these account for over 82% of the edible species, while in square 8 they account for more than 84%. The weights for these four dominant species, however, reveal a different picture – these represent only 32.5% (sq 6) and 57.5% (sq 8) of the shell present (Table 7.9). This is due to the differing shell rugosities: the hairy mussel and limpet are very thin, light shells, while the rock oyster and black periwinkle are medium weight shells. Conversely, shells such as *Cabestana* and *Thais orbita* are much larger and heavier, and this results in their far greater presence when considering weight and dietary estimates.

Table 7.7: Shell species identified at the Great Mackerel site (Names from Dakin 1980).

Latin Name	Common Name	Edible Size	Estuary	Mangrove	Rock Platform
Anadara trapezia	Sydney Cockle	X	X		
Pyrazus ebeninus	Sydney Whelk	X	X		
Velacumantis	Small whelk	X	X		
Crassostrea commercialis	Rock Oyster	X	X	X	
Ostrea	Mud oyster	X	X	X	
Chama fibula	Spiny oyster	X	X		
Trichomya hirsuta	Hairy mussel	X	X		
Cellana	Limpet	X			X
Cabestana spenglerei	Spengler's triton	X			X
Thais orbita	Cartrut	X			X
Nerita atramentosa	Black periwinkle	X			X
Turbo torquata	Large turban shell	X			X
Turbo undulata	Small turban shell	X			X
Haliotis	Abalone	X			X
Australocochlea	Periwinkle	X			X
Bembicium		X	X		
Bittium			X		X
Thalotia comtessi			X		X
Placophora	Chiton	X			X
Scutus	Elephant snail	X			X
Pecten fumatus	Cockle	X	X	X	
Tapes waitlingi	Tapestry cockle	X	X	X	
Littorini unifasci	Oyster borer		X	X	
Morula marginalba	Maroon oyster borer		X	X	X
Agnewia triton	Oyster borer		X	X	X
Cardium racketti	Common cockle	X	X		Beach
Stomatella sp					X
Chamaesipho col'a	Barnacle				X

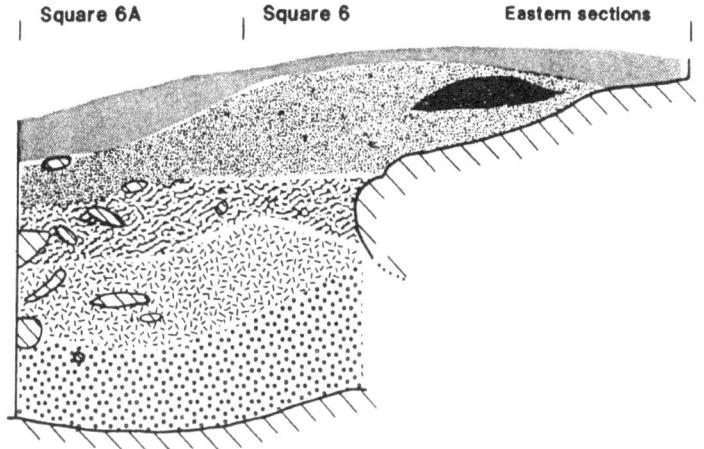

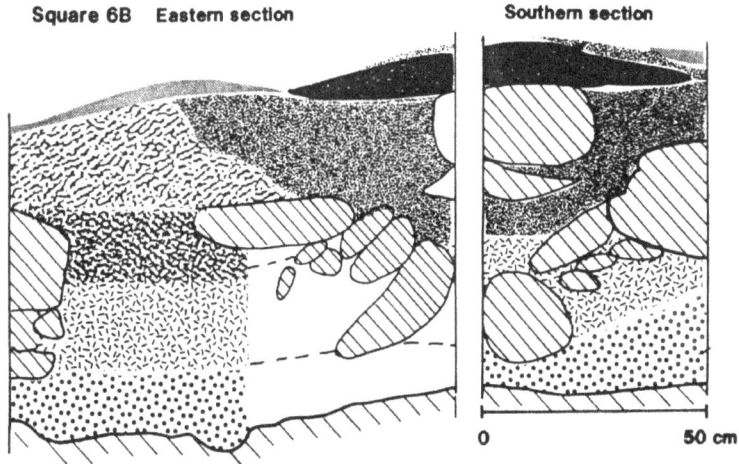

Figure 7.12: Stratigraphic sections, Squares 6 and 8.

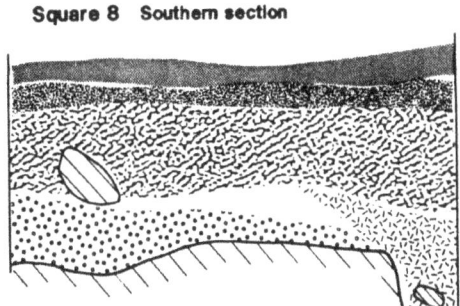

The majority of shell was collected from rock platforms. Muddy estuarine species (*Anadara, Pyrazus, Velacumantis, Ostrea, Bembicium*) accounted for only 11% of the species present. The only identified species which has been collected from an open sandy beach is *Cardium racketti* (the common cockle).

This data suggests that the majority of shellfish collection probably took place below the site, on the rocky platforms on the point at the southern end of Great Mackerel Beach. Today, the nearest muddy estuarine conditions occur at the Basin, over 1.5 km south of the site. In prehistoric times, however, the creek flowing into Great Mackerel and the adjacent flat area behind the beach (where holiday houses have now been built) may well have provided suitable conditions for these estuarine species.

Given the local proximity of a suitable resource and the climb up to the site, it would seem unlikely that shellfish collection took place any further afield. Indeed, it is interesting to speculate upon the reasons which resulted in the collected food been carried up such a steep incline.

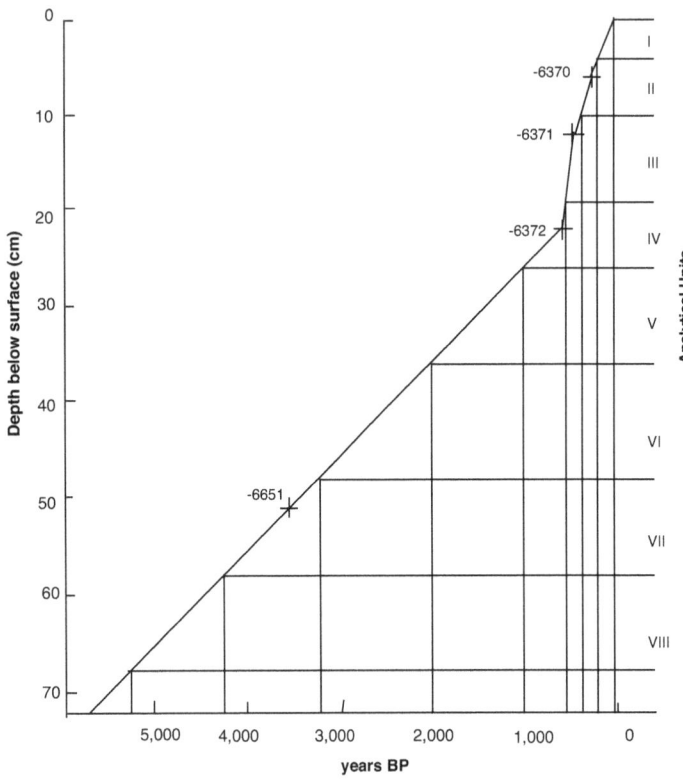

Figure 7.13: Age depth curve based on square 6 spit depths. Dates plotted using ANU lab numbers.

Temporal Variation in Shellfish Exploitation

Dates obtained for the site indicate that the midden layer was produced in a recent relatively restricted time period - between c.560 and c.220 years ago. Because of this, a finer than usual degree of temporal change can be analysed: in many sites, a 340 year time-depth is archaeologically 'invisible'. The environmental and social conditions in the Sydney Region are considered to have been constant during the last 1,000 years, allowing very specific questions to be asked.

What sort of site usage does the deposit represent? Was it a few meals evenly spaced over several hundred years; or alternatively, a more intensive usage for a small period, perhaps interspersed with sporadic use at other times? Does the site represent a base camp focus (Meehan 1982) or is it more likely an occasionally used dinner time camp (albeit some distance from the resource)?

Shell material occurs only in Units I-IV in square 6, while in square 8 some shell also occurs in Units VI and VII (Table 7.9 to Table 7.13). Different taphonomic processes in the two areas of the site are thought to have contributed to this. The general trends in the two squares do appear to be slightly different and these will be discussed separately.

In square 6, the major increase in shell quantities commences in unit IV, peaks in spit III and, to a lesser extent, spit II and then diminishes in spit I. Percentage change in MNI gives a slightly different picture to that achieved by weight, with the former giving a more definite peak in unit III:

Unit	I	II	III	IV	V	VI	VII	VIII
% MNI	6.7	30.3	43.3	16.2	3.5			
%weight	7.3	33.7	34.9	21.3	2.6	0.0		

While the differences between the results achieved by the different methods are minimal, these can be explained in several ways:

1) differential preservation of shells with depth, i.e., the good preservation of shell in spit II has meant that a high number of individuals were identifiable, even though the weight of these shells was not appreciably higher than found in spit IV;

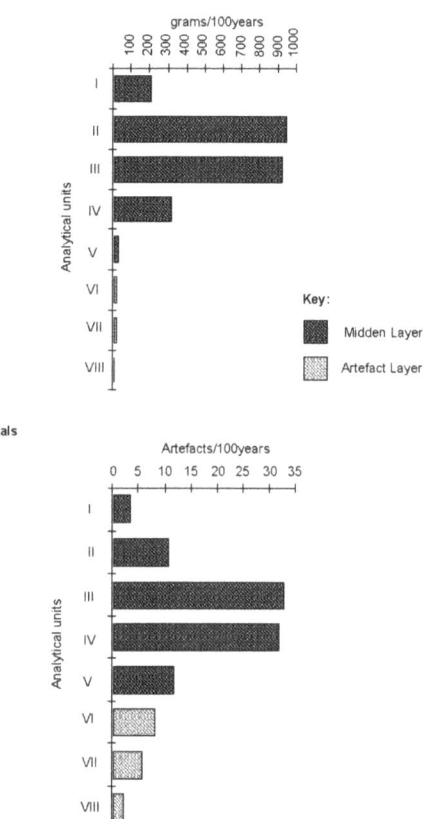

Figure 7.14: Artefact and cultural material accumulation rates per 100 years calculated on basis of age depth curve. Cultural material weights in grams (Figure 7.13).

2) different individual shell weights;

3) that this is too small a time scale to meaningfully subdivide.

Table 7.8: Estimated Number of Individuals (MNI) in Squares 6 and 8.

Species	Square 6	Square 8
Anadara	16	5
Pyrazus	18	-
Velacumantis	39	3
Cellana	167	78
Trichomya	115	9
Crassostrea	326	58
Ostrea	14	7
Cabestana	11	12
Thais orbita	20	2
Nerita	87	57
Turbo torquata	-	4
Turbo undulata	11	-
Bembicium	6	2
Bittium	-	1
Chiton	-	1
Haliotis	-	-
Australocochlea	6	2
Scutus	1	-
Other*	10	4
Total	848	245
Total edible	846	240

*includes *Tapes waitlingi, Chama fibula, Thalotia comtessi, Stomatella, Pecten fumatus, Littorini unifasci, Cardium racketti, Morula marginalba.*

A combination of the first two explanations seems most likely in this case. Recognisable and consistent trends are observable throughout the midden layer. During the 340 years of the site being used as an eating place there are several changes in the dominance of shellfish species. This may indicate either a change in preference of resource zone, or perhaps changing (i.e. declining/increasing) supply.

Size

Most of the minor species remain fairly constant throughout time. This probably indicates that these species were either;

- sparsely (or distantly) distributed,

- relatively undesirable, or,

- species which were not collected and carried away from the resource to be eaten.

Table 7.9: Weight (in g.) of shellfish species in Squares 6 and 8.

Species	Square 6	Square 8
Anadara	139.3	77.2
Pyrazus	270.8	2.5
Velacumantis	93.5	6.9
Cellana	277.5	108.2
Trichomya	207.8	29.8
Crassostrea	897.6	220.2
Ostrea	56.1	35.8
Cabestana	373.0	80.5
Thais orbita	627.5	39.1
Nerita	215.6	157.9
Turbo torquata	-	19.8
Turbo undulata	52.5	-
Bembicium	4.6	1.3
Bittium	-	0.1
Chiton	3.9	5.8
Haliotis	30.1	14.0
Australocochlea	3.8	11.9
Scutus	14.6	-
Other*	40.0	11.8
Unid. fragments	158.9	78.0
Total	3468.1	901.4

*includes *Tapes waitlingi, Chama fibula, Thalotia comtessi, Stomatella, Pecten fumatus, Littorini unifasci, Cardium racketti, Morula marginalba.*

These species include all the muddy estuarine species (*Anadara, Pyrazus, Velacumantis* etc.) as well as many rock platform species (*Cabestana, Thais orbita,* the turban shells, *Australocochlea, Haliotis, Scutus,* Chiton etc.). These species do occur throughout the midden, and indeed in many other middens in the Sydney area and thus the first explanation is probably the appropriate one. Local abundance of the predominant species explains the preference shown in the Great Mackerel midden.

There is a change over time in the proportions of the dominant species. This is most marked in square 6 where *Crassostrea* declines with time and the other three species fluctuate (Table 7.9, Figure 7.15). There appears to be an inverse relationship between *Trichomya* and *Nerita* - while one is present in high proportions, the other is present in lower proportions and vice versa. Conversely, *Nerita* remains fairly constantly around 10% of the shellfish collected, while *Cellana* and *Trichomya* are present much less consistently. This perhaps indicates that *Nerita* was more consistently available (in smaller quantities), while the other species were more sporadic and/or less reliable. Individual meals may well be represented by clusters of same shell species and slight variations on these.

In Square 8 the pattern of shell increase indicates a similar trend although the percentage figures for MNI and weights are slightly different, with the MNI indicating more definite peaks (cf. Table 7.10, Table 7.12).

Here the initial increase occurs deeper in the deposit (units IV and V) and there is a decrease earlier in the sequence, in unit II. This disparity is likely to be taphonomic, square 8 being located in a more exposed part of the shelter and more affected by external soils deposition. Otherwise, the pattern seen in units III - VI (Square 8) is very similar to that of units I - V (Square 6). The small sample sizes (i.e. low shell numbers and weights) in the upper and lower units are acknowledged and the significance of trends is restricted to the units with larger sample sizes (units II-IV; Square 6: units III - V; Square 8).

Spit	I	II	III	IV	V	VI	VII	VIII
% MNI	4.1	1.6	11.8	36.3	40.0	6.1		
% Weight	6.8	3.0	18.4	33.8	31.3	6.3	0.4	

In square 8 the *Crassostrea* declines through time - even in the peak of the midden layer. None of the trends exhibited by the other three dominant species are the same in square 8: *Nerita* is co-dominant with *Cellana* during the peak of the midden layer, while *Trichomya* only increases in the sparse upper layer (units I and II).

Comparing the peak layers within the midden - units IV and V (Square 8) and units II and III (Square 6) further emphasises the dissimilarity of the two squares. The dominant species in square 6 is *Crassostrea* while in square 8 it is *Cellana*. In square 6 the dominance of rock oyster declines,

Table 7.10: Square 6; Minimum numbers: change through time by analytical unit.

Spit	I	II	III	IV	V	VI	VII	VIII
Anadara	4	4	4	4				
Pyrazus		5	10	3				
Velacumantis	2	11	20	6				
Cellana	22	66	57	15	7			
Trichomya	10	78	25	2				
Crassostrea	1	55	171	80	19			
Ostrea	1	6	4	3				
Cabestana	2	3	5	1				
Thais orbita	4	9	2	4	1			
Nerita	7	11	52	14	3			
Turbo torquata								
Turbo undulata		1	6	4				
Bembicium	1	2	3					
Chiton								
Haliotis								
Australocochlea	2	1	3					
Scutus		1	1					
Other	1	4	4	1				
Total	57	257	367	137	30	0	0	0
Total edible	56	257	366	137	30	0	0	0

Table 7.11: Square 8. Minimum numbers: Change through time by analytical unit.

	Analytical Units							
Species	I	II	III	IV	V	VI	VII	VIII
Anadara		1	2	1	1			
Pyrazus								
Velacumantis	2			1				
Cellana	1		2	32	38	5		
Trichomya	3	1	2	1	1	1		
Crassostrea	1	1	14	9	27	6		
Ostrea	1	1	1	2	2			
Cabestana				12				
Thais orbita					2			
Nerita	1		5	28	21	2		
Turbo torquata			1	1	2			
Bembicium	1				1			
Chiton				1				
Australocochlea				1	1			
Other			2		2	1		
Total	10	4	29	89	98	15	0	0
Total edible	9	4	28	89	96	14	0	0

while hairy mussel and limpet increases dramatically. In square 8, the limpet remains dominant while rock oysters decline and black periwinkles increase. Hairy mussel is present in only small quantities in square 8.

There are several possible interpretations for these intra-site differences - temporal, spatial and sampling. As the sequence in square 8 has not been dated the first of these possibilities is not immediately testable (although suitable charcoal samples were collected from this square). Given the distance between the two squares, a spatial explanation is also highly likely. The discard from a single meal (or depositional event) would not be distributed over the full extent of the site, except, perhaps, where there was the clearing of larger material, or some other post-depositional or taphonomic processes. There certainly didn't appear to be any evidence for clearing activity at the site - large and small shells were distributed fairly regularly across the site, and in several instances the remnants of a single meal (particularly limpets) were found stacked inside each other.

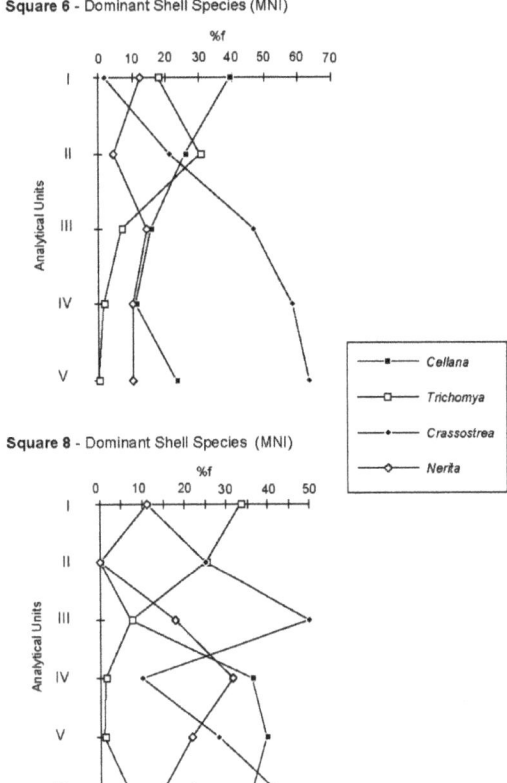

Figure 7.15: Trends in dominant shellfish species throughout the midden layer. Squares 6 and 8.

Table 7.12: Square 6. Shellfish species and other cultural material; change over time by analytical unit. Weight in grams except where indicated.

Species	Analytical Units							
	I	II	III	IV	V	VI	VII	VIII
Anadara	31.3	29.2	41.3	37.4	0.1			
Pyrazus	2.7	83.5	112.1	72.5				
Velacumantis	2.8	31.0	45.6	14.1				
Cellana	35.6	121.3	90.3	17.6	12.7			
Trichomya	14.5	139.7	49.2	4.3	0.1			
Crassostrea	27.6	280.0	390.6	156.1	43.3			
Ostrea	6.0	35.5	6.5	8.1				
Cabestana	39.5	136.7	149.8	47.0				
Thais orbita	38.6	165.5	116.1	299.4	7.9			
Nerita	19.1	40.9	113.1	28.0	14.5			
Turbo undulata	1.8	25.4	14.9	10.4				
Bembicium	0.3	2.5	1.8					
Chiton	0.5	1.1	1.7	0.6				
Haliotis	6.4	17.1	3.7	2.9				
Australocochlea	0.5	2.2	0.4	0.7				
Scutus	1.2	3.0	9.1	1.3				
Other	2.0	17.9	16.2	3.9				
Unid frags	24.9	36.8	47.8	35.4	13.5	0.5		
Fishhooks	0.2	0.8						
Total shell	255.5	1,170.1	1,210.2	739.7	92.1	0.5	0	0
Artefacts	2.5	8.8	41.4	275.4	103.1	144.0	144.5	34.0
Charcoal	32.8	179.1	200.2	116.2	18.6	14.4	3.4	1.2
Bone	4.5	11.9	0.9					
Pigment	0.1	0.4						
Total	295.4	1370.3	1452.7	1131.3	213.8	158.9	147.9	35.2
Deposit (kg)	74.75	80.75	83.75	107.0	84.5	92.35	59.75	86.75

The different material in the two squares may indicate that different people or parts of the social group ate in separate areas of the site (square 8 is on the periphery of the midden, and near the outer edge of the sheltered area). Difference in sample sizes from the two areas (both in terms of areas excavated and number of shells retrieved) is another potential cause of the variability. Indeed, Squares 6A and 6B were excavated to increase the sample size for analysis. The trends demonstrated by analytical square 6 are considered to be more reliable.

Table 7.13: Great Mackerel Square 8. Shellfish species and other cultural components per analytical unit. Weights in grams except where indicated.

Species	Analytical Units							
	I	II	III	IV	V	VI	VII	VIII
Anadara	11.3	7.1	30.0	23.3	4.2	1.3		
Pyrazus	0.3	0.4		1.8				
Velacumantis	5.6		0.8	0.5				
Cellana	0.9		9.5	50.4	42.2	5.0	0.2	
Trichomya	3.2	2.9	15.6	4.3	1.5	2.3		
Crassostrea	3.6	2.9	30.0	61.0	97.3	25.4		
Ostrea		3.4	7.0	18.5	6.9			
Cabestana	6.4	3.6	20.6	25.9	20.0	3.2	0.8	
Thais orbita	7.5	3.2	4.2	3.1	21.1			
Nerita	2.4	0.2	16.7	74.5	57.7	5.3	1.1	
Turbo torquata	3.9		4.2	3.7	8.0			
Turbo undulata				0.3				
Bembicium	0.6				0.7			
Chiton			0.8	5.0				
Haliotis	1.1		7.1		3.7	2.1		
Australocochlea				0.4	5.3	5.8	0.4	
Other			4.8	4.6	1.5	1.0		
Unid frags	14.4	3.4	14.8	27.6	11.7	5.2	0.9	
Total shell	61.2	27.1	166.1	304.9	282.1	56.6	3.4	
Artefacts	0.8		17.8	46.6	32.7	25.0	8.7	1.5
Charcoal	5.0	13.1	14.8	5.7	6.4			
Bone	0.3		0.5	0.2	1.3			
Pigment								
Total	67.3	40.2	199.2	357.4	322.5	81.6	12.1	1.5
Deposit (kg)	24.8	25.3	43.5	39.3	41.3	31.75	21.5	23.5

Table 7.14: Squares 6 and 8. Dominant shell species - peak midden units.

	Square 6	%f	Square 8	%f
Cellana	123	19.7	70	37.8
Trichomya	103	16.5	2	1.1
Crassostrea	226	36.3	36	19.5
Nerita	63	10.1	49	26.5
Total edible	623	(82.6%)	185	(84.9%)

The Bone

Faunal material was found in the midden layer in all excavated squares (Table 7.11 and Table 7.13). It occurred only in very small quantities (total weight 19.6 grams). While several different genera were recognisable - identification to species level was mostly impossible. Fish (bones, vertebrae, spines, two otoliths and one scale), small unidentifiable macropod (fragments), small reptile and bird bones were recovered. The fish was a Snapper (*Chrisophrys auratus*) identified by comparison with Snapper bones/otoliths at Angophora Reserve (Wood 1992). The fish bones appear to represent a single individual. The two otoliths are identical in size; one is from the right and the other from the left side. The reptile recovered was probably a smallish skink.

The presence of bone coincides with higher densities of shell and those spits where greater shellfish variety occurs. From the (small) sizes of the animals present and species of shell involved this may indicate a generalised dependence on a wide range of food resources and resource zones - both in littoral and on the forested ridgetop.

All the food represented by the deposit may have been collected by women, and the site may have been used for food consumption almost exclusively by women – based on ethnohistoric sources (Bowdler 1976, Collins 1798[1975], Phillip 1789[1970], Worgan 1788) and ethnographic analogy (Hiatt 1965, Meehan 1982). Fishhooks were also reported in Sydney region to be women's fishing apparel, and the presence of fishhooks in the deposit supports an interpretation of the site as one used mainly by women. This proposition is discussed further below.

Dietary Estimates

A calculation of the energy content of the excavated food remains informs us about the dietary significance of the deposit. Energy content is calculated using a figure of 30% as a proportion of the total weight of shellfish which was meat - and 65 kcal/100g meat weight (Shawcross 1967). Meehan's figures for these same calculations are 21% (flesh/shell weight) and 80 kcal/100g (Meehan 1982:143). From the site's excavated shell total (4,370g) between 1,000g (following Shawcross) and 918g (following Meehan) of shellfish meat could have been expected. An average daily calorific requirement of 2,200 kcals/day is taken as representing the number of kilocalories each member of a family group requires per day (Bailey 1975; Meehan 1982; Shawcross 1967; Wood 1989). The meat weight from the excavated squares would have provided a total of between 852 kilocalories (Shawcross) and 734 kilocalories (Meehan): insufficient (alone) to support one person for a single day.

The (largely unspeciated) faunal remains at the site were one small snapper (say 0.3 kg live weight; 60% edible; 13,070 kcal/edible kg), a small macropod (say 1.1 kg live weight; 75% edible; 20,700 kcal/edible kg) and a small lizard (say 0.1 kg live weight; 70% edible; 525 kcal/edible kg). These three individuals provided as much as 19,500 kcals (based on figures calculated in Wood 1989:76-83, following Meehan 1982:147). The faunal remains, as insignificant as they seem (weight wise) in the cultural deposit may have contributed more than 26 times as many kcals as the shellfish remains present. These findings concur with previous archaeological analyses [Wood's (1989) findings at the Angophora Reserve site and Bailey's (1975) results from NSW north coast middens]. They are also predictable in light of Meehan's ethnographic data which indicates that shellfish contributed between 0.3% and 14% of the daily calories available for each person (Meehan 1982: 144).

Extrapolating on the basis of the total floor area at the site (i.e. approx 44 sq m) - sheltered and with depth and headroom - and the percentage of the deposit excavated (1.75 sq m = 4.0%), the projected total weight of shell at the site is in the order of 108.6 kg, while the potential edible meat weight from fish and land animals would be in the order of 34 kg. From this much shell, between 32.6 kg/22.8 kg (Shawcross/Meehan) of shellfish meat would have derived. This amount of shellfish meat would have provided between 21,190 and 18,240 kilocalories: sufficient for an adult living only on shellfish for a total of slightly more than nine days. From the projected meat weight from fish and land animals a calculated 628,983 kcal would have been provided; sufficient for a single person for 285 days. A family group of four (two adults, two children), then, could have been supported by the estimated dietary remains at the site for approximately 73 days (2.4 months).

Given that the economic remains at the site would not represent all food consumed in any one day (based on our understanding of hunter gatherer foraging patterns), the fact that we do not know how much shellfish meat was eaten by women as they collected, and the fact that we do not know which other sites in the locality were being used at the same time and were complementary to the Great Mackerel site, such calculations are relatively meaningless. However, the food remains at the site can be interpreted as small scale. The dietary remains at the Angophora Reserve site on the opposite site of Pittwater (McDonald 1992a) were found to represent the intake of 'one person [eating] continuously for 6,303 days (17 years). Alternatively it would take 1,260 days (3.5 years) for a family group of five to consume the amount of energy represented' (Wood 1989: 82). These calculations certainly place the site in perspective, in terms of its relative importance as an eating place.

Stone Artefacts

A total of 511 stone artefacts were recovered from the site. The majority of these, particularly in the upper layers, are undistinguished, split quartz pebbles. Bipolar was the predominant technique used here. No backed or ground implements were present amongst the assemblage. Only thirteen artefacts (2.5%) were found to have evidence of retouch and/or usewear (R/U).

Raw Material

Quartz was the dominant raw material accounting for 55% of the total artefact assemblage. Silicified tuff is the next most important (14.7%) with quartzite, chert, silcrete and volcanic being common (<10%). The remaining materials present are quite rare (<1%). These include fine grained basic (FGB), fine-grained-siliceous (FGS), siltstone and 'other' (Table 7.15).

Table 7.15: Great Mackerel, all squares. Raw materials and artefact totals.

Material	6	6A	6B	8	TOTAL	%
Quartz	32	92	105	52	281	55.0
Silicified Tuff	6	16	41	12	75	14.7
Quartzite	5	6	27	8	46	9.0
Chert	5		24	3	32	6.3
Silcrete		11	12	1	24	4.7
Volcanic	8	7	11	3	29	5.7
FGB	1			1	2	0.4
Siltstone	1		3		4	0.8
FGS	1				1	0.2
Other	3	3	8	3	17	3.3
Total	62	135	231	83	511	(100.1)
%	12.1	26.4	45.2	16.2	(99.9)	

Size

The vast majority of the assemblage (96.5%) was <3cm long (maximum dimension). Of these, 13.5% were <1cm long. Very few artefacts (14 - 0.2%) were between 3-5 cm long, while one only was >5cm in length.

Artefact Types

The assemblage consisted primarily (90.8%) of debitage. Thirteen artefacts had evidence of retouch/usewear (R/U) and the 34 artefacts were identified as cores. No diagnostic artefact types were located in the assemblage. All pieces with R/U consisted of modified flakes (38.5%) or flaked pieces (61.5%).

The absence of backed blades may well be a sampling issue. This artefact type usually represents between 0.2-0.9% of shelter site assemblage (Attenbrow 1987, McDonald 1994) and 0.2-3% of open site assemblages (JMcD CHM 2005a). Given these usual proportions, the expected number of backed blades at Great Mackerel would be between 0.3 and 1.5 in the midden layer and 0.6 and 2.9 in the artefact layer. The absence of this diagnostic artefact type cannot necessarily be used to characterise nor to infer the age of this assemblage. Similarly, little interpretative value should be put on their absence (e.g. in terms of gender specific tool use).

Almost all (94%) of the cores retrieved were quartz and the majority of these (62.5%) were bipolar. The other two cores were quartzite bipolars. There were eight quartz multi-platformed cores, and four quartz single-platformed cores.

The absence of cores in the other raw materials - particularly silicified tuff - suggests that knapping of these other material did not occur on site. Again, the low sample size for the assemblage generally makes such conclusions tenuous. Figures from Mangrove Creek show that between 0.5-1.% is the normal expected number for hand held cores (only one site – Loggers shelter - had a higher percentage frequency - 3%: Attenbrow 1987). Thus at Great Mackerel, given that there were 75 silicified tuff artefacts, the expected frequency of cores here would have been 0.3-0.75 (or 2.25 if at Loggers). More recent investigations of Pre-Bondaian assemblages on the Cumberland Plain – these dominated by silicified tuff assemblages – indicate that core represent only c.1% of overall assemblages. This is taken as evidence for different procurement strategies operating, and different gearing up strategies. Again, the low core numbers retrieved from the lower level is likely to be a sampling issue.

The 13 artefacts with evidence of retouch and/or usewear, represent too small a sample to meaningfully analyse in detail (Table 7.16). Most of these artefacts occurred in the units with the highest artefact frequencies. Quartz was not the most used material (contrary to general assemblage proportions). Silicified tuff and siltstone were the most used material, followed by silcrete, volcanic and chert (Figure 7.16). Only one quartz artefact had R/U.

Most of the R/U consisted of fine scalar flaking, although heavy and notched edges were also observed. Most (61.5%) retouched pieces were on flaked pieces, the others were modified flakes. Most (77%) of these were in the 1-3cm size category, the remainder were in the 3-5cm category.

Five artefacts were observed to have macroscopic residues present (these have not been analysed in more detail). Four of these were otherwise unmodified quartz artefacts, and one was a (macroscopically) unmodified silicified tuff piece. The high proportion of quartz present in the assemblage, but the apparent low rates for use of this material, combined with the presence of residues on what appear to be otherwise unmodified quartz pieces is suggestive of this material being used for tasks which leave no forms of macroscopic usewear evidence. The proliferation of quartz bipolar in later ERS assemblages has in part been attributed to the shattering qualities of quartz (Hiscock 1986). The generally low rates of use for this material have, however, been extensively documented (Attenbrow 2004, McDonald 1985b, McDonald *et al.* 1994). A possible explanation for this pattern is the use of this raw material to process plants (see below).

Table 7.16: Artefacts with Retouch/usewear. Raw Material and vertical distribution.

Analytical Unit	Quartz	Silcrete	Silicified Tuff	Volcanic	Siltstone	Chert
I						
II						
III						
IV			1	1	2	
V						1
VI	1	2	1	1		1
VII			1		1	
VIII						
Total	1	2	3	2	3	2

Change through time

As with the shell material, the excavated spits from all squares were amalgamated into analytical units for the purposes of investigating changes through time. Data from Squares 6, 6A and 6B were grouped here as 'Square 6'. Because of the relatively small number of artefacts, change over time was analysed in terms of stratigraphic phases: the midden and the artefact layers. Changes in raw material, artefact size and type were analysed (Table 7.17).

There are twice as many artefacts in the lower layer as in the midden layer. Given that the vast majority of the identifiable cultural material is located in the midden layer, this variation is considered important. It is notable, however, that the artefact accumulation rates (Figure 7.14) are actually higher in the Midden Layer than in the Artefact Layer.

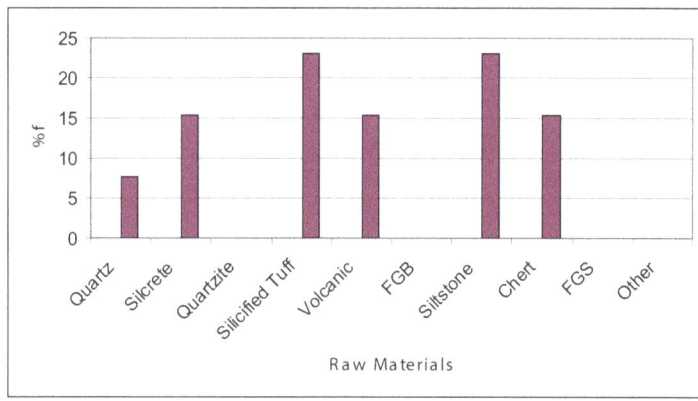

Figure 7.16: Great Mackerel: Artefact with retouch/usewear. Raw material preferences.

Quartz predominates in both layers, accounting for over 50% of each assemblage. The major differences in raw material usage over time are the decrease in usage of Silicified Tuff and silcrete in the midden layer with the increase in usage of chert, and to a lesser extent, siltstone.

The proportions of artefacts with R/U are notably higher in the artefact layer (61.5%) than in the midden layer (Table 7.17). Given sample sizes little can be said on patterns identified in this artefact class. The constriction in the range of raw material used in the midden layer may result from the smaller assemblage.

The assemblage's characteristics (an absence of backed and ground artefact types and the preponderance of quartz and the bipolar technique) suggest it could (in the absence of dates) either be classified as late Bondaian and less than 1,500 years old or early Bondaian in age (between 3,500 and 5,000 years old). In fact, the age determinations received *are* Early and Late Bondaian in age.

The minor changes in raw material preference over time suggest some technological changes are being demonstrated. Similar trends have been demonstrated in larger shelter assemblages [e.g. Loggers, Mussel etc. (Attenbrow 2004); UDM and Mt Yengo (this volume); Cherrybrook (McDonald 1985b); Angophora Reserve (McDonald 1992a)]. The relative absence of stone tools in the midden layer suggests that stone working was not a major activity at the site in its later phase of use (cf. Angophora Reserve). This may relate to the availability of raw material resources (i.e. through trade access: Kohen 1986), or to a preference for shell and bone tools (as suggested by the ethnohistoric literature).

Table 7.17: Squares 6. Changes in Raw Material, per cultural layer.

Material	Midden layer	%	Artefact layer	%
Quartz	93	55.7	136	52.1
Silicified Tuff	14	8.4	49	18.8
Quartzite	15	9.0	23	8.8
Chert	22	13.2	7	2.7
Silcrete	3	1.8	20	7.7
Volcanic	9	5.4	17	6.5
FGB	1	0.6	-	
Siltstone	3	1.8	1	0.4
FGS	1	0.6	-	
Other	6	3.6	8	3.1
Total	167	100.1	261	100.1
%	39.0		61.0	

Table 7.18: Square 8. Changes in Raw Material, per cultural layer.

Material	Midden layer	%	Artefact layer	%
Quartz	10	52.6	42	65.6
Silicified Tuff	4	21.0	8	12.5
Quartzite	-	-	8	12.5
Chert	2	10.5	1	1.6
Silcrete	1	5.3	-	-
Volcanic	1	5.3	2	3.1
FGB	1	5.3	-	-
Siltstone	-	-		-
FGS	-	-		-
Other	-	-	3	4.7
Total	19	100.0	64	100.0
%	22.9		77.1	

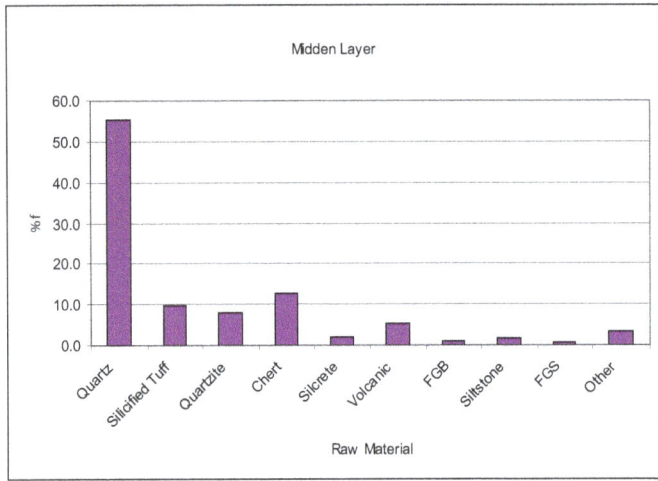

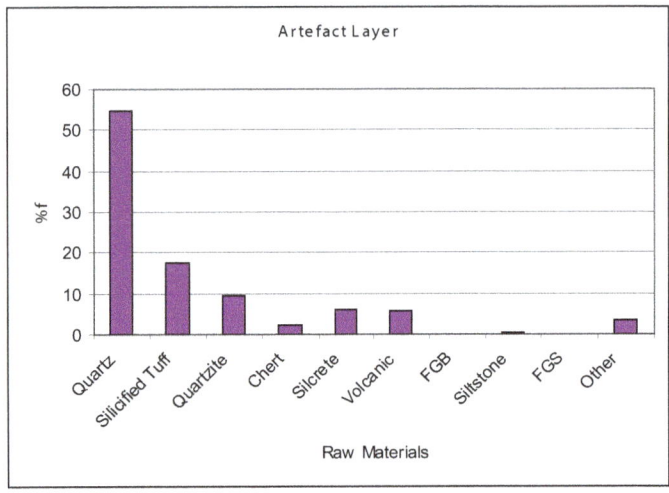

Figure 7.17: Raw Material proportions in the Midden and Artefact Layers. Combined squares.

Shell Artefacts

Five shell fish-hooks were recovered from the midden layer in Square 6 (two from spit 6/3, one each from spits 6/1, 6A/1 and 6A/2). These were made from the small turban shell (*Turbo undulata*). Several pre-forms or blanks were located as well as the complete and broken ones (see Figure 7.18). This artefact type is well known both from ethnohistoric references (e.g. Tench 1793, Collins 1798) and the archaeological literature (e.g. Bowdler 1976, Lampert 1971a, Megaw (ed.) 1974 and Clegg 1979).

As well as these relatively well known artefact types, seven shell scrapers were also recovered from the deposit. These consisted of *Anadara* and *Tapes waitlingi* specimens with usewear along their distal margins (see Figure 7.19). All shell artefacts derived from the upper spits of the midden and were found only in Square 6 (Table 7.19).

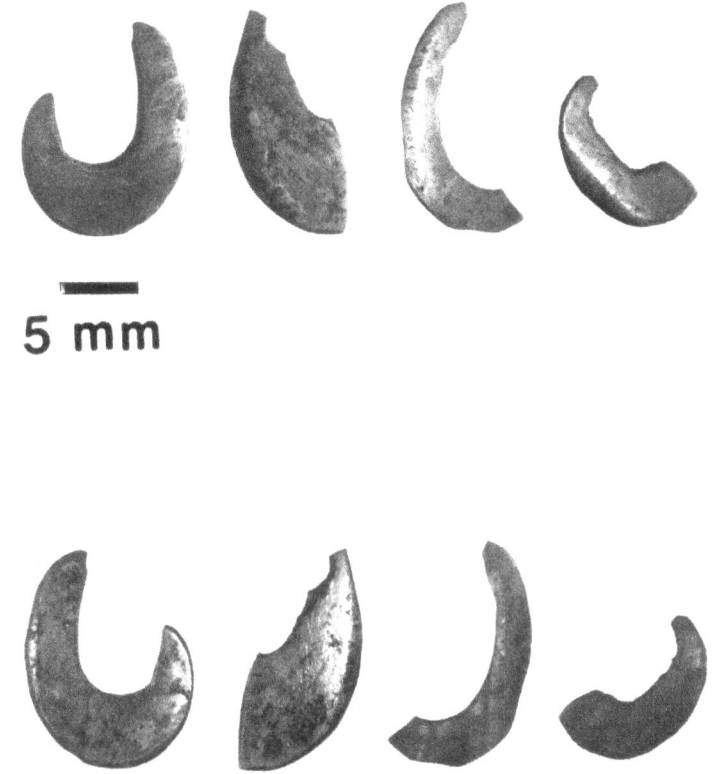

Figure 7.18: The Great Mackerel rockshelter. Excavated fishhooks (finished) from Squares 6 and 6A. Examples shown front and back (from left to right) from spits 6/3, 6A/2, 6/3 and 6/1.

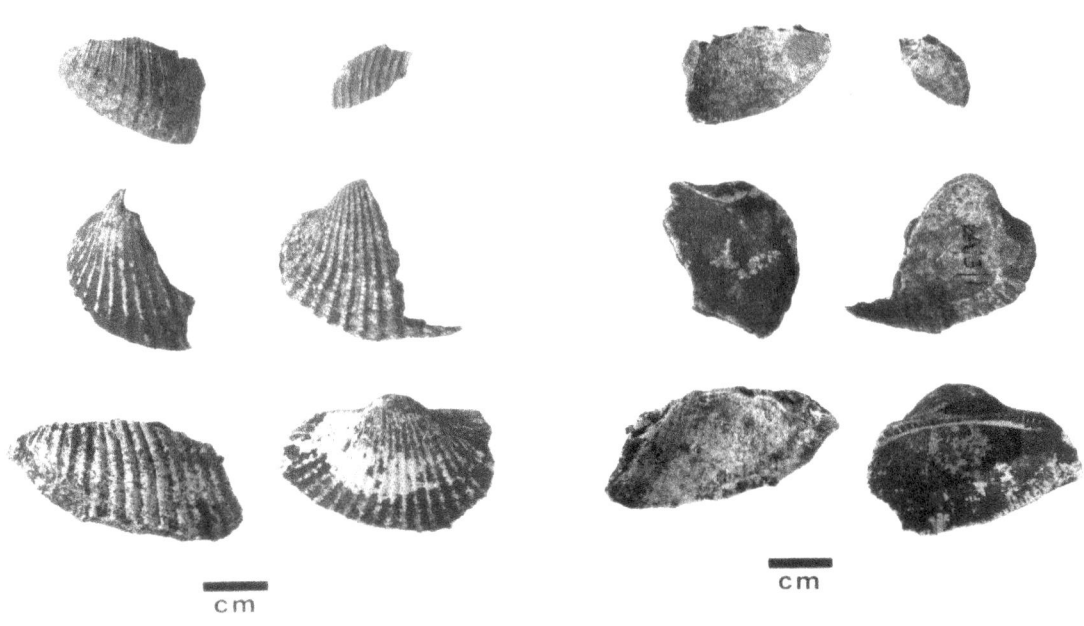

Figure 7.19: Great Mackerel shell scrapers. Examples shown front and back (from left to right, top to bottom) from spits 6B/2, 6A/1, 6B/1 and 6A/3, 6B/2 and 6B/1.

Table 7.19 Great Mackerel, Square 6. Distribution of shell artefacts.

Archaeological Unit	Square		
	6	6A	6B
I	H	HS	2S
II	-	H	2S
III	2H 2B	S	-
IV	S	-	-

H = Hook   B= Fishhook Blank   S = Scraper

Table 7.20: Great Mackerel. Distribution of pipeclay throughout the archaeological sequence (weight in grams).

Archaeological Layer	Square			
	6	6A	6B	8
Midden	0.4	-	0.1	-
Artefact	-	-	-	-

The Pigment

A very small quantity (0.5g) of white pipeclay was located in the excavated deposit (Table 7.20). The pipeclay is very fine grained and of good quality. All fragments are small and there is no evidence of faceting - probably due to their small size. It is unlikely that the pipeclay was used for drawing at the site, since there are no striations on this material. Given the extensive use of white pigment in the art assemblage it is likely that these fragments are evidence of pipeclay preparation for stencilling. As this material does not derive naturally from the shelter's ceiling or surrounds, it has been included in the analysis of the site's cultural material.

Also found at the site was a quantity (130g) of red/brown roof-fall (sandstone and ironstone) which produces a stain when rubbed on paper. This is of particularly poor quality as far as ochres are concerned, being very sandy and gritty. This material is not considered ideal pigment nor is it considered likely that this was used to produce art at the site. It is not faceted (either from direct contact with the wall or during pigment preparation). This material is not the same colour as any of the art i.e. it is a dark brown compared to the (red) stencils recorded at the site. Because of these factors, this material has not been included in the analysis.

The pigment found in the deposit was inconclusive in substantiating the contemporaneity of two separate art phases with the two separate occupation phases at the site. While white pigment was only found in the midden layer (and not below Unit II, Square 6's; Unit V, Square 8), there was an absence of red pigment in the deposit.

White pipeclay is not considered to survive well in archaeological, especially acidic, deposits (Morwood 1979). It could, therefore, have been present in the lower layers but not survived the processes of time. Secondly, the absence of red pigment means that there can be no correlation of the red art at the shelter with either component of the occupation deposit. The following interpretations of this result are possible. Either:

1) the red stencilling predates (or possibly correlates with) the earliest occupation evidence in the shelter, while the phase characterised by white stencils coincides with the midden layer; or,

2) the white stencilling phase took place throughout the entire usage of the shelter as an occupation site; or,

3) it is not possible to correlate the art with the occupation evidence, since the presence of pigment in the deposit is sparse and all the materials employed in the production of the art are not represented in the deposit.

While explanations 2) and 3) are not refuted by the site's evidence, explanation 1 is the preferred option.

Discussion and Interpretation

Discussion here is restricted primarily to the specific aims of the excavation. Further discussion of the site and the interpretations of its evidence have been made elsewhere (McDonald 1992b).

As stated above the aims of the excavation were threefold. These were, to determine:

1) the nature of the occupation evidence at the site;

2) if (as suggested by the art) there were two phases of occupation at the site; and,

3) whether the art was contemporaneous with the occupation deposit.

All three aims have been addressed by this report, the third albeit inconclusively.

Two phases of occupation were identified at the site, an early Bondaian phase with no shell (dating from around $3,670 \pm 150$ BP [ANU-6615]) overlain by a late Bondaian midden layer, dating to the last 600 years [ANU-6373, ANU-6370].

The earlier occupation layer contained stone artefacts only. These were fairly sparsely distributed, mainly undiagnostic quartz and silicified tuff debitage. While little knapping of artefacts appears to have occurred at the site at this time (no cores), stone tools appear to have been used for a variety of tasks. The more recent layer contained shell material, faunal remains, stone artefacts, white pipeclay and shell fish hooks and scrapers. By weight, shell was the predominant cultural component (74%), followed by stone artefacts (15%), charcoal (10%) and pipeclay and faunal material (<1%).

Slightly less than 4.5kg of shell was retrieved from the deposit, yielding a minimum number of 1,093 individual shells. Most of these shells (1,086: 99.4%) were of edible size and species. A total of 29 molluscan species were identified from the assemblage. Six of these were considered to be inedible largely due to their minute size (four varieties of oyster borer, plus worm tubes and barnacles). These species amongst the assemblage reflects the parasitic nature of these smaller species and shellfish collection techniques rather than dietary preference.

Dominant molluscan species at the site include *Crassostrea, Trichomya, Nerita,* and *Cellana. Cabestana, Thais* and *Anadara* are also relatively common. The range of shellfish present at the site reflects the range of species readily available on the rock platforms, beaches and estuarine conditions along the north western shore of Pittwater.

A very small amount of bone (19.6g) was recovered from the site. This included a single Snapper, a small unidentified macropod and a small lizard. Estimates of dietary contributions indicate that these faunal remains while insignificant weight-wise, would have provided more than 25 times the calories of the excavated shellfish remains.

A minute quantity of pigment (white pipeclay) was recovered from the midden layer. This could be interpreted as indicating that the most recent phase of art at the site (predominantly white stencilling) coincided with the most recent occupation – in the last 600 years. While there are problems in firmly correlating the two occupation phases with the two phases of art, the contemporaneity of the midden layer with the more recent art assemblage is posited. The cultural remains in the midden layer, in particular the fishhooks, shellfish, fish and small land animals is interpreted as indicating the presence of women at the site. The more recent art phase also suggests the presence of women. This art assemblage includes stencil compositions of women's and baby's (sized) hands and digging sticks.

The older occupation layer represents sporadic use of the shelter and may well indicate a different pattern of land use. This earlier occupation post-dates the sea-level rise (between 7-5,000 BP) and the environment during all phases of the site's occupation can be considered comparable. The absence of organic remains in the deeper deposits is inconclusive but would appear to relate to the acidity of these layers – in the absence of midden deposit. Differences between the stone artefacts in the two phases, however, are quite clear. Possibly this site was first used as a shelter by men, while hunting on the ridge tops. It is tempting to correlate this earlier occupation which comprises only stone tools with the older phase of art at the site, which consists mainly of very large (men's?) red hand stencils.

But what does the more recent use of the shelter represent? What type of site is suggested by the archaeological remains and the art? How does this site 'fit' into our understanding of prehistoric Sydney at European contact?

The most intensive period of occupation at the shelter took place between 220 and 480 years ago. The archaeological evidence, including calculations of the dietary significance of the shellfish and faunal remains, indicate that the site was probably neither a dinner time camp nor (*per se*) a home base camp (following Meehan's 1982 definitions). While being aware of the archaeological isolation of the Great Mackerel site, it is possible to test Meehan's definitions in a limited fashion. The site is too far from the shellfish collection area and permanent water, and contains the remains of too many species (shellfish and other), for it to be a dinnertime camp. On the other hand, it is too small (physically) to accommodate a territorial group. The occupation deposit does contain a wide range of species in relatively sparse quantities, which could support its identification as a home base. The dietary remains and dates indicate a total of 285 person days of occupation over a period of 340 years. While it is recognised that the food remains present would not represent evidence for the entire daily consumption, the site contains significantly less food debris (both variety and quantities) than the Angophora Reserve Shelter. While the two are not comparable (temporally), the differences in occupation evidence are clear. The Angophora Reserve site has been interpreted as a home base camp (McDonald 1992a; Wood 1989).

The archaeological evidence from the Great Mackerel site does suggest a semi-permanent site for a smaller group. It could represent the base camp for a family group - or a foraging group - over a longer period of time. It could also have been a regular dinnertime camp during times of inclement weather. The fact that no ethnohistoric observations were made with regard to the function of art, or about its production means that this site's function cannot be distinguished on the basis of it being decorated. This research has shown that 65% of occupation shelters in the Sydney region also have art, and that the presence of art need not necessarily imply a non-domestic function. Indeed, in the Ku-ring-gai area most of the shelter art sites are located at the bases of cliffs and hillslopes, adjacent to the maritime resources and the primary resource zones where all members of the local group are likely to have spent most of their time (McDonald 1991).

There is some evidence in the Great Mackerel art for a mixed group (not just women) being present during the more recent art's production; viz. the stencil composition of boomerang held to the wall by a pair of large men's hands. This evidence supports either model: of a shelter for the gatherers during the day, or a base camp for a family group. Ethnographic work (cf. Meehan 1982, 1988) and ethnohistoric references from Sydney both indicate that shellfish collecting parties comprised women, children and sometimes old men 'the Midshipmen met with a very old man and woman and two small children; they were close to the waterside where several more were in their canoes gathering shellfish ' (Beaglehole 1768-71[1955]: 309). Meehan identifies that foraging groups often comprised men, women and children equally as commonly as groups which contained just women and children (23.7% each). The men in these groups were said to be 'usually middle aged or elderly, whose failing eye sight and declining agility prevented them from successfully pursuing normal male hunting activities' (Meehan 1988: 173).

Whatever the function of the site, two forms of archaeological evidence, art and occupation deposit, indicate the presence of women occupants. It is on this basis that the contemporaneity of the most recent art and midden layers is concluded.

8

EXCAVATIONS AT UPSIDE-DOWN-MAN

This chapter details the excavation of Upside-Down-Man (UDM) shelter in the lower Mangrove Creek catchment, north of the Hawkesbury River (Figure 8.1). The site contains an extensive pigment art assemblage with several phases of production. Also present are all three shelter engraving types (pecked, Sydney miniature and abraded), thought to be diachronic indicators in shelter art production (McDonald 1991). Dripline scours initially indicated that there was occupation evidence at the site. The site was test excavated in September 1991. A total of 12 person days was spent testing the site.

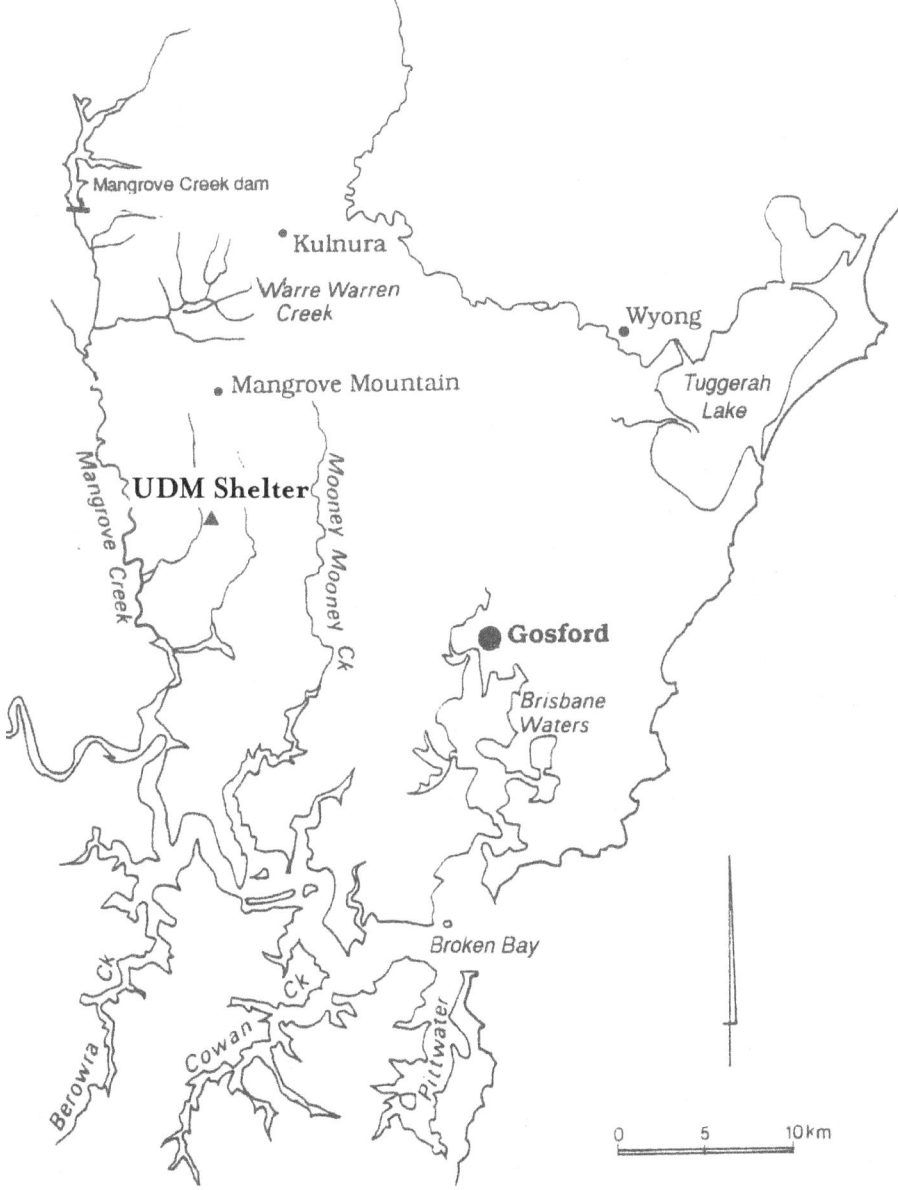

Figure 8.1: Locality Map. UDM Shelter in its local context.

Environmental Context

The site is located on the hillslope 300m from the confluence of Ironbark Creek and one of its minor eastern tributaries. The site is 50m from and 20m in elevation above the tributary creek. There is a spring c.50m upstream of the shelter on the tributary creek and a smaller spring at the southern end of the shelter. While the understorey and surrounding vegetation were extremely dry at the time of the excavation, both of these springs were producing water - in sufficient quantities to feed a waterfall.

The confluence of Ironbark Creek and Mangrove Creek is less than 10km upstream of the junction with the Hawkesbury River. Tidal estuarine mudflats occur in the lower reaches and at the mouth of Mangrove Creek less than 3km from Ironbark Creek (Figure 8.1).

The shelter bedrock is of the Hawkesbury Sandstone formation and the shelter is surrounded by dry sclerophyll forest. Dominant tree species include a variety of *Eucalyptus* spp. and *Angophora* spp., while the mid-storey contains many wattles, Geebung (*Persoonia* spp.) and grevilleas.

The Site

The shelter is large and cavernously weathered, measuring 17m x 6m x 3.5m (Figure 8.2, Figure 8.3; Figure 8.4). It has a westerly aspect. The archaeological components of the shelter include art (engravings, paintings, drawings and stencils) and occupation deposit (including lithics, ochre, faunal and floral remains, shellfish and charcoal). Outside the shelter on a sloping sandstone ramp there are numerous grinding grooves. On the platform roof above the shelter there is an engraved macropod.

The Art

A moderately large assemblage of 274 motifs was recorded within the shelter. Most of these are black, red and/or white drawings (Table 8.1). A complex range of artistic techniques has been employed at the site. One simple-non-figurative motif on Panel 4 is polychrome (black, white and yellow). Of the 58 technique variables identified in detailed diachronic work in Mangrove Creek, 36 are present at Upside-Down-Man (Table 8.2). Of the 66 sites recorded in detail around the Mangrove Creek area, in Upper Mangrove Creek (Attenbrow 2004, Gunn 1979) and in the Warre Warren area (McDonald 1988a), only one shelter has a larger assemblage. That site is Swinton's, with a motif total of 857 including 575 hand stencils.

Stencils

Fifty-five stencils were recorded amongst the UDM assemblage. Of these, 36 were of hands and 18 were of hand variations (hand and wrist, hand and arm, finger manipulation). The majority was white (90%) with the remainder red (7%) or yellow (2%).

Most (67%) of the 36 hand stencils were of left hands, a fair proportion (22%) were indeterminate and four only were of right hands. Only half of the 18 hand stencils variations were of left hands while a larger proportion (39%) was of right hands. Only two of these were indeterminate. While for a fair proportion of the stencils 'handedness' is no longer recognisable (owing to pigment flaking, fading etc.), left hands appear to have been stencilled more often than right hands.

This pattern generally holds for the white stencils [(67%) were of left hands, (18%) were of right hands and (16%) were indeterminate] but not for the minor colours. Half of the four red stencils were indeterminate and there was one each of the right and left hands. The one yellow stencil was of a right hand. While these coloured stencils provide too small a sample to make much of comparative preferences, the nature of the red stencils indicates a greater antiquity for this technique (i.e. red stencilling). The pigment in these examples has bonded with the rock and

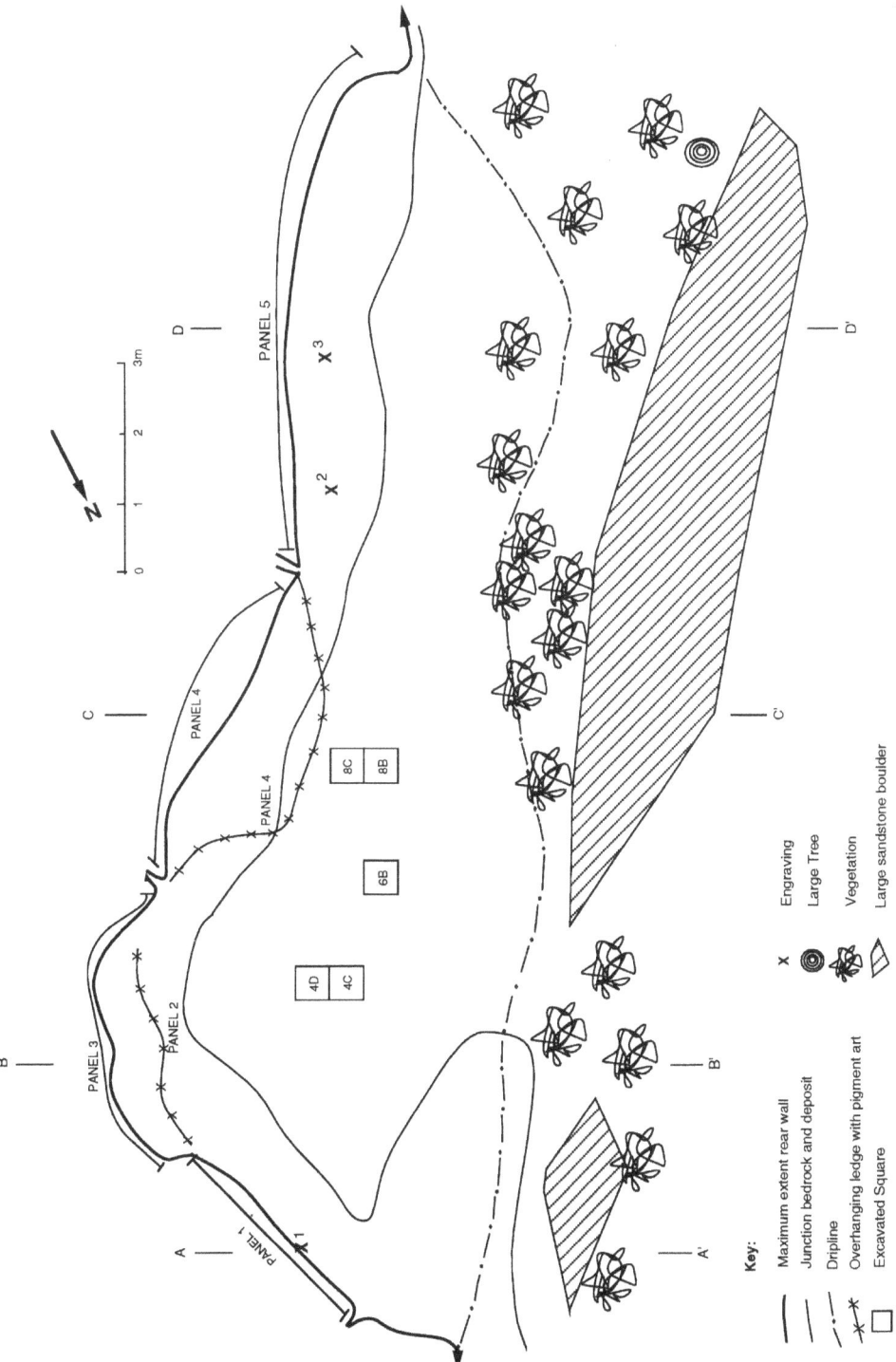

Figure 8.2: UDM Shelter. Site Plan.

there is considerable exfoliation of the rock surface. Red hand stencils are consistently found beneath white hand stencils where superimpositionning occurs.

Most of the pigment art is in good condition, although there is evidence for exfoliation postdating some art production and predating a later phase in Panel 5. There is also deterioration (cracking, spalling and exfoliation) of the case-hardened surface in Panel 1.

Engravings

A number of petroglyphs occur within the shelter. Three of these are engraved male anthropomorphs. One of these is a pecked upside-down-man on the vertical surface, low on Panel 1[23] and beneath

[23] This motif is probably the source of the shelter's name, although there is also a red drawn upside down human

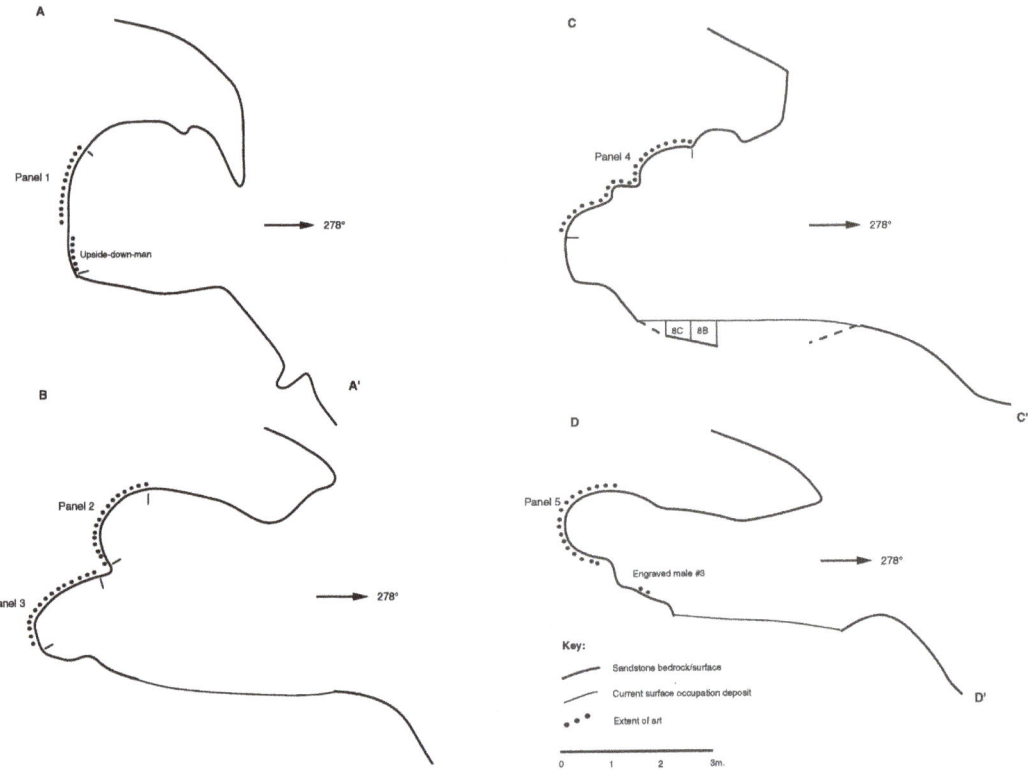

Figure 8.3: UDM Shelter. Site Cross-sections A - D.

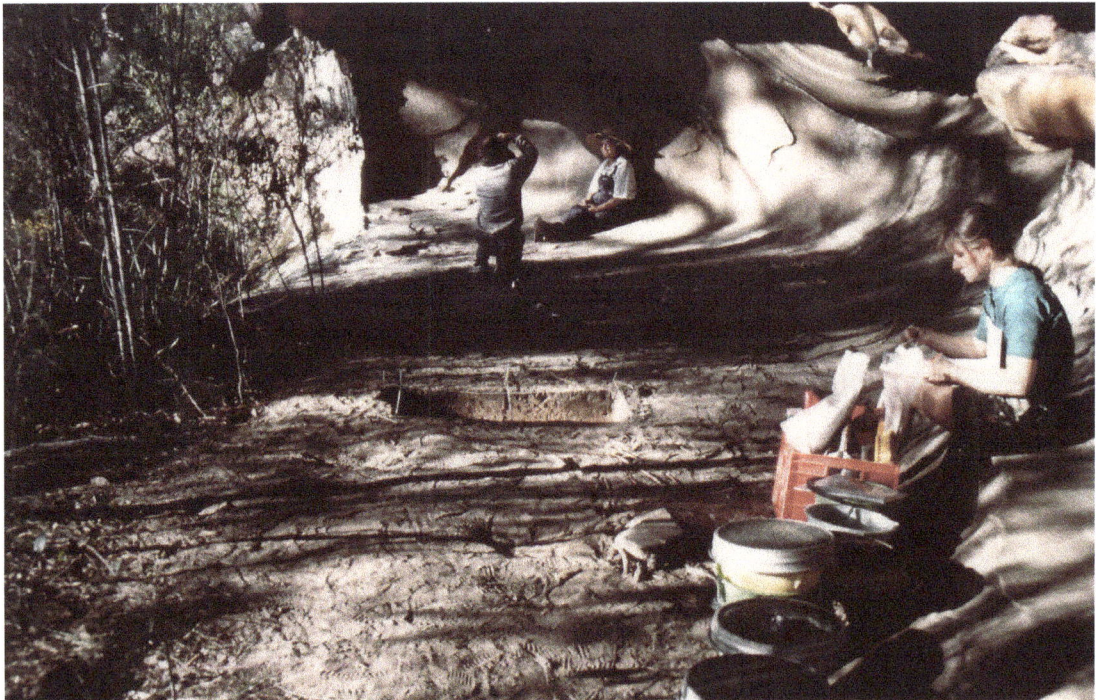

Figure 8.4: View of the internal space of the rockshelter, camera facing north-east. Panel 1 is in the background; Squares 4 and 8 are open.

Panel 5 there is a pecked man and an outline pebraded miniature-Sydney style man, both on a horizontal ledge.

These three motifs (Figure 8.5) are clearly separated spatially from the pigment art. The five abraded motifs have been incorporated into a pigment panel (Panel 4). Superimpositionning analysis reveals that the abraded motifs occurred fairly late in the site's art production, and postdate charcoal motifs drawn on this Panel (Figure 8.6).

figure amongst the pigment art on Panel 1 also.

Table 8.1: UDM Shelter. Motif and Technique Information.

Motif	Colour				Technique				Form			Total
	Red	White	Black	Yellow	Dry	Wet/*	Stencl	Engr'v	Out.	Infill	O/I	
Man	2	1			2	1		3å	2	3	1	6
WomanΩ	2	1	1		3						3	3
Anthrop. Ω	10	5	5	1	9	3/2				4	10	14
MacropodΩ	2	14	11		18	4/2		2ß	12		14	26
SnakeΩ		4	4		5						5	5
EchidnaΩ		1	1			/1					1	1
Reptiles≈	3		1		1	3					4	4
Fish	1				1					1		1
EelΩ		2	1		1	1			1		1	2
Shield			1		1						1	1
Axe	1				1						1	1
O. mat objΩ	1	1				/1			1			1
Hands	2	33		1			36					36
Hand var.	2	16					18					18
Bird tracks		2			2					2		2
Circle			1		1				1			1
Dots		18			18					18		18
CXNF	1				1				1			1
SNFΔ		2	2	1	3			1ß			3	3
Other†	2				2						2	2
Unid lines	1	2	3		6				6			6
Unid solidΩ	34	26	62		111	7/1	1	2ß	24	57	41	122
Total	64	128	92	4	165	40/7	55	8	48	85	87	274

ΩEighteen of these are bichrome (i.e. black and white, red and white) and Δone is trichrome (yellow, white and black)
å 2 pecked, 1 outline (miniature Sydney)
≈ 3 goannas, 1 turtle
* wet and dry
ß abraded
† shields?

The pecked motifs beneath Panels 1 and 5 cannot, because of their locations (Figure 8.2), be directly correlated with the pigment tradition. As stated above, two of these motifs are pecked (intaglio), and the other is a miniature (pecked and abraded) outline engraving. The pecked intaglio male figures would appear to be related to an earlier transitional phase of art production in the region. This has been dated at Yengo 1 to the pre-Bondaian (this volume). There is a small number of sites with this earlier art style recorded in the Mangrove Creek area (at 11 sites; 17%). In all other instances the art consists of tracks (bird and macropod), circles, and pecked pits or dots.

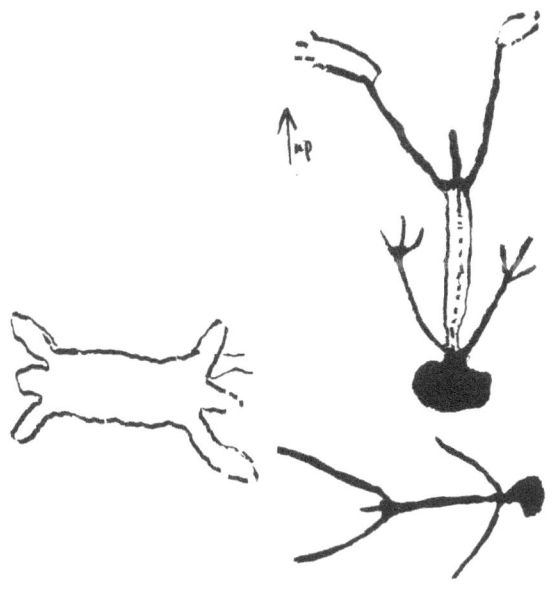

Figure 8.5: UDM Shelter. The three engraved male human figures.

Table 8.2: UDM Shelter. Technique variables. Variables in identified superimposition relationships (Table 8.4) highlighted in bold.

Variable	Technique description		Total motifs
1	black outline	(all dry)	13
2	black infill		34
3	black outline + infill		29
4	red outline		7
5	red infill		16
6	red outline + infill		21
7	white outline		14
8	white infill		2
9	white outline + infill		15
13	white outline	(all wet)	7
14	white infill		23
15	white outline + infill		1
17	red infill		8
19	yellow outline		1
20	yellow infill		1
22	wet + dry, o + i (w/w + b, b/w, b + w/b*)		3
23	red dry outline/wet infill	(wet/dry)	1
24	black + white outline	(bichromes)	1
25	white outline, black infill		4
26	black outline, white infill		1
27	black + white, outline + infill		1
28	black + white outline, black infill		1
31	red + white outline		2
32	white outline, red infill		2
34	red + white, outline + infill		1
35	white outline, red + white infill		2
44	black, white + yellow outline, black, white + yellow infill, white + black outline, yellow infill.	(polychrome)	1
47	linear infill		2
48	white	(stencils)	50
49	red		4
50	yellow		1
54	intaglio	(engravings)	2
55	outline		1
56	incised outline		1
57	incised o/i		4

Variables numbered according to the more extensive classification system used in broader analysis.
*white and black combinations

The UDM intaglio motifs, being figurative, are not within the usual range of motifs found in this earlier style, and it is possible that they represent a transitional phase of art production. The inverted man beneath Panel 1 has an infilled circular head and stick body and limbs (Figure 8.5). While the feet are depicted as open ovals (*mundoes*), both hands have three fingers, reminiscent of bird tracks. Some of the red, early wet infilled anthropomorphic depictions at the site (Panel 5) also have three fingers on their hands. The other intaglio male figure also has a solid circular head, but no digits appended to the limbs (Figure 8.5, Figure 8.7). These motifs will be further discussed in terms of the art phases identified at the site.

Change over time

Ongoing surface exfoliation as well as superimposition evidence at the site indicates several art production episodes. While the pecked and outline engravings cannot be related to any particular phase of pigment art production at the site by direct association, there are several pigment

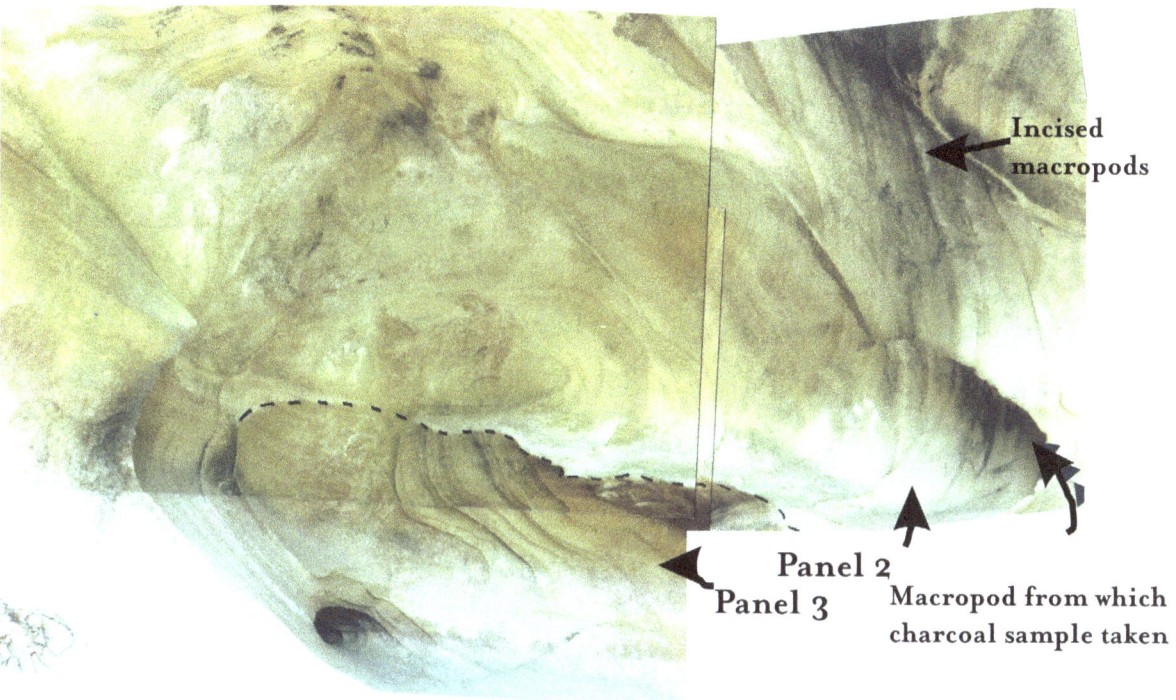

Figure 8.6: Panel 2 (above) and Panel 3. Incised macropods on far right of Panel 2 (detail Figure 8.10). The current floor level is c. 1m below Panel 3.

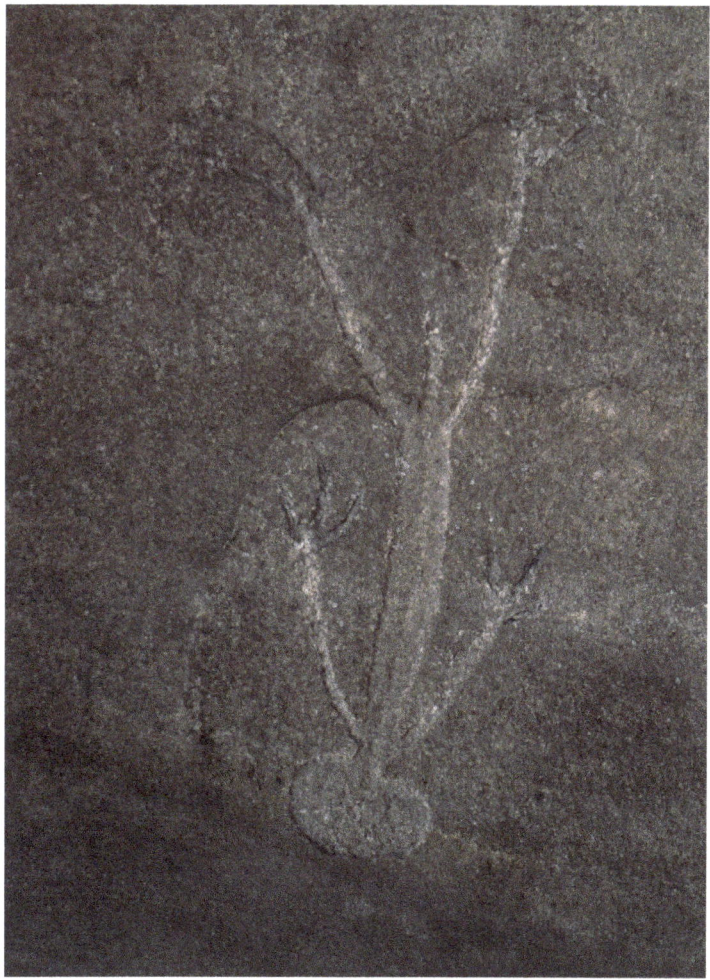

Figure 8.7: Pecked Upside down-man beneath Panel 1. Note the charcoal infill in the hands and feet. This may be recent graffiti. The spine of the figure is more deeply pecked than the rest of its intaglio body.

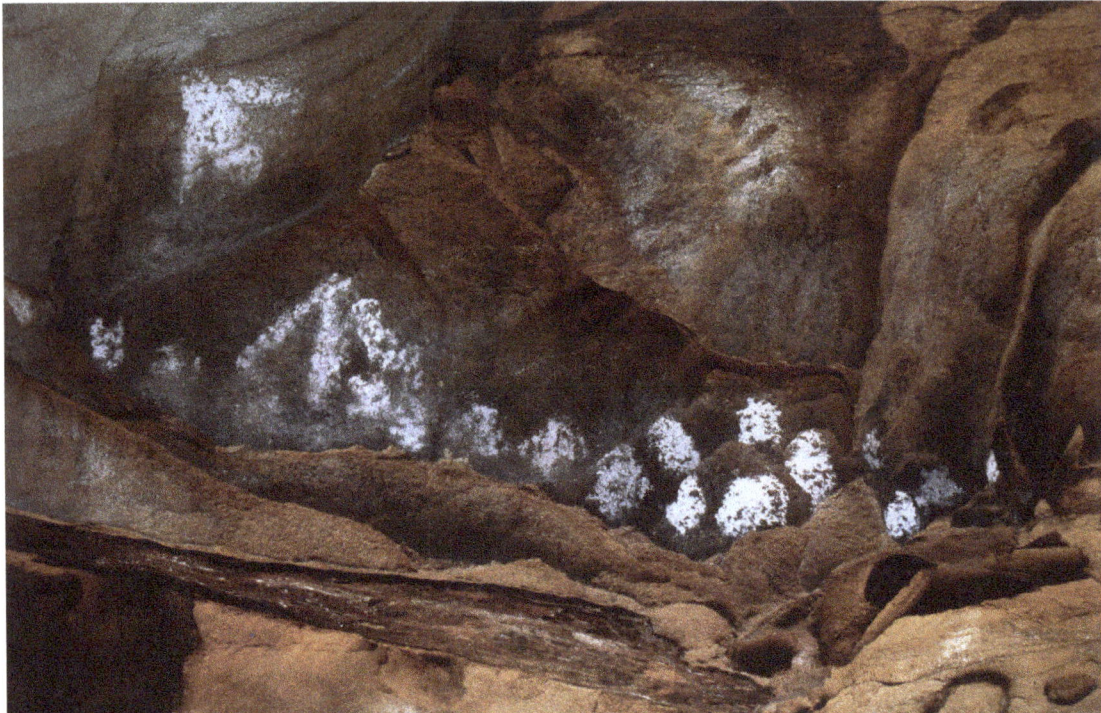

Figure 8.8: UDM Panel 4. White painted blobs, bird tracks and white hand stencils.

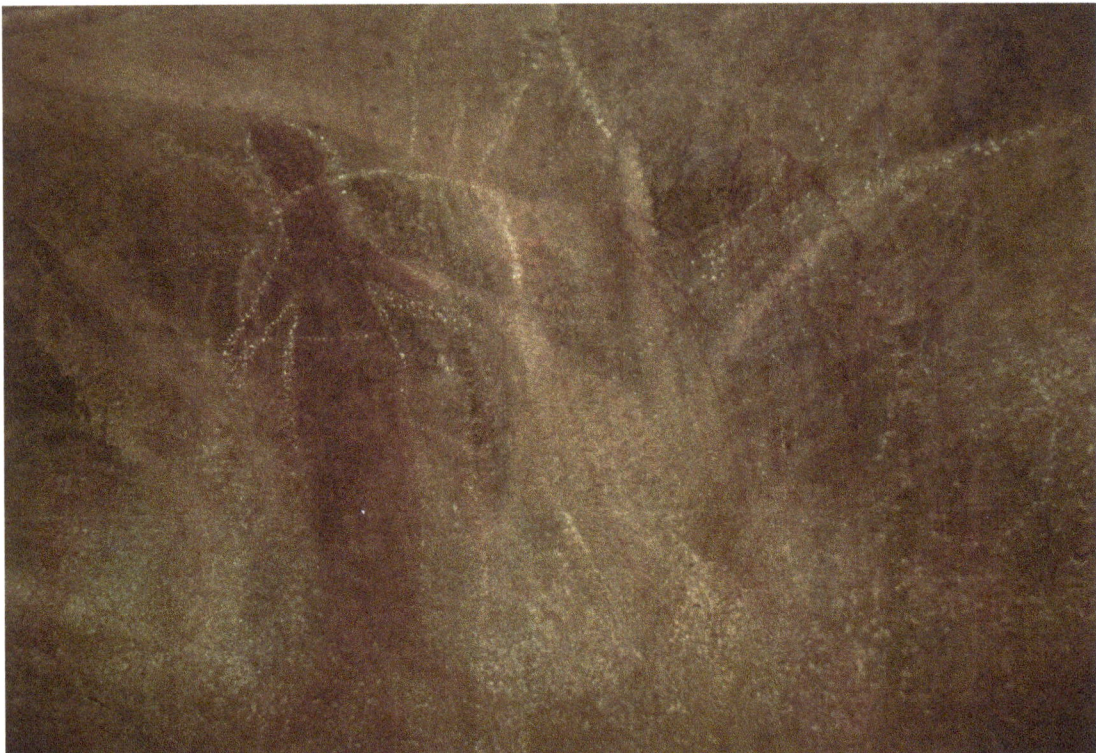

Figure 8.9: UDM Panel 5. Red anthropomorphs (one with white outline) beneath white outlined macropods.

techniques (and motifs) which are consistently beneath others and are interpreted as being older than the majority of the art. Incised motifs also occur in superimposition relationships with other techniques, and as indicated above, appear relatively late in the art sequence (Figure 8.10, Figure 8.11).

The diachronic analysis used with the Mangrove Creek shelter sites was used at UDM to explore intra-site diachronic variability. Eighteen techniques occur in superimposition relationships at UDM (Table 8.2), and the general sequence identified for Mangrove Creek (Table

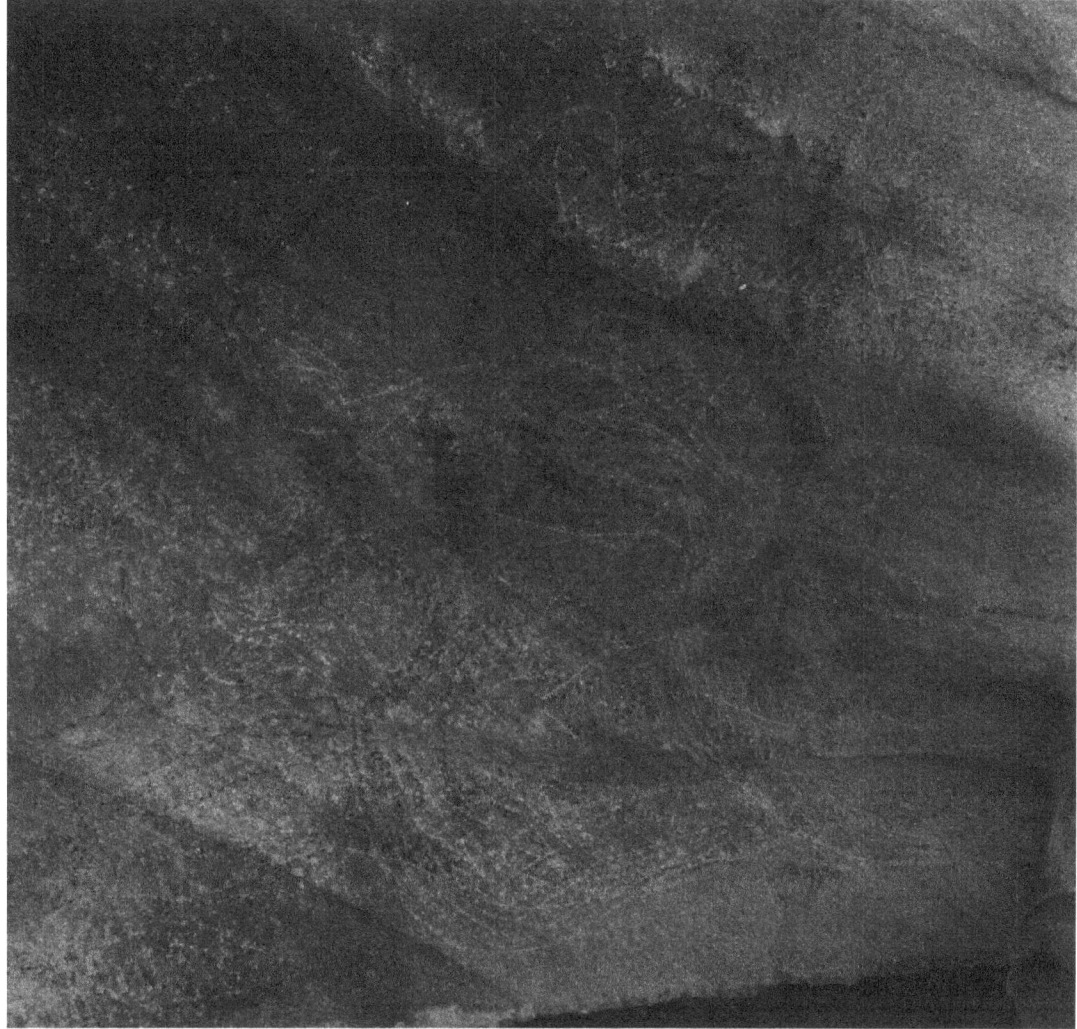

Figure 8.10: UDM Panel 2. Detail of incised macropods over black charcoal macropods.

8.3) was found here also (Table 8.4). The variable numbers used (Table 8.2) relate to the more comprehensive list identified (Table 11.1; chapter 11).

Phases 1 and 2 show distinct separation in motif preferences from Phase 3 (Chapter 9). This is interpreted as indicating that Phases 1 and 2 are temporally discrete from each other and from Phase 3. Phase 3 represents most of the art in the Mangrove Creek area and is characterised by a proliferation of artistic techniques.

Superimposition relationships were recorded at the site (Figure 8.6, Figure 8.8 and Figure 8.9). This was not quantified, i.e. recording every example (cf. Morwood 1979), but indicatively, i.e. recording presence/absence of trends (Table 8.4).

White hand stencils provide the best evidence for the contemporaneity of much of the assemblage and/or the ubiquity of this technique. White hand stencils occur above and below many of the techniques which do not occur in other superimposition relationships. White hand stencils occur over black infilled and outlined and infilled motifs, red o/i, white o/i, wet white outlined and infilled motifs, white + red o/i motifs and red hand stencils. White hand stencils have been drawn over by red infilled motifs, wet and dry black + white o/i motifs and wet white outline motifs.

On the basis of close visual inspection, these appear to be generally contemporanous. All appear to be of roughly the same coloured pigment, and the expertise indicated by the user(s) of the technique across the site is consistent. Weathering varies slightly, mainly as a result of the varying degrees of protection afforded to the different art panels.

At UDM, Mangrove Creek Art Phase 2 is represented by the solid red painted anthropomorphs and goannas and red hand stencils on Panel 5. The red figurative motifs are beneath a series of red

Table 8.3: Proposed Diachronic Sequence; Mangrove Creek Shelter Site Phases 1-3. Those elements observed at UDM in bold.

1. (Earliest)	Intaglio motifs (? usually tracks, circles etc.)
2.	**red hand stencils**
	wet red infill, outline and infill
3.	**dry black outline, infilled, outlined and infilled, dry red infill, outlined and infilled**
	wet red outline, **red and white outline**
	wet white infill, incised motifs
	white, red, **yellow** and pink **stencils**
	bichromes, black outlined and infilled
	dry white infill and white and/or yellow outlined and infill
	polychromes and **wet and dry black and white motifs**
most recent	**dry red outline**, wet red outline, **wet white outline**, most dry yellow outline, **dry black outline**
	contact motifs (white stencils, red outlined and infill and white outlined and infill).

Table 8.4: UDM shelter. Recorded superimposition information.
(Horizontal variables <u>over</u> vertical variables)

	1	2	3	4	5	6	7	9	13	14	17	22	24	31	32	48	49	57
1							1											
2			1		1			1	1	1						1		
3							1	1		1						3		1
4							1		1									
5			1															
6	1		1				2							1		1		
7																		
9																1		
13																1		
14																1		
17			1			1				1					1			
22																		
24																		
31			1															
32			3						1							1		
48					1					1		1						
49																1		
57																		

This analysis revealed the following trends:

<u>Never over other techniques</u>   <u>Never under other techniques</u>

dry black infill   white outline

wet red infill   white + black o/i (wet and dry)

red hand stencils   incised motifs

red and white outline   black and white outline.

and white bichrome anthropomorphs, which, in turn, are below several white outline macropods. The red hand stencils - which are completely bonded to the sandstone matrix - are beneath white hand stencils. In Panel 2, two red and white bichrome anthropomorphs are beneath black drawings and white hand stencils. In Panel 4, the incised motifs are on top of several types of black drawings, which are in turn beneath several white painted motif types.

These superimposition relationships indicate temporal trends and also indicate certain colour relationships. Black and/or red monochrome motifs are never placed over white ones (either wet or dry), although white was placed often over red and/or black motifs. Black was never drawn over white hand stencils, although the converse did occur. Many of these superimposed colour relationships may result from visibility considerations. Colour avoidances, however, may also be operating.

The assemblage at UDM is interpreted as consisting primarily of elements from Art Phases 2 and 3. The pecked motifs may be from Phase 1, although based on motif analyses (chapter 10) these are interpreted as transitional and are therefore conservatively placed in Phase 2. The majority of the UDM art falls in Art Phase 3, with several techniques (namely dry and wet red outline) which occur consistently late in the phase also present.

Given this art sequence, the excavations at the site aimed to determine whether there were multiple occupation phases at the site, and their temporality.

Art Dates

Four samples from two charcoal macropod drawings were collected for AMS dating (Figure 8.12). These samples were processed by John Head (ANU Quaternary Dating Research Centre) and the counting was completed at the ANSTO Facility at Lucas Heights. The following dates were received (Table 8.5).

Table 8.5: UDM Shelter. AMS dates from the two macropod motifs.

Sample No.	Field Number	Motif #	Date
ANU-AMS 773	UDM/1	1	c.480 + 80 BP
ANU-AMS 774	UDM/2	1	indistinguishable from modern
ANU-AMS 775	UDM/3	1	indistinguishable from modern
ANU-AMS 776	UDM/4	2	indistinguishable from modern

These age determinations are inconclusive as to the age of the motifs, and suggest problems with the field sampling procedure (McDonald 2000c).

The three dates which are indistinguishable from modern standard suggest that there has been contamination of both motifs by younger material (lichens or other microscopic organics). Conversely this art was produced either just prior to European contact or up until 1950 AD. This latter scenario would seem unlikely for motif #2 which is faded and partially affected by surface exfoliation (Figure 8.11). Motif #1 was produced using a lump of charcoal and comprises a single line outline. The samples were collected from locations where it was clear that only one stroke had been executed. This makes it unlikely, in archaeological terms, that this motif has been contaminated by a more recent 'touching up' of the motif.

The date c.500 BP may be an accurate representation of the age of motif #1. Conversely, given that the two contaminated dates from this same motif, it could indeed be older. If the contamination is contemporary (i.e. micro-organics still alive at the time of sample collection), then the date returned may be significantly younger than it should be. John Head (Quaternary Dating Research Centre, pers. comm., 1994) suggested that the 500 year old date may indeed be double that - c.1,000 years BP. Resolution of this problem will only be achieved through further fieldwork which aims to identify sources of local contamination on the surface of the rock or growing on the art.

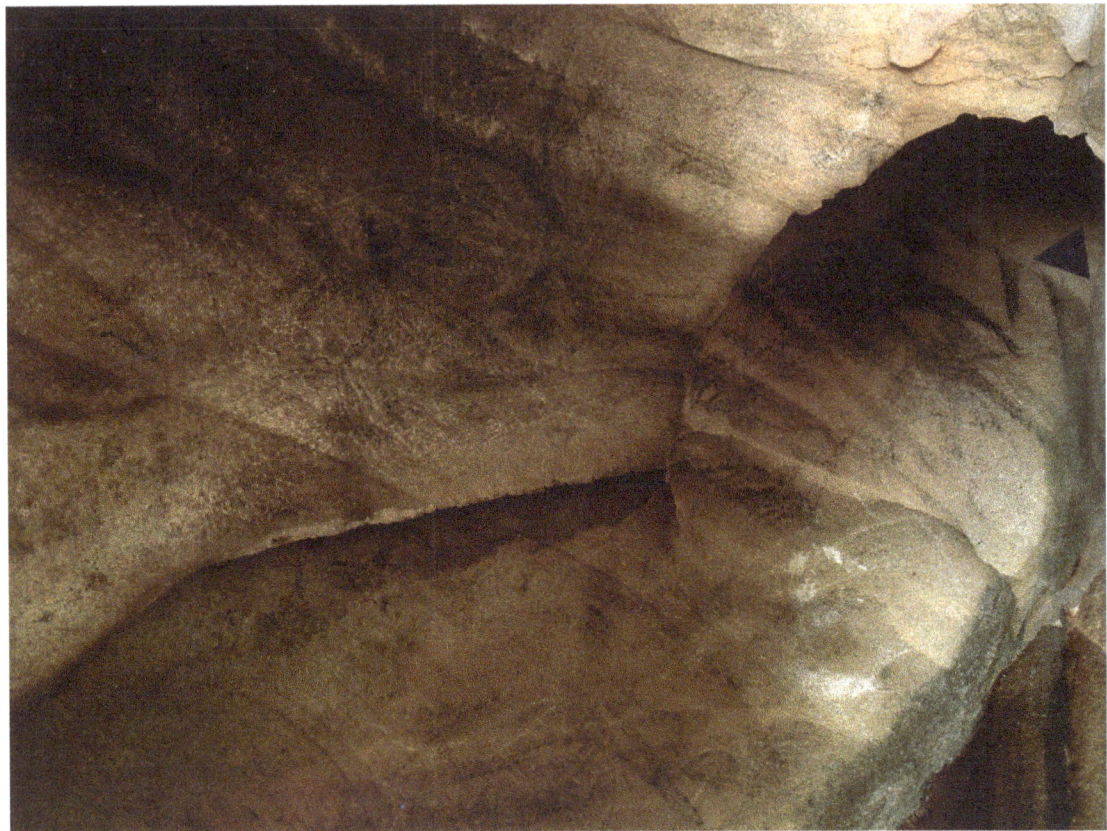

Figure 8.11: Panel 3 showing context of incised panel and the macropod #2 in the alcove (arrowed right) which was sampled for AMS dating.

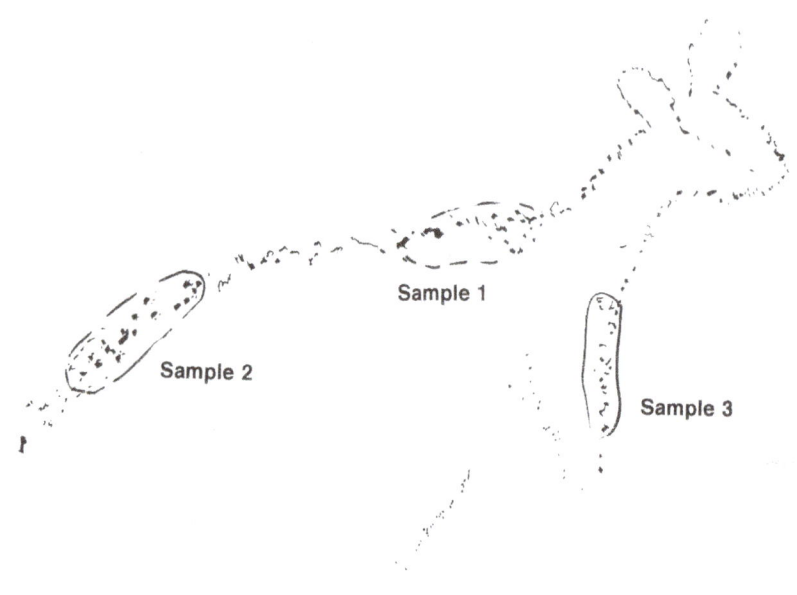

Figure 8.12: UDM Shelter. Macropod motif #1 from which three AMS samples were collected.

These contradictory results make it difficult to make firm conclusions about the production date of the outlined charcoal macropod. On face value, it would appear that art was being produced at this site up until c.500 BP: after other occupation of the shelter had ceased. The fact that contamination cannot be ruled out means that this motif may have been produced in the final or even terminal phase of the site's occupation.

The Excavation

Aims

The aims of the excavation were to determine;

1) the nature and timing of the occupation evidence at the site;

Of particular interest was whether this evidence indicated domestic use of the shelter. Only one site with a larger art assemblage had previously been excavated in the Mangrove Creek Valley (MacIntosh 1965). None of Attenbrow's 31 occupation sites had major art assemblages. It was thus of interest to characterise the occupation debris at UDM.

2) Whether there were two or more phases of occupation evidence at the site (as suggested by the art); and,

3) Whether the art was contemporaneous with the occupation deposit.

All three aims have been achieved by this work, the last inconclusively.

Methods

Conventional archaeological techniques were employed in this test excavation (Figure 8.2). Squares were located based on two criteria:

1) the presence of a relatively flat, protected floor area in relatively close proximity to pigment art panels; and,

2) the likelihood of a reasonable depth of deposit.

The sloping back wall of the shelter suggested that much of the deposit at the site would be shallow. Probing with metal skewers indicated that most of the deposit close to the back wall is <20cm deep. Excavating in immediate proximity to decorated panels (cf. Rosenfeld *et al*. 1981, Morwood 1992b) was not an option here.

A 50cm x 50cm pit (6B) and two 50cm x 100cm trenches (Squares 4 and 8) were excavated. A total of 1.25m^2 was investigated: a 2.8% sample of the floor area with reasonable depth of deposit.

Excavation in all squares was by 5cm spits or in stratigraphic units, whichever were smaller (Figure 8.13). The two 50cm x 100cm trenches were excavated as two consecutive 50 x 50cm pits. Large and/or diagnostic artefacts and charcoal samples were provenanced during excavation. The small size of the excavation units was felt to be sufficient for the purposes of provenancing excavated material.

All deposit was dry-sieved through 5mm and 2.5mm sieves on site. Unsorted residue from the 2.5mm sieve was bagged and retained for later sorting. This has been preliminarily sorted but is retained as a category of cultural material. The residue consists mainly of charcoal, fine gravel, vegetation (leaves and small twigs in the upper most layers, root material lower down) and micro-debitage, although the latter was specifically targeted during sorting.

Stratigraphy

The excavations revealed four stratigraphic layers (Figure 8.16, Figure 8.17). These were identified on the basis of soil colour and texture (Figure 8.14, Figure 8.15). An ashy lens occurred in all squares between layers 1 and 2 (at varying depths). Each of these was considered to be a specific and separate hearth, and so these not given Layer status. This ashy lens is the best indicator of where recent scuffage stops and *in situ* deposit starts. This is generally quite close to the surface (see section drawings). The sandy deposit was extremely compact. The identified Layers were clearly defined on the basis of colour and textural variation. While there may have been some downward movement of smaller artefactual material, the stratigraphic integrity of the site is good.

The Layers identified were;

I   Surface leaf litter, loose grey sandy deposit with high organic/humic and charcoal content. Some recent materials e.g. cigarette filter (Colour 10YR 3/3; pH = 9).

II   More consolidated than Layer I, brown grey, sandy deposit. Reddish brown in square 4 (5YR 5/4, pH = 6.5); more orange brown (Colour 10YR 5/3; pH = 6.5) in Squares 6 and 8.

III   Yellow brown, sandy deposit with low charcoal content. There are again slight colour differences between square 4 and Squares 6 and 8, with the latter being somewhat pinker (Colour 7.5 YR 6/6; pH = 3.5) than found in square 4 (Colour 7.5YR 6/4; pH = 3.5).

IV   Yellow/pink white (Colour 7.5YR 7/4; pH = 4.5) sandy deposit with very low charcoal content. Many small roots. N.B While there is a general increase in acidity with depth, layer III is more acidic than layer IV.

In Square 4 (Figure 8.16):

> Layer II - was quite orange with a red area (Colour 10YR 5/6) at the eastern end of the pit. This was apparently associated with the burning event/ashy lens (very low quantity of charcoal and bone were observed, suggesting an extremely hot fire oxidising the surrounding deposit).

> Layer III - yellow brown sandy deposit in contact with bedrock: immediately adjacent to bedrock there is a darker band (<1cm thick) of deposit.

The following specific notes were made regarding Square 6 (Figure 8.16):

> Layer I - Ashy lens here is associated with two separate hearths with ashy lenses above the charcoal rich features. One of these is in the southern corner, the other along the northern baulk.

> Layer II - extends down to bedrock in southern corner in this square, beneath the intact hearth. This suggests prehistoric disturbance (pit? some other sort of excavation?), which has disrupted the units below. Several largish pieces of sandstone roof-fall were found here and above intact Layer III in eastern corner.

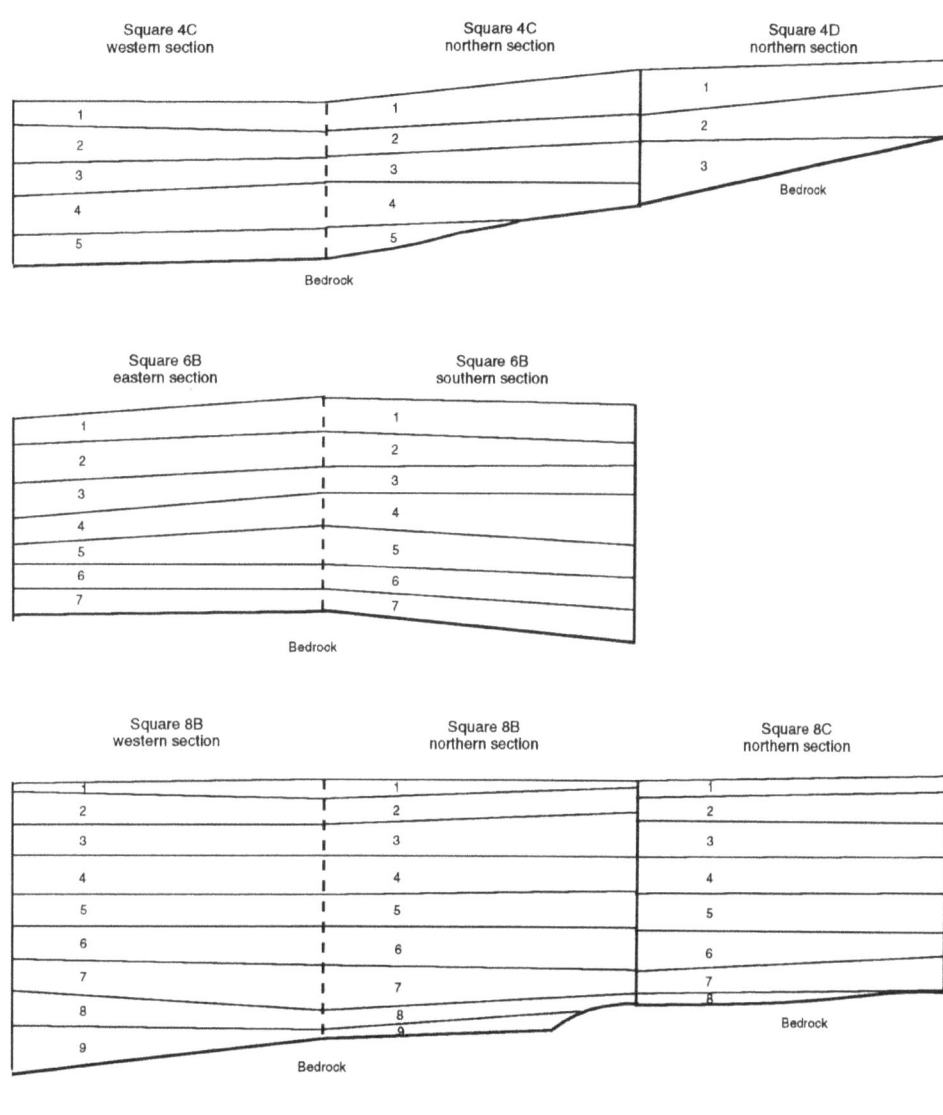

Figure 8.13: UDM Shelter. Excavated spits – all squares.

Layers III and IV - thicker at western side of pit (deeper owing to sloping bedrock).

In Square 8, the following notes were made (Figure 8.17):

Layer I - deeper at the western end of pit (square 8b) and thinner at eastern end (square 8c). Ashy lens found at both ends of pit but not in centre.

Layer II - found only at eastern and western ends of pit on the northern side of the square, but present along entire length by southern side of pit. This Layer is quite deep at the western end of the square.

Layer III - quite pink in this square. Very large extant root in this unit (and below) at western end of pit, may have affected the stratigraphic integrity of square 8B particularly Layer III.

Layer IV - most intact and deepest example of this unit across the three pits.

Correlation of Excavation Layers with Stratigraphic Layers

For the purposes of analysing vertical distribution and diachronic change, the excavated spits have been correlated into analytical and stratigraphic layers. This was straightforward since the shelter floor was flat and adjacent squares were dug simultaneously (Table 8.6; Figure 8.13).

Dates

Four charcoal samples were submitted to the ANU Quaternary Dating Research Centre. These have been counted and the following age determinations obtained (Table 8.7). The proveniences of these samples are illustrated on the stratigraphic sections (Figure 8.16 and Figure 8.17).

Table 8.6: UDM Shelter. Correlation of analytical units with excavation spits and stratigraphic layers (Squares 4, 6 and 8).

Analytical Layer	Squares and excavated spits					Stratigraphic Layer
	4c	4d	6	8b	8c	
1	1	1	1	1	1	I
2	2	2	2	2	2	II
3	3	3	3	3	3	II
4	4	-	4	4	4	II
5				5	5	III
6	5		5	6	6	III
7	-		6	7	7	IV
8			7	8	8	IV
9			-	9	-	IV

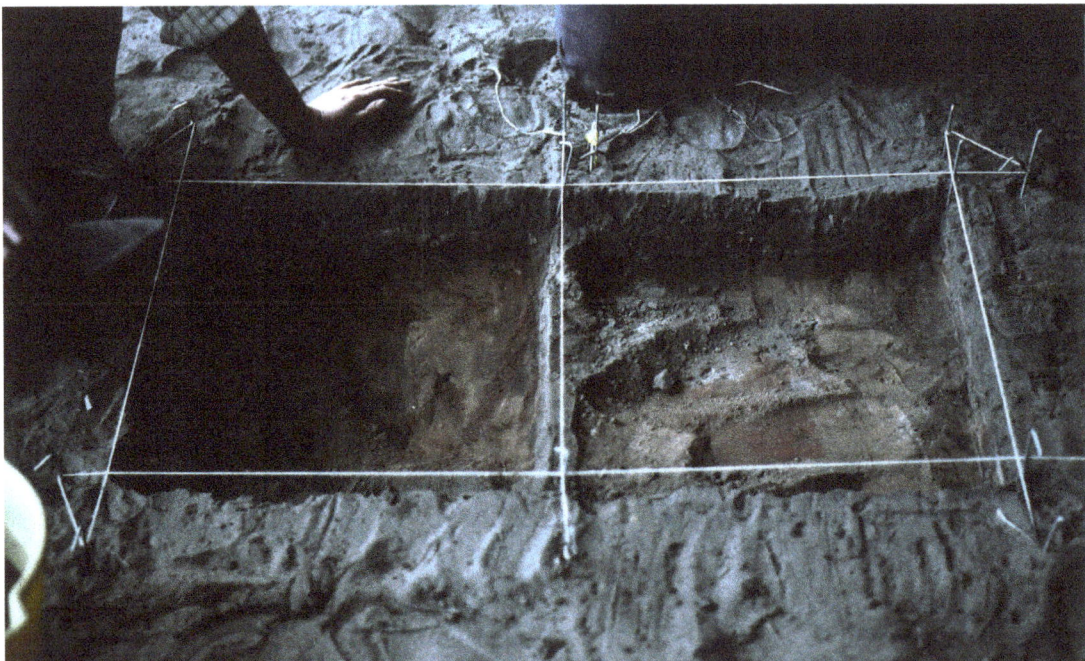

Figure 8.14: Square 4, excavation in progress. Note the marked colour contrast between the reddened ashy lens below Layer 1.

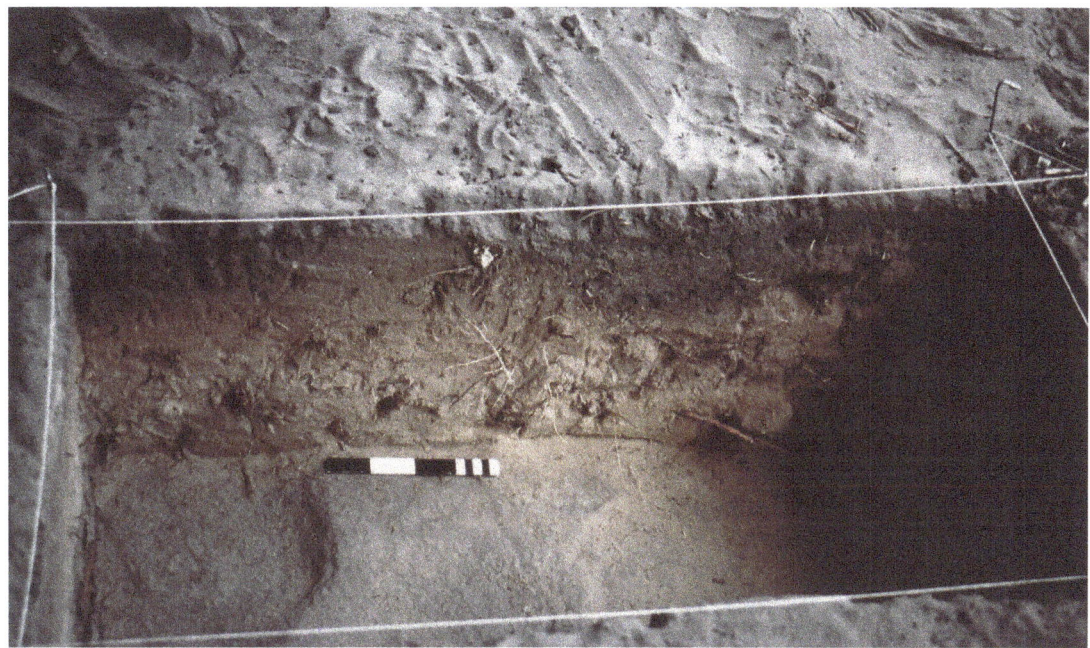

Figure 8.15: Square 8, excavation to bedrock. The colour differences in the different strata are quite clear.

Table 8.7: Radiocarbon dates, depth below surface and association with stratigraphic units.

Sample No.	Field Number	Depth below surface	Stratigraphic Layer	Date
ANU-8134	6B/2/1	6.5 cm	II	1,220 ± 120 BP
ANU-8135	6B/4	15-20cm	II	1,860 ± 70 BP
ANU-8133	8B/5/1	20 cm	III	1,540 ± 60 BP
ANU-8132	8B/8	34-39cm	IV	4,030 ± 140 BP

The samples derived from two squares (6B and 8B). While each square's dates are internally coherent, there is a reversal in the expected outcome for layers II and III. The presence of the large tree root in square 8B may have affected the integrity of this square, although the field notes accompanying sample 8B/5/1 suggest that this was not obvious in the field. The field notes do indicate that there were a number of smaller roots though this spit; hence it is possible then that this age determination has been contaminated by younger carbon. This was, however, a provenanced sample while sample ANU-8135 was collected from the sieves. The *in situ* sample is considered to be the more reliable of the two. The two results are within the same range for the Middle Bondaian.

Table 8.8: Proportions of lithic, charcoal, shell, bone, residue and 'other' from the excavated squares. Weights in grams.

Cultural component	Total (g)	%f
Stone artefacts	1,561	27.6
Shell/bone	63	1.1
Charcoal	1,980	35.0
Unsorted Residue	2,055	36.3
Other (seeds and pigment)	1.1	0.0
Total	5,660	100.0

Cultural Material

A total of 543kg of deposit was excavated from the site. From this derived slightly less than 5.7kg of cultural deposit (Table 8.8) and 4.5kg of roof-fall. The shallowest squares (4D and 4C) had the most recent deposit and the highest densities of cultural material. Square 8C had the lowest proportions and densities of cultural material (Table 8.9).

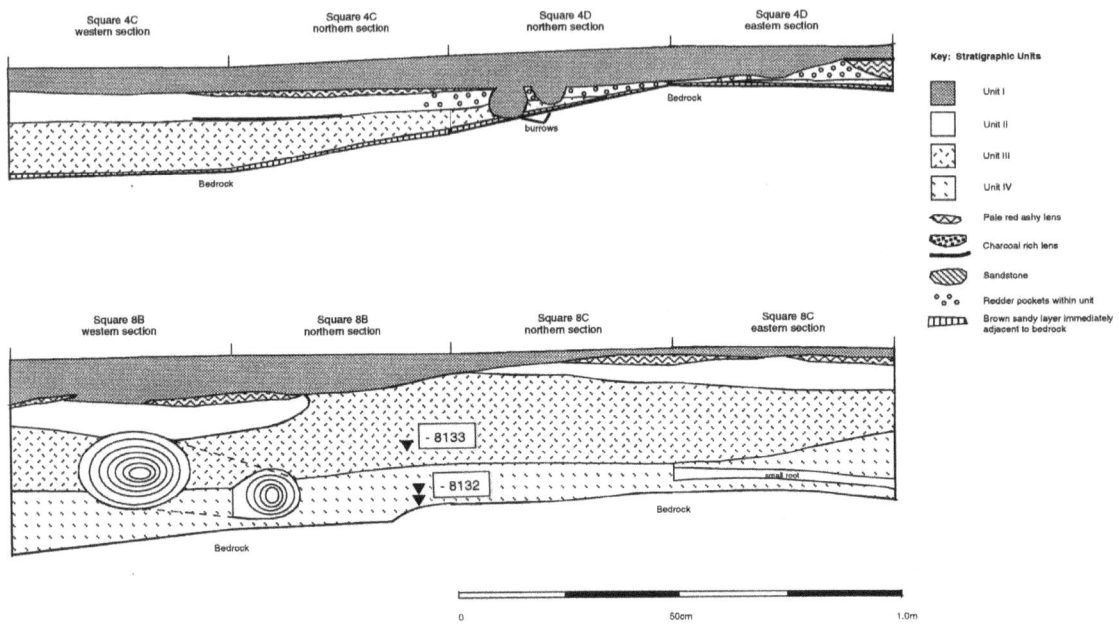

Figure 8.16: UDM Shelter. Stratigraphic sections, Squares 4 and 8.

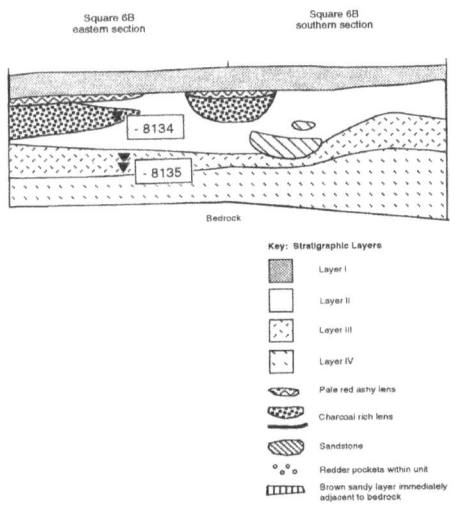

Figure 8.17: UDM Shelter. Stratigraphic Section, Square 6.

Stone Artefacts, Charcoal and Unsorted Residue

Charcoal and unsorted residues represent the bulk of the cultural material by weight. Both of these categories, being largely organic (the latter also contains small gravel fraction, and possibly some micro-debitage), are present in greater quantities in the upper layers than deeper in the deposit.

Stone artefacts represent a slightly lower proportion of the cultural material (by weight). These are present in fairly high numbers.

Table 8.9: Proportions of cultural material retrieved from the excavated squares. Weights in grams except where specified.

Square	Artefacts	Shell/Bone	Charcoal	Residue	Other	Total Cult. Mat	Excavated Deposit [kg]	density/m^3 (%)
4D	96.2	14.5	571.6	466.8	0.8	1,150	85.5	1.35
4C	53	14.9	682.7	417.1	0.1	1,168	56.3	2.08
6B	392.4	28.3	242.8	401.3	0.2	1,065	125.0	0.85
8B	509	2.5	270.1	515.1	0	1,297	146.8	0.88
8C	510.5	2.5	212.6	254.7	0	980.3	129.5	0.76
Tot	1,561.1	62.7	1,979.8	2,055.0	1.1	5,660.3	543.1	1.04%

Shell and Bone

Small quantities of highly fragmented shell [Sydney cockle (*Anadara trapezia*), hairy mussel (*Trichomya hirsuta*) and freshwater mussel (*Hyridella sp.*)] were present in the upper layers: 12g in the five squares. Bone was also present in very small quantities (50.7g) in the upper layers. Most of this is highly fragmented. With the exception of several macropod and small marsupial teeth, rib fragments and a few reptile vertebrae, most is unidentifiable. The bone fragments were found in several conditions: burnt, calcined, weathered and/or stained and fresh. While some of the fresh bone, particularly of smaller species, may represent recent additions, the majority is considered to be culturally derived.

Plant Remains and Pigment

Amongst the cultural remains, in minute quantities, are seeds, white pipeclay and red ochre.

The seeds were analysed by the Seeds Laboratory (NSW Department of Agriculture). These included:

- Geebung (*Persoonia spp.*) 2 seeds;

- Native cherry (*Exocarpus cipressiformis*) 1 seed each in 4C/1 (gnawed) and 4D/1 (cracked);

- Vine lilac or false sarsaparilla (*Hardenbergia violace*) in 4C/1, 1 seed; in 8B/3, 4 seeds; and,

- Senna (*Cassia spp.*) - in 4C/1, 1 seed; 6B/5, 1 seed.

There are also 20 seeds in 8B/3 and two desiccated fleshy pieces which could not be identified. Under a microscope, the interior of these seeds (somewhat like *Callitris* seeds) has a honeycombed appearance, more in keeping with bony structure.

Pigment was found only in isolated spits (pipeclay in 6B/3 and 8B/2, red ochre in 4C/3 and 4D/1). The pipeclay (four fragments in 6B/3; one fragment in 8B/2) is highly weathered and, possibly as a result of this poor condition (and small size), shows no traces of use. The two fragments of red ochre in 4C/1 are of a fine texture and produce a dark red colour (Munsell 5R 3/6) on white paper. These two fragments are extremely small (0.1g). Neither shows striations or other evidence of usewear. This may result from their very small size. The one small fragment (0.2g) of dark red ochre in 4C/3 is very granular and of poor quality.

Change over time

The site's cultural deposit demonstrates obvious changes over time in its composition. In the lower two levels, stone artefacts make up the bulk of the cultural material present. In the upper layers there is significantly more charcoal and a higher proportion of finer residue. Shell and bone are only found in the top two strata (Figure 8.18).

Many of these patterns, particularly those involving the site's organic remains, may be due to unequal preservation: the lower layers are very acidic. This is consistent with the environment usually provided by Hawkesbury sandstone but also reflects an absence of neutralising components in the older depositional period (i.e. ash, shell) which provided a more alkaline (neutralising) environment (see McDonald 1992a).

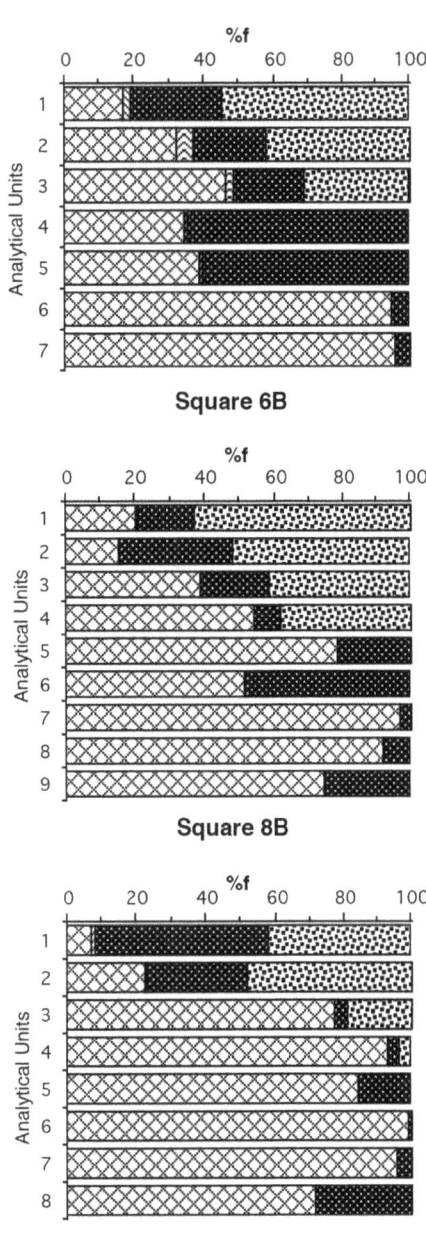
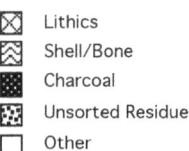

Figure 8.18: UDM. Cultural components in the three analysed squares. Change over time.

Artefact Analysis

When this site was first recorded by Patricia Vinnicombe in 1978, a bifacially flaked hatchet preform (Figure 8.19) was found in the vegetation at the front of the shelter. Four pieces of fragmented shell (*Anadara sp.* and *Hyridella sp.*) and 25 knapped stone artefacts were collected from the dripline (and lodged in the Australian Museum). These artefacts were made of a range of raw materials, silcrete, silicified tuff, quartzite, fine-grained basic (FGB), quartz. Three of the FGB pieces amongst this collection had evidence of previously ground facets, and there was one broken silcrete backed blade. Also found was one piece of thick black/green glass, although this does not appear to have been worked. Many more artefacts were observed in the dripline during the current fieldwork, but these were not collected or recorded.

A total of 3,550 artefacts were retrieved from the five test pits at the site. Only three of the pits (6B, 8B and 8C), with 3,290 artefacts, were analysed in detail.

The artefact material from the three analysed squares is dealt with collectively and then detailed according to each square, separately and in combination, for the analyses of vertical distribution.

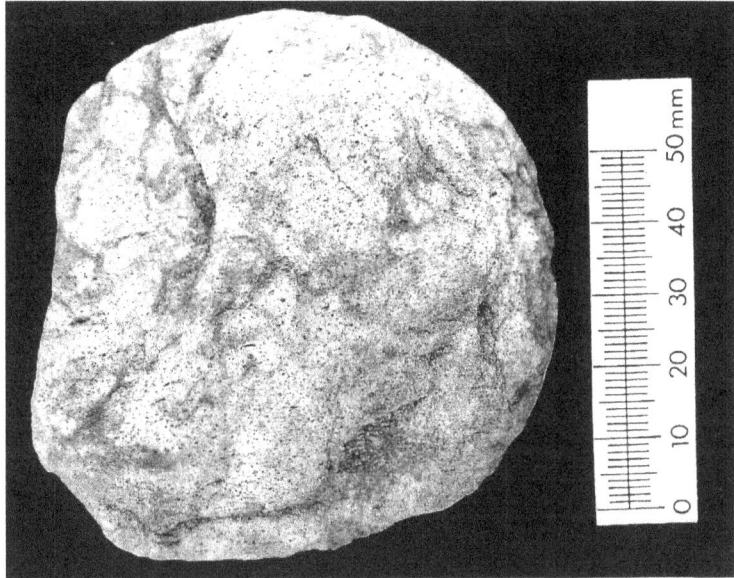
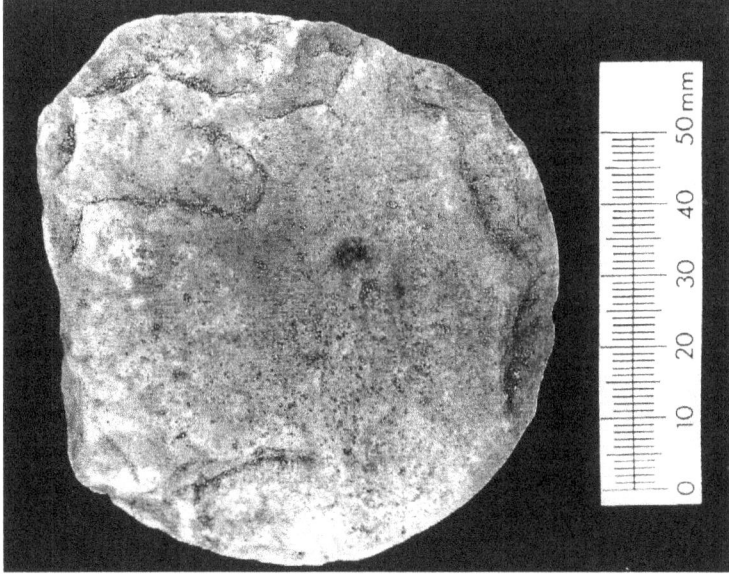

Figure 8.19: The bifacially flaked basalt axe pre-form collected by Pat Vinnicombe in 1978. Pitting on one face suggests that this may have also been used as an anvil.

Raw Material

Quartz is the predominant raw material found at UDM (38%), followed by Silicified Tuff (21%), fine grained basic (19%) and veined pebble chert (15%). The remaining raw material types (silcrete, chert, quartzite and 'other') were present in very small numbers. The general proportions are mirrored in the individual squares (Figure 8.20), although there is more FGB in square 6B and a larger proportion of silicified tuff in Squares 8B and 8C.

When only artefacts >1cm are considered (Figure 8.21) the overall pattern is similar, although there is proportionally more veined chert in square 8B, more quartz in 8C and an equal dominance of quartz and FGB in square 6B. These patterns can be partially explained by the presence of discrete knapping events in various locations across the site, and partially by temporal variability (see below 5.3.4.ii).

Some interesting patterns are observable when the raw material preferences for utilised artefacts are considered. Most of the artefacts with R/U (retouch and/or usewear) are made of silicified tuff (31.8%) or chert (22.7%), with veined chert being the next most used material (13.6%). While these figures show a consistency in preference (i.e. compared with the overall assemblage figures) for silicified tuff and veined chert, chert figures much more highly amongst the R/U than it does in the assemblage generally. Over 17% of the chert artefacts (>1cm) have evidence for retouch and/or usewear (Table 8.9) significantly more than the percentage frequency for any other raw material. The fact that there are no cores made of this raw material suggests either that this was a valued raw material, artefacts of which were transported around the landscape in finished form; or perhaps that valued cores of this material were subsequently removed from the site for continued use elsewhere. Some knapping of this material obviously occurred at UDM; although some of this micro-debitage may have been produced by the usage of several of the chert artefacts (e.g. the tool from 6B/3 has extensive and heavy R/U and some notching). Many (60%) of the R/U chert artefacts are of translucent chert, more common in the Hunter Valley than the Sydney sandstone.

Size

The majority (71.7%) of the artefacts in the assemblage are <1cm long while most of the remainder (26.8%) are between 1-3cm long (Table 8.10). Relatively few artefacts (1.1%) are between 3-5cm, and two only (0.1%) are larger than 5cm[24].

Artefact Types

The majority (98.2%) of the artefacts at the site consists of unmodified debitage (Table 8.10). The remainder include 37 cores (1.1%) and 22 artefacts with retouch and/or usewear (R/U - 0.7%). Of this latter type, eight are backed artefacts. Seven (87.5%) of these are Bondi (i.e. asymmetric) points, while the other is a trapezoidal geometric. Also present amongst the assemblage are numerous flakes (63) with residual evidence of edge-grinding in the form of smooth faceted planes with striations. These artefacts indicate that several edged-ground implements have been broken up at UDM. None of these artefacts shows any sign of further utilisation and thus, while indicative of grinding as a technological component, they are not, strictly speaking, utilised artefacts. While included in the artefact type discussions, they have been excluded from some calculations of artefact type proportions.

Almost half (43%) of the material with evidence of grinding is micro-debitage (<1cm in size): two pieces only are in the 3-5cm range. From the colour ranges in this raw material it would appear that three edge-ground implements were knapped in the vicinity of Squares 6 and 8. None of this material has evidence for further retouch/usewear and these implements appear to have being broken up for some reason other than the production of useable flakes: possibly further flaking prior to additional grinding?

Of the eight backed artefacts at the site, four are incomplete. Two are butt ends probably broken during manufacture. The other two are the distal tips of backed blades, one of which has R/U on the chord. These may have broken during use: two of the eight have fine flaking along their chords, indicating some sort of use. The geometric backed blade has macroscopic residue, suggesting this artefact may have been hafted. There is no evidence for usewear or damage to the chord or point of this artefact.

There is an increase over time in the density of artefactual material deposited (Figure 8.22). This has been calculated on the basis of artefacts per kilogram of excavated deposit.

Vertical Distribution

Clear stratigraphic divisions in the deposit were observed in the field. Analysis of the excavated assemblage indicates that some of these differences are cultural and that there is change over time in the nature of the artefactual material deposited.

---

[24] At the time that this analysis was undertaken (1991) the artefact size was usually measured in these size categories. The current standard is 5mm intervals Rich (1992), McDonald *et al.* (1994), JMcD CHM 2005a.

Table 8.10: UDM Shelter. Artefact Types according to raw material and size. The three analysed squares (6B, 8B and 8C).

Raw Material	Unmodified Debitage				Cores	BA	R/U	Total
	<1	1-3	3-5	>5				
Quartz	953	261	15	0	27	1	1	1258
Silcrete	27	26	2	0	1	1	0	57
ST	499	184	6	0	2	2	7	700
FGB	434	127	4	1	0	0	64	630
Quartzite	10	23	0	0	1	0	1	35
Veined chert	345	133	4	0	5	1	2	490
Chert	48	23	0	0	1	2	3	77
Other	15	22	3	1	0	1	1	43
Total	2331	798	34	2	37	8	79	3290
%f	70.9	24.3	1.0	0.1	1.1	0.3	2.4	100.1
Total (R/U minus ground artefacts)								
	2358	836	35	2	37	8	14	3290
%f	71.7	25.4	1.1	0.1	1.1	0.3	0.4	100.1
Type		Size						
	<1	1-3	3-5	>5				
Unmodified*	2358	836	35	2				3231
R/U	1	18	3					22
Cores		27	10					37
Total	2359	881	48	2				3290
%f	71.7	26.8	1.5	0.1				100.1

* includes ground material

Artefact Density

Artefact densities in the earliest two units (III and IV) are the lowest, with a minor peak being demonstrated across all squares within unit IV. Unit II consistently contains the highest density of artefactual material with a decrease over time within this unit. Square 8B is anomalous: it contains the highest artefact density in the most recent spit.

Table 8.11: UDM Shelter. Modified artefacts and raw material types.

	Backed Artefacts	R/U	Grinding	Total	% of R/U	Artefacts >1cm	% R/U
Quartz	1	1	0	2	2.4	303	0.7
Silcrete	1	0	0	1	1.2	30	3.3
ST	2	5	0	7	8.2	161	4.3
FGB	0	1	63	64	75.3	629*	10.2
Quartzite	0	1	0	1	1.2	24	4.2
Volcanic	1	2	0	3	3.6	141	2.1
Chert	2	3	0	5	5.9	28	17.9
Other	1	1	0	2	2.4	26	7.7
Total	8	14	63	85		1342	6.3
%f	9.4	16.5	74.1				

*FGB total raw material has been included here since 43 % of the FGB material with evidence of grinding is in the <1cm size category.

Changes in artefact density seem to indicate that there were two main phases of occupation at the site, with peaks in these depositional periods coinciding first with unit IV and then later with unit II. The earlier period is characterised by a sparse artefact density; the later period was much more intensive. The occupation represented by Unit III indicates either a hiatus in shelter use, or at least a very low level of occupation. Interpretation of this occupation can be refined by more detailed artefact analyses.

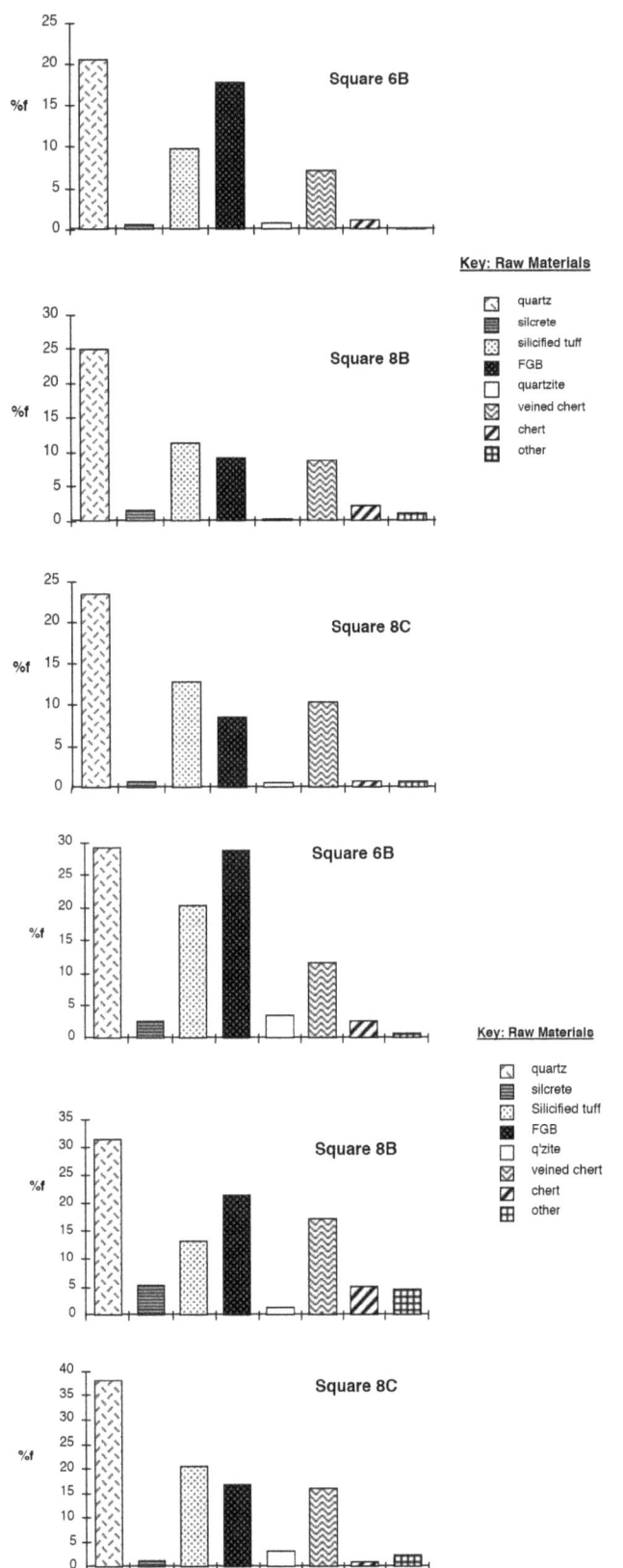

Figure 8.20: UDM. Raw material proportions in the three analysed squares. All artefacts.

Figure 8.21: UDM. Raw material proportions in the three analysed squares. Artefacts >1cm.

The age-depth curve and estimates of artefacts deposited/year produce a somewhat different picture[25]. While the two more recent stratigraphic layers exhibit the similar trends to those indicated by density calculations, the earliest strata have extremely low artefact deposition rates (see Table 8.12, Figure 8.22). The main period of occupation at the site is definitely defined by this reworking of the data, as is the sporadic nature of the earlier occupation.

[25] The middle date of 1,540±60 BP (ANU-8133) was used here as this was a provenanced sample, while ANU-8135 was a sample collected from the sieves.

Table 8.12: UDM Shelter. R/U artefacts and cores and raw material types.

Raw material	Backed Artefacts	R/U	Total R/U	% of R/U	cores	Artefacts >1cm	% R/U	% cores
Quartz	1	1	2	9.1	27	305	0.7	8.9
Silcrete	1	0	1	4.6	1	29	3.4	3.4
ST	2	5	7	31.8	2	201	3.5	1.1
FGB	0	1	1	4.6	0	169	0.1	-
Quartzite	0	1	1	4.6	1	25	4.0	4.0
Veined chert	1	2	3	13.6	5	145	2.1	3.4
Chert	2	3	5	22.7	1	29	17.2	3.4
Other	1	1	2	9.0	0	28	7.1	-
Total	8	14	22	(100)	37	931	2.4	4.0

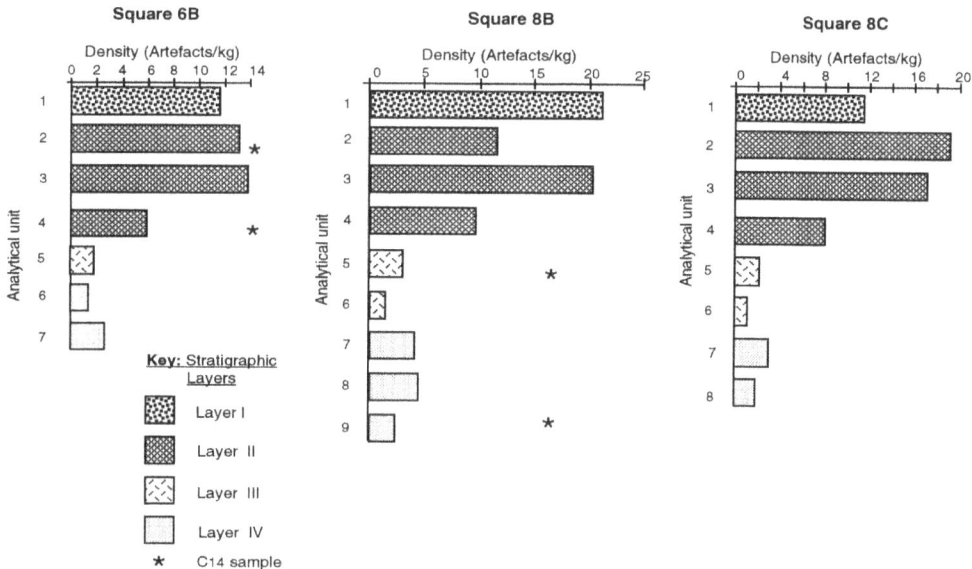

Figure 8.22: Artefact densities in the three analysed squares showing locations of the four radiocarbon dates.

Table 8.13: UDM Shelter. Age-depth calculations: artefacts per year.

Analytical Unit	Number of Artefacts	Number of Years	Artefacts/year
1	522	90	5.80
2	898	110	8.16
3	949	110	8.63
4	441	90	4.9
5	104	800	0.13
6	77	800	0.10
7	149	800	0.19
8	115	750	0.15
9	35	600	0.06

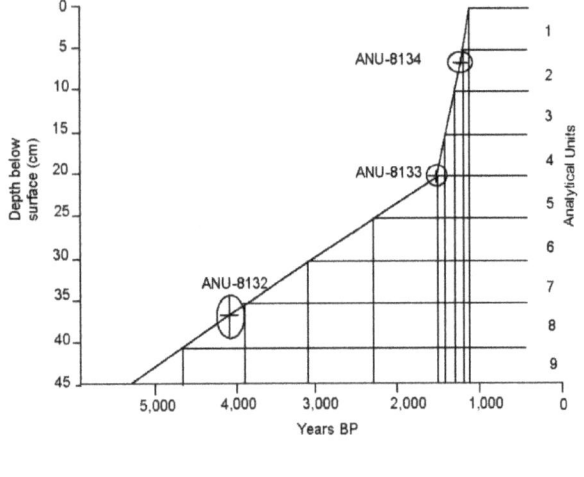

Figure 8.23: UDM shelter. Age depth curve and artefact accumulation rates by analytical and stratigraphic units.

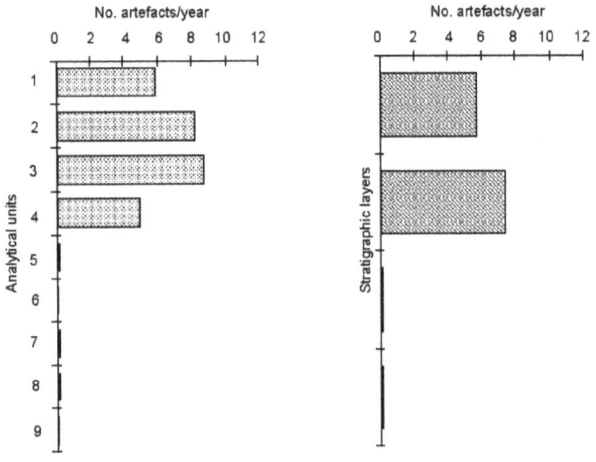

Figure 8.24: Correlation between sample size and number of raw material types per analytical unit.

Raw Material

There are definite changes in raw material preferences over time. While there are differences in sample size between the earlier and the later assemblages, the sample sizes in the earlier units are sufficiently large (i.e. >100 artefacts per stratigraphic unit) to make meaningful comparisons. Given the biases inherent in smaller sample sizes, a correlation between sample sizes and raw material ranges was sought (Figure 8.24). There is no direct positive correlation between sample size and the range of raw material present – so sample size will not affect the interpretation of these results (James 1993). The data below are based on calculations for both total artefact numbers (Figure 8.25) and artefacts >1cm (Figure 8.26).

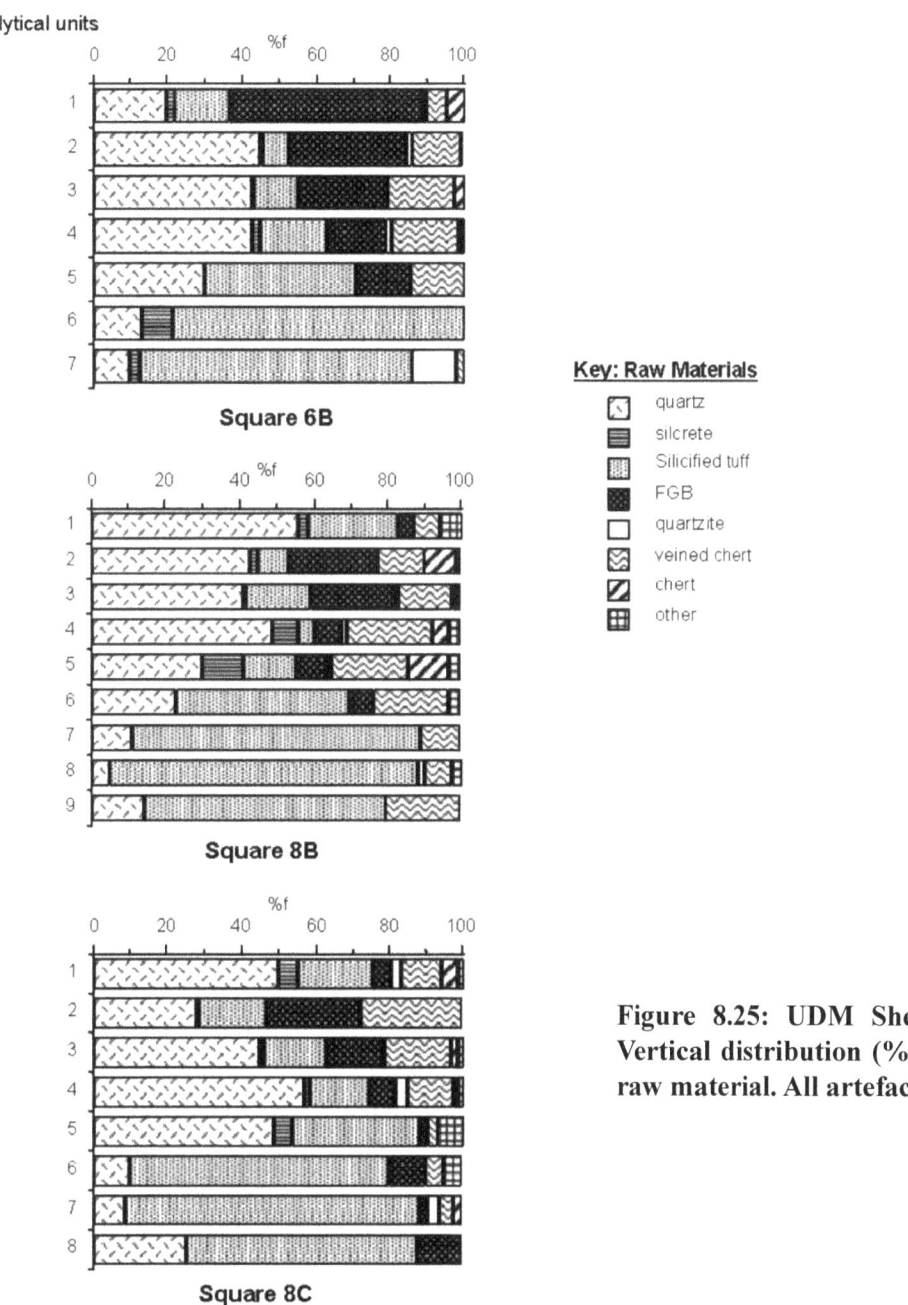

Figure 8.25: UDM Shelter. Vertical distribution (%f) of raw material. All artefacts.

In the earlier period of occupation, silicified tuff predominates and there is a more restricted range of raw materials generally. In the later period of occupation quartz gradually increases in use, while silicified tuff declines. There is a proliferation in the range of raw material being used. Some of these are not found at all in the earliest phase of occupation. The relationship between quartz and silicified tuff (Figure 8.27) is contrary to the divergent pattern which would be the expected result if sample size were responsible for raw material proportions (James 1993).

FGB material is notably absent from the earlier layers of deposit. While some microdebitage of this material occurs in square 8C, it is virtually absent from the earliest layers. Based on this evidence, the earlier phase is presumed to predate edge grinding as a technique. Sample sizes in Layers III and IV are comparable, justifying this conclusion. This evidence, supported by the dates received, suggests that the earlier occupation is Early Bondaian or older (Attenbrow 2004).

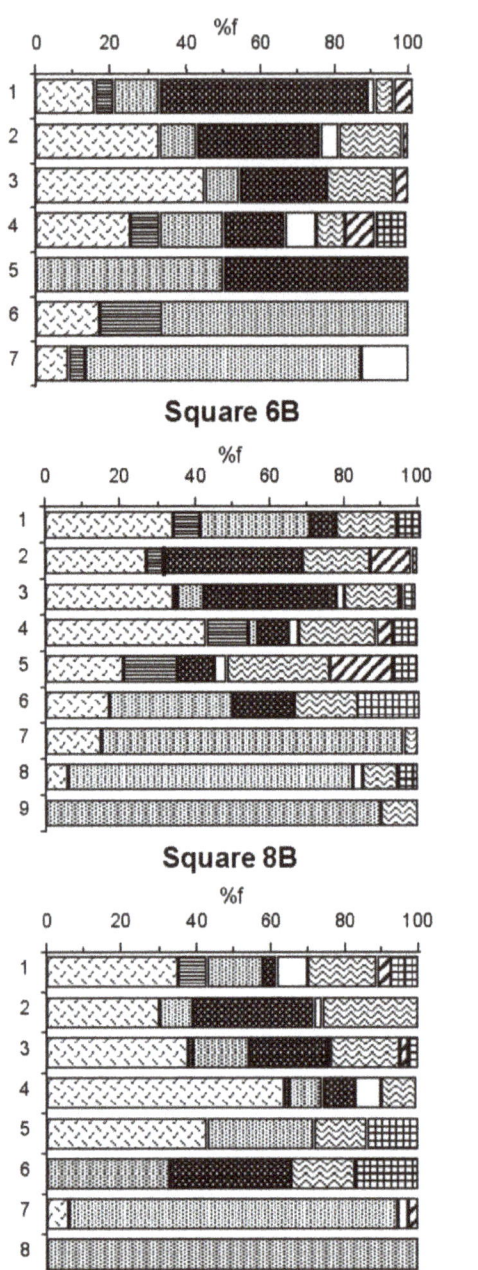
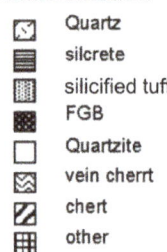

Figure 8.26: UDM Shelter. Vertical distribution (%f) of raw material per analytical unit. Artefacts >1cm.

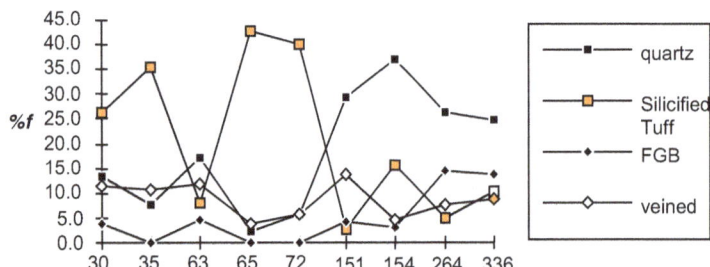

Figure 8.27: UDM. Raw material proportions plotted against sample size.

Table 8.14: UDM Shelter. Size ranges for backed blades and artefacts with retouch/usewear per stratigraphic layer.

Stratigraphic layer	<1cm	1-3cm	3-5cm
I	1	3	-
II	-	11	-
III	-	1	-
IV	-	3	3

Table 8.15: UDM Shelter. Size ranges for cores per stratigraphic layer.

Stratigraphic layer	<1cm	1-3cm	3-5cm
I	-	2	-
II	-	23	5
III	-	-	2
IV	-	2	3*

*includes core with R/U

Size

As noted above, the majority of the lithic material at the site is very small. There are definite indications, particularly amongst artefacts with R/U and cores (all tending to be >1cm) that there is a general decrease in the size of the assemblage over time.

In stratigraphic layer IV, only larger artefacts with R/U occur; in Layers II and III all of the retouched artefacts are between 1-3cm, while in Layer I, most of these artefacts are 1-3cm long and the other is <1cm long (Table 8.14). Because of small sample sizes these items are discussed in terms of stratigraphic rather than analytical units. Cores show a similar general reduction in size with time. In Layer I, all are in the 1-3cm category; in Layer II the majority are in the 1-3cm category, while in Layer IV less than half are in this size category (Table 8.15).

The debitage indicates a similar trend of decreasing size with time (Table 8.16 and Table 8.17). While there is a general decrease in size with time, the two artefacts in the excavated assemblage which are >5cm occur in layers II and III. The artefact in Layer II consists of a porphyry (BP) flake, the only piece of this material at the site. The large artefact in Layer III is an FGB flake (with a broad platform and platform preparation). Both of these larger artefacts were found in square 8C.

Table 8.16: UDM Shelter. Raw material per analytical unit. All Artefacts.

Unit	Quartz	Silcrete	IM	FGB	Q'zite	V Cht	Chert	Other	Total
1	198	16	98	137	5	39	18	11	522
2	342	7	106	242	7	166	23	5	898
3	408	7	149	194	4	160	18	9	949
4	224	15	55	41	10	77	10	9	441
5	39	9	23	6	1	14	7	5	104
6	17	0	39	8	0	11	0	2	77
7	16	2	117	1	2	10	1	0	149
8	9	1	90	1	6	6	0	2	115
9	5		23		0	7			35
Total	1,258	57	700	630	35	490	77	43	3,290

During the earlier occupation periods at the site around half of the material was larger than 1cm, while in the more recent phases only 20-30% of the assemblage was this big. Micro-debitage may well indicate intensity of knapping activity. It has also been suggested (Hiscock 1986) that raw material and knapping technique may contribute to the amount of shatter in an assemblage, and that the change to bipolar technique is heralded by the very high proportion of small material in the later phases of the Eastern Regional Sequence. At UDM both bipolar quartz and indurated mudstone (knapped used a hand held technique) reveal a similar pattern in artefact size reduction between the earlier and later phases at the Upside-Down-Man Shelter. This tends to support the hypothesis of a greater degree of knapping *per se* in the upper levels.

Table 8.17: UDM Shelter. Raw material per analytical unit. Artefacts >1cm.

Unit	Quartz	Silcrete	IM	FGB	Q'zite	V Cht	Chert	Other	Total
1	31	7	22	28	3	14	3	5	113
2	60	3	12	71	4	41	9	3	203
3	111	3	32	75	4	50	8	7	290
4	81	10	11	15	8	23	4	6	158
5	9	4	2	3	1	9	5	3	36
6	1	0	5	4	0	2	0	2	14
7	8	2	60	0	1	2	0	0	73
8	4	1	48	0	4	3	0	2	62
9	0	0	9	0	0	1	0	0	10
Total	305	30	201	196	25	145	29	28	959

Artefact Type

Debitage

Platform characteristics of the 424 flakes and the shape of the 393 flaked pieces (i.e. all debitage >1cm) were analysed to investigate possible changes over time. Four types of platforms were distinguished on the flakes; broad ('B'), broad with evidence for platform preparation (i.e. platform remnant on dorsal surface; several scars on platforms - 'BP'), focalised ('F') and bipolar ('Bip'). Two types of flaked pieces were identified; amorphous ('Am') and lamellate ('Lamm') (see McDonald 1994: Appendix A3.1 for raw data; Appendix 8 for glossary of terms used).

The proportions and the changes in the proportions of these characteristics were investigated (Table 8.17; Figure 8.28). Detailed technological analysis was not undertaken, but the approach was based on the findings of the (then) recent work of Baker (1992), following Hiscock's (1986) approach to technological change during the Holocene. Baker's work had refined Hiscock's earlier model and determined that certain flake characteristics are sensitive indicators of technological strategies and change over time. He demonstrated that:

> the Sandy Hollow pre-Bondaian level SH1/5 is characterised by core rotation, a lack of platform preparation and ... low platform angles ... By contrast Phase I levels SH1/3 and SH1/4 are characterised by platform preparation and high platform angles typical of a blade based technology. Phase II levels SH1/1 and SH1/2 are characterised by focalised platforms which represent continued attempts to remove thin flakes or, more probably, blades from the core. (Baker 1992: 84)

Baker's analysis validated Hiscock's analytical approach at the Sandy Hollow site. However, he encountered difficulties using the approach to test for chronological phases in the Narama open site assemblages from the central Hunter lowlands (Rich 1992). There was no consistent association between any of the Sandy Hollow levels and the Narama assemblages, which, from their associations with backed artefacts and other flake and core characteristics, were clearly Bondaian in age (Baker 1992:84). Baker's analysis indicated that an abbreviated set of variables may as effectively allow different technological phases to be differentiated. The current analysis was attempted on the basis of that finding.

At UDM, focalised platforms are the most common form (43.6%), followed by bipolar flakes (30.9%) and broad platforms (23.1%). Platform preparation is relatively rare and was observable on only 10 flakes (2.4%). Lamellate flaked pieces (56.5%) are slightly more common than amorphous flaked pieces (43.5%: Figure 8.28).

The most consistent indicator of change over time is flake platform shape. The general pattern is one of a decrease over time in broad platformed flakes, with a concomitant increase in bipolar flakes. Focalised platforms are present in roughly the same proportions throughout the sequence although percentage figures for this platform type are consistently highest in the upper layers (Table 8.18). Platform preparation appears to be present only in the early to mid occupation of the shelter.

Table 8.18: Platform characteristics over time. Frequency and %f per analytical unit.

Unit	Broad		BP		Focal		Bipolar		Total
		%		%		%		%	
1	8	19.5	0	0	27	65.9	6	14.6	41
2	11	12	0	0	44	47.8	37	40.2	92
3	25	19.5	0	0	51	39.8	52	40.6	128
4	9	13.6	1	1.5	27	40.9	29	43.9	66
5	2	13.3	2	13.3	7	46.7	4	26.7	15
6	4	40	1	10	5	50	0	0	10
7	17	47.2	5	13.9	13	36.1	1	2.8	36
8	20	69	0	0	7	24.1	2	6.9	29
9	2	28.6	1	14.3	4	57.1	0	0	7

With the exception of analytical unit 9 (where n=2), the proportions of lamellate to amorphous flaked pieces is generally consistent over time (Figure 8.28).

Cores

The 37 cores were analysed, with multiplatformed ('M'), single platformed ('S') and bipolar ('B') types being distinguished. The general proportions and changes in these characteristics were investigated and the results plotted (Figure 8.28).

Only bipolar (86.5%) and multiplatformed cores were found. The fact that no single-platformed cores were observed possibly reflects a conservation of raw materials by the knappers at the site: i.e. cores were used until exhausted, or at least more than once.

The sample size for cores is relatively small for considering temporal change at the site. However, with the exception of one bipolar core found in spit 8B/7, all the cores in the lower levels are multiplatformed while all those in the upper layers are bipolar (Figure 8.28). This pattern reflects the trend observed in the debitage for an increase in dominance over time in the bipolar flaking of (particularly) quartz.

Using the mosaic of platform characteristics over time, the assemblage is best interpreted as demonstrating two phases of knapping over time.

1. The earliest phase is characterised by the presence of mainly broad-platformed flakes although focal platforms are used; platform preparation and bipolar techniques are rare.

2   The subsequent phase shows a sharp increase in platform preparation and an increase in focal platforms. The bipolar technique becomes increasingly important. This phase represents the most intensive occupation (in which the highest discard rate occurred). Broad platforms occur in much smaller proportions. No platform preparation is in evidence. In the terminal phase of site occupation a decreased artefact frequency (and sample size) results in fewer bipolar pieces and an increase in focal platforms.

Backed Artefacts

Eight backed artefacts were found at the site. There are no raw material preference trends over the short period of their use (Figure 8.30). These were only found in the top five units: in Layers I, II and the top of unit III (Figure 8.29). The presence of this artefact type coincides with the period of maximum artefact density (and larger sample sizes), and the increase in quartz and the bipolar technique. Flakes with focalised platforms also predominate here, as expected with a technology geared to the production of backed artefacts.

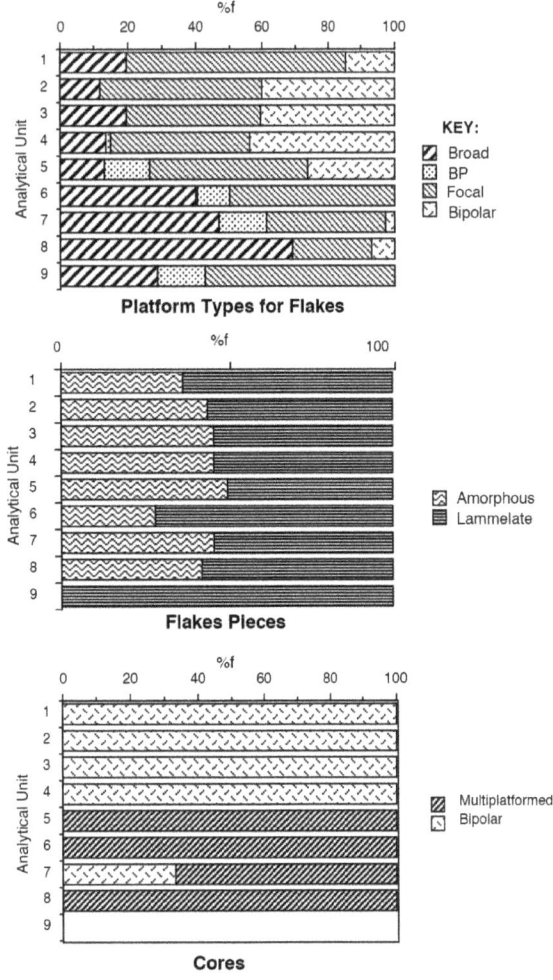

Figure 8.28: UDM Shelter. Platform, flaked piece and core types over time.

The presence of this tool type postdates the introduction of focalised platforms and flakes with platform preparation. This pattern is consistent with other excavated assemblages (e.g. Sandy Hollow), and would appear to indicate either a time lag in the appearance of this tool type after the appropriate technology had evolved, or that the production of backed blades was initially only a low frequency occurrence, leading to very low deposition rates for this rarer artefact type (i.e. a sampling effect).

Amorphously Retouched Flakes

Fourteen artefacts with retouch/usewear (R/U) were recorded from the three analysed squares. These were found in all analytical units except 5 (Figure 8.29). While there is a general decrease in the size of retouched artefacts with time (Table 8.14) and a trend in raw material preferences (i.e. from exclusively indurated mudstone to a proliferation of raw material types: Figure 8.30), there are few other artefact characteristics which are easily seen as changing throughout the deposit. With the exception of one core tool found in unit IV (spit 6B/7) the artefacts in this class can only be described as amorphously retouched pieces; there are no scrapers or any other diagnostic tool types present.

Table 8.19: UDM Shelter. Artefacts with R/U per stratigraphic layer. Platform and flaked piece types.

Layer	Broad	BP#	Focal	Bipolar	Amorph's	Lammel
I	-	-	1	-		1
II	1	-	2	1	-	-
III	2	-	1	-	-	-
IV*	2	1	2	-	1	-

*excludes core with r/u
broad platform with evidence of previous platform preparation

There are suggestions of trends in flake platform and flaked piece shapes over time, but the sample is too small to make much of these (Table 8.19). While the sample is extremely small, the trend indicated by the retouched material is similar to the general pattern found in the whole assemblage (Figure 8.29); i.e. a change over time from broad platform in the earlier deposit, to focal platforms and bipolar technique in the later deposit.

As with artefact density (Figure 8.22) and platform shapes (Figure 8.28) the vertical distribution of the artefacts with macroscopic R/U suggests two phases of site occupation. There are two peaks in the use of artefacts (Figure 8.29). The very low number of artefacts with R/U in Layer III suggests that this period represents some sort of hiatus in site usage, rather than a period of very low density occupation. Again, low accumulation rates and smaller sample size in the lower units makes this interpretation difficult.

Grinding

As well as the ground edged hatchet collected in the dripline and the many grinding grooves located outside the sandstone shelter, grinding activity is documented by the presence of 63 fragments with ground facets in the deposit. They are all unmodified debitage and the distinctive colour ranges indicate that at least three edge-ground implements were broken up at the site. The three colours are black, blue-grey and brown-grey. While the blue-grey is found in both pits, black occurs only in square 6B, but the brown-grey is restricted to square 8 (Table 8.20). Only a small proportion of the blue-grey occurs in square 8, and it would appear that the focus for the knapping of the first two colours was in square 6B.

The vertical distribution of this material includes Layers I-III (Table 8.20), with the main concentration in the upper layers of Layer II (Figure 8.29).

Table 8.20: Colour distribution of ground FGB artefacts; as per stratigraphic unit.

Stratigraphic layer	black	blue-grey	brown-grey
I	3	3	
II	12	10	33
III	-	2	1
IV	-	-	-
	< Square 6B >		< Squares 8B and C >

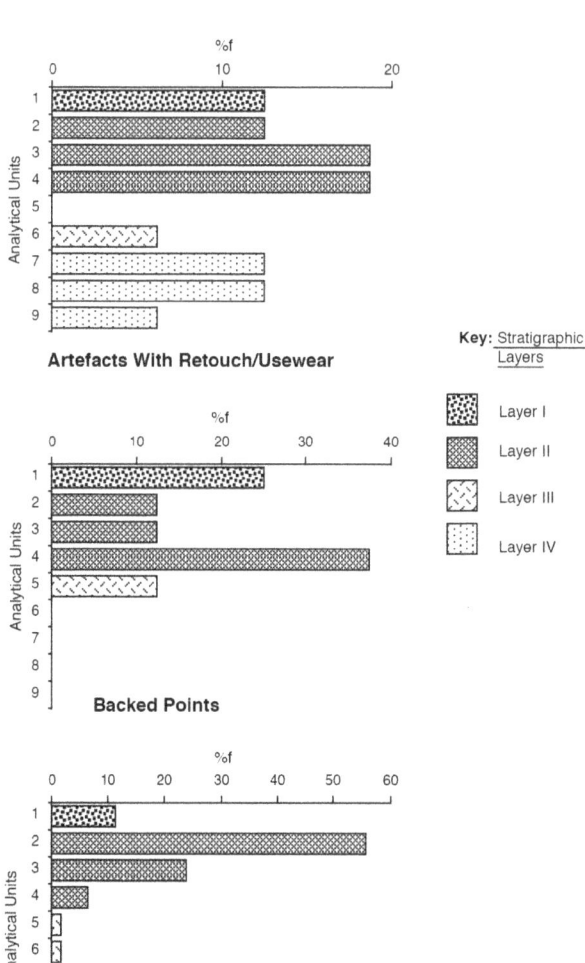

Figure 8.29: UDM Shelter. Vertical distribution of artefacts with retouch usewear, grinding and backed blades.

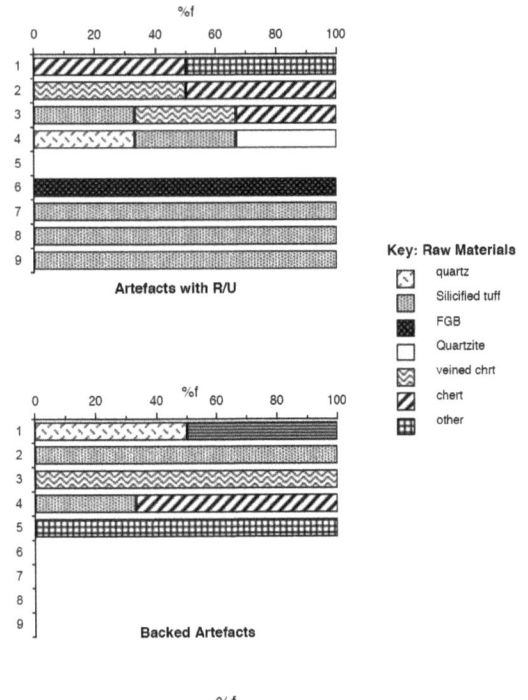

Figure 8.30: UDM Shelter. Modified artefacts and raw materials shown by analytical units.

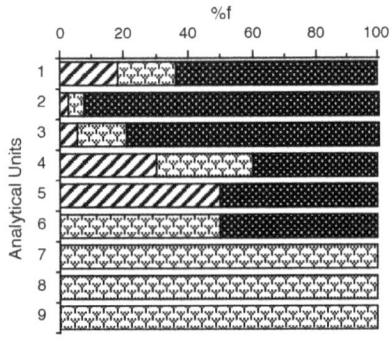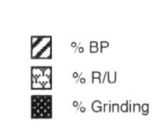

Figure 8.31: UDM Shelter. Proportions of modified material per analytical unit.

Discussion

The aims of this excavation and analysis were to determine:

1) the nature and timing of the occupation evidence at the site;

2) whether there were two or perhaps three phases of occupation evidence at the site (as suggested by the art);

3) whether the art was contemporaneous with the occupation deposit.

The first two aims were achieved by the excavation.

Contemporaneity has thus been argued on the basis of excavated and associated evidence. It had been hoped to resolve this issue more specifically – and charcoal samples were collected from two drawn macropod motifs at the site in the hope of dating these using Accelerator Mass Spectrometry (AMS). Four samples were collected (Figure 8.11, Figure 8.12) and submitted to the ANU Quaternary Dating Research Centre for processing. The dates returned on these samples were inconclusive and indicate sampling problems (McDonald 2000c).

Only one site with a large art assemblage, *Dingo and Horned Anthropomorph* (DHA), has been excavated previously in the Mangrove Creek valley (MacIntosh 1965). The analysis of this excavated material was cursory and not comparable to more recent work. The finds associated with this art assemblage were reported as 'sparse'. It should be noted that the opportunity afforded by DHA for camping, in terms of flat, protected floor area, is limited.

None of Attenbrow's 31 excavated sites in Upper Mangrove Creek (UMCC) contained a major art assemblage. The largest UMCC art assemblage (Site #17) contained 61 motifs. The issue of differing site functions, then, could not be addressed without the excavation of a major art site with accompanying occupation debris. Characterising the occupation evidence at UDM was of additional interest.

Of further interest was a comparison of the UDM site with the results of Attenbrow's (1987, 2004) UMCC research. Attenbrow's defined patterns of site usage - over the last three millennia in particular - are of considerable relevance to the analysis of diachronic variability in the shelter art in this valley. All her sites were located in the upper reaches of Mangrove Creek, as was *Dingo and Horned Anthropomorph*. Patterns in the middle to lower reaches of this valley system test the wider applicability of Attenbrow's results.

As well as the specific UDM excavation aims, there was the broader aim of identifying how shelter occupation evidence associated with a large art assemblage fitted into Attenbrow's scheme for the prehistoric use of Mangrove Creek valley. Significant differences in art site content in the Mangrove Creek Valley have been identified (cf. Gunn 1979; Attenbrow 1987; and McDonald 1988a) raising questions about the stylistic variability inherent in any particular catchment. The diachronic art analyses undertaken for this research used the art from the wider Mangrove Creek catchment. Because Attenbrow's excavated data was derived almost entirely from sites with 'minor' art assemblages, there was a need to investigate occupation evidence from a major art site. The UDM shelter fulfilled both criteria of art assemblage size and geographic location.

The Archaeology of Mangrove Creek

A Sydney-wide regional perspective of the archaeological context is discussed in detail in chapter 4. Here the archaeology of the Mangrove Creek valley, particularly in its upper reaches (Attenbrow 1981, 1987, 2004) is discussed. The art from this valley has also been the focus for considerable previous interest and research (Gunn 1979; McDonald 1987, 1988a; Smith 1983; Vinnicombe 1984). Because of this prior research focus, the art of this drainage basin is used to exemplify diachronic change in the Sydney region. More than 80 shelter art sites have been recorded from this catchment, 65 in sufficient detail for further analysis. Many of these sites were located as a result of a stratified random sampling procedure (Attenbrow 1987, McDonald 1988a). This sample of shelter art sites represents one of the most systematically collected in the region. With the detailed archaeological context provided by Attenbrow's excavations, the shelter art of this valley appeared ideally suited to answering questions of the context and contemporaneity of shelter art with general occupation evidence. The exact nature, direction and timing of changes in the archaeological record, particularly occupation indices, were of considerable interest.

This current analysis has used a combination of (the modified) Hiscock, Baker and Attenbrow approaches to determine the nature and timing of change at Upside-Down-Man. Before discussing the specifics of artefact change at UDM, however, the general pattern of change in the Upper Mangrove Creek (UMCC) are outlined.

UMCC

To measure quantitative change, Attenbrow converted into indices the number of habitations (excavated shelter sites) used and the artefact accumulations within these. Indices used were the rate of habitation establishment and use over time and the rate of artefact accumulation. Rate was calculated as frequency per millennium for each factor.

Attenbrow identified highly variable patterns of individual habitation (site) usage and thus averaged the catchment's data to demonstrate generalised patterns and to quantify local characteristics. She found that while there was a continuing increase in the number of habitations established and used in successive periods of time, there was a substantial decrease in the local artefact accumulation rates during the most recent phase and last millennium. She also identified that the onset of quantitative changes and the introduction of changes in typology, technology and lithology did not necessarily coincide. Two scales of temporal change were achieved by using millennial increments as well as typological phases. Attenbrow concludes that:

> using the millennial increments as the basis for the temporal sequences, results in a more accurate representation of the trends in the [intensity] indices, than does the use of the typological phases. ... The main implication is that each aspect of the archaeological record is likely to have its own trajectory and temporal sequence, and therefore should not be presented in terms of previously constructed sequences designed for other purposes. (Attenbrow 1987: 214)

Sampling Issues

Attenbrow's habitation indices were based on artefact analysis and radiocarbon dates from eleven sites. The identification of phases within the assemblages at the remaining 20 sites was achieved on typological grounds (i.e. the presence/absence of particular artefacts etc.). The typological data that Attenbrow analysed are in summary form in her thesis but vertical distribution data was provided only for Loggers and Mussel shelters (Attenbrow 2004: Table 4.6).

Data for Attenbrow's Phase 4 sites (Attenbrow 2004: Table 6.2) indicates that certain sampling problems are inherent. The classification of assemblages in this time period was based both on the depth of deposit and the typological characteristics of the assemblages (Attenbrow pers. comm.). Small sample sizes are an issue.

Should the absence of backed implements be considered sufficient evidence for Phase 4 occupation? It would seem that the absence of this artefact type most clearly distinguishes Attenbrow's Phase 3 from Phase 4. The designated Phase 4 sites all contain less than 25 artefacts *in toto*. The percentage frequency of backed artefacts at sites with large assemblages is between 0.2-0.9% (at the five sites with >1000 artefacts; 2004: Table 4.6). Thus the probability of locating backed artefacts in such small assemblages would be extremely low (~0.025 backed blades would be expected in an assemblage with 25 artefacts). The absence of backed artefacts in the assemblages for the vast majority of Attenbrow's most recent sites may be due to no more than sampling bias.

Because the majority of UMCC Phase 4 sites have extremely small assemblages, doubt is also cast upon the calculations of artefact accumulation rates over time and the conclusion that this rate dropped significantly in the last millennium. Similarly, the rates of shelter establishment could also be affected by this unfounded placement of sites into the later parts of the sequence.

Rates of artefact accumulation were based on the estimated total number of artefacts in an archaeological deposit and on the estimated total number within each spit/phase/millennium. Estimated artefact totals were (quite reasonably) used 'to avoid problems associated with inter-site variability' (Attenbrow 1987: 202) and in an effort to make comparable disparate data sets. Estimates were based on mostly very small and very different sized samples. At most sites (68%) between one and four 50cm x 50cm test pits were excavated, representing <3% of the total floor area. At only two sites (Loggers and Black Hands) was more than 10% of the floor area excavated (Attenbrow 2004: Table 3.5). The number of stone artefacts retrieved from sites also varied enormously, ranging from 1-30 artefacts at 15 sites to more than 7,000 at another site.

The index for artefact accumulation rates appears to have the most problems, since the method for calculating this is subject to more sources of potential error than that for habitation establishment. However, the classification of assemblages into typological phases on the basis of extremely small sample sizes, for the majority of Attenbrow's sample, also has the potential to have skewed the results towards a proliferation of sites with low artefact numbers (and therefore low accumulation rates and late establishment dates) in the most recent millennium.

In order to establish if these sampling problems have affected Attenbrow's patterns, her data were re-analysed using only the seven larger sites (i.e. >100 excavated artefacts).

This testing confirmed the general pattern of Attenbrow's results for artefact accumulation rates particularly in the most recent millennia (N.B. however the reversal indicated between Phases 1 and 2). However, it revealed very different patterns in terms of the habitation indices (Table 8.21, Table 8.22; Figure 8.32, Figure 8.33).

Table 8.21: UMCC Sites with >100 excavated artefacts. Reworked estimated artefacts totals in successive periods of time (Attenbrow 2004: Table 6.7).

Site	Phase 1	Phase 2	Phase 3	Phase 4	Total
Loggers	63,000	5,600	22,950	36,450	128,000
Uprooted tree	800	3,250	4,650	22,500	31,200
White figure	500	2,100	19,250	15,250	37,100
Sunny		3,600	38,550	46,050	88,200
Emu Tracks 2		14,500	199,000	36,100	249,600
Black Hands		950	41,700	34,350	77,000
Delight SH			3,200	1,800	5,000
7 UMCC habitations	64,300	30,000	329,300	192,500	616,100
UDM	17,588	10,677	165,294		193,559
Mangrove Ck Tot	81,888	40,677	494,594	192,500	809,659

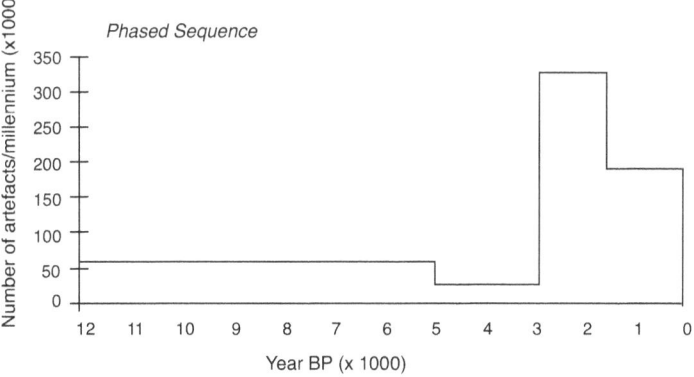

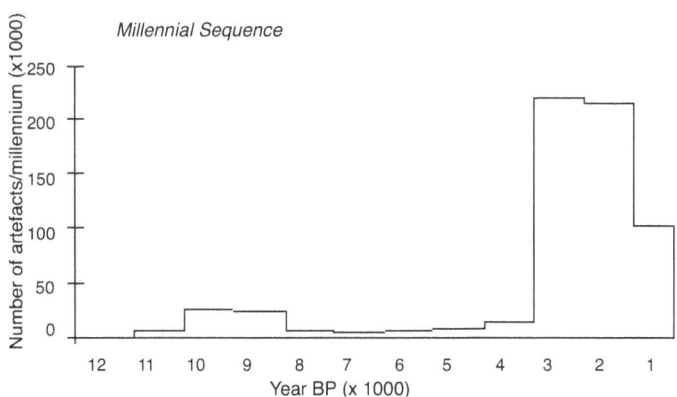

Figure 8.32: Reworked UMCC data. Local rates of artefact accumulation in successive periods (cf. Attenbrow 2004: Table 6.9).

As indicated by Attenbrow's results, artefact accumulation peaked in Phase 2 (particularly between 3,000 and 1,000 years ago), followed by a decrease in the last millennium.

The re-calculation of habitation indices, albeit based on a considerably smaller number of sites, indicates that there was a low establishment rate up until c. 4,000 years ago. After the third millennium, no new shelters were established. The pattern revealed by number of habitations in use over time is like that achieved by Attenbrow, although there is no increase, by these re-calculations, in either the last 2,000 or 1,000 years. This index shows a consistent or stable use of the same shelters over the last three thousand years.

Table 8.22: UMCC Sites with >100 excavated artefacts. Reworked estimated totals of artefacts in successive periods of time: subsequent millennia (Attenbrow 2004: Table 6.13).

Site	12th	11th	10th	9th	8th	7th	6th	5th	4th	3rd	2nd	1st
Loggers	150	3850	24,100	23,200	4,300	3,650	3,550	4,150	5,850	7,000	11,800	36,450
Uprooted tree					200	300	850	1,150	1,300	2,350	13,050	12,050
White Figure				50			300	450	1,550	11,550	17,250	6,000
Sunny								300	2,100	13,950	55,850	16,050
Emu Tracks 2									3,650	165,200	66,300	14,450
Black Hands									100	16,050	46,300	14,600
Delight Shelter											2,650	1,100
Total UMCC	150	3850	24,100	24,100	4,500	3,950	4,700	6,050	14,550	217,350	213,200	100,700
UDM									17,588	10,588	165,294	0
Mangrove Creek Total	150	3850	24,100	24,100	4,500	3,950	4,700	6,050	32,138	227,938	378,494	100,700

Discussion

One of Attenbrow's most significant findings was the marked contrast between:

i) the substantial decrease in the local artefact accumulation rates during the most recent phase and millennium; and,

ii) the persistent increase in both the rates of occupation establishment and numbers of habitations used over time.

The re-testing of Attenbrow's data using only sites with reasonable sample sizes does not support this contrast. The pattern would appear to indicate stability over the last 3,000 years in the habitation indices, while supporting a decrease in artefact accumulation in the established shelters in the last millennium.

This pattern does not suggest the same degree of mobility or increase in the territorial range over the last 2,000 years as the occupation mosaic proposed by Attenbrow. It does however emphasise the decrease in shelter site usage - or at least the deposition of stone artefacts - in the last millennium.

Another of Attenbrow's important conclusions was the unsynchronised nature of changes in the occupation indices. She argued there was no coincidence in the timing of the typological phases with either artefact discard rates or the numbers of shelter being used. The re-calculations done here do not support this finding. Rather they show a major coincidence in typological changes (to the Middle Bondaian) with an increased artefact discard rate and increasing number of shelters being used. The peak millennium for habitation establishment appears to be the fourth, predating the typological change and increase in artefact discard rate. But these different sources of evidence all appear to support a proliferation of activity and possibly a major social change at the beginning of the third millennium.

On the basis of these results the following model for the catchment is proposed.

The period of most intensive shelter usage in the valley appears to have been between 3,000 and 1,000 years ago, when enormous numbers of artefacts were deposited within shelters. The

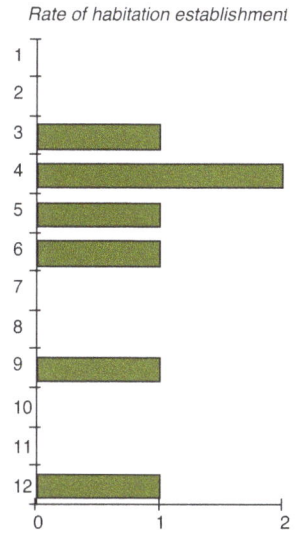
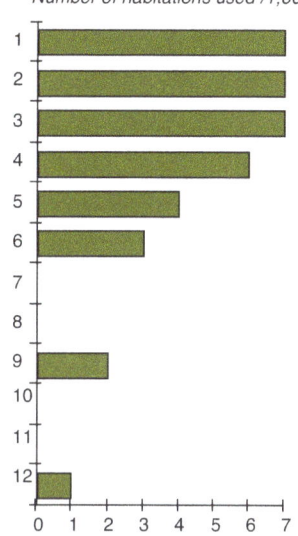

Figure 8.33: UMCC reworked data. Rates of habitation establishment and habitation use in the seven sites with >100 excavated artefacts (cf. Attenbrow 2004: Figure 6.2).

beginning of this major increase coincides with the beginning of Phase 3, when backed blade production was at its peak and the assemblages began to be dominated by bipolar flakes of (particularly) quartz. This peak period slightly postdates an increased habitation establishment rate, but coincides well with the beginning of a relatively stable period of shelter occupation.

The local artefact accumulation rate drops significantly in the last millennium, while the rate of habitation usage remains consistent. People continued using their established shelters but they deposited less artefacts within these locations. All three indices suggest that over the last 3,000 years there was an increased use of shelters. On the basis of artefact accumulation rates it would appear that the use of shelters as foci in the landscape - for stone tool manufacture or discard at least - declined in the last millennium.

Morwood (1986) identified a similar pattern at Gatton in SE Queensland in the last millennium. As well as decreasing artefact discard and the disappearance of backed blades and 'barbs', he noted that;

> increased rates of faunal, charcoal and sediment deposition indicate that the site was being used more intensively, while late increases in faunal diversity show a broadening of the resource base, with more extensive exploitation of a wider habitat and species range. ... general changes in the technology of predation are suggested. (Morwood 1986: 117)

He concluded that;

> Given the integrated nature of subsistence settlement systems, there are likely to have been associated changes in other components such as group size, frequency of site occupation, duration of occupation and inter-site distribution of activities. (Morwood 1986: 117)

Two of Attenbrow's shelters (Loggers and Mussel) contained significant faunal remains (Aplin 1981, Attenbrow 2004: 70, 92). Neither of these sites showed the same pattern as that found at Gatton.

Loggers revealed a contrary pattern both to the Gatton shelter and to general trends in the UMCC: an increased deposition rate for artefacts in the top spits - dated to the last millennium[26]. While sedimentation was fairly constant over time, there is an increase in organic matter in the top 30cm of deposit, coinciding with increased artefact density (Hughes and Sullivan 1979: Figure

[26] Three geometrics were found in spit 2 which was dated to 780 ± 80 BP (SUA-1124). Attenbrow has classified the two upper spits at Loggers as Late Bondaian on the basis of the date and not on artefact typology. [Attenbrow has reported that reanalysis of artefacts originally identified in spit 2 as eloueras, has revealed that these had been incorrectly identified; pers. comm. 1994]. The Logger's artefact density and typological data (Attenbrow 1981: Table 6.6) suggests that spit 2 could more properly be assigned to the upper level of the Middle Bondaian layer, supporting a later changeover date for these two Phases (like UDM and Yengo 1). With a reclassification of spit 2, Loggers shelter would no longer anomalous in terms of increased artefact accumulation rates in Phase 4.

10). Charcoal was collected only for dating purposes at Loggers (Attenbrow 1981: 64) so no analysis of changing proportions of this component was undertaken.

Aplin described a proliferation of faunal remains and increase in the number of species in the most recent unit (319 individuals; 23 species) compared with the preceding, Middle Bondaian, unit (66 individuals; 14 species: Aplin 1981: 25; Table 4). Unit I contained new species from both 'wet' and 'dry' environmental conditions. While he does not draw the conclusion, it could be inferred that these species indicate a broader resource range in the most recent period. Aplin does point out that there is a decrease in the relative abundance of large macropods in the most recent level, this being 'compensated by an increase in the abundance of smaller macropods as well as in the introduction of several smaller species not previously present amongst the assemblage e.g. bandicoots, wombats etc.' (Aplin 1981: 23-5). Thus while the artefacts do not conform, patterns identified by Morwood (increase in organic content, late increase in faunal diversity) are suggested at Loggers.

At Mussel, where the top unit is clearly (typologically) late Bondaian there is a sharp decrease in artefact densities. Hughes and Sullivan (1979: Appendix I) identified an increase in organic matter in the top spits. This shelter then would appear to support Morwood's conclusions: a decrease in artefacts in the most recent deposit during a period of increased sedimentation. Aplin, however, determined a concomitant decrease in faunal remains between the middle and top units (1981: Table 7). He also identified that the species in the middle unit are predominantly from 'dry' environments while those from the upper unit are from 'wet' environments.

Interpretation of the Upside-Down-Man site

The initial occupation of the UDM site took place around 4,000 BP. This was relatively ephemeral with low artefact deposition rates. Raw material preferences were restricted, with silicified tuff (known as 'indurated mudstone' in 1994 and 'chert' in Attenbrow's classification) predominating. Later occupation was more intensive and is marked by the introduction of backed implements and a focus on bipolar knapping of quartz. Backed implements were in use throughout the later occupation of the shelter, which appears to have been abandoned before the start of the last millennium.

Based on the assemblage's typological characteristics, and compared with Attenbrow's diagnostic traits, the earlier occupation appears to have taken place during Phase 1, while the later occupation would appear to be Phase 3 (compare Table 8.21 with Figure 8.22 to Figure 8.25). A period of even lower occupation between these two 'phases' (analytical units 5 and 6; Stratigraphic Layer III) may represent a hiatus in site usage, but could also represent Attenbrow's Phase 2: the absence of backed and ground material may be the result of low sample sizes (Table 8.16 and Table 8.17). FGB does occur within Layer III but ground fragments are found only at the top of this Layer and these are demonstrably part of the knapping event focussed within Layer II. There is no Phase 4. Backed artefacts are present in spit I and the terminal date for the site (at 6.5cm depth) was 1,220 ± 120BP.

Using the modified Hiscock approach (Baker 1992), the assemblage characteristics throughout the site's usage are similar to that found by Hiscock at Sandy Hollow (SH1). At UDM there is the additional use of the bipolar technique. The earliest UDM phase is characterised by a lack of platform preparation (viz. SH1 Pre-Bondaian levels). The assemblage in Unit III is characterised by platform preparation (viz. SH1 Phase I Bondaian levels), while the most recent UDM assemblage is characterised by focalised platforms (Figure 8.28).

While the typological analyses indicate that the site was occupied by knappers during Attenbrow's Phases 1-3 (Hiscock's Pre-Bondaian and Phase I and Phase II Bondaian) the dates for UDM are not in accord with the dates from Sandy Hollow, nor within the time frames set by Attenbrow for these Phases in Upper Mangrove Creek. The inherent inaccuracies of the SH1 dates (Hiscock 1986: 42) means that this discord in dating is not problematic. The differences with the general trends outlined by Attenbrow, however, require further discussion.

The UDM Phase 1 material is more recent - by at least a millennium - than any Phase 1 material excavated by Attenbrow. Similarly, the Phase 3 assemblage fits better into the time frame for Attenbrow's Phase 4, starting as it does before Attenbrow's transition date of c. 1,600 BP and continuing until c.1,200 BP.

Assemblage Size and Characteristics

In comparing UDM with the UMCC sites, assemblage totals only will be used. In terms of lithic assemblage size, UDM is a major site. A total of 3,550 artefacts was retrieved from the five test pits, making it the third largest assemblage excavated in the Mangrove Creek Valley (Attenbrow 2004: Table 4.6). A projected artefact total of 193,559 artefacts is calculated for UDM based on the 2.8% sample excavated and 1.7% sample analysed (Table 8.21). On this basis the site is the second largest in the catchment after Emu Tracks 2 with a projected 249,600 artefacts (Attenbrow 2004: Table 6.13).

Artefacts with retouch/usewear and backed artefacts are relatively rare at UDM (0.4% and 0.3% of the assemblage respectively: Table 8.11). These figures are indeed extremely low when compared with Attenbrow's data (i.e. only those five UMCC sites with >1,000 artefacts). Artefacts with retouch/usewear represent between 0.6-1.9% of the UMCC assemblages, while backed artefacts represent between 0.2-0.9% (Attenbrow 2004: Table 4.7). At UDM, then, while backed blades fall towards the low end of the expected range, artefacts with retouch/usewear are notably lower. Conversely, ground fragments commonly occur only as very small percentage frequency of the larger assemblages (between 0.1-0.7%). At UDM this artefact type represents 1.9% of the artefact assemblage, significantly more than any other site.

Artefact Accumulation Rates

Artefact accumulation rates at UDM were calculated on the basis that Unit II material accumulated in roughly 1,000 years, while the deposit in units III and IV accumulated in roughly double that time, 2,000 years (Figure 8.23: ignoring sample ANU-8133 a constant sediment accumulation rate prior to unit II is assumed based on the age-depth curve). The floor area estimated to have depth of deposit is 45 square metres. The excavations represented a 2.8% sample of this deposit while the analysed data represents a 1.7% sample. Based on these figures and on the artefact totals retrieved (three analysis squares), the estimated artefact accumulation rates for UDM were calculated (Table 8.23).

Attenbrow's analyses identified a highly variable pattern of individual site usage amongst her sample of 31 sites. Comparison of UDM with her 31 sites in terms of artefact accumulation rates reveals that this site is not the same as any other site in UMCC. The site is immediately differentiated on the basis that it has no Phase 4. It would appear to be the only shelter in the catchment which, once established, was not occupied into the last millennium.

Table 8.23: UDM: Rates of Artefact accumulations in successive phases.

Layer(Phase)	Artefact total	Estimated Total	Rate of accumulation*
Units I+ II (Phase 3)	2,810	165,294	165,300
Unit III (Phase 2?)	181	10,677	10,700
Unit IV (Phase 1?)	299	17,588	17,550
Total	3,290	193,559	

*Rate calculated /1,000 years; rounded to nearest 50 (following Attenbrow 1987: Table 7.8)

Comparing this site with the re-calculated data, it in fact accords well with the general UMCC trends. The site is established in the fourth millennium, the peak identified in the other main shelters (Figure 8.34), and the major period of artefact deposition was in Phase 2, albeit later than predicted by Attenbrow's dated sequence.

A re-working of the artefact accumulation rates and habitation indices, including the UDM assemblage, provides a slightly different pattern for the Mangrove Creek Valley (Figure 8.35). A later date for the Phase 3/4 transition is proposed on the basis of the UDM shelter. This re-worked habitation pattern is tested, in the diachronic analysis of the Mangrove Creek shelter art sites (Chapter 10).

Hatchets and grinding technology

A notable feature of the site's artefact assemblage is the evidence for the breaking up of three edge-ground hatchets. These fragments (63), as evidence for this technological component, are present in far higher proportion than elsewhere in the Mangrove Creek catchment. The small size of this material (43% < 1cm in size) indicates that these ground items were knapped in the vicinity of Squares 6 and 8. From the absence of use-wear and /or retouch on any of the larger broken up pieces, it would appear that this knapping was completed for the purpose of breaking up or reworking these implements.

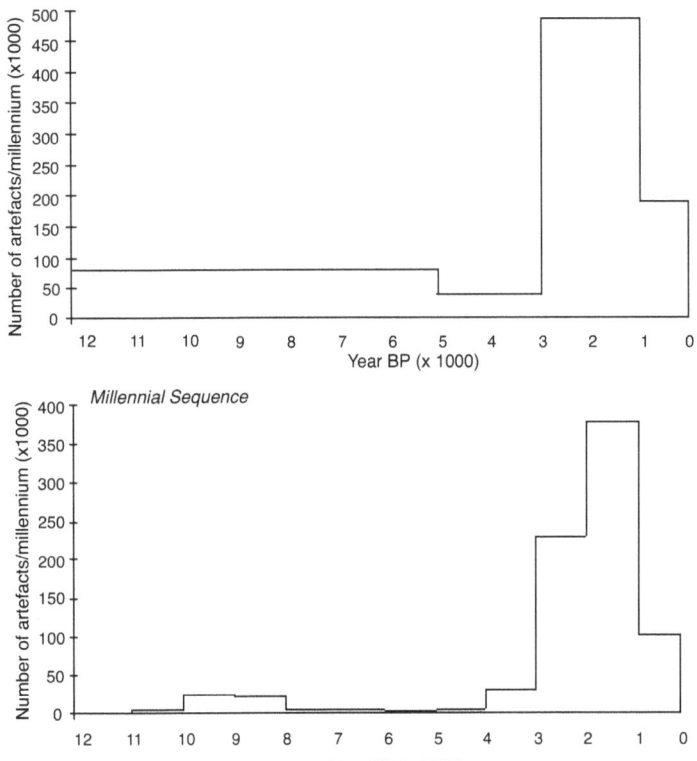

Figure 8.34: Mangrove Creek Valley. Rates of artefact accumulation in successive periods. Attenbrow's seven large sites plus UDM (cf. Attenbrow 1987: Table 7.9 and Table 8.22).

A ground edged hatchet (Figure 8.19) was found at the site and the sandstone 'ramp' outside the shelter has a large collection of grinding grooves. Several more groups of grooves (with up to 20 in each group) were observed around a number of potholes on sandstone surfaces within 100m of the site.

Clearly, edge-grinding technology was used by the inhabitants of Upside-Down-Man during the most recent phase of its occupation. This use included the maintenance or breaking up of previously ground implement.

While stencilled hatchets are found elsewhere in UMCC, no stencilled hatchets occur in the UDM art assemblage. On Panel 1, however, a red anthropomorph is drawn with a red axe (Figure 8.7). An association can be drawn on this basis between the more recent phase of occupation deposit and with the production of the red outlined and infilled motifs on Panel 1.

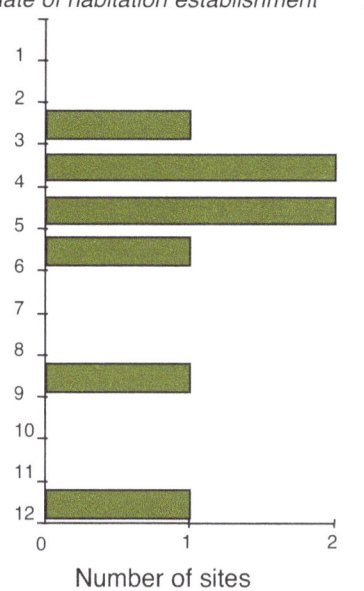
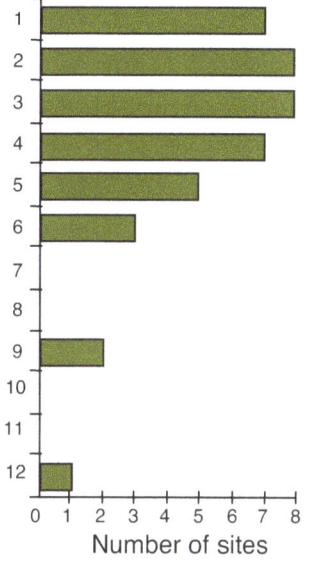

Figure 8.35: Mangrove Creek Valley. Rates of habitation establishment and habitation use in the eight sites (including UDM) with >100 excavated artefacts (cf. Attenbrow 1987: Figure 7.11).

The art and the occupation evidence

Very little ochrous material was retrieved from the site. And none of the red ochre or pipeclay retrieved bore evidence of use. The absence of faceted usewear evidence on this material makes definite correlation between the art and deposit tenuous. Certain points can be made about this material, however.

1. None of the excavation squares were directly below art panels. The presence of art producing material in the general floor area is considered to be fortuitous and less likely than in close proximity to the art panels.

2. These materials do not occur naturally in the shelter, and therefore must have been brought to the shelter during its occupation history.

3. All possible art producing material is confined to the top three spits of the deposit.

4. Ochre was recovered from only one of the shelters in UMCC, and this site had no art (Attenbrow 1987: Tables 5.5; A3/1).

While there is no evidence that the pipeclay and red ochre were used in the production of art, this conclusion is not unreasonable. Both colours are found amongst the art. While the colour of the pigment recovered is quite distinctive, the red-coloured motifs at the site vary considerably. On Panel 1, the red is mainly tomato red, with some more cherry hues (Munsell colours were used in the recording of the art: see McDonald 1994; Appendix 3.3). On Panel 2, the red is mainly dark cherry red, while the large red shields on Panel 3 are more tomato coloured. Panel 5 contains the widest range of reds. The 'oldest' wet infilled motifs are a dark brown red; the red anthropomorphs with head-dresses are dark cherry red; the CXNF motif is dark carmine; while the red outline macropod is dark crimson.

White pipeclay appears to have been used late in the art production sequence at the site. It is found as an outlining material to both black and red bichrome motifs; in the white hand stencils, the white outline paintings and in the white painted blobs found in Panels 4 and 5 (see photographs). The four fragments of this material found in square 6B/3 are from below an intact hearth (dated to 1,220 + 120 BP). The use of this ochrous material is firmly placed within the most intensive period of site usage.

The two pigment fragments in 4D/1 produce the same colour red (5R 3/6) as the painted boomerang and stick figure on Panel 1 - and the red outlined macropod on Panel 5. The boomerang motif appears to be associated with a dry outline/infilled anthropomorph (i.e. is positioned as if in the hand of the anthropomorph: Figure 8.4), while the red stick figure is superimposed by the red woman motif. On Panel 5, the motif in this colour is beneath a painted white outlined macropod.

The depiction of a hafted hatchet in the hand of an anthropomorph on Panel 1 and the presence of edge-ground fragments in the upper units at the site provides another correlation between elements of the art and features of the occupation deposit.

The faunal remains at the site provide no such correlation. These are too highly fragmented to allow identification, but are almost without exception from very small species. There are not, for instance, any large, medium or even small macropod bones present amongst the faunal remains, such as would 'match' the depiction of such species amongst the art.

A direct link to the people who created the art is provided by the hand stencils. These were not measured (and engendered information is thus difficult of ascertain: see the Mount Yengo excavation chapter). During recording of the art assemblage it was noted that these filled a range of sizes which included children, adolescents and adults.

Other than in terms of stone tool production, the deposit provides scant insight into the lifestyle of the site's occupants. The plant remains suggest medicinal use (senna, sarsaparilla) as much as food sources (native cherry, Geebung), and these remains are too few to be other than highly speculative. Both the native cherry fragments indicate breakage and/or gnawing, which suggests that these two items may have been introduced to the deposit by small herbivores. The other plant remains appear to have been humanly introduced into the deposit (mainly on the grounds that none of the other species concerned was observed in the immediate vicinity of the shelter mouth at the time of excavation). Those remains in 8B/3 and 6B/5 (particularly) had a highly carbonised appearance under the microscope, and looked different from the comparative specimens at the Seeds Lab. It would seem unlikely that these have been introduced into the deposit recently, and had worked their way down into the deposit (and this scenario is not supported given the interpretation of stratigraphic integrity of the deposit). Of course this does not discount their natural introduction into the site some time during its prehistory. The small amount of shell in the deposit also provides a glimpse of the ranges covered by the shelter's occupants in their quest for food. The nearest estuarine conditions necessary for *Anadara* are around four kilometres from UDM, while the hairy mussel may have come from rocky shores of the Hawkesbury River. Freshwater mussel could have been collected from closer to the site, in deep pools on Ironbark Creek.

The AMS samples

The black outline macropod motif on Panel 1, from which three AMS samples were collected, is clearly superimposed over the red art on this Panel. It would appear from the superimpositionning analysis of this Panel, that this was one of the final motifs created on Panel 1. It was hoped that the dates from this motif would provide a minimum date for the final period of art production at the site. The other macropod motif, from which one sample was collected, is not in direct association with any other techniques/motifs. It is however one of series of similar motifs, which occurred relatively early in the art sequence, beneath incised motifs, white hand stencils and white outlined motifs. It was hoped that a date from this motif would provide an age estimate for the main period of art production at UDM.

Unfortunately, the age determinations received are inconclusive, and suggest problems with the field sampling procedure (McDonald 2000c). One sample from the outlined macropod (ANU-AMS-773) returned a date of c.480 ± 80 BP. The two other dates from this motif, as with the date for Motif #2, were indistinguishable from modern.

The three dates which are indistinguishable from the modern standard suggest that there has been contamination of both motifs by younger organic material.

Several difficulties arise from these results and several interpretative scenarios are possible.

1) The date c.500 yrs BP may be an accurate representation of the age of motif #1. The two more recent dates suggest either than this motif was partially contaminated, or that part of it was redrawn around contact. This scenario is unlikely on archaeological grounds.

    The modern date for motif #2 suggests that this too was produced around contact. This seems unlikely, however, given its location in the superimposition sequence, that it is faded and affected by surface exfoliation.

2) Both motifs were created more recently, using an older piece of charcoal found lying on the surface. Contamination of this charcoal may have resulted from the artist having some form of organic material (e.g. animal fat) on his/her hands. This material may have encouraged the growth of micro-organics, thus affecting, unequally, the radiocarbon signal of the charcoal in the drawing.

3) It is possible that this motif could be older, if the contamination is younger than contact, i.e. more recent European interference or ongoing micro-organic growth on the surface of the rock. If this is so, then all the dates returned would be younger than they actually are. A modern contamination of the prehistoric charcoal could mean that the 500 year old date is realistically double that received; c.1,000 years BP.

The contradictory nature of these results makes it unwise to rely on them and impossible to make firm conclusions about the production date of the outlined charcoal macropod. On face value, it would appear that art was being produced at this site up until c.500 yrs BP: after other occupation of the shelter had ceased. As contamination cannot be ruled out, it is possible that this motif was produced in the final phase of the site's occupation.

The use of AMS at UDM shelter did not provide confirmation on the terminal date of art production as had been hoped. Thus, the association of the excavated evidence and the art assemblage is used. Associations suggest that the main art production phase coincided with the main occupation phase at the site. The presence of red pigment in the top spit of square 4D suggests that art production continued beyond the main occupation phase and possibly after use of the shelter for habitation ceased. The one (apparently reliable) AMS date suggests that use of the shelter for art production may have continued into the last millennium.

Upside-Down Man in the local context

The occupation of UDM reveals a markedly different pattern from the habitation pattern proposed by Attenbrow 1987 (2004).

1. The earliest UDM stone tool assemblage, which falls into Attenbrow's typological Phase 1, is more recent (i.e. 4,030±140 BP ANU-8132) than is defined by Attenbrow for this Phase (i.e. c 11,200 to c. 5,000 years BP);

2. The most intensive occupation of the site (as seen in artefact accumulation rates) took place between 2,000 and 1,000 years ago - late within Attenbrow's sequence for Phase 2 (Figure 8.34);

3. The site was not occupied during Phase 4. The site's occupation ceased prior to the last millennium and it was not used in the contact period. The occupation evidence is supported by the art assemblage, which has no contact motifs (which are found elsewhere in the catchment);

4. There is evidence for a slight decline in artefact densities in the top spit at the site. This occurs within one typological Phase and probably reflects taphonomic processes (i.e. recent visitor treadage and scuffage) at the site.

UDM follows the general pattern of the reworked UMCC data, with the only anomaly being the later date for the main (Phase 2) occupation. When the UDM material is combined with the data from the other large sites in the catchment, the habitation indices show a peak in habitation establishment rate in the fourth and fifth millennium. The rate of habitation usage peaks slightly later, between the second and third millennium, and there is a decline in the most recent millennium.

These combined data from the Mangrove Creek sites suggest that shelter occupation in the catchment peaked in the third and second millennium, but that there was a decrease in shelter usage in the most recent millennium. They also suggest that the timing for the changeover between the Middle and Late Bondaian should perhaps be pushed to c.1,000 years BP, and this date is proposed as an alternative (see below).

9

THE CONTEMPORANEITY OF ART AND DEPOSIT

The problem, previous approaches, current aims

> We can ... base a dating system on clearly defined stylistic conventions which encompass a whole system of reference and through which secure bonds may be established between stratigraphically dated mobile art objects and parietal works, the latter generally lacking a datable context. (Lorblanchet 1977:56)

Here, the contemporaneity of different archaeological elements in shelter art sites is explored. The assumption that occupation evidence and art were produced at the same time, and that the two are complementary forms of evidence for the group(s) which produced them, is addressed. This assumption was not a new one when this research was originally undertaken, having formed the basis for many analyses in Australia and overseas (e.g. in the USA - Geib and Fairley 1992; Geib *et al.* 1986, Gunnerson 1969, Lipe 1970, Lister 1964, Schaafsma 1985, Talbot and Wilde 1989; in Europe - Bahn and Vertut 1988, Begouen and Clottes 1985, Cartailhac and Breuil 1906, Conkey 1978, Gonzales Echegaray 1974, Lorblanchet 1977). In various Australian regions the assumption has often been explicit (or even implicit) in more generalised analyses (e.g. Chaloupka 1977, 1994; Morwood 1979, 1992b; Taçon 1989; Taçon and Chippindale 1993) with the age of art phases being tied to phases of occupation evidence. While a common assumption, the contemporaneity of art and occupation evidence has rarely been investigated on a regional scale.

A number of Australian shelter sites have been excavated previously with the intention of indirectly dating the art. Specific chronologies have been extrapolated to other sites or broader regions on the basis of stylistic criteria and comparison (e.g. Beaton 1991b; David 1994; Flood and Horsfall 1986; Frost *et al.* 1992; Morwood 1979, 1986, 1992b; Rosenfeld *et al.* 1981; Ward *et al.* 2006). Some shelter art sites have been excavated in order to characterise their lithic assemblages, and it has been assumed that the art and the deposit were contemporaneous phenomena (e.g. Quinnell 1975, Wright 1971). Other art sites have been excavated and little or no effort has been made to marry the art with the deposit (e.g. Attenbrow 1987; Attenbrow and Negerevich 1981, 1984; Beaton 1991a, 1991b; Cox *et al.* 1968; Mulvaney and Joyce 1965; White and Weineke 1975).

While the assumption of contemporaneity and occupation evidence is broadly based, there can be no definitive correlation between art production and the domestic use of shelters.

In some social contexts, art and its production take place in the ritual sphere. In such situations, a nexus between domestic evidence and art would not be expected, e.g. art associated with mortuary practices in central Queensland (Morwood 1979). Other models have proposed an inverse relationship between art and occupation evidence. Morwood (1986) suggested that the proliferation of art in south-eastern Queensland was linked to declining use of shelters for habitation as art in the region began to function as broad scale communication. There are instances where the contemporaneity of art and deposit would not be expected.

Several Sydney shelters have been described as 'important ritual galleries' (Elkin 1949, McCarthy 1961) with the implication that these probably were restricted (or 'closed') to parts of the social group (i.e. based on age, level of initiation and gender). There is no ethnohistoric evidence about the social context of art production in the Sydney region at contact (Chapter 3). In the absence of traditional knowledge about this art's function and production, the possibility of art having either ritual and/or secular functions requires consideration.

Over 65% of shelter art sites also have recognisable occupation deposit (i.e. from surface evidence[27]) which provides strong support for pigment art having a domestic function, if contemporaneity is a valid assumption.

The only way of fully testing contemporaneity would be through a comprehensive effort to obtain direct dates for the pigment art (using Accelerator Mass Spectrometry ('AMS') radiocarbon and companion techniques, e.g. Rowe 2001) at a number of excavated (and dated) occupation sites. There is no other way to verify that the two sorts of evidence were produced at exactly the same archaeological time (see Chapter 10).

For this research project a number of motifs were dated using AMS. At the time, progress was slow, due to the pioneering nature of this technique, restricted funding and technical problems in setting up facilities in Australia. The myriad potential sampling problems and interpretative pitfalls introduced by the experimental nature of these techniques also meant that limited time could be spent on this aspect. A small number of sites have been dated by this process in the greater Sydney region (McDonald 2000c; McDonald *et al.* 1990). One of these sites was the UDM shelter excavated for this research (chapter 8). In the absence of sufficient uncomplicated dating results, this chapter demonstrates a conventional methodological approach to investigating the assumption of contemporaneity of habitation and art production. This is done at the site specific and regional levels.

It is assumed that the engraved and pigment art in Sydney were produced by the same group(s) of people over the period that the region was occupied. The concept that occupation evidence and art are different components of the same culture is implicit to this assumption. Based on the Sydney Basin's occupation indices, it can be assumed that the majority of the Sydney Basin art dates to within the last 4,000 years. Sheltered art sites are the focus of this analysis since these are the locations which provide the evidence of art and occupation in close proximity.

Testing the assumption of contemporaneity at a regional scale presents different challenges. Many sites have no datable evidence, either in the pigment sequence or the archaeological deposit. Often the context of the art in a site does not allow for its relative age to be judged, e.g. engravings located on sloping shelves not covered by deposit. Some art assemblages create analytical difficulties in broader terms. These include sites which have:

- only one phase of occupation evidence but several phases of art present, either stylistically distinct motifs or obviously episodic art production;

- several phases of occupation evidence but only one art phase present;

- artistic evidence which demonstrably postdates the occupation evidence;   or,

- obviously ancient art which is accompanied by more recent occupation evidence.

Archaeological phases or cultural periods by their very nature 'collapse' periods of time in terms of the individual or even generations. Direct correlations between 'an individual site occupation' and 'a specific artistic production episode' could rarely be expected [cf. Cosquer Cave provides such a unique opportunity (Clottes and Courtin 1993, d'Errico 1994)]. Artistic trends over centuries or even millennia are often the usual scale of prehistoric art analyses, although the successful use of AMS dating may alter this situation (Clottes and Courtin 1993, Cole *et al.* 1994, Geib and Fairley 1992, McDonald 2000c, McDonald *et al.* 1990, McDonald and Veth 2008).

[27]These figures are based on 1994 NPWS (now AHIMS) site records. CHM test excavation programmes in the Sydney region have demonstrated that more than 80% of shelters with Potential Archaeological Deposits (PAD) and open PADs i.e. with no visible (surface) archaeological remains are indeed archaeological sites once excavated (Attenbrow 1987, Attenbrow and Negerevich 1981, Koettig 1985, McDonald 1985b, McDonald *et al.* 1994).

Chapter 9: The contemporaneity of art and deposit

Regional patterning operates multifariously and 'many important processes and institutions operated at and should be understood at a variety of scales of inclusion' (Conkey 1987:70). We need to understand the complexities of intra-site patterning from individual sites, as well as regional patterning which results from viewing the material at a broader perspective.

The assumptions of this research were intentionally 'telescopic'. The assumptions of contemporaneity are first explored at a finer scale with the potential for anomalies being identified prior to exploring the regional trends. The sites purposively excavated for this research represent a miniscule sample for the region. These excavation results, however, combined with evidence from other excavated and dated art shelters, can be used to predict how reliable this approach is.

The four sites excavated for this research (chapters 6-8) all contained evidence for the episodic production of the pigment art. Yengo 1 also contained an engraved panel partially covered by occupation deposit. The two Yengo sites, located only 10m apart, contained very different art assemblages. Different degrees of direct and indirect association between the art and the deposit were demonstrated in these four sites.

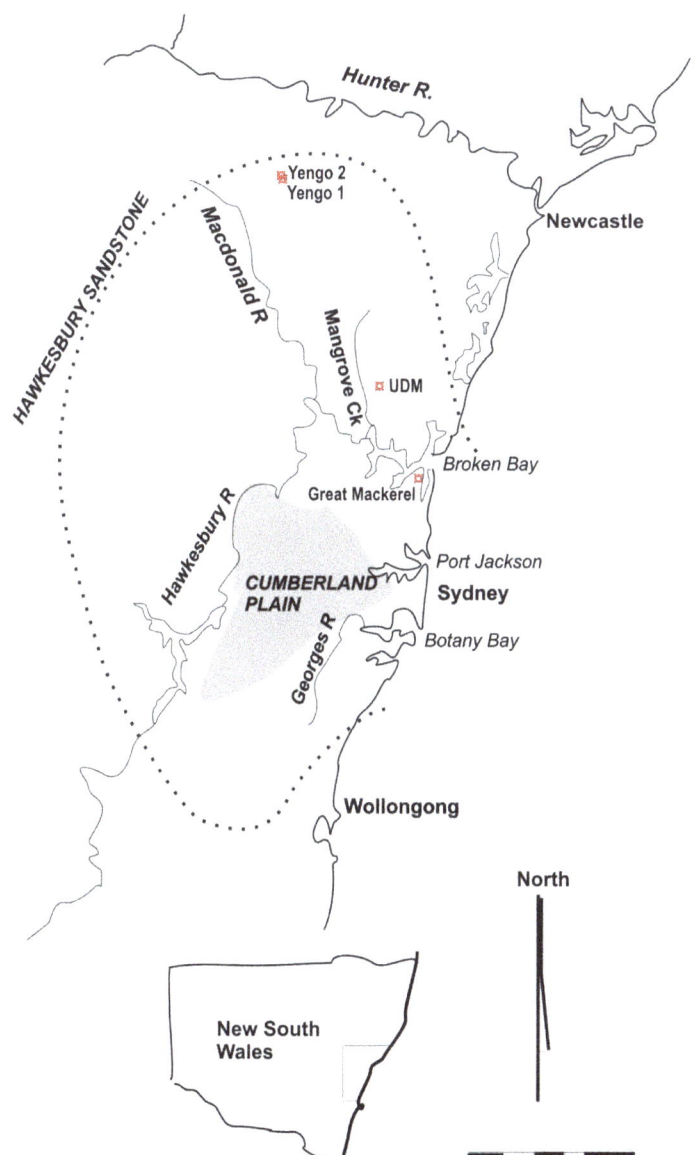

Figure 9.1: The four shelter art sites excavated for this research.

This chapter reviews and analyses the previously excavated shelter art sites in the Sydney region. Regional patterns in shelter site usage are identified and the ramifications of these discussed. A discussion regarding the validity of assumptions made regarding the contemporaneity of art and occupation evidence concludes this chapter.

Regional patterns for the Sydney region

The problem of demonstrating contemporaneity between art and deposit relates primarily to the fact that parietal art, until recently, could not be directly dated. Theoretically AMS radiocarbon dating is the solution to this issue, but this technique is still in its adolescence (Keyser 2001) because of the relative newness of the techniques and the lack of theorising about applicability of these techniques to art assemblages generally (see Beck *et al.* 1998; Bednarik 1996; Hyman and Rowe 1997; McDonald 2000c; McDonald *et al.* 1990; Rosenfeld and Smith 1997).

Direct dating techniques involve the collection and dating of small samples from art (e.g. pigment, charcoal, beeswax) or from crusts and/or deposits overlaying (or underlying) art motifs (e.g. oxalate crusts, desert varnish, and mud-wasp nests). Accelerator Mass Spectrometry (AMS) is the most widely used technique because it requires much smaller samples than conventional radiocarbon (ca.0.0005 gram versus 5 grams: see Rowe 2001). AMS counts the number of radiocarbon ($_{14}C$) molecules (as a ratio to carbon) in any organic material. The main difference between this and conventional radiocarbon dating, is that AMS counts the actual $_{14}C$ atoms – as opposed to the number of atoms that decay over a given time period (Rowe 2001). Charcoal is the most common archaeological material used for dating and, although there are certain identified caveats, i.e. potential contamination and the old wood/fossil charcoal problems, the techniques for dating charcoal are reliable and well tested.

Researchers have experimented with a number of other materials and techniques. These have included plasma-chemical extraction or organic carbon from inorganic pigments (Hyman and Rowe 1997), fibres found in paints (Watchman and Cole 1993), beeswax (Nelson 2000), blood residues (Loy *et al.* 1990, although see Nelson 1993, Gillespie 1997), oxalate crusts (Watchman 1993a) and optically stimulated luminescence dating (OSL) of mud wasp nests over or beneath rock-art (Roberts *et al.* 1997).

Focussed dating programmes have resulted in chronological control in a number of countries – but only in the order of only c.100 radiocarbon art dates have so far been published (Rowe 2001: 148). In Australia, other researchers have dated charcoal, beeswax, oxalate crusts, Bradshaw figures and plant fibres in paint (David *et al.* 1999, Nelson 2000, Roberts 1997, Watchman *et al.* 1997). A current ARC Linkage Project on the Canning Stock Route aims to target a range of pigment styles in an ambitious attempt to provide a chronology for the recent art and dreaming stories (Tjurkurrpa) in the Western Desert (McDonald and Veth 2005, 2008; McDonald and Steelman 2008).

Unless both the occupation evidence and the art have been dated, the association between the two types of evidence is not easily proved. Direct associations, e.g. where art is covered by dateable deposit or has been detached from the wall and is buried, are rare - and these mostly only indicate minimum dates.

Occupation sites providing a dateable context for art in the Sydney region are extremely rare. So far, only engravings have been found in suitable contextual locations – and the art covered in these situations is not representative of most of the engraved art in the region. Pigment art is not generally preserved below deposit (because of the acidity and moisture). There is one shelter art site on Cowan Creek (Mega Midden: McDonald 1987) in which several metres of midden deposit has built up to within a metre of the roofline. White hand stencils occur across the back wall very close to the current surface of the deposit and clearly must have been produced before the deposit accumulated to its current level. There is a possibility that pigment art exists below the deposit in this shelter (preserved by the alkaline deposits) and that an association could be achieved in this location.

In the Sydney region the problem is further exacerbated by the fact that the drawing technique predominates. Drawing involves the use of dry pigment as a crayon – not the preparation of paints, which involve grinding, use of binders, mixing palettes and so on. This lack of pigment preparation suggests that the likely residues of such activity might not be expected in the deposits to the same degree as in regions where painting predominates (e.g. the Australian arid zone, Kimberley, Arnhem Land and Laura). Faceting of ochres would not be expected – nor would preparation palettes. Crayon ochres would occur in a much less predictable manner, i.e. as a result of being dropped or fragments breaking off during the act of drawing. Further, the ubiquity of charcoal in Sydney pigment art production means that any large pieces of charcoal in the deposit could potentially have been used for art production.

Whether lack of pigment preparation or taphonomic factors is the cause, very few excavated shelters in the region have provided evidence which assists in the indirect dating of pigment art production. Of the 31 shelters excavated in Upper Mangrove Creek, only one had ground ochre: and this site had no parietal art (Attenbrow 2004: Table 4.5). It is possible that excavated ochres have not been recognised by archaeologists who are excavating without an art research focus, especially since these types of remains when they are found, are usually highly fragmentary. Further, much of the red and yellow pigment which is found in shelter sites is only gradationally finer in grain size and friability than natural ironstones which are ubiquitous across the Sandstone Formation. It is interesting that the shelters excavated for this research all contained buried pigments (either pipeclay or ochre). Perhaps this can be attributed to the fact that these sites all had major art assemblages, hence increasing the probability for this type of material to be deposited, and recovered.

Prior to this research, 35 art shelters in the Sydney region had been excavated (Attenbrow 1987; Attenbrow and Negerevich 1984; Clegg 1979; Cox *et al.* 1968; Koettig 1985; MacIntosh 1965; McDonald 1992a; Menses (in) Miller 1983; Moore 1970, 1981; White and Weineke 1975). While many of these authors discuss the art located in these excavated shelters, only one of these excavations was undertaken with the expressed aim of contextualising the art (MacIntosh 1965).

Most (78%) of these art sites were excavated for cultural heritage management purposes (Table 9.1). Eighteen of the 25 salvaged art shelters were excavated in the Upper Mangrove Creek catchment (Attenbrow 1981, 1987, 2004; Gunn 1979). Two other shelters were investigated for management purposes i.e. prior to the installation of protective cages (Clegg 1979, Menses 1970's in Miller 1983). No detailed excavation report exists for either of these sites. Charcoal samples, however, were submitted from both of these shelters and dates for these deposits are available (McDonald 1992a, Miller 1983, Mackay and White 1987). Five of the excavated art sites had broader research questions: contact between the Hunter and the Hawkesbury (Moore 1981) and characterising coastal stone tool assemblages (e.g. Cox *et al.* 1968).

Here the data from some of these excavated shelter art sites is discussed to highlight the potential for interpreting the contemporaneity of art and deposit. By analysing the occupation patterns demonstrated by these sites, general patterning across the region can be defined.

Table 9.1: Excavated shelters with art in the Sydney region.

AHIMS #	Site Name	Published / Reference	Dates (years BP -uncalibrated)	Lab-ID	Art details	Sequence?
[Southern Sydney]						
52-2-37*	Bull Cave	Miller 1983	1,820 ±90	SUA-2106	white stencils,	b/r/w
			<1,050 ±90		blk and red drwg	
52-3-30*	Audley	Cox *et al.* 1968	no		w and r stencils	st/r/w
			(midden/occ)		b,r and w drwgs	/b/b+w
	BC1	Attenbrow and Negerevich 1984	no		b drwg	no
52-2-771*	BC2	'	no		bl, r drwg + y paint + prints	y/red
52-2-774*	BC5	'	no		blk drwgs	no
52-2-778*	BC9	'	1,630 ±90	SUA-1746	blk drwgs	no
52-2-1031*	M11	Koettig 1985	480 ±70	SUA-2255	w stencils	no
			1,520 ±70	SUA-2256		
			2,220 ± 70	SUA-2257		
45-6-602	HLD	White and Weineke 1971	5,240 ± 100	SUA-60	red h stencils	no
			870 ± 95	SUA-59		

AHIMS #	Site Name	Published / Reference	Dates (years BP -uncalibrated)	Lab-ID	Art details	Sequence?
	Bindea Rd	Attenbrow and Conyers 1983	2,340 ± 100	BETA-5887	?	?
[Northern Sydney]						
37-5-1*	Yengo1	McDonald 1995	5,980 ± 290	ANU-6059	engravings	
			4,590 ± 300	ANU-6055		
			2,750 ± 220	ANU-6215	w,r,b drwg+ paint	yes
			1,950 ± 400	ANU-6054	w,y,r,b stencils	
			540 ± 180	ANU-6058		
37-5-2*	Yengo2	McDonald 1995	still to come		w,b,r +y drwg paintings + stencils	yes
37-6-349	Big L	Moore 1970	2,495 ± 105	SUA-756	w stencils+paint	no
			930 ± 50	ANU-648/2		
45-2-39?	MR1	Moore 1981	2370 ± 100	SUA-387	R paint	no
			5,820 ± 110	SUA-564		
45-3-317*	Dingoand Horn	MacIntosh 1965	581 ± 120	GX-70	r/w, r, y and b drwg	
			144 ± 125	GX-69	w stencils	yes
45-3-787*	Black hands	Attenbrow 1987	3,040 ± 85	SUA-932	b, w, r + w stenc + polychrome	yes
45-3-1207	Dingo	Attenbrow 1987	1,840 ± 60	SUA-2166	?	
45-3-1528	Elonga'd Fig	Attenbrow 1987	1,810 ± 80	SUA-2170	?	
45-3-789*	Roo+Ec'idna	Attenbrow 1987	6,700 ± 150	SUA-2172	b drwgs	No
45-3-776*	Loggers	Attenbrow 1987	530 ± 85	SUA-1124	b, r,w, b+w,	yes
			2,480 ± 60	SUA-2165	incised engrav	
			11,050 ± 135	SUA-931		
45-3-1165*	White Fig	Attenbrow 1987	5,230 ± 70	SUA-2167	?	
45-3-1159*	Wolloby Gully	Attenbrow 1987	400 ± 60	SUA-2168	b, w	yes
45-3-1179*	Emu tracks2	Attenbrow 1987	3rd - 1st mill		engravings	no
45-3-1164	one tooth	Attenbrow 1987	2nd?-1st mill		?	
45-3-1161	low frontage	Attenbrow 1987	1.5 mill		?	
45-3-1168	Elongarrah	Attenbrow 1987	1.5 mill		?	
45-3-1201	McPherson	Attenbrow 1987	1st mill		b drwg	
45-3-1174*	Bird Track	Attenbrow 1987	1st mill		b, w, b+w (out, o/i)	no
45-3-1160	Boat Cave	Attenbrow 1987	1st mill		?	
45-3-1170*	Venus	Attenbrow 1987	<1st mill		b, r drwg	
45-3-1210*	Ti-tree	Attenbrow 1987	<1st mill		b drwg	
45-3-1196	Mangrove mansion	Attenbrow 1987	<500 yrs		?	
45-3-1204	Firestick	Attenbrow 1987	<500 yrs ?		?	
45-6-150*	Milligans	Clegg 1979	5,340 ± 105 BP	SUA-?	b dr'wg, engrav	yes
			800 ± 90			
45-6-72*	Angophora Reserve AR1	McDonald 1992a	2,000 ± 150	ANU-6584	b + r drw'gs	no
			1,750± 90	ANU-6923		
			1,150 ±100	ANU-6583		
45-6-1614*	Great Mackerel	McDonald 1992b	3,670±150	ANU-6615	red stencils	yes
			560±160	ANU-6372	b,w drw'gs +	
			220±120	ANU-6370	w stencils	
45-3-1114	UDM	McDonald 2000c	1,220±120	ANU-8134	white stencils	yes
			1,860±70	ANU-8135	bl + wh drw'g	
			1,540± 60	ANU-8133	red stencils	
			4,030±140	ANU-8132	engravings	

Henry Lawson Drive

The art assemblage at this site is described as:

> Rock paintings on the back wall of the shelter above the shell midden area. They were situated about one metre above floor level and comprised at least seven red stencilled hand-prints (sic) which are only just visible. Some red ochre elsewhere on the walls indicates that other art had previously been present. (White and Weineke 1975:7)

Two dates were obtained for the site. The older (5,240 ± 100 BP) derived from the excavation outside the shelter and predates the main midden occupation phase (White and Weineke 1975: Figure 7). A backed blade was associated with this dated charcoal sample. The base of the midden inside the shelter was dated to 870 ± 95 BP.

The artefact assemblage retrieved from the 36cm of deposit below the midden was small (n=246). It included backed artefacts but no bipolar or use-polished material. The date returned is slightly older than expected and its derivation (outside the dripline) means that this may have been subject to a number of taphonomic agencies (see also Hiscock 2003). The possibility of this being a faint signature of early to mid Holocene occupation in this part of the region is posited by Hiscock (2003) - a possibility reinforced by the discovery of an early Holocene open site in Tempe (JMcD CHM 2005c).

The rock art in this shelter could relate to either occupation period. However, the fact that the art is very faded, and that it occurs only 1m above the present ground surface, suggests that the stencils predate the midden period and relate to the lower units which were more than 45cm below the current surface level. Adult hand stencils rarely occur low on shelter walls, and certainly, stencilling one's hand below waist-height (approx. 1m) would be extremely awkward. Given the build-up of the floor level over time, it seems reasonable that the stencils predate the midden and the roof-fall period and could be contemporaneous with the earlier occupation period, c. 5,800 years ago.

Mill Creek

The M11 site contains faded white stencils of human hands and macropod feet (Koettig and McDonald 1984, Koettig 1985).

The deposit at this site was dated to between c.500 - 2,200 BP (see Table 9.1). Two main phases of Middle and Late Bondaian occupation were found. The earlier phase of occupation was the main one, with artefact densities three times higher than recorded in the upper assemblage. No excavated pipeclay was reported, although the poor state of preservation of the faunal remains suggested that the survival chances for this would be low. The preservation of faunal material is best in the earlier layer, and the absence of bone from the upper levels is stated as 'noteworthy' (Aplin 1985:82).

The production of the art here may have taken place during the more intensive middle Bondaian phase. Occupation at this time was more intensive, and a correlation between the faunal remains (including macropod) and the art (stencilled macropod feet) is possible. It is unlikely that a macropod's feet would have been stencilled without the beast also being consumed (or at least deceased!).

It is argued that the production of white stencils at this site took place between c.2,200 and 1,500 years ago.

Barden's Creek

Only one of the four excavated art shelters in Barden's Creek was dated (BC9; 1,630 ± 90 BP). Based on an age-depth curve, occupation in this site is estimated to have commenced between

3,500-3,000 years ago (Attenbrow and Negerevich 1984:143). The stone tool assemblage was characterised as mainly Middle Bondaian, with a sparse Late Bondaian phase.

The small (<20 motifs) art assemblage comprised faded black drawings of anthropomorphs and a quadruped. Charcoal was found in greatest quantities throughout the dated occupation layer.

While a correlation between charcoal rich layers in the deposit and charcoal drawings is not conclusive, it is not an unreasonable assumption. As this shelter had several phases of occupation, concluding contemporaneity of the art with one particular phase of deposit is difficult. The production of charcoal drawings at BC9 probably dates to somewhere in the last 1,600 years.

Bull Cave

The art from this shelter was recorded in detail by the Sydney Prehistory Group (1983). The art here includes several contact motifs amongst its complex assemblage of black and red drawings and white hand stencils. The contact motifs are large bulls executed in traditional techniques; outlined and infilled drawing. One complete bull is red; the other black. Both have white eyes.

Schematically, while obviously being bulls, their feet are not bovid and their heads are bird-like. Clegg (1981) argued that these bull drawings depict the cattle that went missing from the early days of the colony (in 1788) since these cattle were polled, unlike their 61 progeny which were eventually found in 1795. He also argues for a developmental schema whereby the earlier (red) drawing of the two is more bird-like than bull-like, while the later black version is more schematically 'correct'. The bull's feet were obviously problematic for the Aboriginal artist, because these were unlike any native fauna. From a distance these would have been concealed by grass: a similar argument was mounted by Percy Trezise (at a conference in 1988) in relation to depictions of diprotodons).

The analysis of the excavated occupation deposit did not attempt to correlate the art with the deposit (Miller 1983). A cited date of c.1,800 years BP (Table 9.1) for the main phase of this silcrete rich assemblage may well be associated with the majority of the art present in this shelter. The bulls obviously post-date the majority of the art – and can be taken of evidence that rock art was still being produced at contact (McDonald 2008). Koettig's (1985) re-analysis of the Bull Cave material indicates that an undated Late Bondaian assemblage is in the top spit. The contact art may correlate with this the terminal (undated) phases of occupation at this site. Otherwise this art has inconsistent dated excavation evidence.

Mangrove Creek

The Mangrove Creek sites (Attenbrow 1987, Gunn 1979, MacIntosh 1965) are discussed in more detail in Chapter 10. Two of these, however, are discussed here. The Dingo and Horned Anthropomorph shelter is discussed as this site was excavated with the expressed purpose of dating the art. Emu Track 2 is discussed as this site demonstrates a situation of potential ambiguity.

Dingo and Horned Anthropomorph

The art at this shelter is located on two separate panels. The main panel, after which the site is named, is in a large circular alcove and contains an aesthetically pleasing composition of two red culture heroes (horned anthropomorphs), two dingoes and two echidnae. The other art panel has a complex panel of drawings and stencils. There is an engraved fish on the sloping rock panel at the base of this panel and a number of grinding grooves on a platform outside the dripline.

Two dates were obtained from a pit, excavated near the horned anthropomorph panel (Table 9.1). No lithic or faunal assemblage is described from this particular excavation, but clear stratification of charcoal (some hearths) as well as the two 'red powdery' strata amongst the yellow sand was found. MacIntosh argues that these powdery strata represent the two artistic episodes at the sites. The older date was collected from a sample below the darker red lens in which one piece of faceted pigment (the same colour as the dingo and horned anthropomorphs) was found. The pigment from between 24-27cm depth was described as having 'three of its eight sides 'rub-polished'' (MacIntosh 1965: Plate V).

The other dated sample came from above the upper lighter red layer, only 4cm below the surface (MacIntosh 1965:92-3). This more recent date cannot be correlated with the art at the shelter (see discussion Chapter 7).

Emu Tracks 2

At Emu Tracks 2, 15 engraved emu tracks were recorded as being level with the current surface of the deposit. It is generally assumed (Clegg 1978; Morwood 1979, 1992; McDonald 1993b; Rosenfeld *et al.* 1981) that engravings on vertical shelter walls are contemporaneous with living surface 35-50cm below the engravings.

On the basis of artefact typology, the occupation of this shelter commenced sometime c. 4,000 years ago (Attenbrow 2004: Table 6:2). The major occupation of the shelter is thought to have occurred in the 3rd millennium (in the Middle Bondaian). The basal levels at Emu Tracks 2 were c.80cm below the current surface, and thus it could be assumed that these engravings relate to the main, Middle Bondaian occupation of this shelter.

However, the morphology of the shelter and nature of the engravings (residual Panaramitee) suggest that this may not have been the case. The back wall and bedrock in the shelter is steeply sloping. The single (50 x 50cm) test pit excavated here had contracted significantly at its base, and it is possible that evidence of the site's earliest occupation, towards the front of the shelter, was undetected by this very small sample (Val Attenbrow, pers. comm., 1991). While a minimum age of Early to Middle Bondaian is suggested by the dated deposit in this site (Chapter 10; Table 10.7), the art may indeed be older.

Based on this shelter's art evidence, this art is placed at older than 5,000 years BP (Chapter 7).

Angophora Reserve

The main occupation phase at this site was between c. 2,000-1,200 years ago. The art consists of small assemblage of faded black and red fish and macropod drawings. There is no obvious association between the deposit and art here, mainly because these are spatially separated by more than 10m (McDonald 1992a: Figure 2).

The art has the potential to be dated, using AMS on the charcoal pigment or on oxalates within the silica coating which has covered several of the motifs. The art has been severely vandalised in recent years, and the risks of charcoal contamination are high. The silica coating, however, offers good dating potential (Watchman 1994).

Audley

The 38 motifs at this site were described by Maynard as being 'not all done at the same time, but on several occasions, to which different fashions in colour and form were appropriate' (Cox *et al.* 1968:99). The art sequence here was described as white stencils and red followed by white, and black and white bichrome.

Several phases of occupation deposit were observed, although these were not dated and appeared to 'offer no index of chronology ... ' (Cox *et al.* 1968: 99). The earlier phase of occupation included artefacts, with charcoal and a few shells. The later phase comprised tightly packed shell midden (Cox *et al.* 1968: Figure 1).

The shells weren't analysed but were apparently 'estuarine species readily comparable with those recovered from a rock shelter some three miles downstream at Gymea Bay'. Bone material was in quantities 'insufficient to allow any useful identification'. Human skeletal remains were found in this site (Cox et al. 1968: 94, 97).

Other excavations in shelters across the region have revealed similar patterns of occupation (Attenbrow 1992; Clegg 1979; Glover 1974; McDonald 1992a, 1992b; Megaw 1974) and, from the limited information provided, the Audley shelter seems to have a similar occupation history to a number of other decorated shelters in estuarine locations.

The phases in the art are also suggestive of similar trends noted, particularly in the Great Mackerel shelter (Section 6.3). In both shelters there is an earlier phase of red pigment followed by a later phases comprised of white stencils and black and white bichromes.

The absence of dates, shell or lithic analysis makes drawing conclusion difficult. Based on similar trends identified elsewhere (and particularly the absence of backed artefacts) the upper midden unit (possibly associated with white stencils and black and white bichromes) at Audley may be late Bondaian, while the earlier occupation unit (possibly associated with red paintings) may be early Bondaian.

Daley's Point (Milligan's)

The Daley's Point shelter (north of Broken Bay) was excavated by John Clegg prior to regulators constructing a metal grid to protect the art from vandals. The art consists of charcoal drawings and pecked outline engravings (fish?/echidna? and macropod) low on the sloping back wall.

The site is stated to have been occupied 'for between 200 and 600 years from 900 to 700 years ago, or 1100 to 500 years ago' (Clegg 1979:2). This is interpreted as meaning a date obtained for the occupation level was 800 ± 100 BP, although this interpretation is clearly only one of many. A basal date is firmly stated as being $5,340 \pm 105$ BP (SUA-?: not referenced).

Clegg observed change over time in the occupation material 'the stone artefacts are low down. All the fish-hook stuff is right on top. The lower shells are larger than the upper shells' (Clegg 1979:3). Attenbrow's (1979) analysis of the molluscan remains confirms that a vast quantity of shell (69kg) was removed from one square. The majority of this (82.6%) was *Anadara trapezia*.

The pigment and engraved art is interpreted here as being associated with the main, midden occupation phase. Its production is thus tentatively placed within the last millennium.

Conclusions

Of the 35 shelter art sites excavated in the region, 22 have been dated (Table 9.1). A careful reading of the original excavation reports allows an inferred age for the art in many instances - assuming that the art generally correlates with more intensive occupation. No temporal trends in art styles are suggested based on this data. Diachronic change is explored in chapter 10, at which times these trends are further discussed.

The length of time that sites with art have been occupied and the main phases of occupation at each (Figure 9.2) confirm the general occupation trend for the region (chapter 4). The fact that this occupation pattern is mirrored by shelters with art is considered significant. These decorated shelters document a very similar pattern of behaviour to habitation sites generally. This similarity supports the contiguity of use of these locations for general occupation and for the production of art.

Three sites excavated specifically for this research demonstrated different approaches to testing and establishing the contemporaneity of art and occupation deposit. In all cases it is possible to conclude that particular elements of art production were contemporaneous with particular phases of the occupation evidence.

At Great Mackerel, the more recent occupation phase was established as contemporaneous with the more recent art phase. Both art and occupation suggested the presence of women at the site and on this basis it was concluded that the two were recent and contemporaneous. The more recent occupation layer contained white pipeclay, a material present only in the more recent art phase. The earlier occupation at this site seems to represent short, sporadic visits, possibly by hunting parties. While it was not possible to conclusively demonstrate the contemporaneity of the earlier occupation with the older rock art, it is possible that these may also have been produced contemporaneously.

At Upside-Down-Man shelter, occupation occurred between 4,000 and 1,200 years ago. Two main phases of Pre-Bondaian and Middle Bondaian (Attenbrow's Phases 1 and 3) occupation were documented interspersed with sporadic occupation, possibly Early Bondaian (Attenbrow's Phase 2).

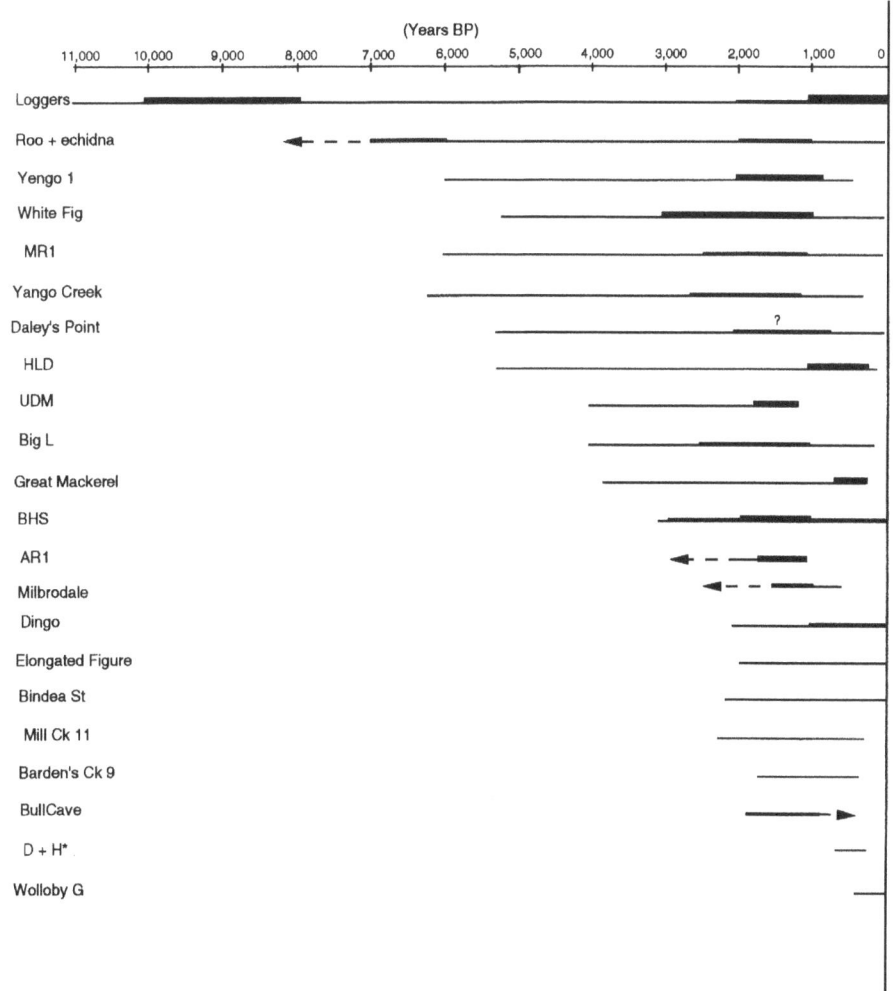

Figure 9.2: Dated shelter art sites showing length of occupation and period of most intensive artefact accumulation.

The art here provides evidence for episodic production. Small quantities of pigment and pipeclay were found throughout the most intensive period of the shelters usage (c.1,500-1,200 years ago). This suggests that most of the more recent pigment art was produced during the later, and main, habitation of the shelter.

AMS dating was used in an effort to date one of the terminal motifs of the assemblages. Unfortunately, contamination issues means that this was inconclusive. Assessing the age of the earliest art production here is even more difficult. It is argued that the earlier pigment art is older than middle Bondaian in age. The figurative schema of the pecked engravings represents a conceptual and stylistic development from iconic to figurative. The pecked figurative motifs are a transitional form of the regional Panaramitee style (which is thought to be pre Bondaian in age).

A transitional form would be expected to be younger than the Yengo 1 engravings (a minimum of 5-6,000 years) but older than the Bondaian art which predominates in the region.

On this basis, it is argued that two of the UDM pecked human figures are contemporaneous with the earliest use of the shelter for occupation. The seemingly late date (i.e. 4,000 years BP) for the oldest lithic assemblage found here suggests that this may also be a 'transitional' assemblage. The anomalous nature of both the art and the stone assemblage is again suggestive of contemporaneous events.

The Yengo 1 site provided a more conventional set of data for the establishing the age of its art. The partially buried boulder at the front of the shelter provided the opportunity to indirectly date the regional Panaramitee style engravings by association with deposits. It is argued that these are pre-Bondaian in age (dated here to between 5-6,000 years). The occupation evidence, including pigment and ground edge fragments, established the contemporaneity of the most recent phase of occupation with the production of pigment art and the sharpening of ground edged implements. The morphology of the shelter floor assisted in this interpretation (given the ceiling height prior to deposition).

In the Great Mackerel and UDM sites, the earlier pigment art consisted of red stencils and red paintings. At Yengo 1 and UDM, the earlier art forms were pecked engravings.

At UDM use of the shelter for habitation appears to have ceased around the turn of the last millennium. After use of the shelter for camping ceased, it is possible that art production continued. There is no evidence in any of the sites excavated for this research that art production continued until European contact, i.e. there are no post-contact motifs. At the Great Mackerel site, the terminal phase of the shelter's occupation probably coincided with European contact. Use of the shelter *per se*, may have finished abruptly. The terminal dates in Yengo 1 suggest usage up until just prior to contact.

Only the ambiguous AMS evidence from the UDM site suggests that art production in shelters might have continued once these were abandoned as places of occupation - as does possibly Bull Cave.

In all shelters tested, it is argued that the main phase of pigment art production coincided with the most intensive occupation period. In multi-phased art sites, earlier low intensity occupation appears to have had an artistic component: also of low intensity. The assumption that art production and shelter occupation coincide, is argued as being valid.

These results broadly demonstrate evidence for diachronic change in pigment art over the period of the region's occupation. The clearest evidence for this is in the presence of engraved motifs. While the majority of these are not in datable contexts, the Yengo 1 excavation supports the contention that this style of art predates the bulk of the art in the Sydney region; as has been demonstrated elsewhere in Australia (e.g. the arid zone, central Queensland, north Queensland, Victoria River Downs, etc.).

The results support the model (McDonald 1991:83) of an earlier, low density artistic tradition predating the main occupation and main artistic period of the region. Only a small number of shelter sites contain residual Panaramitee engravings: this earlier low density art phase matches other forms of early occupation evidence (e.g. Attenbrow 2004; JMcD CHM 2005a, 2005b). This suggests a continuing tradition over time for the contemporaneity of art and occupation in shelter sites of the region.

This is highly suggestive about the role this medium played in terms of information exchange theory. If art is being produced in shelters where the community group is spectator – where there is an open social context, then this art is functioning very differently to art being produced 'in private' or closed social contexts.

10

DIACHRONIC VARIATION IN THE ART OF THE SYDNEY BASIN

This chapter investigates the role of time as a significant factor accounting for variability in the Sydney rock art. It is assumed that the majority of the region's art coincides with the recent archaeological evidence in the region - the late Holocene Bondaian periods. The Bondaian has three phases of production based on changes in stone tool technology. At issue here is whether these phases correlate with the different kinds of social changes and relations which might also be reflected by the region's art.

There are few indications of clear-cut diachronic change in the Sydney Basin art style (*contra* Arnhem Land, the Kimberley or Western Desert art sequences; Chaloupka 1985, 1994; Lewis 1988; McDonald 2005a; McDonald and Veth 2007; Taçon and Chippindale 1993; Walsh 1994). Social change in Sydney may not have been as dramatic as was evidenced in Arnhem Land or the Western Desert over the last 10,000 years.

The relationship between lithic technology, typology and social change is a highly complex one. The 'intensification debate' (Lourandos 1985, Lourandos and Ross 1994) characterises late Holocene increases in the number of sites and changes in their uses. A number of researchers have explored the archaeological evidence for Holocene demographic and technological changes in a range of different environments (e.g. Attenbrow 2004, Barker 2004, Beaton 1985, Boot 1996, David *et al.* 2006, Lampert and Hughes 1974, Lourandos 1985, Ross 1985, M.A. Smith 1982, Thorley 1999; Ulm 2006; Veth 2006; Williams 1985). Such studies have produced a variety of behavioural explanations (see Attenbrow 2004:185-186; cf. Bird and Frankel 1991).

If the social ramifications of changes in lithic technology are poorly understood, then the relationships between lithic technologies and artistic traditions have been even less transparent for the Sydney region.

In other parts of Australia, the burgeoning of regional art bodies has been explained in terms of greater demographic pressure, increased social complexity and amplified territoriality (David 1991; David and Cole 1990; McDonald and Veth 2006; Morwood 1984, 2002; Rosenfeld 1993, 2002). It has been argued that patterns of change in symbolic behaviour (including rock art) were functionally interrelated with changes in resource structure, technology and economy (David 2002, Morwood 1987). A similar model has now been developed for the Sydney region. This model views art as the material manifestation of peoples' value systems and ideologies and as having a role in negotiating social relations. Through the art, inherent conflicts and stressful situations are mediated.

The expectation of this model is that increased social pressure would result in increased art production. Social pressure and increased interaction is likely to be a consequence of increased population pressure. Cultural phases with the greatest amounts of archaeological evidence (Attenbrow 2004, Lourandos 1985, Ross 1985) are likely to be correlated with phases of peak art production providing *in situ* evidence for symbolic behaviour (Conkey 1978, 1980; Gamble 1982; Morwood 1987). During such cultural phases there would be increased levels of territoriality, with amplified inter-group competition and imposed 'political' control necessitating group action and re-action. During such a period the conditions would be ideal for demonstrations of local and regional group identity (viz. Wiessner 1989).

While group identity and territoriality are imposed at a local level, increased social complexity also necessitates broad scaled social cohesion.

> The emergence of regionally distinctive systems for encoding ... may therefore document the emergence of more standardised ceremonial, trade, and mating networks, which were bounded by emphasising stylistic differences...(Morwood 1984:370)

Regionally distinctive styles are seen to not only distinguish one culture area from another; they can also demonstrate social cohesion at the regional level. Shared ceremonial commitments and widespread stylistic similarities in items of material culture and body design are other signs of regional cultural cohesion.

In the preceding chapter the contemporaneity of art production with other forms of occupation evidence in sheltered locations was demonstrated. The art produced in shelters is considered to have had highly visible social context and should reveal localised stylistic patterning. In this chapter I explore how changes in the lithic production and occupational evidence may be correlated with changes in pigment art production over time.

Sheltered art is the better medium for exploring diachronic variability at several levels. Open engraving sites generally lack any superimpositionning. Sheltered sites often contain superimposed sequences. As indicated above, they are also often associated with dateable deposits. Pigment art also has the potential to be directly dated, using AMS techniques. This technique was in its infancy when this research was undertaken, and so only limited testing was attempted (McDonald 2000c, McDonald *et al.* 1990). Since then, techniques have improved (McDonald and Veth 2005, 2008; Rowe 2001) and the Sydney region offers great potential for further direct dating analysis (e.g. Taçon *et al.* 2006).

A previous regional approach to diachronic variability

Previous regional art studies have analysed the possibility of diachronic stylistic variability, but few in the detail of Morwood's (1979) Central Queensland Highlands work. Working on a sample of 92 art sites with 17,025 motifs, he analysed stylistic variability over time, and correlated this with excavated data for the region.

> [S]ome of the changes in the art through time, are clear. Others are less so, while others are still tenuous. However, taken in combination, they suggest that there was a general pattern of change which included colours, techniques, compositions, and context. Relative dating techniques ... used to outline these differences, are differential weathering, superimpositions, spatial analysis, subject and style. (Morwood 1979:278)

Morwood detailed the technical difficulties associated with the recognition of superimpositions and discussed the problems of demonstrating comparative bias (i.e. non-random variations) in the distribution of artistic variables in any sequence. He concluded that superimposition sequences can only deal with major changes in the most common variables. Morwood argued that much of the inter- and intra-site variability in his region reflected the episodic nature of art production. He identified that small assemblages commonly reflect a single artistic event and successfully demonstrated that spatial analysis of these could refine a regional sequence.

A threefold sequence was defined for the Central Queensland Highlands based on technique and colour and a change in motif emphasis. Seriation of the three art phases with the archaeological context was done using relative chronologies from excavated site data, the introduction of stone axes (in deposit and art) and stencilled contact motifs (particularly trade axes). Elements of Morwood's approach are of relevance here. The following assumptions and methodology direct the current analysis:

1) It is assumed that association (and direction of association) between contemporary art categories is random. Therefore:

    a) the number of superimpositions that an art type is involved in is proportional to the frequency of that art type and,

    b) there being no inherent bias in the direction of association between pairs of categories;

2) small art assemblages provide better information about contemporaneity of stylistic variables than large complex sites. Small assemblages are likely to represent a single artistic event;

3) large complex sites are most likely to produce superimpositionning information;

4) colour, context, composition and technique are all aspects which are sensitive to change over time. These need to be analysed individually and in combination, by the processes of superimpositionning, and through quantitative analyses to investigate trends in spatial patterning and motif preference.

An Earlier Model for Diachronic Change in the Sydney Basin shelter art
McCarthy (1967[1979], 1988) identified a temporal sequence for pigment art in the Sydney Basin. This was based on the introduction (and decline) of certain colour preferences and artistic techniques, e.g. stencilling, bichrome, polychrome. The sequence was not correlated with the engraved component, nor was it synchronised with the ERS. McCarthy's art sequence was based on several sites from across the region (McCarthy 1988:18) and was as follows:

(1) Stencil phase. Stencils in red and white, also yellow, of human hands and feet, and artefacts, in wet paint, together with the imprints of human hands and feet, and an occasional outline figure. This is the earliest phase.

(2) Red and white phase. Drawings in dry pigment in outline, solid and various infilled styles, of culture heroes, humans, animals and artefacts.

(3) Black phase. Drawings in dry charcoal in a wider range of subjects than Phase 2, in outline, solid and various infilled styles, with an important series of black and red, black and white, black and yellow bichromes, red, white and black trichromes; the richest phase of shelter art in the region.

(4) Polychrome phase. This is known in only one figure, a culture hero in four colours, associated with a large red bora initiation ground figure.

(5) White stencil phase. A very rich phase of stencils of human hands and feet, animals' paws, a wide variety of artefacts parts of plants and other subjects.

McCarthy's (1988:18) sequence does not withstand large scale field testing (McDonald 1988b). The major methodological, theoretical and practical problems with the sequence include:

- stencils at the beginning and end of the sequence provide an untestable hypothesis; how do you differentiate between an early white stencil and a late one?;

- the sequence omits several techniques which are quite common across the region - white, red and yellow paintings and the three distinct engraving techniques;

- the sequence is not correlated with any recognisable archaeological phases. While not invalidating the sequence, this limits its usefulness. Several of the phases were thought to be 'probably contemporaneous' (McCarthy 1988: 18): hence these are not technically temporal phases.

In the north of the region, Sim's work in a large rockshelter west of the Macdonald River provided a detailed superimpositionning analysis (1969: 168-70). He proposed the following sequence at this site:

1. red infill;

2. white stencils, black infill, red infill/white outline, white outline;

3. black infill/white outline, white infill, red outline;

4. black outline.

At Maroota (south of the Hawkesbury River and north of 'Canoelands') a complex medium sized assemblage provided a different superimposition sequence (McDonald 1986a (Vol. 2): 60-70). The art in this shelter included white stencils (hands, material objects and twigs) and drawings, black and red drawings and cream drawings and paintings. One black and white bichrome motif was present. The superimposition relationships here indicated that the art was executed in a single phase. A charcoal outline and infill ship motif amongst the assemblage, led to the assumption that this assemblage was relatively recent. The Maroota assemblage challenges McCarthy's sequence, both because a contact motif (in charcoal) appears over white stencils (Phase 1 or 5?), and because of the contemporaneity of techniques which McCarthy suggests may be separated into different phases.

At the Audley site in the south of the region, a sequence was discerned which fitted into the early and intermediate phases of McCarthy's model (Cox *et al.* 1968). However, Officer's work in the Campbelltown region indicated that there is no support for a chronological sequence of colours or support for a stylistic division according to colour (Officer 1984: 32).

McCarthy's sequence has been found to have major limitation in its application to sites across the Sydney region. In part this is because it does not accommodate the full range of artistic traits known to exist; but also because it subdivides the art into unworkable and untestable divisions. Individual sites, including no doubt his original type sites, support parts of the sequence but at a broader scale there is limited applicability.

Diachronic variability in the art of Mangrove Creek
The aim of this analysis was to synthesise trends in rock shelter occupation indices and artistic traits. This analysis was based on superimposition analysis in large assemblages, analysis of motif preferences and a quantitative analysis of artistic variables in small sites.

The Sample and the Technique variables

This test of diachronic variability involves shelter art sites from the Mangrove Creek drainage basin. The 65 shelter art sites used included 26 sites recorded in UMCC (Attenbrow 1981, 1987; Gunn 1979) and 39 sites from Warre Warren in the mid-reaches of the valley (McDonald 1987, 1988a). These sites provide both sufficient detail of recording in a local context where there is an excavated and dated local archaeological sequence. Seven of the 65 shelter art sites included in this analysis have also been excavated (Attenbrow 1981, 1987, 2004; MacIntosh 1965: Upside-Down-Man, this volume).

A localised assemblage was selected for this analysis given synchronic variation across the region. The resultant sequence may not, therefore, be broadly applicable on a regional basis, but require modification to accommodate localised variability.

A comprehensive list of 58 variables (colour, form and technique combinations) was initially used in this analysis (Table 10.1). Two levels of analysis were undertaken:

1. sites were counted for the presence and frequency of these 58 variables, i.e. the each assemblage was treated as a single entity and the presence of variables was counted; and,

2. the motif classification (used elsewhere in this research) was counted using these 58 variables to investigate motif/technique preferences and possible trends in preferences over time.

The frequency and distribution of these technique variables at the sites was initially investigated with the aim of determining which of these had the potential to provide the necessary information.

Assemblage sizes in the Upper Mangrove Creek sample are generally smaller than those found in the Warre Warren area. The 26 UMCC sites contained 424 motifs, while the 39 Warre Warren sites produced 2,371 motifs: 5.6 x the number of motifs from 1.5 x as many sites (Table 10.2). A twin column chi squared test on the motif assemblages from these two samples revealed that the differences between these were not statistically significant (at .05 level).

Generally speaking, the extended technical variables (cf. Table 10.3 and Table 10.4) occur in very low percentage frequencies, i.e. 17 variables account for 91.3% of the motifs, while the other 41 variables account for only 8.7% of the motifs.

The most commonly used techniques are white stencilling (26%), black outline and infill drawing (16%) and black infilled drawing (10%). These occur in a great number of sites, but their distribution does not correlate with the relative frequency of their use. The technique which represents the highest frequency of motifs does not occur in the greatest number of sites - white stencils are found in 23% of sites; black outline and infilled motifs and black infilled motifs are each found in c.60% of sites. Black outline (accounting for 6% of motifs) is found in 57% of sites. Most of the red and white monochrome drawings (consistently a low-average proportion of motifs present) are found in between 30-40% of the sites. Engraved intaglio motifs represent only 1.7% of the motifs counted, but are found in 15.4% of sites. Linear infill is found in 1.3% of motifs, but in 13.8% of sites.

To determine how this proliferation of variability correlates with overall assemblage size, technique occurrence and assemblage size was investigated. This suggests that much of the variability is not of temporal or spatial significance - since these techniques occur rarely, as one-off production episodes.

Of the 41 variables which occur infrequently (<1%: see Table 10.3):

- 8 (20%) are found at small sites;

- 17 are found at medium sized sites;

- 10 are found at large sites;

- 16 (37.6%) are found only at very large sites.

Table 10.1: Techniques and/or colour. Artistic variables counted in the diachronic analysis.

Variable	Colour and form	Technique
1	black outline	Dry
2	black infill	
3	black outline + infill	
4	red outline	
5	red infill	
6	red outline + infill	
7	white outline	
8	white infill	
9	white outline + infill	
10	yellow outline	
11	yellow infill	
12	yellow outline + infill	
13	white outline	Wet
14	white infill	
15	white outline + infill	
16	red outline	
17	red infill	
18	red outline + infill	
19	yellow outline	
20	yellow infill	
21	black infill	
22	wet + dry, outline + infill (w/w+b, b/w, b+w/b)	Wet/Dry combinations
23	red dry outline/wet infill	
24	black + white outline	Bichrome
25	white outline, black infill	
26	black outline, white infill	
27	black + white, outline + infill	
28	black + white outline, black infill	
29	black + red outline, black infill	
30	white outline, black + white infill	
31	red + white outline	
32	white outline, red infill	
33	red outline, white infill	
34	red + white, outline + infill	
35	white outline, red + white infill	
36	red outline, black infill	
37	yellow outline, black infill	
38	yellow outline, white infill	
39	black + red, outline + infill	
40	yellow + white, outline + infill	
41	red and white outline, black infill	Polychrome
42	red + black + white/yellow, outline + infill	
43	black + red + white outline, black + white/red infill	
44	black, white + yellow outline; black, white + yellow infill; white + black outline, yellow infill.	
45	black + red + white outline, black infill	
46	black, red, white and yellow outline	
47*	linear infill	

Table 10.2: Mangrove Creek valley: different sized assemblages located in Upper Mangrove Creek compared with the midstream Warre Warren sites.

Assemblage sizes	UMCC sites	%	WW sites	%
Small Sites (< 20 motifs)	18	69.2	23	59.0
Medium Sites (20-50 motifs)	6	23.1	7	17.9
Large Sites (50-100 motifs)	2	7.7	3	7.7
Very Large (>100 motifs)	0		6	15.4
	26	(40%)	39	(60%)

Most of the rarer technique combinations occur only in a small number of sites.

- 30 variables (52%) are not found at more than three sites (i.e. they occur in <5% of the sample);

- 42 variables (72%) are not found at more than 6 sites (they occur in <10% of the sample).

Of the 30 variables which occur at <u>very</u> few (<5% of) sites:

- 3 are found in small sites;

- 10 are found in medium sized sites;

- 4 are found in large sites;

- 14 are found only in very large sites.

More than half (57%) of these variables only occur at medium, large or very large sites. This suggests that these variables result from chance development rather than specific stylistic phasing, i.e. they are likely to result from greater frequency of artistic activity at large sites (and random development) rather than as a result of changing artistic trends over time.

The three rare techniques which occurred only in small sites were analysed although none of these are found in superimposition relationships with other techniques. These are:

#26 dry black outline, white infill

#37 dry yellow outline, black infill

#45 dry black, red and white outline, black infill

The conclusions which can be reached on the basis of these analyses:

- Most of the art in the Mangrove Creek valley can be accounted for by an abbreviated list of 17 technique variables. These include the monochrome dry variables (excluding yellow), white paint, red infilled paint, and combinations of red, white and black dry bichrome techniques. White hand stencils are common; red and yellow hand stencils are less common. The engraved intaglio technique is also quite common.

- There is no direct correlation between most frequently used techniques and the numbers of sites at which these techniques are used. This suggests that a few large sites may be foci for both the development of certain technical combinations and for a proliferation of these.

- The most common techniques are rarely present in more than 50% of the sample sites. This suggests that the art is relatively diverse and that there is no standardised (i.e. culturally prescribed) technical formula for art production.

- The small sites contain a range of common, average and rare techniques. These sites have the potential for testing the combination of techniques which may have temporal or spatial significance.

Table 10.3: Mangrove Creek shelter art sites. Number and % frequency of motifs for each technique variable.

Variable	Total	%f techniques
1	169	6.0
2	289	10.3
3	440	15.8
4	106	3.8
5	87	3.1
6	96	3.4
7	137	4.9
8	51	1.8
9	130	4.7
10	24	0.9
11	6	0.2
12	9	0.3
13	34	1.2
14	48	1.7
15	9	0.3
16	8	0.3
17	59	2.1
18	1	0.0
19	7	0.3
20	2	0.1
21	4	0.1
22	9	0.3
23	6	0.2
24	7	0.3
25	9	0.3
26	3	0.1
27	18	0.6
28	9	0.3
29	3	0.1
30	6	0.2
31	3	0.1
32	5	0.2
33	1	0.0
34	5	0.2
35	2	0.1
36	2	0.1
37	2	0.1
38	1	0.0
39	13	0.4
40	4	0.1
41	2	0.1
42	4	0.1
43	5	0.2
44	3	0.1
45	2	0.1
46	1	0.0
47*	36	1.3
48	718	25.7
49	76	2.7
50	69	2.5
51	12	0.4
52	20	0.7

More can be made of these findings on the basis of the superimposition and multivariate analyses.

Superimposition Analysis

Thirty-six of the 58 technique variables (62%) occur in superimposition relationships (Figure 10.1). Assemblages with superimposition relationships were found in 19 of the 65 sites (29%). A total of 189 superimpositions relationships were recorded. This was not intended to be a numerically-oriented recording of superimpositionning (cf. Morwood 1979) since the total number of relationships at each site was not always recorded. Rather, the existence of certain trends at each site was recorded[28], as was the reversal of such trends if these occurred. Superimposition relationships between two techniques would thus only be mentioned twice at any site if the relationship was found to be reversed, i.e. indicating contiguity in the use of those artistic variables.

Many of the 36 variables were not recorded in more than one superimposition relationship. This paucity is due to the relative infrequency of certain technique variables and lack of superimpositionning generally, rather than deliberate avoidance of certain technique combinations.

As expected, the most common technique variables occurred most frequently in superimposition relationships. The greater frequency of examples was not taken to be necessarily significant, given the methodology involved presence and/or absence data (Figure 10.1). This analysis revealed the following:

- Intaglio motifs occur very rarely in superimpositions (at only 10% of sites with such motifs) as the majority of these are spatially separated from the pigment assemblage in most shelters. Where they do co-occur with pigment motifs, they are always underneath the pigment art.

It has already been argued (chapter 6) that these motifs are residual artistic elements, predating the Bondaian. The spatial separation of this technique from pigment assemblages generally supports a change in focus.

- The earliest pigment techniques include red paint. Red (and sometimes white) hand stencils also appear low in many superimposition relationships.

[28] At many sites the superimposition relationships noted represented repeated trends. In other sites, these were single occurrences.

Variable	Total	%f techniques
53	7	0.3
54	37	1.3
55	2	0.1
56	6	0.2
57	6	0.2
58	1	0.0
Total	2795	99.7

Table 10.4: Mangrove Creek shelter art sites: Sites at which different technique variables are present.

Variable	Number of sites	%f sites with technique
1	37	56.9
2	27	41.5
3	39	60.0
4	20	30.8
5	22	33.8
6	26	40.0
7	26	40.0
8	7	10.8
9	24	36.9
10	3	4.6
11	3	4.6
12	2	3.1
13	5	7.7
14	7	10.8
15	5	7.7
16	3	4.6
17	6	9.2
18	1	1.5
19	4	6.1
20	2	3.1
21	3	4.6
22	3	4.6
23	4	6.1
24	6	9.2
25	3	4.6
26	3	4.6
27	8	12.3
28	4	6.1
29	3	4.6
30	4	6.1
31	2	3.1
32	4	4.6
33	1	1.5
34	6	9.2
35	1	1.5
36	1	1.5
37	2	3.1
38	1	1.5
39	5	7.7
40	2	3.1
41	2	3.1

There is a subsequent proliferation of techniques, with wet and dry pigment, a variety of hand stencil colours (including bichrome). Outline motifs occur in the uppermost layers of many superimposition sequences (wet or dry red, dry yellow and wet or dry white) finishing several sequences. Contact stencils (e.g. metal axes) occur only in white pigment. While stencilling was obviously being practised at contact, although there is some evidence for a decline for this technique in the last production phase.

- McCarthy's phased sequence is not supported by the data. Stencils (particularly of hands) occur throughout the entire production period of the pigment art. However, a stencil-only phase does not predate the depictive art – and nor does the depictive art replace stencilling in a terminal phase (Figure 10.1). There is no evidence for a red and white bichrome phase predating the predominant 'black phase'.

The following superimposition sequence is identified (Table 10.5). This was achieved by initially clumping the variable techniques into 12 gross classes (e.g. all three dry black variables, all three dry red variables, polychromes, stencils etc.: Figure 10.1) to identify general trends. A more detailed ordering within these general trends was then explored. Given the number of variables involved and the variety of superimposition relationships, this was the only rational way to manually attempt the task[29].

On the basis of this sequence it appeared that three technical phases of production could be identified. The main pigment art phase contains a proliferation of techniques. While these are generally contemporaneous (archaeologically speaking), there are apparent trends in sequencing which are indicated by the order that the variables shown. The general scarcity of superimpositionning makes more definitive division of the main phase impossible. To further explore this phased sequence for the catchment, an analysis of motif preference was undertaken.

Changes in Motif Preference over Time

The motif range in each of the three potential art phases was examined to determine whether there was any major change in motif preference between these. This would support the supposition that these were discrete artistic assemblages.

[29]Enquires were made about seriation programmes (Computing Services Section, ANU). At the time this research was completed, none were known which could have dealt with this specific problem.

Variable	Number of sites	%f sites with technique
42	3	4.6
43	3	4.6
44	3	4.6
45	2	3.1
46	1	1.5
47*	9	13.8
48	15	23.1
49	5	7.7
50	7	10.8
51	1	1.5
52	2	3.1
53	1	1.5
54	9	13.8
55	2	3.1
56	5	7.7
57	3	4.6
58	1	1.5
Total	65	

Stencils were removed from the analysis, since this technique occurs throughout the pigment sequence. This analysis was restricted to identifiable motifs (Table 10.6). This analysis involved 24 motif classes, a total of 853 motifs and a more restricted range of techniques (n=48: variables #16, 18, 19 and 35 were excluded on the basis because only unidentified motifs were recorded in these; variable #47 was excluded because it was not independent; and variables #48 - 53 were excluded because they were stencils).

This analysis demonstrates clear differences between the three proposed phases (Figure 10.2, Table 10.6).

Phase 1 has a very restricted range of motifs - tracks predominate (95%) and there are circles and 'other' motifs (dots). A Simple-non-figurative motifs (SNF) and male anthropomorph (one each) were also recorded (<3% of this assemblage).

Phase 2 has a slightly less restricted motif range, with 'other' motifs (dots) dominating (61%). Macropod tracks, SNF, circles and men are present (15%), but new elements are introduced. Most important amongst these are anthropomorphs (13%) and goannas (7.5%), but a snake, a quadruped, and a complex non-figurative motif are also present (2% each).

Phase 3 has a proliferation in the motif range and amongst these a dominance of macropods (30%). Tracks, circles and 'other' motifs diminish significantly, and this shift is towards a greater focus on figurative motifs (and a characteristic *Darkingung* motif assemblage: see chapter 9).

As the Phase 3 assemblage represented a much larger sample than the other two, a Student's t-test (designed to test whether differences between two populations may be the result of random chance) was run to assess the statistical significance of their similarities and/or differences.

This revealed the following statistics ($df = 23$):

	Statistic (chi-)	Degree of Significance
Phase 2:	-0.2	0.88
Phase 3:	-3.0	0.006
Phase 3:	-3.1	0.005

This test indicated that Phases 1 and 2 are both very different to the later Phase 3. This result shows that the differences between Phases 1 and 2 - as well as the differences between the earlier two phases and the later one - are unlikely to be the result of chance. The differences between the two earliest phases and the third are statistically significant.

The main phase (designated Mangrove Creek Art Phase 3: Table 10.7), contains a proliferation of techniques. The production of new art appears to have been cumulative. New techniques have not superseded older ones. It does not appear possible to archaeologically subdivide this major artistic period into discrete temporal events.

Prior to this, however, there was an engraved phase and a pigment phase, both of which may also have been temporally discrete. It is possible that pigment component accompanied the engraved form, although tying these two together is difficult. The phased motif analysis shows that these two assemblages are sufficiently different to warrant calling them separate populations.

This is supported by a low correlation between the two techniques occurring in the same site (Table 10.9). There are seven sites which have engravings (Phase 1) but no pigment art; three

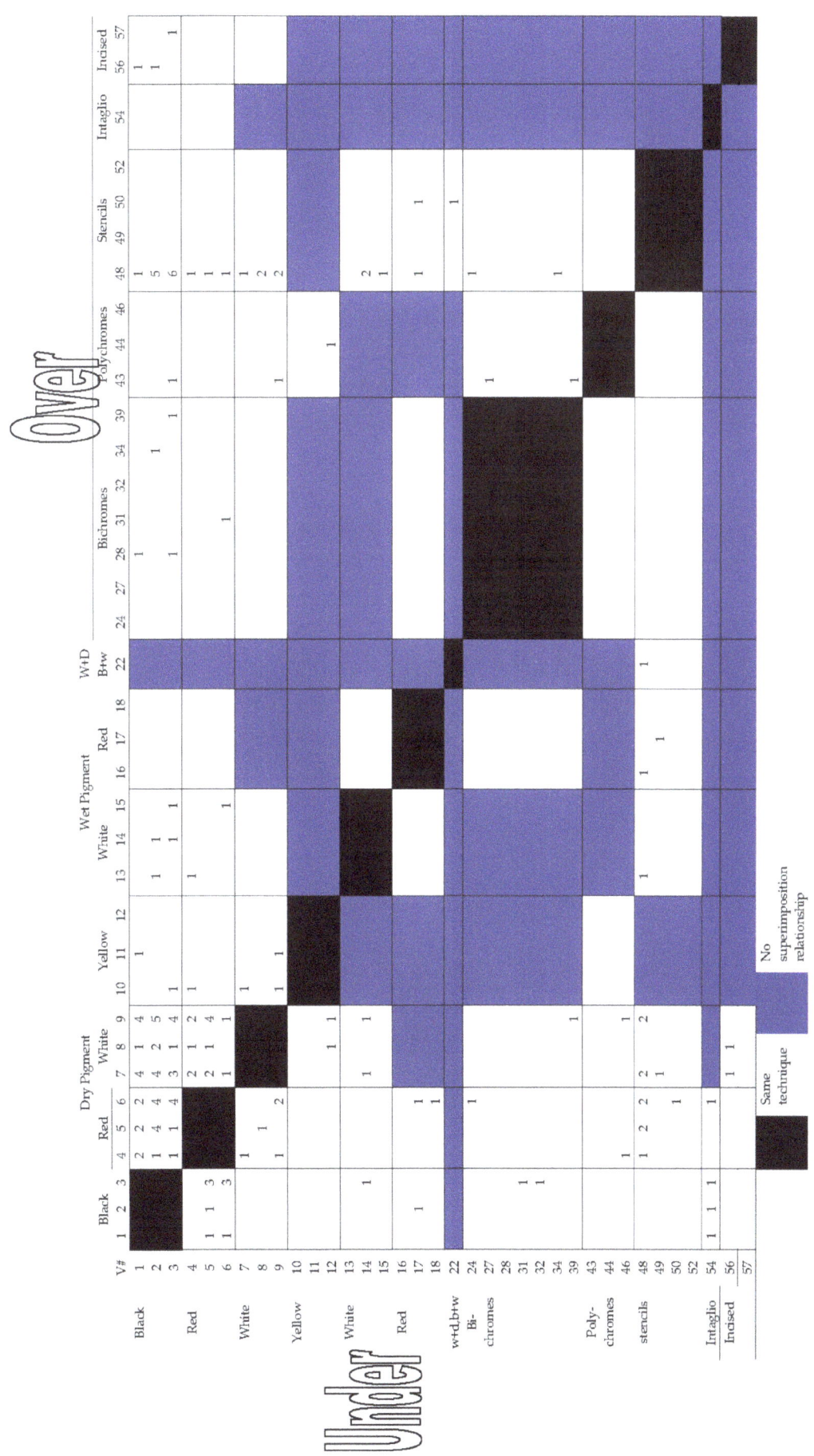

Figure 10.1: Results of the superimposition analysis.

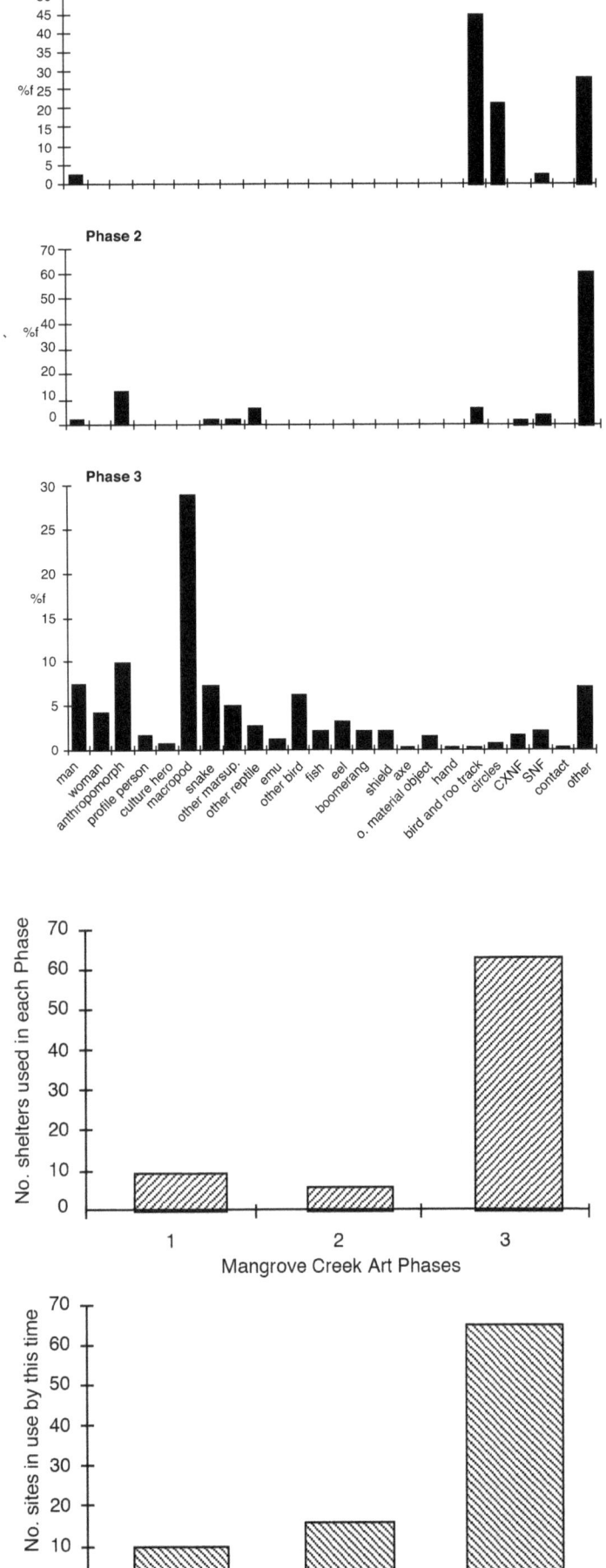

Figure 10.2: Mangrove Creek shelter sites. Motif preferences in the three identified art phases.

Figure 10.3: Number of sites demonstrating the three identified phases and cumulative frequency of sites with particular phases in use.

Table 10.5: Mangrove Creek Shelter Art sites. Superimposition sequence.

Earliest	Intaglio motifs
	white and red hand stencils, wet red infill (solid), wet red outline and infill
	dry black outline, infilled and outlined and infilled motifs, dry red infill and outline and infill motifs, wet red outline, wet white infill, white, red, yellow and pink stencils, incised, bichromes, black outlined and infilled, dry white infill and white and/or yellow outlined and infill, polychromes and wet and dry black and white motifs
	dry and wet red outline, wet white outline, dry yellow outline. Contact motifs occur in white stencils and red and/or white outlined and infill drawings
Most Recent	

sites which have Phase 2 but no Phase 1 engravings, and three sites where both art forms are present.

To view diachronic patterning in the Mangrove Creek pigment sites a cumulative frequency analysis of these three phases was applied (following Attenbrow 2004, Morwood 1987). The art in the 65 sites was categorised into the three identified phases (Table 10.9).

The sites with more than one phase were weighted to demonstrate where the focus of art production lies. The number of shelters used in each Phase varies. Phase 3 has the largest number of sites. Phase 2 has the smallest number of sites (Table 10.8).

Multivariate Analyses

The sample sites were subject to multivariate analysis. The variance displayed by the small sites was considered important, given the assumption that these should be more sensitive to temporal or spatial trends. Correspondence Analysis (CA) was employed for this purpose (Chapter 11 contains a detailed description of this technique).

All variables which were present at <5% of the sites were clumped into generalised categories and a reduced variable list of 35 attributes was analysed. This avoided the possibility that the rarer techniques would force 95% of the data into a homogeneous indistinguishable mass - thus masking any patterning therein. The revised variable list is shown (Table 10.9).

The CA results showed that the first two components accounted for 47% of the variance in the data set. The scree slope plot (Wright 1992) demonstrates that the variance is well described by this CA (Figure 10.4).

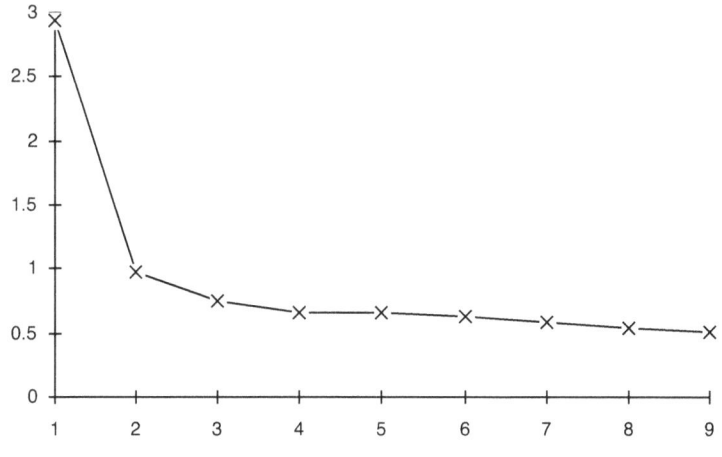

Figure 10.4: Mangrove Creek shelter CA results. Scree slope plot of latent roots.

Table 10.6: Mangrove Creek shelter art: Motif totals in the proposed art phases.

Variable	man	woman	anthrop	prof pers	Cult hero	roo	snake	Marsup.	reptile	emu	bird	fish	eel	boom'g	shield	axe	mat obj	hand	tracks	circles	CXNF	SNF	contact	other	Total
Phase 1																									
Total	1	0	0	0	0	0	0	0	0	0	0	0	0	0	0	0	0	0	19	9	0	1	0	12	42
%f	2.4	-	-	-	-	-	-	-	-	-	-	-	-	-	-	-	-	-	45.2	21.4	-	2.4	-	28.6	
Phase 2																									
Total	1	0	6	0	0	0	1	1	3	0	0	0	0	0	0	0	0	0	3	0	1	2	0	28	46
%f	2.2	-	13.0	-	-	-	2.2	2.2	6.5	-	-	-	-	-	-	-	-	-	6.5	-	2.2	4.4	-	60.9	
Phase 3																									
Total	57	33	76	13	6	222	56	39	21	9	48	17	25	16	16	2	12	2	2	6	13	16	3	55	675
%f	7.5	4.3	9.9	1.7	0.8	29.0	7.3	5.1	2.7	1.2	6.3	2.2	3.3	2.1	2.1	0.3	1.6	0.3	0.3	0.8	1.7	2.1	0.4	7.2	100

A bivariate plot of the attribute scores revealed a tight cluster around the origin with few outliers (Figure 10.5). The only variable to fall out significantly was intaglio pecking (variable #32). A surprise combination which also separated were two black + white bichrome variables (#s 18 and 21).

Focussing on the main body of variables (Figure 10.6), certain patterns are clear:

- there is a tendency for colour and technique characteristics to cluster: the dry reds, blacks and whites are similarly and separately distributed. The painting techniques are inversely correlated on the second component with the drawing techniques.

- the monochrome techniques, particularly the dry red and black separate (and cluster internally), while the polychrome and most of the bichromes cluster in close proximity to each other.

Table 10.7: Proposed Diachronic Sequence in the Mangrove Creek sites. Mangrove Creek Pigment Phases 1-3.

1 (Earliest)	intaglio motifs
2	white and red hand stencils, wet red infill (solid), wet red outline and infill
3	dry black outline, infilled and outlined + infilled motifs, dry red infill and outline + infill motifs, wet red outline, wet white infill, white, red, yellow and pink stencils, incised, bichromes, black outlined + infilled, dry white infill and white and/or yellow outline + infill, polychromes and wet + dry black + white motifs dry and wet red outline, wet white outline, dry yellow outline. Contact motifs occur in white stencils and red and/or white outline + infill drawings.

In terms of colour and technique correlations:

- Red and black drawings appear to be negatively correlated, while black and white bichromes are strongly and negatively correlated with black monochrome art.

- White stencils appear to be negatively correlated with the remainder of the stencil combinations, suggesting an inverse relationship in colour usage.

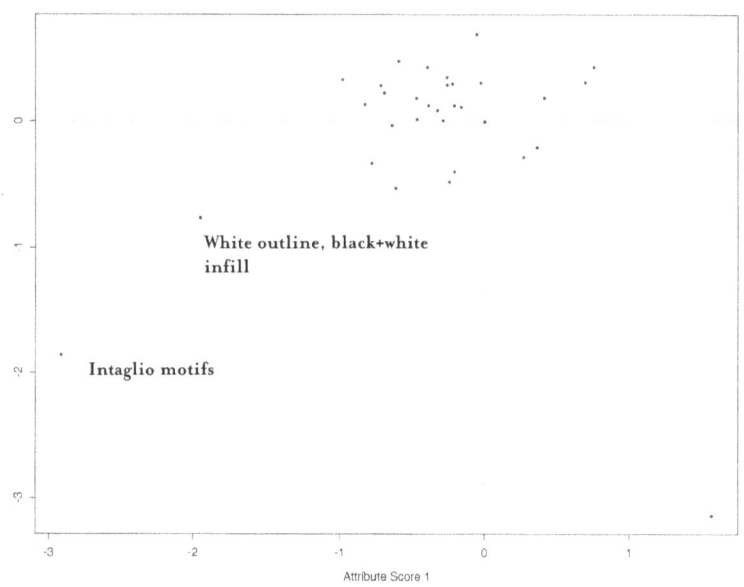

Figure 10.5: Mangrove Creek CA Results. Bivariate plot of component scores. Technical variables.

Table 10.8: Phased Sequence: Mangrove Creek Shelter Art sites.

Site No	Site Name	1	2	3
26	Emu Tracks 2‡	x		
51	Drought breaker	x	+	+
7	Emu Tracks	x	x	x
35	Corroboree	x	x	xx
65	Upside down man‡	?	x	xx
14	Black Hole	x		x
43	Black echidna	x		xx
47	Tic Alley	x		xx
36	Three emus	x		x
58	Break-a-leg	x		x
54	Wasps and women		x	x
27	Swinton's		x	xx
18	Fallen Rock		x	xx
5	Kangaroo Head			x
6	Red Eel			x
9	Roo and Echidna*			x
10	Second Look			x
16	Black Figure			x
19	Bracken			x
20	Owl Figure			x
21	Eleven Cranes			x
22	Damsite 3			x
24	Bird Tracks‡			x
25	Lizard's Leap			x
29	Solitary Kangaroo			x
32	Little end shelter			x
37	The nook			x
38	Macropod and eel			x
42	Eel shelter			x

These results appear to offer only limited support to the temporal sequence with the exception of an earlier pecked phase. Wet red infill (Phase 2) does not separate distinctively from the other phase 3 variables, although it is inversely correlated with red drawings. Red paintings (Phase 2) are inversely correlated with pecked engravings, supporting Phases 1 and 2 as separate phases rather than pigment and engraved versions of the same style.

The CA component score diagrams (plotted according to site size: Table 10.2) reveal some interesting patterning (Figure 10.7). The distribution of sites across the plot (cf. Figure 10.5, Figure 10.6) is similar to that shown by the variables.

Small sites

Analysis of the small sites showed some clustering of sites, although this appears to have little temporal significance. Two sites separated strongly on the basis of their pecked assemblages.

Site No	Site Name	1	2	3
45	Scumball			x
49	The cranny			x
52	Follow the bouncing ball			x
53	Matrigaggle			x
56	Hunting site			x
63	Headwater			x
64	Blunt Instrument			x
8	Black Hands‡			xx
23	Dairy Arm 8			xx
62	Fishmonger			xx
34	Flannel flower			xx
55	Lion's mouth			xx
50	Swain's surprise			xx
48	Wave Rock			xx
1	Dingo and Horned Anthrop‡			x
3	Loggers‡			x
4	Echidna			x
13	Frogman			x
15	Red Figure			x
17	Sandy cave			x
28	Metal Axe			x
31	Eel headed men			x
33	Roos and snakes alive			x
40	Warre Warren Ck			x
41	Rain Gott			x
46	Formal macropod			x
59	Big valley site			x
60	Red Ned			x
61	Black mac			x
11	Candelabra			x
12	Sailing Boat			x
57	White ones			x
39	Banksia			x
2	Red rat kangaroo			x
44	Waratah 1			x
30	WW Corroboree			x
	No. of shelter used for each phase	10	6(+1?)	63 (+1?)
	No. shelter used by this time	10	16	65

+ white stencils only - phasing indistinguishable
‡ excavated sites with radiocarbon dates (see Table 10.10)
* excavated site

With the exception of the two outlier pecked sites, the small sites are spread diagonally across the top of the plot (Figure 10.7). These sites can be divided into four main groups. In the top right quadrant, the sites contain almost exclusively black motifs. This is the largest group (16 sites: 39%). Another large group (11 sites: 27%) is clustered around the origin. These sites contain black, red and/or white assemblages. In the top left quadrant the art consists of predominantly white motifs. In the lower left quadrant the site assemblages contain predominantly red infill.

Only two of the small sites contain white stencils. Only four of these sites (10%) contain bichrome motifs (one of which also contains a polychrome motif). Seven multicoloured techniques are used at these four sites, and, as identified earlier, three of these techniques occur only in small sites. These represent unique combinations rather than temporally significant markers.

Medium sites

The medium sized sites show a much more restricted distribution with a cluster to the top right. These sites contain mainly black and/or red drawings and white hand stencils. The sites close to the origin have a mixture of techniques and colours. The two outlying sites contain red, black and/or white bichromes (site #31) and black and red drawings and pecked motifs (site #58: see Figure 10.7). Most of the sites in the top right quadrant in this size group contain white hand stencils (unlike the smaller sites).

Large and complex sites

The large and very large sites show a very different distribution. Almost all of the very large sites are located in the top left quadrant, containing a much wider variety of techniques (stencils, bichromes and polychromes) and colours (particularly yellow and white).

Table 10.9: Techniques and/or colour combination variables used in Correspondence Analysis.

Variable	Colour and Form	Technique
1	black outline	Dry
2	black infill	
3	black outline + infill	
4	red outline	
5	red infill	
6	red outline + infill	
7	white outline	
8	white infill	
9	white outline + infill	
10	dry yellow	
11	white outline	
12	white infill	
13	white outline + infill	Wet
14	red infill	
15	yellow outline	
16	red dry outline/wet infill	Wet/Dry
17	rare wet combinations	
18	black + white outline	Bichromes
19	black + white, outline + infill	
20	black + white outline, black infill	
21	white outline, black + white infill	
22	white outline, red infill	
23	red + white, outline + infill	
24	black + red, outline + infill	
25	rarer bichromes (combined)	
26	polychromes (combined)	
27	linear infill	
28	white	Stencils
29	red	
30	yellow	
31	rarer stencils (combined)	
32	intaglio	Engravings
33	incised outline	
34	rarer engravings (combined)	
35	incised o/i	

Conclusions

These combined analyses demonstrate:

1) intaglio motifs (Phase 1) separate significantly from the remainder of the techniques;

2) the wet red infill (Phase 2) does not separate well from many of the more complex colour and technique combinations. The fact that few sites are found with this phase alone may be masking this characteristic;

3) red and white hand stencils separate well on the first component;

4) monochrome colour usage (in a variety of techniques) distinguishes well between art assemblages. This suggests that art production using a single colour occurred in an episodic fashion;

5) the strong separation between the black and red use was not indicated by the superimpositionning analysis (Table 10.6; Figure 10.1), although this hinted that black might precede red (i.e. red drawings consistently covered black drawings: dry black infill is only recorded over red infill, and none of the black techniques were recorded over red outline motifs);

6) assemblages with more complex uses of colour and technique and those with of rarer colour usage cluster together, along with the use of wet pigment;

7) small sites, representing probably single artistic events, demonstrate clustering on the basis of specific colours. Four main groups are distinguished:

   i) predominantly black;
   ii) predominantly white;
   iii) predominantly red; and,
   iv) dry black, red and white combined.

8) hand stencils and bichrome/polychrome motifs rarely occur in sites with <20 motifs;

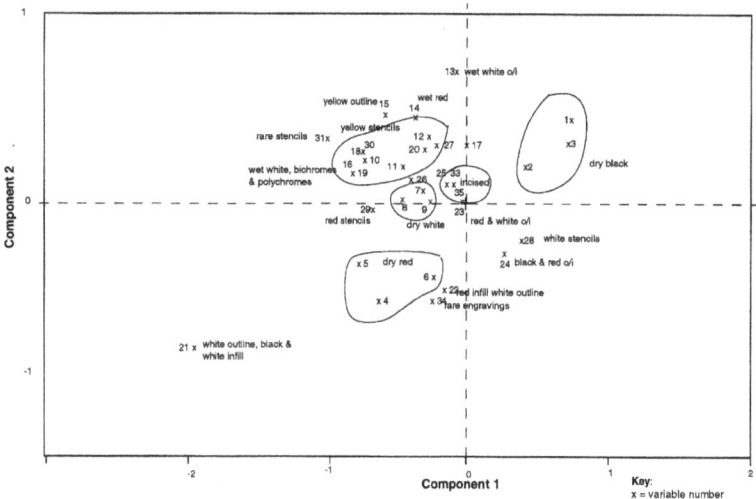

Figure 10.6: Mangrove Creek CA Results. Bivariate plot of component scores. Technical variables (excluding variable #32, intaglio).

9) medium sized sites contain assemblages which use predominantly black and red drawings and have white hand stencils. These sites rarely have other coloured hand stencils present;

10) large sites contain a range of (mainly monochrome) techniques with white hand stencils, and or pecked engravings;

11) very large sites contain the full range of techniques employed in this valley; most of the rarer techniques and colours are only found in these sites.

The CA has demonstrated clustering of the sites based of colour and technique preferences. These groupings do not indicate temporal trends – as determined by superimposition analysis.

These results appear to demonstrate how the art body in this valley system was produced - namely that small sites contain limited artistic events created with a range of similar raw materials. These appear to be *ad hoc* episodes producing art with whatever raw material was to hand (i.e. the ubiquitous charcoal). Sites which represent more conspicuous art foci contain a wider range of techniques and raw materials. The sites which have been used extensively for the production of art demonstrate the full range of techniques and colours available to the artists in the Valley. Increasing artistic complexity - while the result of repeated artistic events – does not demonstrate an evolution of the style.

Two earlier art phases are supported by motif preference and superimpositionning analyses. The separation of the pecked motifs from other techniques is also strongly supported by the CA. No strong trends in colour preference or technique are indicated in the main art phase. Complex sites appear to have been used for longer periods of time or for more intensive episodes of artistic activity. The proliferation of techniques in these sites represents an accumulation of technical options rather than an evolution of traits. The fact that the largest and most complex sites cluster cohesively (Figure 10.7) demonstrates an overriding stylistic homogeneity in the sites of this size. A similar pattern is identified with the medium sized sites. These results are suggestive of different types of art activity locations being identified.

Mangrove Creek Art Sequence

Basis on superimpositionning, motif preference and multivariate analyses the following phases of art production are discerned within the Mangrove Creek valley:

Mangrove Creek Art Phase 1   pecked engravings of tracks and circles.

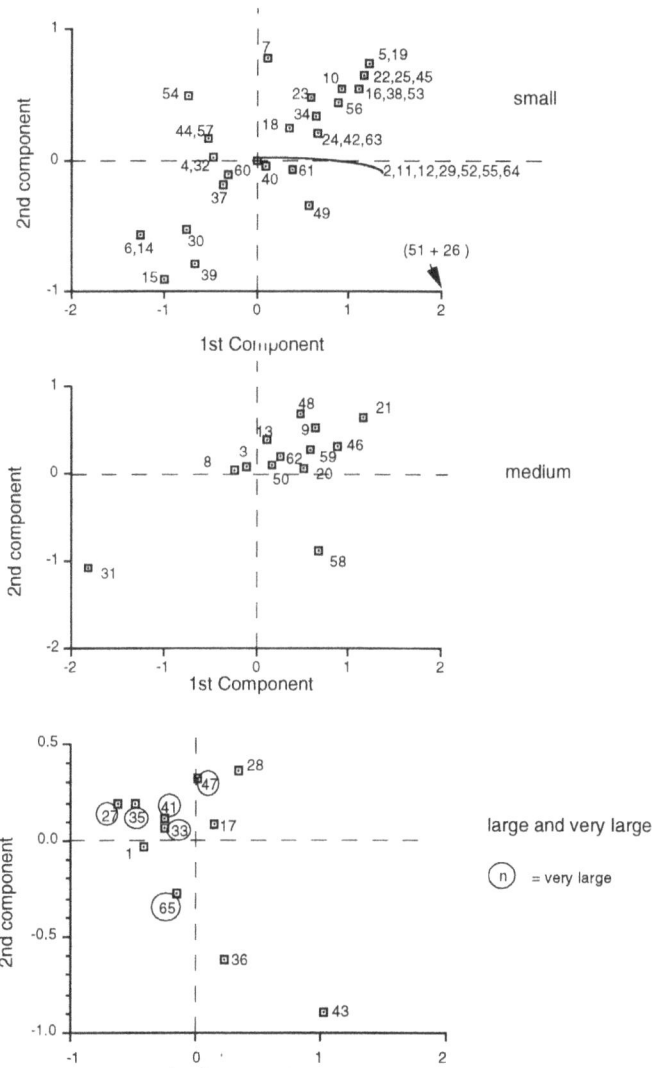

Figure 10.7: Bivariate plot of component scores, Mangrove Creek. Small, medium and large-very large sites plotted separately.

Mangrove Creek Art Phase 2   red paintings and hand stencils, and white hand stencils (red and white hand stencils do not co-occur).

Mangrove Creek Art Phase 3   a proliferation of techniques and colour use, perhaps starting with plain dry black and dry red motifs and then developing into a range of paints, dry bichromes, stencils of varying colours, polychromes and incised motifs. Outline-only motifs end the sequences of many shelter, although contact motifs have also been recorded in white stencils and drawn red and white outlined and infill forms.

How does this art sequence correlate with occupation indices? Is it possible to propose an absolute chronology based on the data obtained here? Definite comments about the earliest and terminal phases are possible.

At Yengo 1 a date of c.6,000 years BP is given for Phase I (associated with a Pre-Bondaian lithic assemblage). A pre-Bondaian association can also be inferred for Emu Tracks 2 in upper Mangrove Creek, although this is not demonstrated by the excavation results.

At Emu Tracks 2, the 15 engraved emu tracks were recorded as being level with the current surface of the deposit. Here the first major occupation of the shelter began sometime around 4,000 years ago (Attenbrow 1987: Table 7.2). Regardless of the exact timing of the shelters first usage, it would appear that the art's position on the wall indicates that this was contemporaneous with earliest usage of the shelter, predating the Early to Middle Bondaian occupation (see Table 10.10).

At the most recent end of the art sequence, there is evidence that pigment art was being produced at European contact. Drawings of sailing ships in a range of colours and the stencilling of metal axes indicate that the art in the Mangrove Creek valley was being produced as late as 1788, although 1770 (the arrival of Captain Cook at Botany Bay), is a possible *pro terminus* (McDonald 2008).

Both Mathews (1897c) and McCarthy (1939) cite evidence of an Aboriginal person in Wollombi Brook producing art as late as 1843, but there is no evidence that this was produced in a culturally prescribed fashion. The ethnohistoric literature offers no evidence of people observed producing art in the early days of white settlement. The fact that contact motifs are relatively rare in the Sydney area suggests either that the production of art at this time was either sporadic, or that the use of art as a cultural medium quickly diminished after contact.

There is an absence of associated evidence for Phase 2 art and the question of timing between Phases 2 and 3 is problematic.

A correlation between art and deposit can be investigated at the seven excavated shelter art sites in the Mangrove Creek valley, five of which have been dated[30]. From these dates and the archaeological sequencing at each site, correlations between art and occupation phases can be made (Table 10.10). This is not a simple exercise as the sites excavated by Attenbrow were not excavated for this explicit purpose.

Of the five art shelters excavated by Attenbrow, four have multiple phases of occupation and most have Phase 3 art. The site that has a single phase of both art and deposit is Bird Tracks (Table 10.10). The art here is designated Phase 3, the deposit (undated) Phase 4 (Attenbrow 1987: Table 7.2). If one assumes that the two were produced contemporaneously, Art Phase 3 at this site dates to the last 1,600 years.

Table 10.10: Excavated shelters with radiocarbon determinations, estimated archaeological phases and designated art phases.

Site	Dates	Occupation Phases	Art Phases
Emu Tracks 2	-	2, 3, 4	1
Upside Down Man	4,030 ± 140BP	1,	1?, 2
	1,5440 ± 60 BP	3	
	1,220 ± 120 BP	3	3
Roo and Echidna	6,700 ± 150 BP	1	
	-	2, 3, 4	3
Bird Tracks	-	4	3
Black Hands	3,040 ± 85 BP	2	
	-	3, 4	3
Dingo and Horned Anthrop	581 ± 120 BP	not analysed	3
Loggers	11,050 ± 135 BP	1	
	7,950 ± 80 BP	2	
	2,480 ± 60 BP	3	3
	530 ± 90 BP	4	3

Three of the art shelters with multiple phases of occupation (Roo and Echidna, Black Hands and Loggers) had their most intensive occupation during Lithic Phase 3 (2,800-1,600 years ago). There is no direct evidence from these shelters to correlate the two forms of evidence. If it's assumed

[30]While 18 shelter art sites were excavated by Attenbrow, the art in only five was recorded in sufficient detail for this analysis.

that the production of pigment art coincided with the most intensive period of occupation, a Middle to Late Bondaian association is suggested for the Phase 3 art in these sites.

More direct associations are possible from the remaining excavated art shelters.

The presence of at least three artistic episodes at the Upside Down Man (UDM) shelter has already been argued (above), while the dating and analysis of the deposit indicates two distinct phases of use. The contemporaneity of the main phase of occupation (Lithic Phase 3) and the majority of the art (Art Phase 3) is indicated on the basis of excavated evidence. It is also argued that a late artistic episode, including a suite of outline motifs, one of which was AMS dated, may well indicate that art production continuing after the abandonment of the shelter for habitation, c. 1,200 years BP.

The two pecked motifs at this site are figurative, unlike the majority of the pecked assemblage in this creek valley. It is possible that these represent a 'transitional' form between Phases 1 and 2. The relatively recent basal dates and the nature of the pre-Bondaian Phase 1 assemblage, support such an argument. The early pigment art (with painted red anthropomorphs that are stylistically similar to the engraved anthropomorphs) may be more closely related to Phase 1 and 2 motifs. The UDM Phase 2 motifs are correlated with the earlier occupation of the site.

The main phase of Middle Bondaian occupation at UDM ends later than proposed by Attenbrow's (2004) sequence (i.e. between 2,800 -1,600 years ago). Similarly, the UDM early Bondaian (Unit III) also returned a comparatively recent date (c.4,000 years BP).

Using UDM's excavated assemblage, the following dates are suggested for the art phases:

Art Phase 2	<4,000 -> 1,600 years BP
Art Phase 3	c.1,600 - 1,200 years BP

The other shelter excavated expressly to date its art was Dingo and Horned Anthropomorph (MacIntosh 1965). Two dates (144 ± 125 BP (GX-0069) and 581 ± 120 BP (GX-0070)) were obtained from one pit at this site. This site has the most conclusive excavated evidence for ochrous art production in the Sydney region[31]. The faceted red ochre, of the same colour as the dingoes and horned anthropomorphs, associated with the earlier date establishes the likely production date for these motifs (see discussion above). The remainder of the art at this shelter (in a separate panel) is classified as Phase 3. Its condition suggests that much of its production may predate the dingo and horned anthropomorph composition. The horned anthropomorphs[32], dingoes and echidnae are classified, as late Art Phase 3. The dated evidence for this site indicates that these motifs were produced midway through the last millennium, long after occupation in UDM ceased.

Based on Emu Tracks 2 and these two shelters with good correlations between art and deposit, and given to the presence of contact motifs in the Valley, the following chronology for the Mangrove Creek Art sequence is proposed:

Art Phase 1	Pre- or Early Bondaian	> 4,000 years BP
Art Phase 2	Early Bondaian	<4,000 – c.1,600 years BP
Art Phase 3	Middle to Late Bondaian	c.1,600 - European contact

Difficulties were encountered in accurately pegging this chronology because of the scarcity of sites with art in dateable contexts. There are inconsistencies in dating the stone tool phases both in Mangrove Creek and in the broader region (see chapter 4). Art Phase 3 could have commenced a millennium earlier than the dates proposed here: and the Late Bondaian transition may have occurred later.

[31]Attenbrow 1987 (Sunny), Glover 1974 (2CU/5), Tracey 1974 (4cU/5) and Megaw and Roberts 1974 (WL/-].
[32]This particular motif has an extremely restricted distribution in the Mangrove Creek catchment and near Mogo Creek just to the west (Chapter 12).

A correlation of art phases with broader lithic phases appears to be the most judicious categorising of the material. Assuming that the main art production period in most sites is contemporaneous with the most intensive period of stone tool production, the Middle Bondaian is likely to be the peak art producing period. Art production certainly continued through the late Bondaian and indeed up until contact. Art was produced in shelter contexts throughout this period without appreciable stylistic change. The significance of this finding is discussed (Chapter 13), in terms of the model for stylistic behaviour in the region.

11

SYNCHRONIC VARIATION: SYDNEY BASIN ENGRAVED ART

Introduction

This chapter looks at regional stylistic variability in the engraving component. The investigation is restricted to motif depictions, given that technical variation here is minimal. Analysis was aimed at identifying whether broad scale patterns could be interpreted culturally, not just environmentally (McMah 1965). A comparison of these results with those achieved in the shelter art assemblage is made later (Chapter 12).

There was initial concern that focussing on motif would not investigate style *per se*, but mere compositional variety[33]. Given the overall aim was to analyse the engraving and shelter art components at a comparable level (despite their technical differences and the variability that these introduce to motif form), motif taxonomy appeared the most judicious approach to the problem.

Motif has been successfully employed at a regional and localised level, investigating a range of stylistic questions (Clegg 1987, Officer 1984, Franklin 1984, Smith 1989). Sackett (1990) cites various examples of how the combination of motifs and compositional features may indicate high levels of ethnically significant patterning (e.g. Glassie 1975, Longacre 1981) and suggests that 'themes may well be the [things] that give congruence to isochrestic choices in non-material aspects of cultural life' (Sackett 1990:41). While this type of classification had not been previously attempted on an archaeological rock art assemblage, the approach used here tested Sackett's proposition. Correspondence Analysis (CA), which seeks patterning in the combination of variables (motifs) in the data set, was considered an ideal tool for this analysis.

The CA indicated that the region's engraving assemblage was largely homogeneous (see below). The results were viewed according to the language boundaries in evidence at white contact and to major drainage basins within the study area. These contexts provide an explanatory framework for variation across the region.

The posited style boundary to the south of the region was further investigated. This was initially identified by McMah (1965). The Rock Art Project determined the likely location for this boundary was the Georges River - and that this boundary existed for both art components (McDonald 1985a, 1990a). This research refines further the extent of the stylistic differences on either side of this boundary. Style clines are also defined and described elsewhere across the region. These are manifested as increasing and/or decreasing amounts of homogeneity in localised areas and varying motif foci.

Engraving sites in different topographic locations were investigated to explore the possibility of different social contexts for information exchange. Vertical engraving sites around the estuarine foreshores (in very public locations) were compared with open engraving sites on the ridgelines (where it is assumed that the audience was more restricted).

The distribution of uncommon motifs is also explored. Rare motifs were thought to have the best potential to demonstrate the influence of local (or even individual) stylistic traits. Rare motifs

---

[33] A qualitative approach to individual motifs (e.g. body proportions, angle of macropod's tails, orientation of motif and presence of eyes and other internal features, etc.) may also have revealed stylistic patterning. Given the lack of success of this approach to demonstrating ethnicity on a limited scale (Smith 1983; and see Clegg 1981), and the generalised outcome achieved by such an approach at a regional level (Franklin 1984), this additional type of analysis was not attempted.

are swamped in large-scale analyses, and it was hoped that this type of analysis would provide additional stylistic information. Whales, for instance demonstrate a fairly restricted geographic range. This, however, could be an environmental range as much as a cultural one. Other motifs, such as profile people, culture heroes, certain material objects and Complex-non-figuratives (CXNF's) were considered to be better gauges of cultural choices. The restricted distribution of contact motifs is also discussed.

As well as the regionally based quantitative analyses, several geographically restricted qualitative analyses of motif depiction and preference (Sackett's 'compositional features') were undertaken. Certain attributes on human figures (particularly gender and items of material culture), CXNF's and culture heroes are the focus of this analysis.

Defining a regional style: methodology

The aim of these analyses was to provide a statistical description of each art component (i.e. average assemblage size, motif frequencies etc.) and to determine the amount of variation within the assemblages on the basis of multivariate analysis. The general approach described here was used for both art components.

As with most exercises of quantification, logical steps are required to code the data so it can be read by a computer, analysed and then interpreted. As well as logically ordering the data, it is necessary to justify the selection of variables as meaningful and relevant to the questions being asked. The classification and selection process was discussed fully in the original research (McDonald 1994a: Appendix 1). The procedures followed in quantifying the two art components can be broadly defined as:

- identification of the sample;
- selection of variables;
- collection (counting) of variables for the sample;
- input of counted information into a (computerised) data base;
- selection/clumping of variables for analysis;
- analysis of data base; and,
- interpretation of analysis results.

Having identified the sample (Chapter 5), the selection of variables for analysis proceeded. A motif classification was applied to both art components (these were fundamentally the same with addition of several motifs for the more diverse shelter art component) and a taxonomy accounting for technical variations for the shelter art sites (see McDonald 1994a: Appendix 1). Topographic, grid reference and site association information for both site types was also collected (see McDonald 1994a: Appendices 5 and 6).

Motif Variables

While comparability between the two components was an overriding factor in the selection of motif variables, it was recognised that inherent differences in the two components would necessitate some variation in the motifs identified and counted.

The motif classification was initially devised for the Rock Art Project, specifically for the engraved assemblage (McDonald 1985a). The field recording exercises undertaken in subsequent stages of the Project revealed greater motif variability, particularly in the shelter art component (McDonald 1987, 1990a) which necessitated two additional motif categories (hands and axes) for both components, and the addition of two exclusive categories (hand stencil variations and 'other') for the shelter art sample (see Tables 5.1 and 5.3, Chapter 5).

Counting

Once the motif and technique classifications had been devised, counting proceeded[34]. The counted data was initially recorded on accounting broadsheets. Each site had one column; variables were recorded by row. As well as counting motif and technique variables, site card information (for those sites not visited by me) provided the topographic and site association variables. AHIMS (then NPWS) site identification was used for each site.

All data were analysed on the mainframe computer at the Australian National University.

Analyses

Analysis commenced with the motif count information, i.e. the raw data which had been entered into the computer. This involved the analysis of 27 engraved motif variables.

All data were put through a GENSTAT Correspondence Analysis (Version 3.1), designed to investigate variance within large multivariate data populations (the multivariate analyses used were defined and discussed in McDonald 1994a: Appendix 1). This method found that both art components were largely homogeneous populations, which clustered tightly and showed no underlying structure. The pattern of distribution, both for variables and sites, was affected largely by the presence of a few outliers. The analysis was re-run removing the outliers, in an effort to seek the underlying patterns within the greater data set. This was done a few times, until it became clear that the method was imposing structure upon the data - in much the same way as a cluster analysis.

Two further steps were taken in the treatment of the data. This involved clumping the motifs and converting them to binary data. The motif taxonomies for both components were reduced to seven variables (Table 11.1).

Table 11.1: Engraving Sites. Clumped motif variables used in Correspondence Analysis.

Variable No.	Motif/Variable description
1	Anthropomorphic
2	Terrestrial
3	Birds
4	Marine
5	Material Objects
6	Tracks
7	Other

Clumped variable 1 includes individual variables 1 - 5; 2 = v 6-8; 3 = v 9,10; 4 = v 11-14; 5 = v 15-8; 6 = v 21-24; 7 = v 25-27 (see Table 5.1). Unidentifiable motifs have been excluded from this level of analysis.

Converting to binary data was a simple process of using presence and absence rather than raw count data: if a variable was present it received a value of 1; if it was absent its value was 0. The CA of the engraving sites was based on a reduced sample of 705 sites (i.e. sites with only unidentified motifs were excluded from the analysis).

Correspondence Analysis (CA): data, results and interpretation

The aim of a CA was to investigate sources of variance within the data set, to identify groups of similar and dissimilar objects (i.e. sites). The advantage of this technique over other multivariate tools is that the variables (i.e. motifs) which contribute to these groupings can be identified. It is not so much the presence of individual motifs which creates the variance, but the <u>combination</u> of variables (Baxter 1994, Benzecri 1992, Shennan 1988).

To explore the geographic variability within this assemblage, first a CA was run using grid location as a factor. This revealed no internal groupings or structure, nor any strong positive correlation between geography and motif. This result supported the absence of strong patterning or major divisions within the assemblage.

---

[34]All sites including those which had been counted previously (McDonald 1985a) were recounted for this research.

The data base for this component is too large for the plot of the site distribution to be meaningfully interpreted (Figure 11.1). The results were thus replotted using subdivisions of the data (Table 11.2: see McDonald 1994a: Appendix 7; Figures A7.1 - A7.7). These sub-plots are based on exactly the same results, but the smaller sample sizes enabled more useful interpretation.

Analysis was undertaken on a variety of scales. The first analysis involved an arbitrary division of the sites into map sheet provenance. The data were subdivided into eight groups which could be interpreted broadly on a geographic basis. The resultant groups were unequal in size, indicating e the geographic biases of the sample (Table 11.2). While being arbitrary in terms of archaeological context, this division of the data base gave control over north-south and east-west clines in the data (see Figure 11.2; McDonald 1994a: Appendix 8).

A more detailed subdivision of the region's data was contemplated, but given concerns about the relevance of the documented linguistic boundaries (Chapter 3) and obvious sampling issues (i.e. unequal sample sizes), division of the data into archaeologically meaningful zones was a vexed question. To investigate possible cultural divisions across the region more closely, several localised areas (with good sample sizes) were chosen to investigate linguistic boundaries.

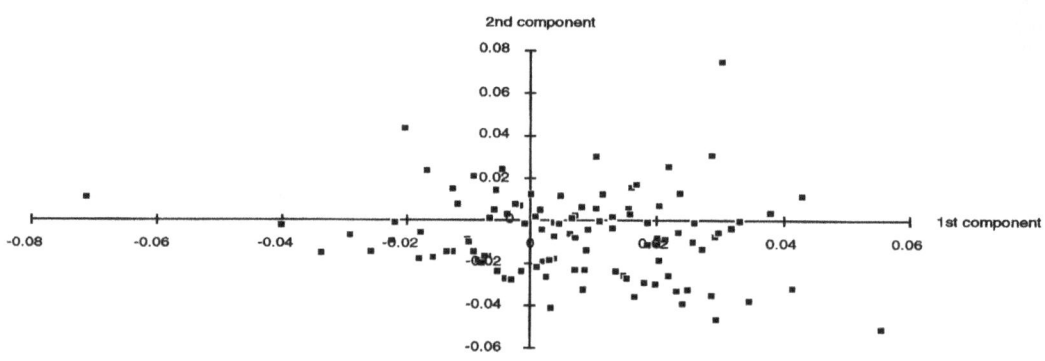

Figure 11.1: CA Scores: entire Engraving component (705 sites). Each dot represents many sites.

Table 11.2: Analytical grouping of engraving sites according to AHIMS numbers. These groups were used in the regional interpretation of the CA results.

Group	Map numbers	1:250,000/1:100,000 maps	Sample size
Group 1	37 - 6 -'s	Singleton/Cessnock	4 sites
	45 - 1 -'s	Sydney/Wallerawang	
Group 2	45 - 2 -'s	Sydney/St Albans	37 sites
Group 3	45 - 3 -'s	Sydney/Gosford	234 sites
Group 4	45 - 4 -'s	Sydney/Blue Mountains	10 sites
	45 - 5 -'s	Sydney/Windsor	
Group 5	45 - 6 -'s	Sydney/ Sydney	377 sites
Group 6	52 - 2 -'s	Wollongong/Wollongong	19 sites
Group 7	52 - 3 -'s	Wollongong/Port Hacking	35 sites

Regional Analysis

The first two components account for 64% of the variance in the sample and these components discriminate well. The first component accounts for the greatest amount of variance in the data base (Figure 11.2), and the scree slope plot (Wright 1992) demonstrates that this component describes considerable structure in the data. Less variance is accounted for by the second component, and the slope then tails off.

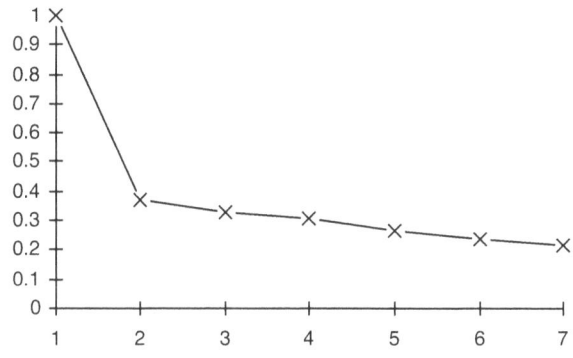

Figure 11.2: Engraving CA results: Plot of the latent roots indicating that the variance in the data set is well accounted for by the first two components.

While no major internal groupings were identified, certain sites were identified as outliers (Figure 11.1). The plot of the first two co-ordinates (Figure 11.3) reveals that three of the variables (2, 3 and 4) are good discriminators. In the first co-ordinate, marine animals (3) and birds (4) are negatively correlated, while in the second co-ordinate, birds (3) are negatively correlated with land animals (2). Thus sites which contain large numbers of bird motifs would have very few marine depictions, while sites with a large number of land animals would contain small numbers of bird depictions (and vice versa). Anthropomorphs are relatively weakly positioned on the first co-ordinate, but have a relatively good discriminating effect on the second co-ordinate. Material objects, tracks and 'other' motifs are poor discriminators being situated close to the origin.

The CA results reveal no evidence for strong or distinctive localised variability across the region. Certain stylistic clines can be observed in the region's subdivisions as shown by the bivariate sub-plots (McDonald 1994a: Appendix 7: A7.1-A7.7) in terms of the distribution and/or clustering of sites relative to the origin (i.e. $X + Y = 0$ on the bivariate plot). Sites close to the origin are poorly discriminated by the axes in question and are stylistically homogenous. Those site distributed away from the origin are well discriminated by their motif assemblage i.e. are stylistically differently. 'Common' sites fall close to the origin; unique and/or more unusual sites are located away from the origin. The identification and distribution of outlier sites in localised areas (according to quadrants on the graph) is the key to investigating thematic variety across the region.

The bivariate plots for each group were analysed. The number of sites within a defined and consistent radius of the origin was noted[35], allowing for a calculation of the percentage of 'common' and outlier sites in each area. This was necessary for comparability given the disparate sample sizes. As the computer generated plots sometimes generated the two axes at different scales, the radius is sometimes described by an ellipse rather than a circle. The distribution of the outlier sites according to the four quadrants was also investigated (Figure 11.4). Variations in the distribution of sites within the quadrants identified different compositional foci across the region.

The sites in the north-west of the region were found to be relatively heterogeneous. Outlier sites occurred predominantly (86%) on the positive side of the vertical axis, indicating that they are more strongly (and positively) discriminated by the first component. Sites in these groups contained many tracks (bird and macropod) as well as a definite preference for macropods and other

---

[35]This was drawn on each of the bivariate plots as a heuristic device.

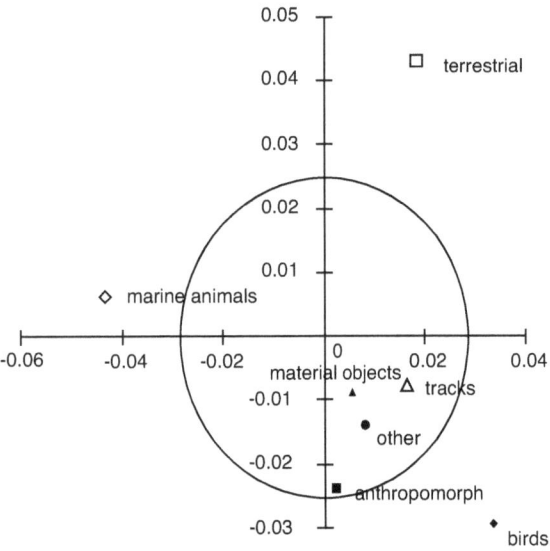

Figure 11.3: CA results: Engraving Sites. Bivariate Plot of Variable Scores.

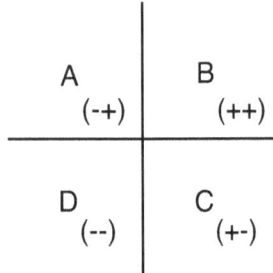

Figure 11.4: Quadrant labels used in the following discussion of the CA results.

land animals, birds and anthropomorphs.

The large group of sites (234) in the north-east of the Basin (north of the Hawkesbury River) were more heavily clustered around the origin (45%), while the outliers showed a strong tendency (76%) to be located on the positive side of the vertical axis. There was a preference in these sites for tracks (particularly *mundoes*), material objects (shields) and marine depictions. Anthropomorphic depictions were also very common.

The 377 sites in the centre of the Basin (located between the Hawkesbury River and Botany Bay; the Cumberland Plain and the coast) probably represent the core Sydney Basin engraving assemblage. Most of the sites (54%) in this group were tightly clustered around the origin. The majority of the sites in this group contained marine depictions, material objects, tracks (particularly *mundoes*) and anthropomorphs. Land animals were quite common, as were birds. There are several outlier groups within this sample. One major cluster (53 sites) contains combinations of exclusively marine depictions. In another major cluster (of 25 sites) each contains a single macropod.

South of the Georges River, there was a marked increase in heterogeneity. South of Port Hacking (group 8) the sites are well dispersed away from the origin (only 26.5% homogeneity). These sites contain a predominance of marine depictions and other material objects. Some of these sites included anthropomorphs; others included terrestrial animals and 'other' motifs. The outlier sites in this group indicate a major difference from preceding groups in subject preference. When the percentages of homogeneous sites across the region are compared (Figure 11.5), this patterning is clear.

The most homogeneous assemblages are in the centre of the Sydney Basin (Groups 5 and 3). This homogeneity decreases as you moving north-west and south. Sites to the west of the central core are also fairly homogeneous.

There was also marked variation in the distribution of outlier sites (compositional focus) across the region. The emphasis on certain combinations of motifs varies across the region, and it is the changes in these combinations that characterise the stylistic clines across the region. These are sometimes explicable in terms of economic/geographic factors (e.g. proximity to the sea). Not all variations, however, were so easily explained.

Language Areas. Searching for boundaries and between-group distinctiveness

Five language areas (following Capell 1970) are recognised to have existed within the study area at European contact (chapter 3). These languages were mutually intelligible, although ethnohistoric evidence suggests that the locations of neighbouring 'tribal territories' were recognised and respected by the various groups. Anthropological work elsewhere on the continent suggests that the boundaries between language groups would not have been impenetrable barriers. Such studies also suggest that such boundaries may have been fluid over time. The territorial distribution of the contact languages may not have extended back more than a few generations.

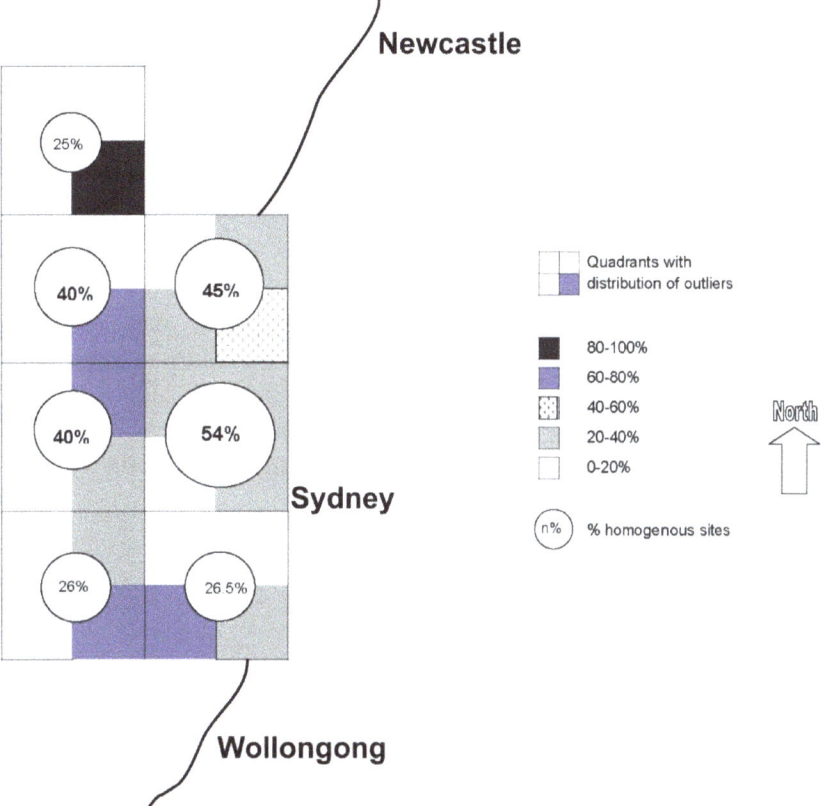

Figure 11.5: Percentage of homogeneous engraving sites in each analytical Group.

Archaeological evidence also casts uncertainty onto the longevity of these boundaries: the cultural change which occurred at around 1,000 years BP (with the introduction of fishhooks and the decrease in the use of rockshelters) may be so significant that the contact language boundaries are meaningless throughout the full extent of the region's art's production (i.e. the last 3-4,000 years).

Capell's (1970) language group boundaries were major rivers and creeklines. I argue (following Tindale 1974, Peterson 1976) that the boundary of any group's range is more likely to be at the periphery of its economically viable area. Boundaries in topographically dissected areas (such as the Sydney Basin – as with the Pilbara) are likely to be along ridgelines. The ethno-historically reported use of ridgelines for access routes around the region supports this argument since it is likely these would have traversed the periphery of any particular group's territory – not bisected the centre.

The model proposed in this thesis suggested that stylistic behaviour which reinforces group distinctiveness should be observable on the basis of drainage basin catchments, with boundaries between groups along ridgelines and not creeklines. The region was thus subdivided by means of drainage basins.

Within the five documented language areas, 25 drainage basins were defined across the region.

As identified earlier, a potential problem with these analyses is the disparate sample sizes (Table 11.3). This distribution of sites represents in part an archaeological 'reality' (Chapter 5), with decreasing site numbers at the periphery of the Sydney Basin. This may in part reflect the geological reality, although this has never been quantified. Much of the bias with this component results from the geographic focus of certain recorders in locations closer to Sydney city (McDonald 1985a). Over 70% of this assemblage was recorded by W.D. Campbell, Fred McCarthy and Ian Sim means that the distribution of this sample largely reflects their areas of interest and recording

Table 11.3: Language areas, codes and sample sizes.

Language Group	Code	No. of sites
Darkingung	1	137
Guringai	2	434
Sydney (*Eora*)	3	32
Dharug	4	49
Tharawal	5	52

focus [Sim (1966a) represents the only focus in the north-west of the Basin]. More recent recording work (e.g. Tacon *et al*. 2006) indicates that there are still many engraving sites to be found in the west of the region.

Several systematic EIS surveys, in the north and south of the region have revealed a relatively low number and density of engraving sites. Likely explanations for this could be geological or cultural. In the Mill Creek valley (south of the Georges River), the ridgelines on either side of the drainage basin are characterised by shale laterite, some of which had been extensively mined by the local Municipal Council for road construction (Attenbrow and Negerevich 1981; McDonald 1990b). Systematic survey here (covering c. 45 sq km) revealed only one engraving site.

In the north of the region, the Mangrove Creek valley, both in its upper catchment and its middle reaches around Warre Warren Creek, has been systematically sampled over an area of approximately 200 sq km (Attenbrow 1981, McDonald 1988a). Engraving sites here were relatively few (n=4: 3% of recorded sites) although large numbers of suitable rock surfaces were located and inspected. These results are in stark contrast with those achieved at Maroota south of the Hawkesbury River in the central-west of the Basin, where 12 engraving sites were located in one square kilometre (McDonald 1986a).

A correlation of sample size with language area and motif categories was made (Figure 11.6), to determine whether patterning in these analyses results from sampling inequities (James 1993).

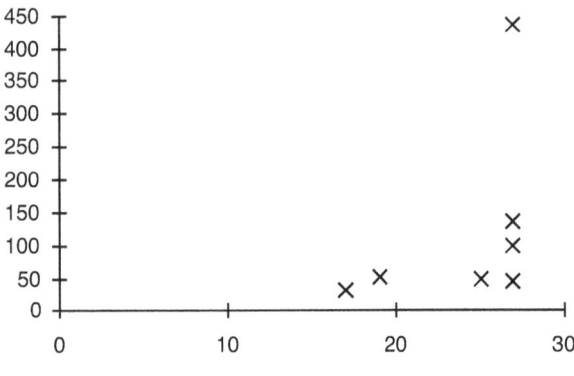

Figure 11.6: Bivariate plot of sample size and number of motifs recorded per sample area. The five language areas and two randomly generated *Guringai* samples.

This demonstrated no direct correlation between site size and motif numbers. Further, the *Guringai* sites were subdivided into random 50 and 100 site samples, for the comparison of assemblage composition (and CA results) according to language area. These different steps indicate that the unavoidable sampling inequities in the data base do not appear to produce significant interpretive issues.

Motif Assemblage Differences across the Basin

Before analysing the CA results, basic assemblage details were investigated for each of the language areas. The division of the sites into language area was based on Capell's (1970) defined boundaries (see Figure 3.1).

Darkingung

This area had 2,127 motifs (1,803 recognisable) from 137 sites. Two of the region's four largest sites (>100 motifs) are found in this area, and average site size is 15.5 motifs/site. The motif focus here is on tracks (bird, human and roo's respectively). The macropod is the next most common motif. Whale motifs are not represented in this assemblage.

Men are the most commonly depicted human figures, followed by non-gendered anthropomorphs. Profile depictions are quite common. Relatively few (six only) culture heroes are found here. Boomerangs are the most commonly depicted material culture items (Figure 11.7).

Darug

This inland area has less of a focus on tracks than its more northerly counterpart. *Mundoes* and bird tracks still figure strongly, but macropods and land animals represent a large component of this assemblage (Figure 11.8). Human figures are again focused on males, and here there is a greater emphasis on profile figures than non-gendered anthropomorphs. Three culture heroes (at two sites) are located in this area. The average site size here is eight motifs/site.

Guringai

This sample is located on the coast and represents the largest sample in the region. The 434 sites in this area produced 4,699 motifs. The average site size is 11 motifs/site. The other two sites with >100 motifs are located in this language area.

To examine the effect of sample size, two random sub-samples were generated (one with 50; the other with 100 sites) to see what effect this may have on the results (Figure 11.9). All three histograms reveal the same focus on *mundoes* and fish, followed by macropods, other land animals and men. While the peaks and troughs of these graphs vary slightly according to sample size, the results are basically the same. Only the sample of 50 resulted in a reduction of motif variables (roo tracks and contact motifs). These were two of the least common motif types in the total sample. While this sample is located on the coast, its motif focus is not entirely explicable in terms of environment (cf. the Sydney group).

Sydney (Eora)

This language group is located on the coast, south of Port Jackson and north of the Georges River. This has the smallest sample size (due partly to the focus of European settlement, but also because the Cumberland Plain comprises a large proportion of this area).

The 32 sites in this area produced 245 motifs (an average of eight motifs/site). The motif focus in this area is on fish (46%), with whales and other marine animals also common (Figure 11.10). The reduced motif classification here possibly is a result of sample size.

Tharawal

This area is also located on the coastal strip, but south of the Georges River. A sample here again was small with 245 motifs recorded at 51 sites. The assemblage sizes here (on average) are the smallest recorded in the Basin (5 motifs/site). There is a focus on marine depictions (30.7%), but the most commonly depicted motifs are men and non-gendered anthropomorphs (20.3% in total). Macropods and land animals are also common, while *mundoes* are the most frequently depicted tracks (Figure 11.11).

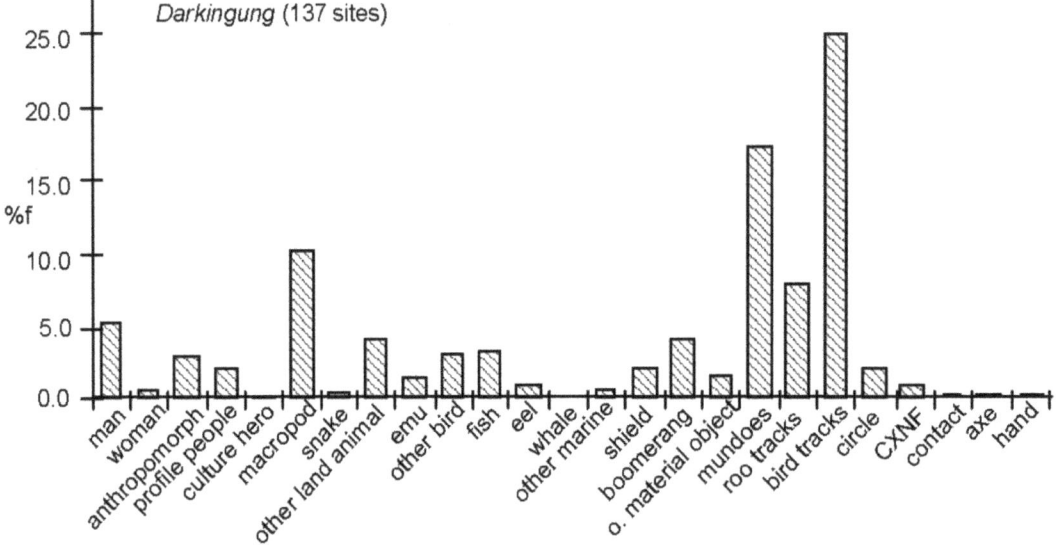

Figure 11.7: Darkingung Language Area. Motif Assemblage.

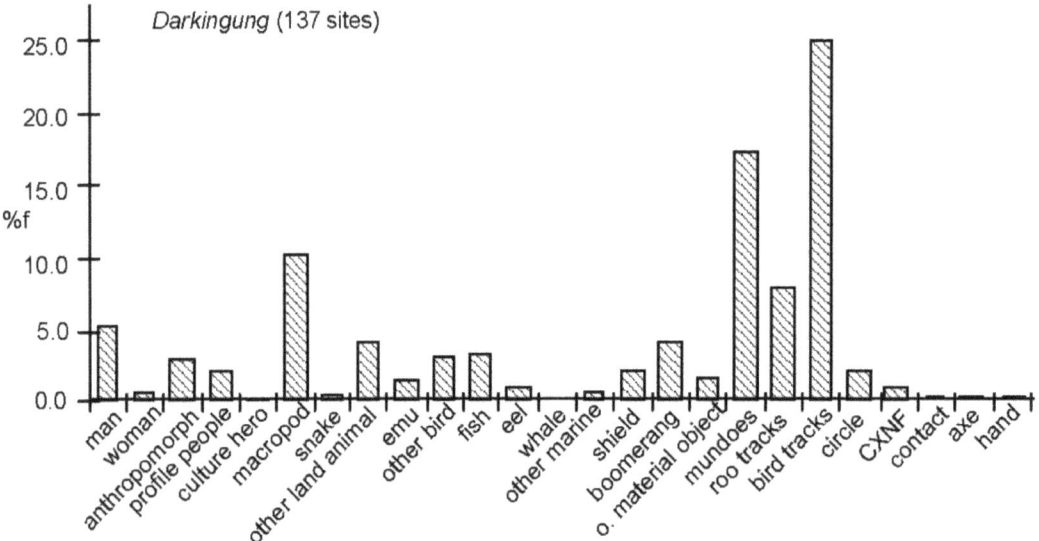

Figure 11.8: Darug Language Area. Motif Assemblage.

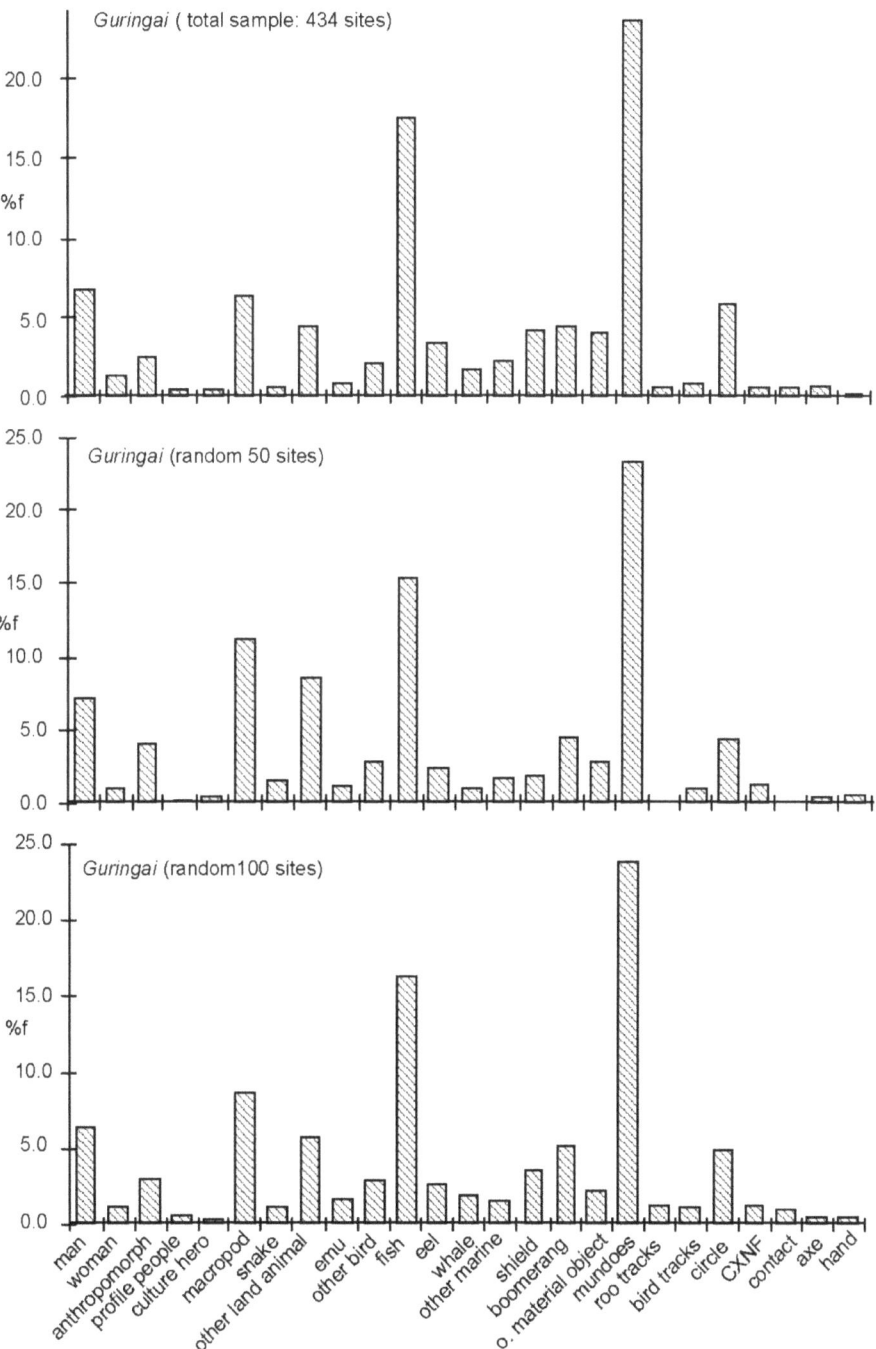

Figure 11.9: Guringai Language Area. Motif assemblage.

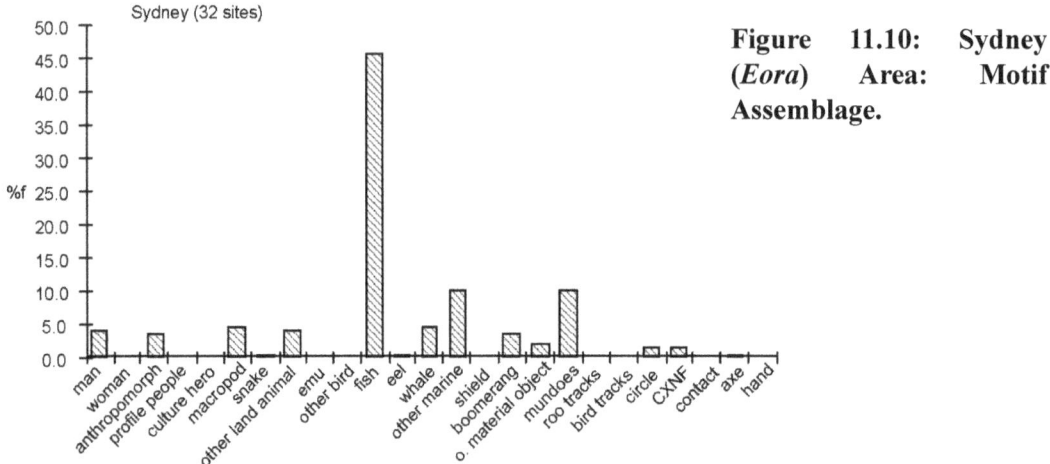

Figure 11.10: Sydney (*Eora*) Area: Motif Assemblage.

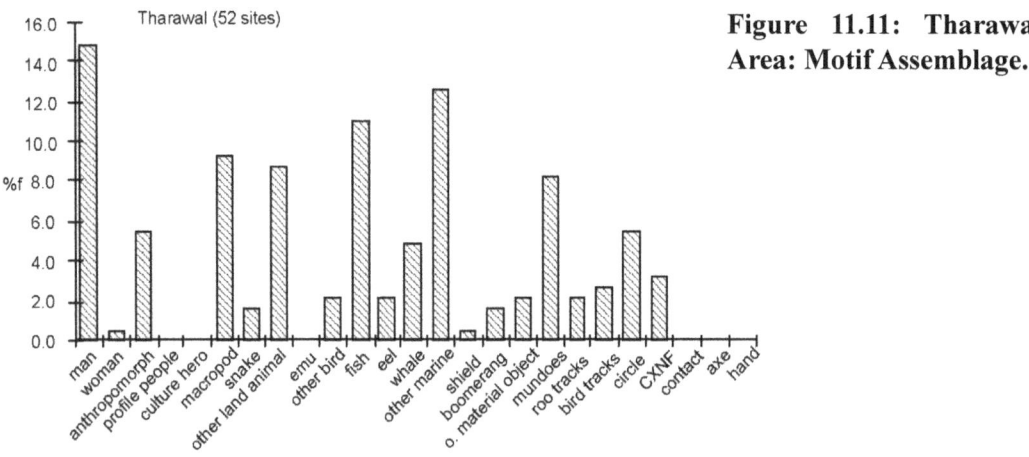

Figure 11.11: Tharawal Area: Motif Assemblage.

Summary

Site size and motif focus vary across the region. While some of these trends are environmental; i.e. more fish and marine depictions on the coast and higher proportions of land animals inland, there are other foci which cannot be explained so simply. There are varying proportions of animal and human tracks in different parts of the region, and a dominance of human figures in coastal sites south of the Georges River:

The *Darkingung* sites are generally large with several very large sites found along access routes. The density of sites appears lower than in other areas although average assemblage size is higher. The motif focus is on kangaroo, bird and human tracks and macropods.

The average *Darug* site size is roughly half that found in the *Darkingung* language area. Macropods and other land animals dominate this assemblage, but tracks appear less important.

The Guringai area has the largest number of sites and seems to represent the region's core engraving assemblage. Average site size is larger than *Darug* but smaller than *Darkingung*. Dominant motifs are *mundoes* and fish, followed by macropods, other land animals, and men.

Only a few sites have been recorded from the *Eora* area, and these are relatively small. The motif focus here is on fish, *mundoes*, whales and other marine animals.

The *Tharawal* engraving assemblages are the smallest on average. The focus here is on men, other marine depictions, and fish, followed by macropods and other land animals.

Correspondence Analysis and Language Areas

While trends in the motif assemblages across the Basin are quite clear, the CA results help determine compositional differences and foci in the different areas, and to demonstrate internal variability in the regional assemblage. The language areas were analysed as were drainage basins. This work also tested several of Capell's boundaries along creeklines.

Drainage Basins

A total of 25 drainage basins were defined across the region (Figure 11.12; Table 11.4). These vary considerably in size and very large drainage areas have been defined for areas with low site numbers (e.g. Blue Mountains and Colo). Not all basins have engraving sites; some contain only shelter art sites (chapter 12).

Because of the disparate sample sizes, language boundaries and drainage basin boundaries were investigated in three locations with good sample sizes of both site types. These analyses tested the possibility of language boundaries and explored the degree of intra-language area patterning. The areas tested were:

1) Drainage basins 1, 5 and 6 within the *Darkingung* language area, north of the Hawkesbury River (107 sites);

2) Drainage basins 10-13 to test east-west patterning across the purported *Guringai/Darug* language boundary, south of the Hawkesbury River (316 sites);   and,

3) Drainage basins 18 – 21 to test east-west patterning across the purported *Tharawal/Darug* language boundary, south of the Georges River (51 sites).

In the first two of these areas, extensive rock art recording work has been completed (Gunn 1979; McCarthy (see references); McDonald 1986a, 1987, 1988a, 1990a; Sim 1963a, b, 1966a, b; Smith 1983; Vinnicombe 1980) as were the four excavations completed for this research. The third area has been studied in more detail by other researchers (Officer 1984, Sefton 1988, SPG 1974) and although a smaller sample, this was seen as a useful test area south of the Georges River style boundary.

CA according to drainage basins

This analysis viewed sites according to language areas and drainage basins[36] (Figure 11.12). The seven clumped motif classes were again used here (i.e. 'anthropomorphs' include men, women, non-gendered and profile anthropomorphic figures and culture heroes). Detail on thematic focus results from re-inspection of the site recordings.

1) Darkingung language group (drainage basins 1, 5 and 6).

This group of 107 sites is located north of the Hawkesbury River and includes the major drainage basins of the Macdonald River and Mangrove Creek. The Upper Macdonald and Central Macdonald groups were divided at the Bala Range, a geographic barrier at the centre of the valley. Drainage Basin 4 was excluded here as it contained only one site (although now see Taçon *et al.* 2005, 2006).

---

[36] All plots are based on the same CA results which are sorted according to location (language area, drainage basin, etc.). The plots have a manually drawn circle or ellipse (depending on the scale of the axes) indicating the arbitrary cut-off for the homogenous 'zone'. This procedure provides a visual aid in the interpretation of the bivariate plots. NB. Each dot symbol may represent one or many sites.

Table 11.4: Drainage Basins, Language Areas and Sample sizes.

Drainage Basin	Basin Code	Language Group	No. of sites
Upper Macdonald	1	1	12
Wollombi	2	1	3
Wyong	3	1/2	3
Colo	4	1	1
Central Macdonald	5	1	55
Mangrove Creek	6	1	40
Mooney Mooney	7	1/2	59
Brisbane Water	8	2	88
Kurrajong	9	1/4	-
Cattai	10	1/4	6
Berowra	11	2/4	42/28
Cowan	12	2	113
Pittwater	13	2	27
Middle Harbour	14	2	107
Lane Cove	15	2/4	9/2
Port Jackson	16	3	22
Botany Bay	17	3	1/-
Port Hacking	18	5	35
Woronora	19	5	11
Mill/Williams	20	4/5	-/1
Georges	21	4	1/3
Nepean	22	4	2
Burragorang/Blue Mountains	23	4*	9
Cataract	24	4?/5	-
Avon/Cordeaux	25	4?/5	-

*May be mixture of Darug and Gandangara Language areas

DARKINGUNG   Outliers

Core: 45% 6.8%	**42.4%**
5.1%	**45.8%**

The 107 sites in this language grouping are relatively homogeneous, with a heavy emphasis in positive quadrants B + C (on tracks, anthropomorphs, terrestrial animals and birds).

Upper Macdonald   Outliers

Core: 42% 0%	14%
0%	**86%**

This group of 12 sites is relatively homogenous with a strong emphasis on tracks (quadrant C). Many sites here include combinations of tracks, anthropomorphs, other material objects and birds. There are no marine depictions in these sites.

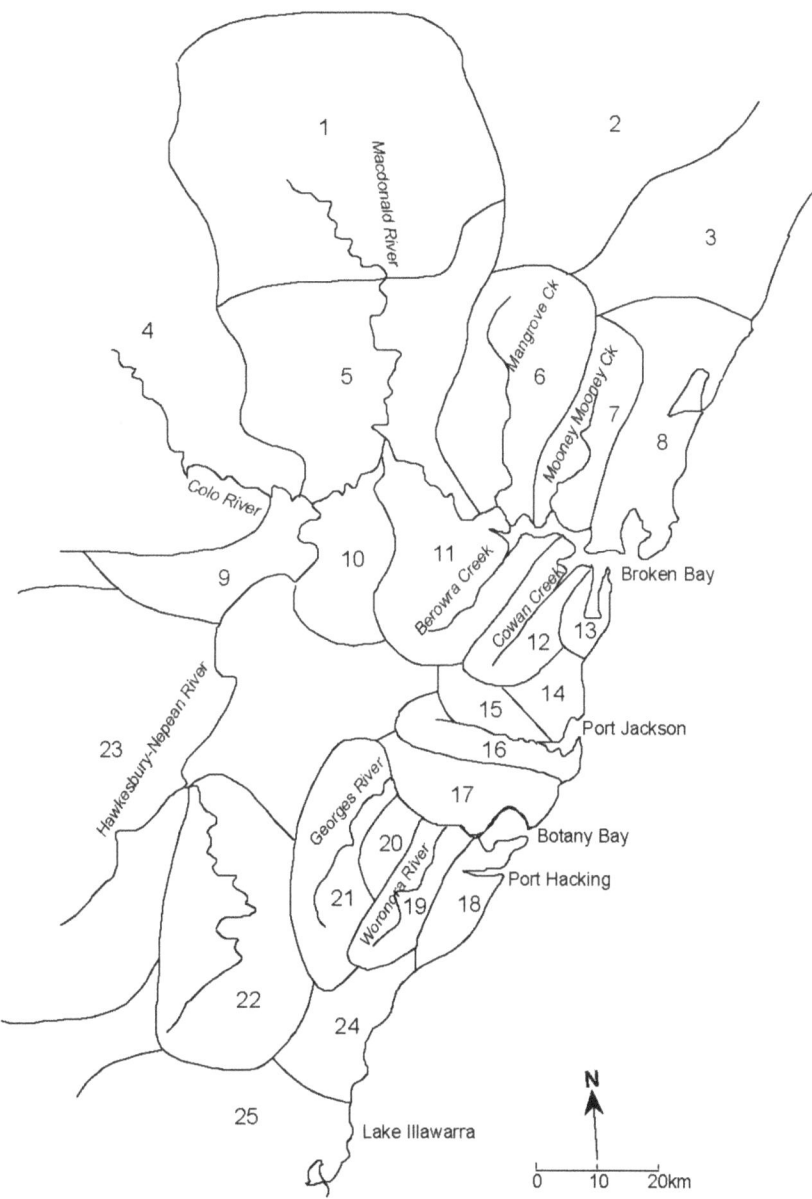

Figure 11.12: The 25 drainage basins defined across the Sydney region (refer Table 11.4).

Central Macdonald   Outliers

	Core: 49%	7.1%	32%
		3.6%	**57%**

This group of 55 sites is the most homogenous. It includes the sites along the Boree Track (a known access route). There is again a strong emphasis on tracks and anthropomorphic figures (the positive quadrants), and a number of sites which have birds only. Material objects (spears, clubs) also figure strongly. There are very few marine depictions. At one site (#172) there are plain fish, eels and turtles.

Mangrove Creek   Outliers

                              Core: 37.5%   8% | **60%**
                                                   |
                                             ——————+——————
                                                   |
                                             8% | 24%

This group of 40 sites are the least homogenous in this language area. Outlier sites here are mostly in quadrant B, with a strong focus on terrestrial animals and material objects. Anthropomorphs and birds also figure strongly.

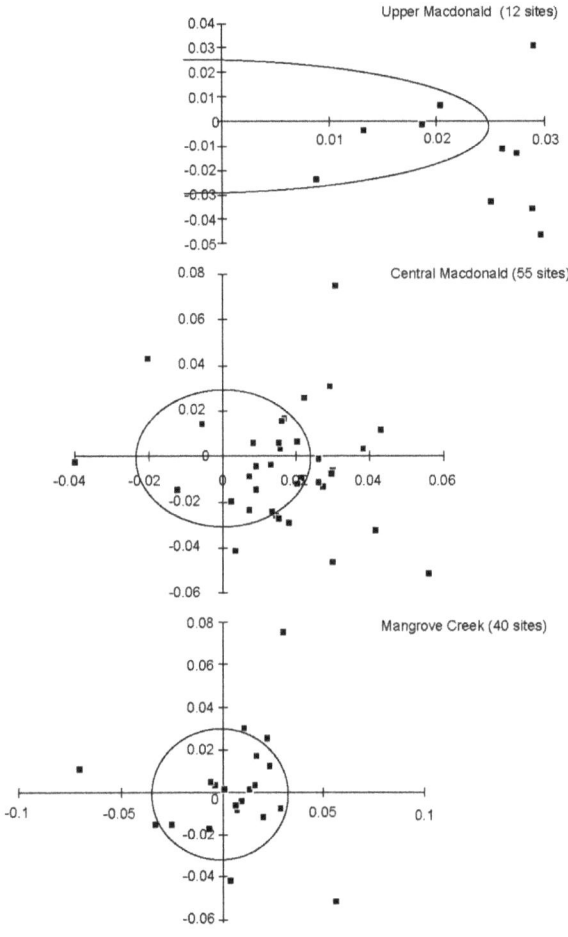

Figure 11.13: Bivariate plots, CA results. Darkingung language area - Upper Macdonald, central Macdonald and Mangrove Creek drainage basins.

2) the Guringai/Darug language boundary (drainage basins 10 - 13)

The purported boundary between these two language areas is Berowra Creek (Figure 3.1). Both banks of this estuarine waterway were surveyed during the Rock Art Project (McDonald 1990b). Vertical engravings were located on both banks of this creek, and the art on both sides (within 40m distance and 10m elevation) was observed to be very similar. This analysis tested the Berowra Creek boundary and found that the sites on either site of the creek do demonstrate differences.

DARUG   Outliers

| | Core: 35% | 18.2% | **45.5%** |
| | | 13.6% | 22.7% |

GURINGAI   Outliers

| | Core: 45% | 27% | 19% |
| | | 16% | **38%** |

The 34 *Darug* sites are less homogeneous than the 182 *Guringai* sites. There is also a general change in focus between the more unusual sites in the two areas (cf. quadrants B + C). More land animals are found in the *Darug* sites and more tracks, birds and marine compositions occur in the *Guringai* sites. The drainage basins reveal considerable variability within these groups.

Cattai   Outliers

| | Core: 50% | **33.3%** | **33.3%** |
| | | 0% | **33.3%** |

This area has only six sites, and thus the results are treated tentatively (Figure 11.14). Half of these sites fall in the core zone and there is one each in three of the quadrants. The outlier sites have fish and eels, land animals and anthropomorphs and material objects.

(*Darug*) Berowra   Outliers

| | Core: 32% | 15.8% | **47.4%** |
| | | 15.8% | 21% |

This group of 28 sites are on the left bank of Berowra Creek. These sites are quite heterogeneous with the motif focus on terrestrial animals (quadrant B: Figure 11.15). There are many sites with single macropods. Anthropomorphs and material objects (particularly shields) are a common combination. Culture heroes and profile anthropomorphs are also present. Vertical engravings are common.

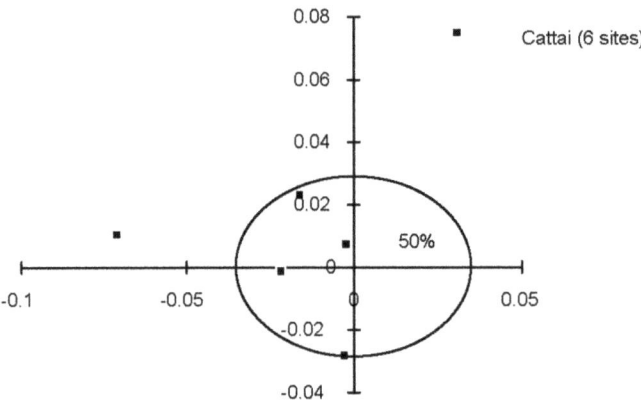

Figure 11.14: Cattai Drainage Basin. Bivariate plot of CA results.

(Guringai) Berowra   Outliers

Core: 46%	21.7%	13%
	17.4%	**47.8%**

This group of 42 sites is more homogeneous than those found on the western side of this drainage basin. The outlier sites here are different to those on the left bank of Berowra Creek (quadrant C: Figure 11.15), with a focus on anthropomorphs, mundoes, shields, culture heroes and other birds. There are more marine depictions (including whales) here. Vertical engravings are common.

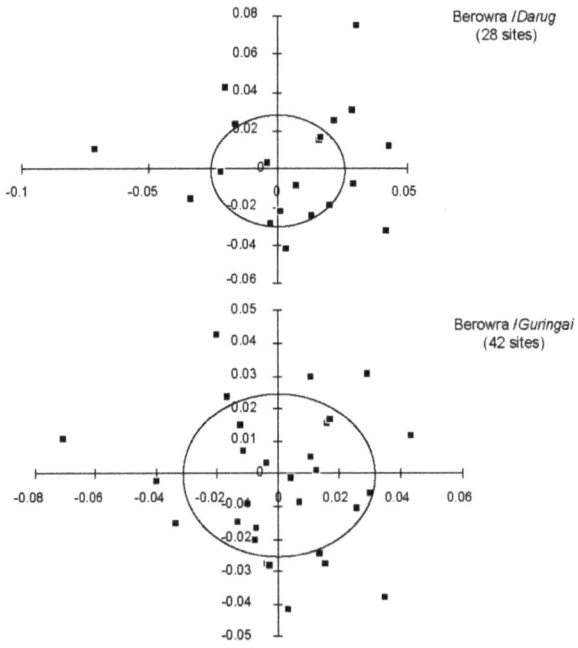

Figure 11.15: Berowra Drainage Basin (*Darug* and *Guringai* Language Areas). Bivariate plot of CA results.

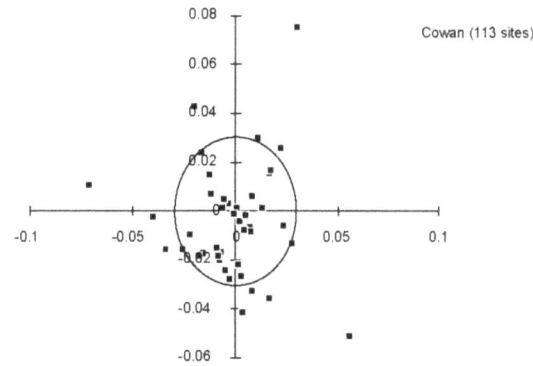

Figure 11.16: Cowan Drainage Basin. Bivariate plot of CA results.

Cowan   Outliers

	Core: 42%	21%	22.4%
		13.4%	**40.3%**

This group of 113 sites is quite homogeneous (Figure 11.16). Themes include a variety of anthropomorphs, birds and other material objects (particularly axes). There is a more even spread between quadrants A + B, with fish and whales as common as kangaroos, shields and boomerang combinations.

Pittwater   Outliers

	Core: 55.5%	**66.7%**	8.3%
		25%	0%

This group of 27 sites is the most homogenous of all those analysed (Figure 11.17). The focus here is heavily on marine animals (quadrant A), and many sites have fish only. There are also many whales.

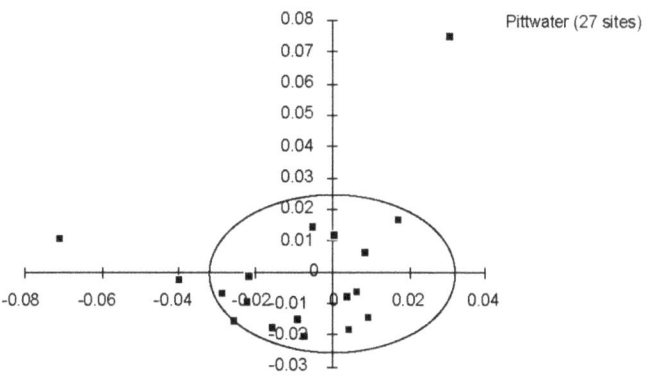

Figure 11.17: Pittwater Drainage Basin. Bivariate plot of CA scores.

3) The Tharawal language area (Basins 18 – 21)

This group of 50 sites south of the Georges River fall within the *Tharawal* language area. There are only a few sites from the western part of this area. The 11 Woronora sites represent a slightly more 'inland' focus, although this creekline is only 10km from the coast and its lower reaches are estuarine.

THARAWAL   Outliers

	Core: 20%	**44%**	30%
		10%	16%

This group of 50 sites is the least homogenous of those analysed in the region.

Mill Creek/Georges River: Outliers

	Core: 75%	0%	0%
		0%	**100%**

The four sites in this drainage basin grouping represent too small a sample for these results to be meaningfully discussed. These results are included in the larger language group discussion.

Woronora: Outliers

	Core: 9%	0%	**70%**
		0%	30%

Sample size may also affect the results of this group (Figure 11.18). The 11 sites here are highly heterogeneous. There is a strong focus (in quadrant B) on kangaroos and other terrestrial animals and material objects. There are less anthropomorphs and tracks (human, roo and bird). There are no marine depictions.

Port Hacking: Outliers

	Core: 17%	58.6%	10.3%
		17.2%	13.8%

This group of 35 sites is also highly heterogeneous and has a different outlier focus (in quadrant A) from the preceding *Tharawal* group. The main compositional focus here is with marine depictions (including whales) and material objects are also common. Most assemblages in this group are fairly small as is the motif range. The largest assemblage has 15 identifiable motifs; the greatest variety of motifs at any one site is six.

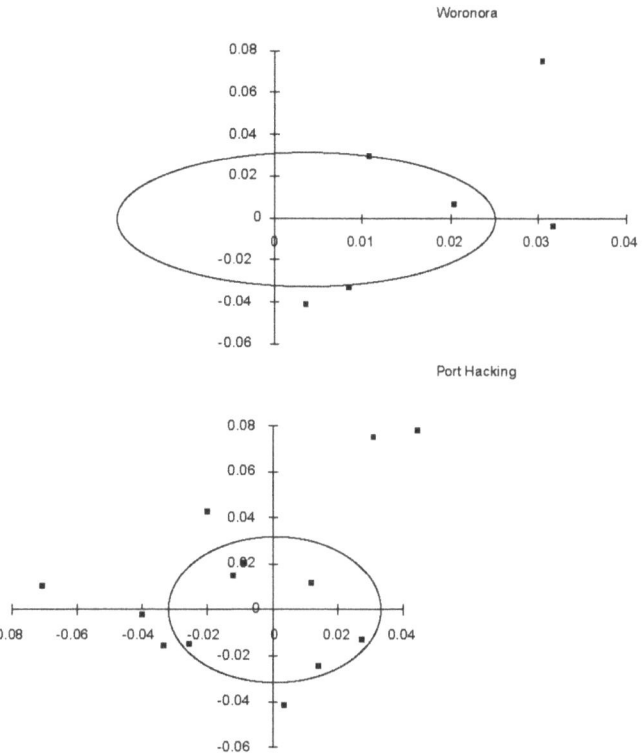

Figure 11.18: *Tharawal* Language Area (Woronora and Port Hacking Drainage Basins). Bivariate Plot of CA Scores.

Conclusions

The *Darkingung* and *Guringai* language areas have the most homogenous engraving assemblages, followed by the engraved *Darug* assemblage. The engravings in the *Tharawal* language area are the least homogenous.

Internal variations across the *Darkingung* language area are relatively small. The upper and central Macdonald River sites are similar in terms of homogeneity and outlier focus (tracks). These sites are, however, different from the Mangrove Creek sites: which are slightly less homogenous and have a focus on terrestrial animals.

The *Darug* sites are more heterogeneous than either the *Darkingung* or the *Guringai* sites. Capell's language boundary along Berowra Creek is supported by this analysis. There are marked differences in the levels of variability, and dissimilar motif preferences.

The *Guringai* area is relatively homogenous, but there is still evidence for localised variability here. The Cowan sites are less homogenous than those from the Pittwater sample. The drainage basin analysis in this central part of the region demonstrates a clinal increase in variability as one moves west away from the coast.

The Mangrove Creek sites appear to be most like the *Darug* sites. Both groups are less homogenous than the *Darkingung* and *Guringai* sites generally. All three sets of sites, however, have different motif foci.

The *Tharawal* sites are markedly dissimilar to all other language groups. The sites are considerably more heterogeneous and the outlier foci are different. Comparison of two drainage basins with reasonable samples within this language area suggests differences between coastal and more inland sub-groups. While having different outlier foci, these two groups demonstrate the most variance of all groups analysed.

These analyses demonstrate a complex network of stylistic variability as defined by engraved motif preference across the region. Several contact language boundaries are supported by these analyses: the east-west *Guringai-Darug* boundary and the northern *Tharawal* boundary. Sites within the *Darkingung* language area, however, show some variability: with sites from Mangrove Creek more like those from the *Darug* language area.

This patterning is discussed after compositional features and the distribution of rare motifs are described.

Ridge top versus vertical engraving sites
This research has identified a number of vertical engraving sites around the foreshores of Broken Bay and its main estuarine tributaries (Figure 11.19). The ethnohistoric literature indicates that this zone as a highly public one, and the sort of location where stylistic bounding behaviour is likely to be demonstrated (Wiessner 1990).

On the ridgelines and plateaux above these waterways are vast numbers of open engraving sites. The art sites at the bottom of cliffs and steep hillslopes provide a different social context from the open engraving sites on the less (economically) productive plateaux above. The sites close to the estuarine resources would have been accessed either by canoe or on foot around the foreshore (Figure 11.20, Figure 11.21).

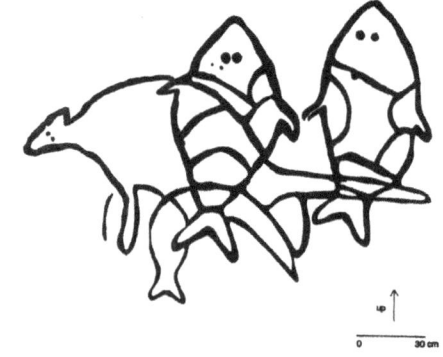

Figure 11.19: Two examples of vertical engraving sites from Berowra Creek (top) and Cowan Creek (bottom). Note the complexity of composition and shared line designs.

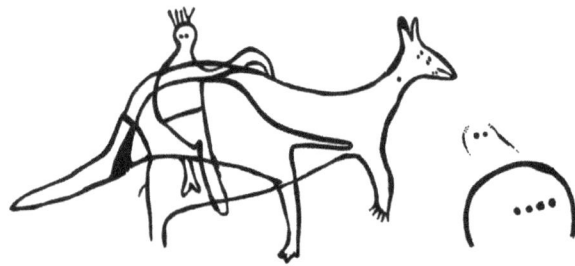

Many of the ridgelines around the region are documented access routes (e.g. the Boree Track, Kulnura Ridge). Sites in these locations will provide a different kind of information, one which promotes social cohesion. Any sites which had ritual significance are also likely to have been located away from the main centres of subsistence economy in any particular social group's territory.

Analysis was undertaken in the *Guringai* language area (Cowan and Berowra drainage basins), comparing the ridgetop, hillside and valley bottom engraving sites. This area was selected because of the large sample of engraving sites generally, and because many vertical engraving sites have been found here too. Motifs were compared (Figure 11.22), and the CA results were re-sorted to determine the varying degrees of homogeneity of these locations. Hillside sites were included as these locations include a high number of the engraving sites.

There are 78 sites in ridgetop locations in this area. The average site size is 12 motifs/site. The average distance to permanent drinking water from these sites is 570m. The predominant motif in these locations is the *mundoe* (c.35%), followed by fish (15%) and men (8%).

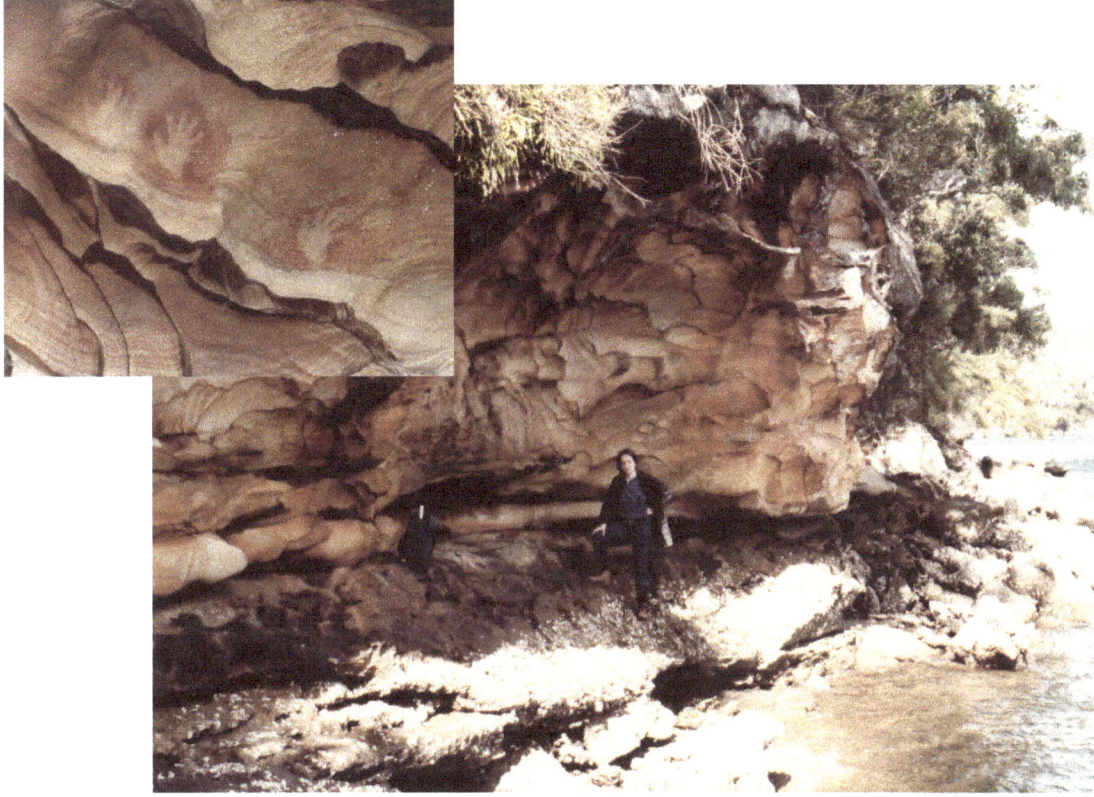

Figure 11.20: Smith's Creek, Ku-Ring-Gai Chase National Park. The red stencils (arrowed and inset) in this shelter must have been produced by artists standing in a canoe at high tide. Photo taken at low tide.

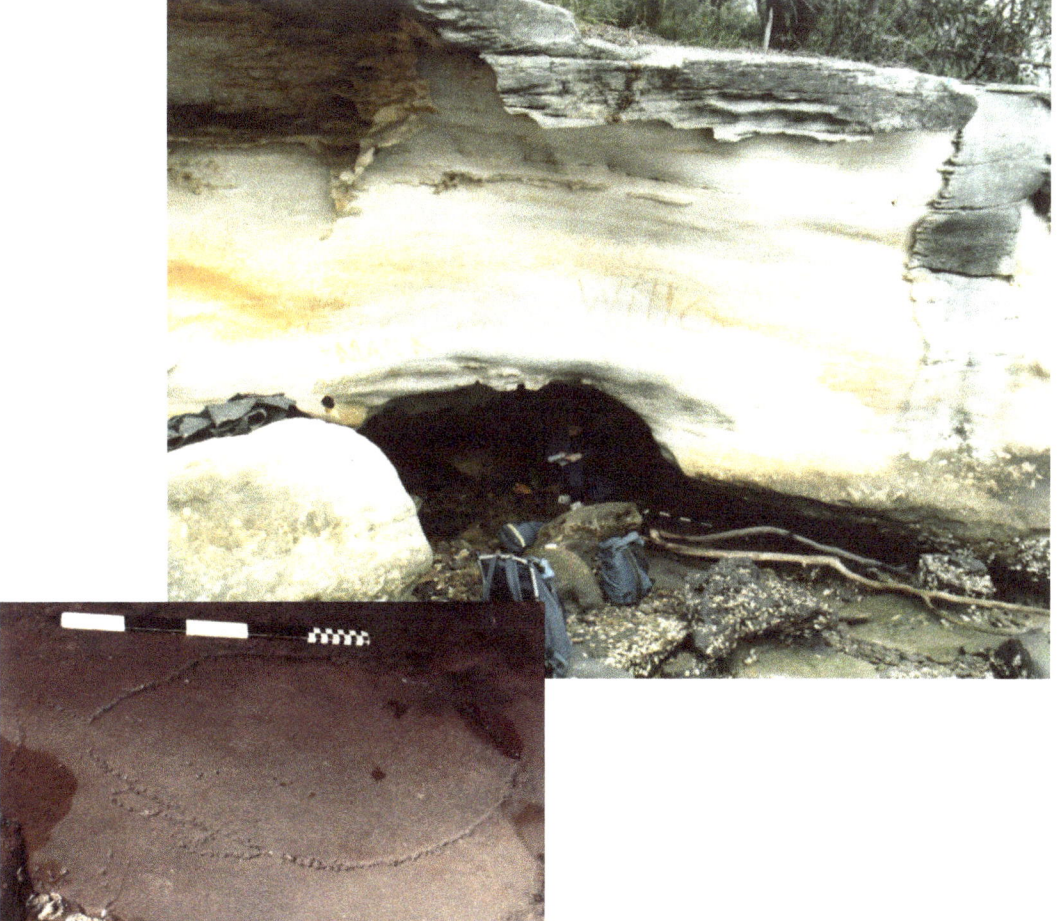

Figure 11.21: Shelter on Cowan Creek, the floor of which is in the littoral zone. An engraved outlined fish (inset) is located on the interior floor surface (arrowed).

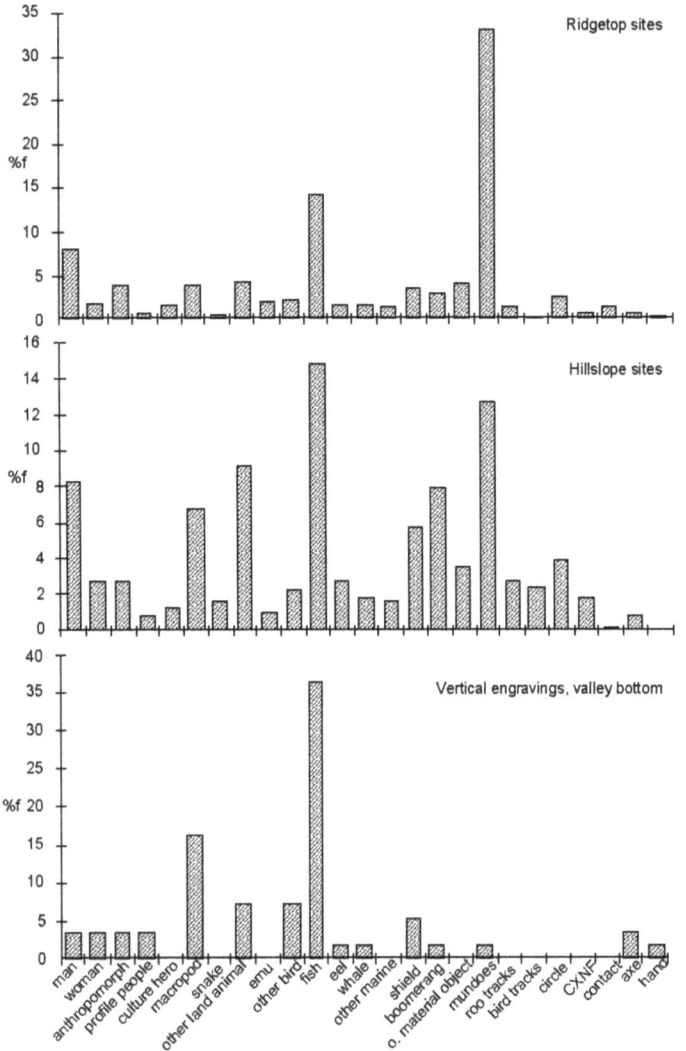

Figure 11.22: Motif histogram for engraving sites in ridgetop, hillslope and vertical engravings in estuarine valley bottom.

In the 63 hillslope sites the predominant motif is the fish (15%) followed by *mundoes* (12.5%), other land animals (9%), men and boomerangs (c.8%), macropods and shields (c.7%). The average distance to permanent drinking water from these sites is 460m.

There are 14 vertical engraving sites and these are generally smaller assemblages (av. 7.5 motifs/site). The average distance to drinking water from these sites is greater than from either of the other locations, for, while these are located next to the water's edge, Berowra and Cowan Creeks are saline and tidal. Fish motifs predominate (35%) in these locations, followed by macropods (15%). Missing motif classes include culture heroes, circles and contact motifs. *Mundoes* are present in a very small number of these sites (c.2%).

The differences in motif preferences on ridgetop and hill side locations are quite striking and cannot be explained in terms of sampling (Figure 11.22). The CA results reinforce engraving site differences in these landscapes (Figure 11.23).

Ridgetops   Outliers

                              Core: 45%   26.2% | 21.4%
                                             ─────────────────
                                             9.5% | **42.9%**

This group is quite homogeneous, with its outlier focus in quadrant C. There is a slight subsidiary focus in quadrant A.

Hillslopes   Outliers

Core: 48%	18.8%	18.8%
	18.8%	**43.8%**

This group of 62 sites has a similar level of homogeneity to the ridgetop sites, and the same outlier focus (quadrant C). The equal distribution between the three other quadrants indicates a broader range of subject combinations in these sites.

Vertical sites, valley bottoms   Outliers

Core: 36%	33.3%	11.1%
	0%	**55.6%**

This group of 14 sites is the most heterogeneous, but has the same outlier focus as the other two groups.

The sites on the ridgetops and hillslopes <u>are</u> more homogenous than those around the water's edge. While there are demonstrated differences in degrees of homogeneity, however, the outlier foci in these three locations are very similar. In other words, the graphic vocabulary of people operating in these different landscapes, is the same. This result is as would be expected within the one language area and suggests that these sites being used by the same group(s) of people in a range of different social or information contexts.

Rare Motifs

Rare and unique motifs were analysed to establish their geographic distributions. It was hoped that this type of investigation would elucidate localised stylistic traits. Analysis concentrated on non-economic motifs in an effort to reduce environmental influences.

The analysis of how many times an individual motif occurred at any engraving site in the region demonstrated some interesting results (Table 11.5; Figure 11.24).

The motifs which occur at the most sites in the region are fish and macropods (c.35%) followed by men and other land animals. Mundoes – while the most numerous engraved motif in the region - are only present at 22% of engraving sites. This result indicates that certain motifs are concentrated in a few sites - while other motifs are more widely dispersed (relatively fewer motifs are placed on many more sites). This analysis focussed on concentrated rare motifs (i.e. ones which are relatively rare and which occur on few sites), and on dispersed motifs (i.e. relatively rare but fairly widely distributed).

The distribution of these motifs was plotted (Figure 11.25 to Figure 11.28), the percentages of sites with each motif type were calculated by language area (Table 11.6). These results were compared with the percentage results for each language area so that relative significance could be determined. To test the statistical significance of these differences, an approximate randomisation method (Noreen 1989, Wright 1991) was used on the figures (see Table 11.7).

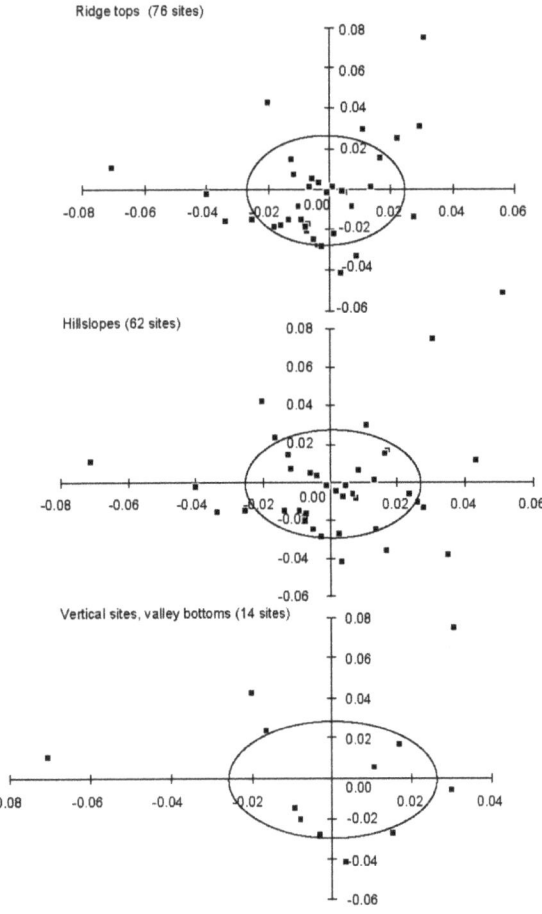

Figure 11.23: CA bivariate plots according to topographic location. *Guringai* language area: Berowra and Cowan drainage basins.

Table 11.5: Engraving sites. Motif total, maximum incidence at any particular site, number of sites in the region with motif present, and % of sites with motif.

Motif	Total	Max incidence	Sites with Motif present	% of Sites with Motif
Man	422	14	199	27.8
Woman	79	5	56	7.8
Anthropomorph	182	15	115	16.1
Profile Person	79	5	53	7.4
Culture Hero	36	2	29	4.1
Macropod	543	13	247	34.5
Snake	56	3	45	6.3
Other Land Animal	312	10	170	23.7
Emu	76	7	51	4.1
Other Bird	166	11	101	14.1
Fish	905	47	250	34.9
Eel	182	7	97	13.5
Whale	101	7	71	9.9
Other Marine	156	9	97	13.5
Shield	232	23	103	14.4
Boomerang	303	15	144	20.1
Axe	45	5	28	3.9
Other material object	218	13	106	14.8
Mundoe	1,360	99	157	21.9
Roo track	186	18	43	6.0
Bird track	541	95	71	9.9
Circle	309	17	96	13.4
CXNF	70	3	53	7.4
contact	36	11	11	1.5
hand	19	3	11	1.5

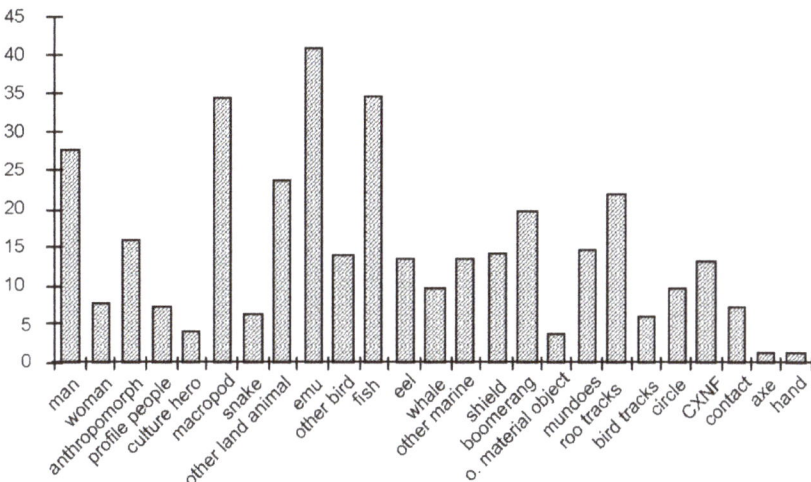

Figure 11.24: Engraving component. Percentage of sites at which particular motifs appear.

This analysis confirmed some of the disparities revealed by the previous analyses:

- There is a significantly <u>higher</u> proportion of profile figures and roo tracks and a significantly <u>lower</u> proportion of shields and axes in the *Darkingung* area.

- Shields and axes occur in significantly <u>higher</u> proportions in the *Guringai* area, while there are significantly fewer profile people and complex-non-figurative motifs in this area.

- In the *Darug* area, there are significantly higher numbers of profile people and complex-non-figurative motifs, and contact motifs and hands occur often.

- In the *Tharawal* area, there are significantly lower proportions of women, snakes, shields and axes.

- Significant differences are identified between language areas (Table 11.7). The differences and similarities between the *Guringai* and *Darkingung* groups are statistically significant. The differences (and similarities) between the *Guringai* sites and the *Darug* and *Tharawal* sites are also significant.

Table 11.6: Rare Engraving Motifs. Distribution per Language Area (outstanding results in red and bold).

Motif	Number (and %) of Sites with motif in each Language Area									
	Darkingung		*Guringai*		*Eora*		*Darug*		*Tharawal*	
Woman	12	21.8	40	72.7	0	0	2	3.6	*1*	*1.8*
Profile person	*26*	*49.1*	22	41.5	0	0	5	9.4	0	0
Culture hero	5	17.2	22	75.9	0	0	2	6.9	0	0
Snake	11	24.4	27	60.0	1	2.2	4	8.9	*2*	*4.4*
Shield	*11*	*10.7*	88	85.4	0	0	3	2.9	*1*	*1.0*
Axe	*3*	*10.7*	23	82.1	0	0	1	3.6	*1*	*3.6*
Roo tracks	*30*	*69.8*	7	16.3	0	0	4	9.3	2	4.7
CXNF	*17*	*32.1*	24	45.3	2	3.8	*6*	*11.3*	4	7.5
Contact	3	18.2	6	72.7	0	0	2	9.1	0	0
Hand	2	18.2	8	72.7	0	0	1	9.1	0	0
Total sample	2127	(19.5)	4699	(61.9)	245	(6.5)	360	(4.6)	245	(7.4)

Table 11.7: Engraving sites. Significant values for rare motifs in the five language areas.

Language Areas compared		Significance value
Darkingung	Guringai	<.001
Guringai	Darug	.013
Guringai	Tharawal	.027

These results confirm the localised character of the engraving assemblage in different areas of the Basin. The distributions of the rarer motifs also demonstrate some interesting connections.

Sites with culture heroes occur mostly in the western part of the *Guringai* territory and into the *Darkingung* territory. This design link is not suggested by the CA results which show the overall foci to be quite different from these two areas. Profile anthropomorphs, axes and contact motifs have very similar distributions to the culture heroes. This will be discussed further below.

Composition

Difference in composition were explored to provide further evidence of the types of cultural and/or stylistic choices (Sackett 1990) being made across the region. Shields and culture heroes were selected for this analysis.

Shields

Previous archaeological analyses of shield designs have demonstrated stylistic patterning explicable in terms of trade and overall alliance systems (Dickens 1992, Hatte 1992; Morwood 1987). Local ethnohistoric evidence suggested that this motif type may provide ethnically-sensitive patterning, with commentators stating that the coastal peoples from around Sydney and further north carried distinctively patterned shields (Bellinghausen and Rossiyisky in Barratt 1981; Enright 1900; Threlkeld in Gunson 1974). It was hoped that patterning in shield design distributions may indicate an interrelatedness of contact around the region.

Threlkeld described the construction of the region's wooden shields from around the Lake Macquarie area (*Awabakal/Guringai* language areas). These were:

> three feet long by eighteen inches ... lozenge shaped, pointed at top and bottom, and pigeon breasted rather than flat. ... The shields are always painted with white pipeclay and are generally ornamented with a St George's Cross, formed by two bands two or three inches wide, one vertical the other horizontal, coloured red ... [Gunson 1974:68].

Rossiyisky describes the wooden shields from the Sydney area similarly, although observing that 'they are daubed with *various* red and white figures' [(in) Barratt 1981:23, emphasis mine]. Bellinghausen's description confirms the colour usage as 'dry white colouring substance over which was painted red stripes' (in Barratt 1981:41; and see Enright 1900; Cave in Brayshaw 1986).

The engraved shields are fairly rare (232 total: 3% of the assemblage) and they are quite dispersed (found at 103 sites; 14%). A few sites have concentrations of this motif type: site# 45-3-376 has 23 shields; #45-6-705 has 15 (Figure 11.29). Most sites have one or two shields (average is 2 shields/site). These motifs are usually found on moderately large sites (average assemblage size = 22 motifs/site). The shield from one site (#45-6-689) was excluded as the motif was incomplete. The sample for this analysis came from 102 sites.

Chapter 11: Synchronic variation: Sydney Basin

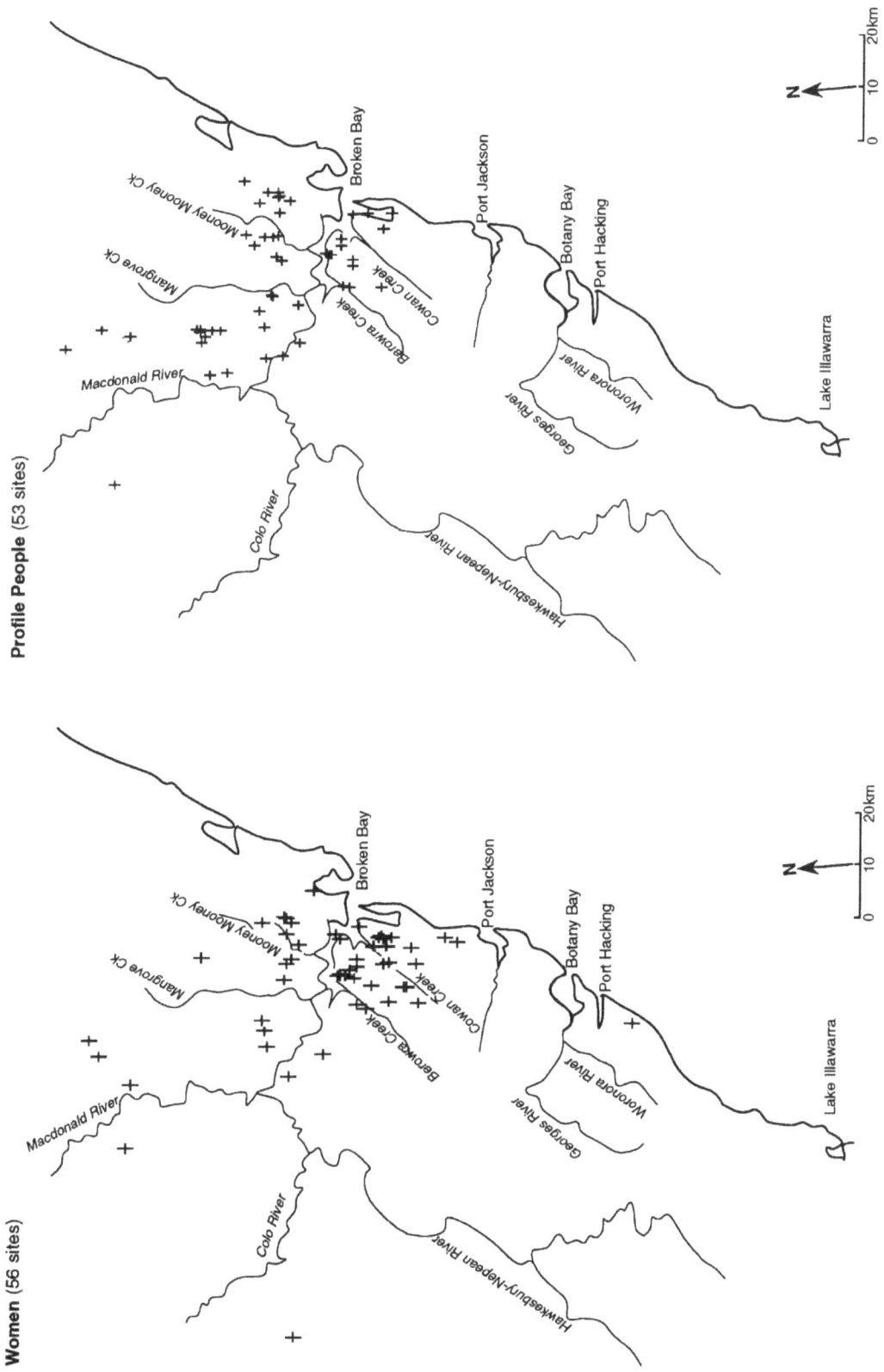

Figure 11.25: Distribution of sites with engraved women and profile people.

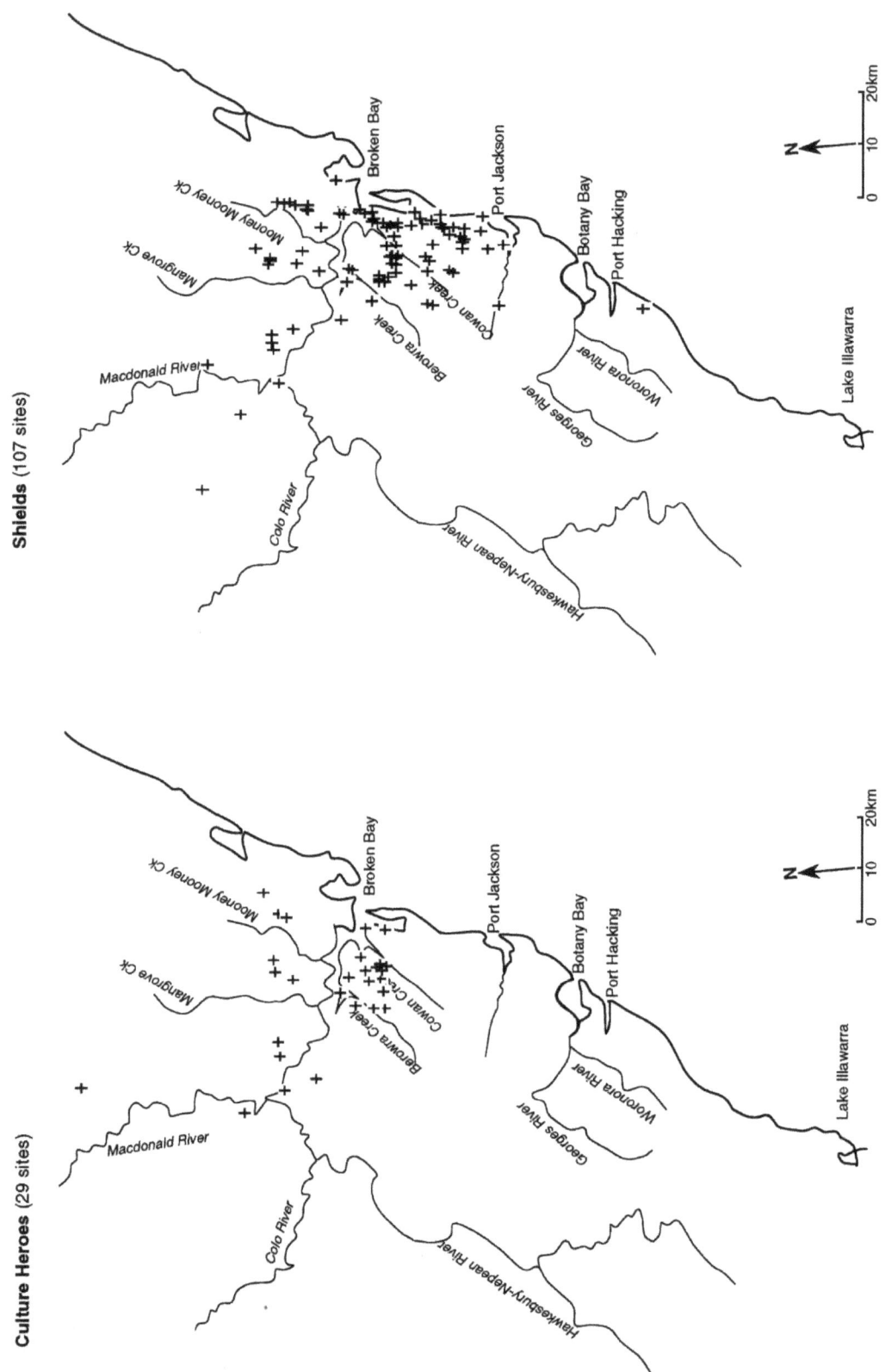

Figure 11.26: Distribution of engraving sites with culture heroes and shield motifs.

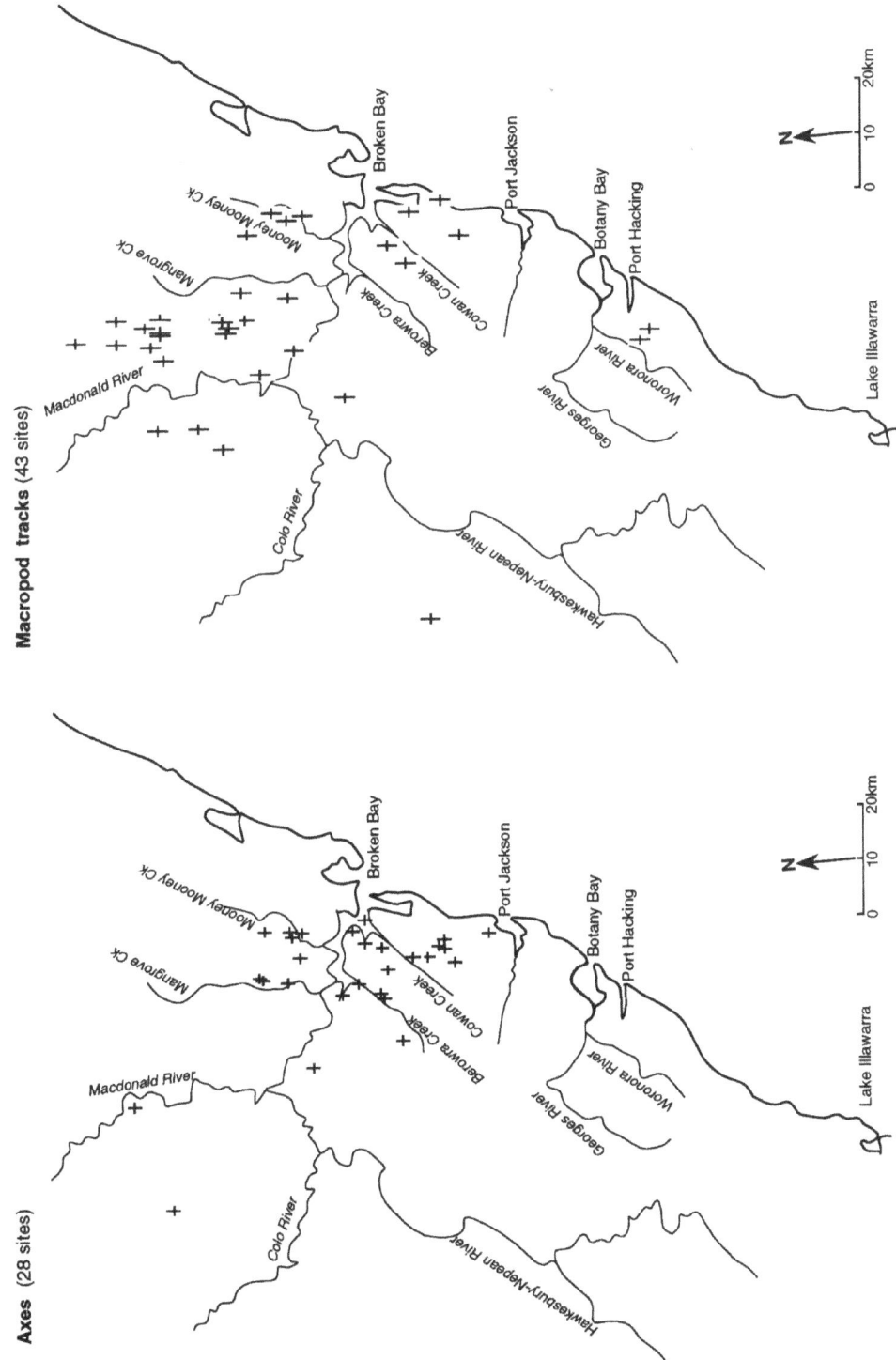

Figure 11.27: Distribution of sites with engraved axe and macropod tracks.

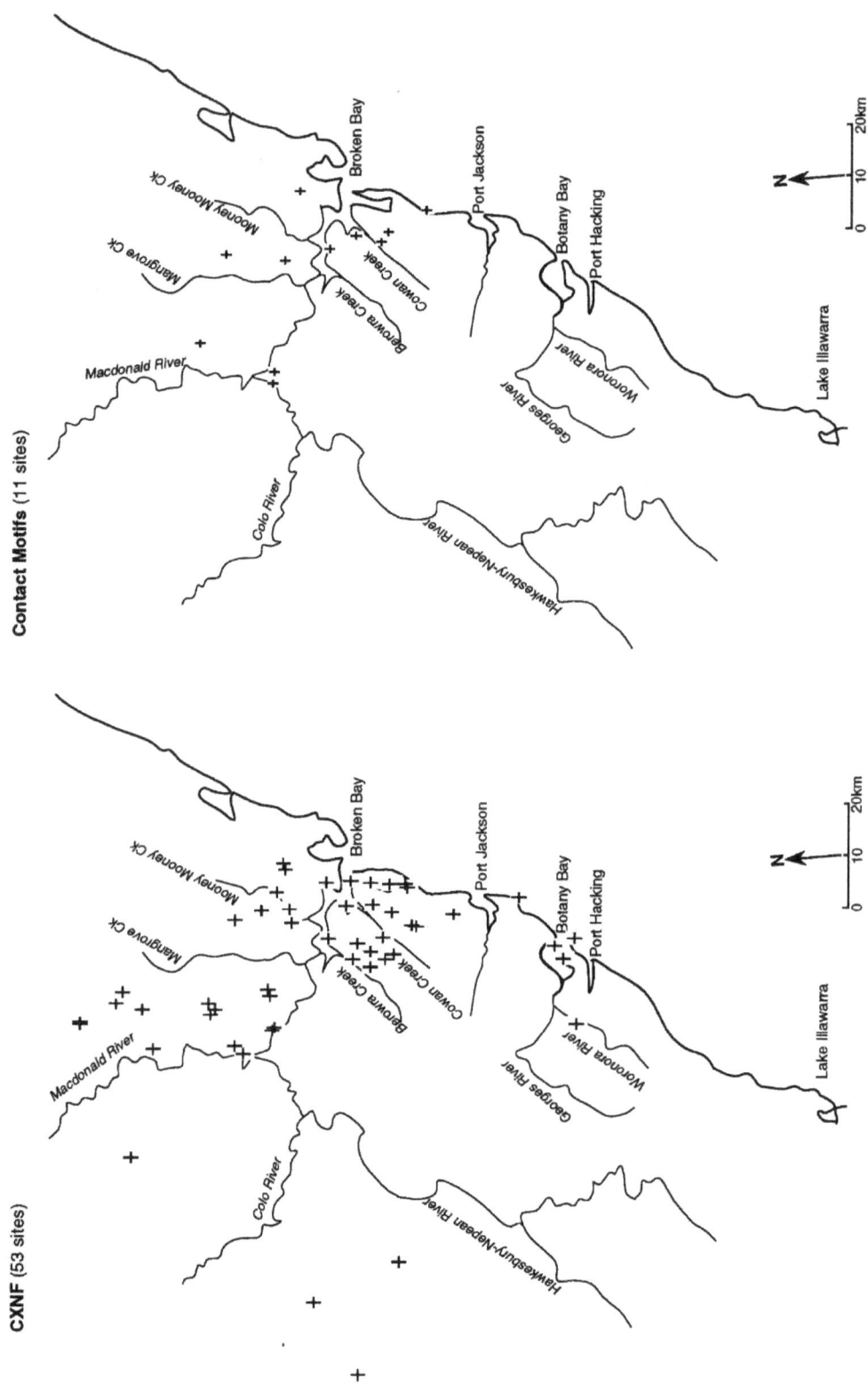

Figure 11.28: Distribution of engraving sites with complex-non-figurative and contact motifs.

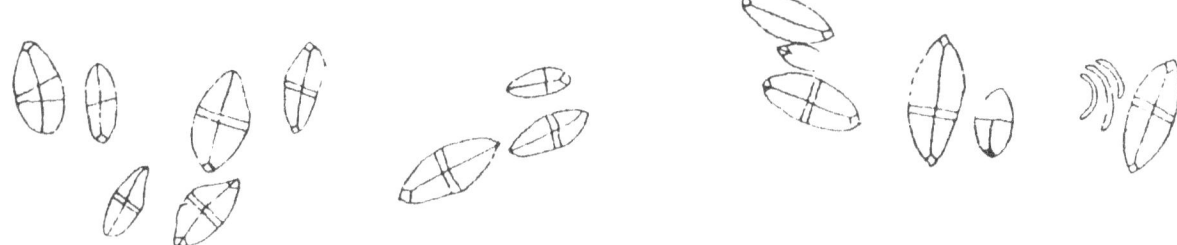

Figure 11.29: Site 575 (NPWS # 45-6-705) with 15 shield motifs and three boomerangs.

Most shields (85.4%) are in the *Guringai* language area. *Darkingung* has the next highest number (10.7%) followed by *Darug* (2.9%) and *Tharawal* (1.0%). No engraved shields are recorded in the *Eora* language area.

Given the paucity of design options provided by ethnohistoric accounts, the engraved shield assemblage was inspected to determine design variety. Analysis revealed this to be considerably greater than suggested by Threlkeld's description. Twenty-six design categories (including undecorated) were identified (Figure 11.30). The designs consist mainly of horizontal and vertical line variations. A significant design element is a diamond shaped component at either or both of the shield's pointed end(s).

There are several unique varieties (designs 6 A-C) which are variations on design themes 2B and 2C. Two of these appear to be shields punctured by many spears (see Megaw 1993).

The two sites with multiple shield motifs (#'s 45-3-376, 45-6-705) are separated by considerable distance. One is north of the Hawkesbury River (Mangrove Creek drainage basin), the other is at Mosman (Middle Harbour drainage basin: Figure 11.31). The *Darkingung* site includes only design varieties 1, 2B and 2C. The *Guringai* site has seven design types (4A, 4D, 4E, 4F, 5B, 5E, and 2C) most of which have the diamond point decorative element at one or both ends.

First, sites with only a single design type present were analysed. It was hoped that this would facilitate clear focus on localised patterning in the design elements and reduce synchronic 'noise'. Threlkeld's description suggested that the 'St George cross' form (2B) would be the most common. This was not the case. Sites with multiple design types were excluded from this initial sort. Four major design themes were identified and these account for 62% of the sites with shields (Table 11.8).

Table 11.8: Engraved sites. Shield Design Types according to language areas and drainage basins. Sites with single design types only.

Language Area and Drainage Basins	Plain	%	2B	%	2C	%	4+5 variation	%	%f
Darkingung cent Macdonald	1	6.7	1	11.1	2	7.1			6.3
Darkingung Mooney	3	20.0							4.8
Darug Berowra	1	6.7			1	3.6			3.2
Darug Lane Cove					1	3.6			1.6
Guringai Mooney	1	6.7			4	14.3			7.9
Guringai Brisbane Water	2	13.3			5	17.9			11.1
Guringai Cowan	3	20.0	3	33.3	7	25.0	7	63.6	31.8
Guringai Pittwater			1	11.1	2	7.1			4.8
Guringai Middle Harbour	3	20.0	4	44.4	5	17.9	4	36.4	25.4
Guringai Lane Cove					1	3.6			1.6
Tharawal Port Hacking	1	6.7							1.6
Totals	15	(23.8)	9	(14.3)	28	(44.4)	11	(17.4)	63

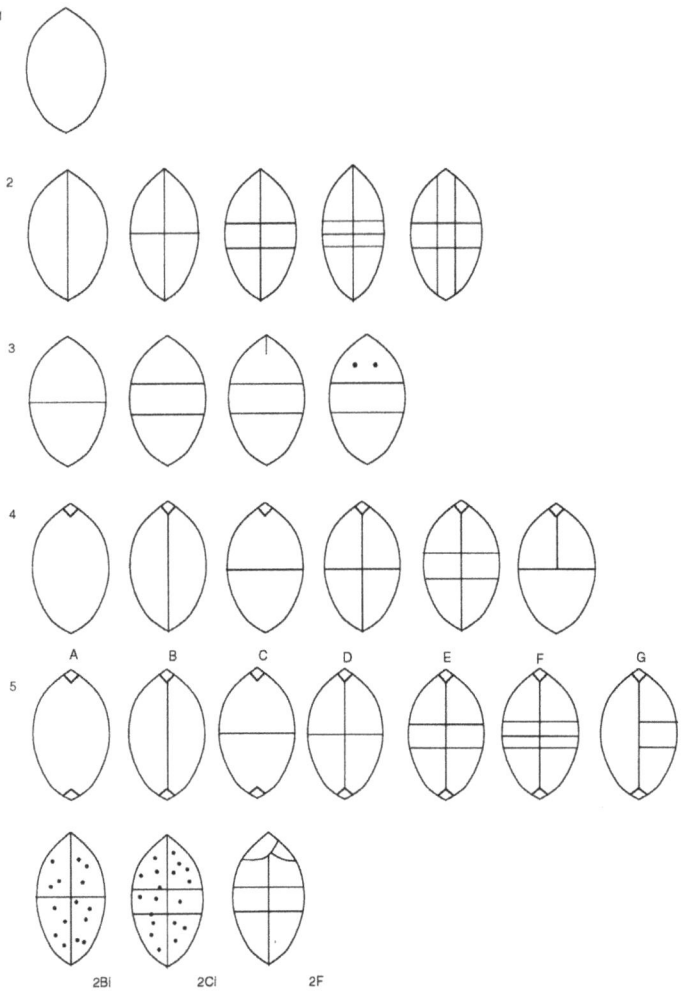

Figure 11.30: Range of shield designs present in the engraved component.

Plain shields, as might be expected, are ubiquitous. They occur on both sides of the Hawkesbury River in *Guringai, Darkingung, Darug* and *Tharawal* sites. The distribution of the St George cross design (2B) was quite restricted. With the exception of one *Darkingung* site west of Mangrove Creek (on Flat Rocks Ridge), this design is confined to the Cowan, Pittwater and (particularly) Middle Harbour drainage basins: all *Guringai* language areas.

The double (horizontal) cross design (2C) is the most common design in the region. It occurs at 28 sites (in isolation) and at another 13 sites in combination with a variety of design forms. It occurs in *Guringai, Darkingung* and *Darug* areas.

The design with diamond elements at either end of the shield has a very restricted distribution. The eleven sites with this design element (in isolation) occur only in the *Guringai* area, in the Cowan and Middle Harbour drainage basins.

The two designs which appear to have the most potential for indicating inter- and intra-language contact were the St George (2B) variety and the diamond-end varieties (design types 4 and 5), both of which are restricted predominantly to the *Guringai* area. Next the distribution of these design elements in sites with combination shield designs was analysed. Again, a restricted distribution was found (Table 11.9). More than 90% of the sites with these designs were in the *Guringai* area, with the Cowan and Middle Harbour catchments containing 68% of these sites.

Shield designs exhibit highly localised characteristics. The *Guringai* language area contains the largest number of shields in the region and exhibits the greatest degree of design variability. There are localised design traits within this area. Cowan and Middle Harbour catchments contain the most shields and this is the design focus for both the diamond infill and double-cross varieties.

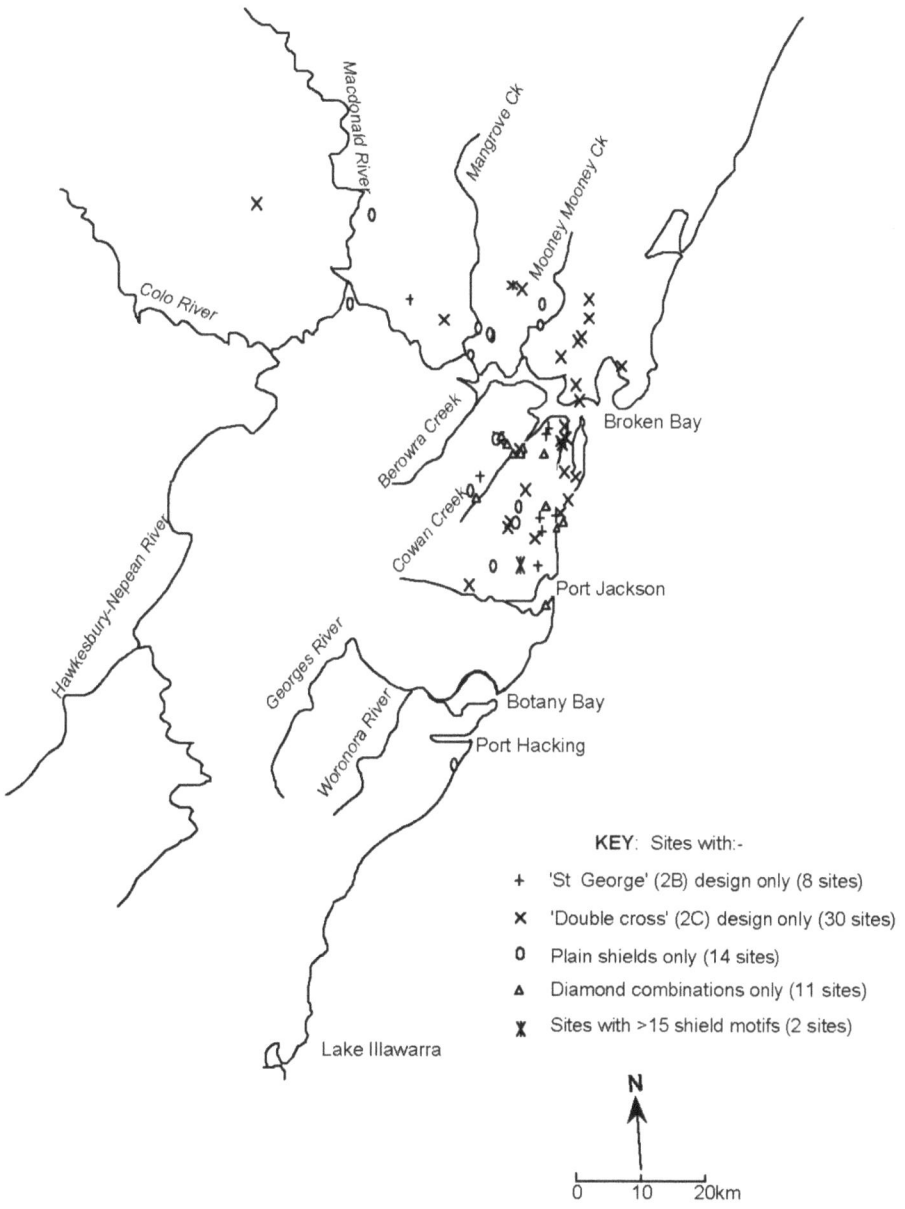

Figure 11.31: Distribution map of main shield designs.

Table 11.9: Sites with engraved shield. Shields Distribution of mixed design types according to language areas and drainage basins. Type 2C, 4 and 5 varieties.

Area	No.	%
Darkingung central Macdonald	1	4.5
Darug Berowra	1	4.5
Guringai Mooney	1	4.5
Guringai Brisbane Waters	3	13.3
Guringai Cowan	7	31.8
Guringai Pittwater	1	4.5
Guringai Middle Harbour	8	36.4
Totals	22	

In this instance, the ethnohistoric literature is highly inadequate with respect to both the design variability present and the distribution of these design elements. There may have been a major constriction in design elements used on shields at contact (and the St George Cross may have predominated as reported). Conversely, the subtleties of design variability were overlooked by early observers. If the former explanation is correct, then the ethnohistoric 'present' has little applicability in the prehistoric past, particularly in terms of ethnically significant designs. This would appear to be another example of the inadequacies of European observation regarding the richness of Aboriginal culture in the region.

Culture Heroes

At 29 sites across the region there are 36 culture heroes (Table 11.10). This motif differs from ordinary anthropomorphs on the basis of extreme size (mean length is 5.2m; standard deviation = 1.4m) and by the amount of infilled decoration. Some of these motifs are therianthropes (Flood 1987; McDonald and Veth 2006b) with animal features including bird's and/or snake's heads. The two main forms of culture hero are the '*Daramulan*' type (partially or fully in profile) and the '*Biaime*' type (in plan, with limbs akimbo). These types were named by McCarthy (1959a; following Mathews 1904, and see Clegg 1981) and a more recent analysis (Higgs 2003) has explored attribute variability in these motif forms.

A feature of this motif form suggests its cultural importance - or at least its continuation of use over time: evidence for multiple episode of engraving. More than half of these motifs have added features or altered outlines. The *Daramulan* figures appear particularly susceptible to change: many having a second leg added, and sometimes an arm or extra penis (Figure 11.33).

These motifs occur amongst a range of assemblage sizes. Some sites are extremely large (97 motifs), but there are also isolated examples e.g. with only a single culture hero present. The average size of sites with this motif present is 25.4 motifs/site. The average distance to drinking water from sites with these motifs is c.500m.

As indicated above, this motif is restricted to the *Guringai* (75%), *Darug* (8.3%) and *Darkingung* (16.7%) language areas. Almost half of the motifs in the *Guringai* area are located in the Cowan drainage basin (Table 11.11).

Most of these motifs occur singly at sites, but are often paired. Some of these pairings are clearly male/female; others are male/male. Some sites include paired and/or transitional *Daramulan* and *Biaime* types (Figure 11.34).

The *Daramulan* type is heavily focussed (86%) in the *Guringai* area, while the *Biaime* form is distributed more widely (still 67% *Guringai*). The two *Daramulan*-type culture heroes located outside the *Guringai* area occur on major engraving sites (Devil's Rock Maroota, Flat Rocks Ridge) in places which could be argued are aggregation locales (Conkey 1980) - or at least on access routes where groups cohesion is being demonstrated.

Table 11.10: Engraving sites. Culture heroes: compositional details.

Site Number	Max length (m)	Type Biaime/ Daramulan	Animal features	Infill	Other features
37-6-8	2.7	B		dot	breasts, penis and foot added, fingers 1 hand, 2 eyes, headdress, 47 dots
45-2-16	6.8	D	bird head	heavy dot/linear	penis + leg added, ornate headdress, bird head
	5.0	B	-	heavy dot/linear	pointy ears headdress, male, dotted linear infill
45-2-45	3.3	B	-	heavy	two deep eyes, body linear infill waist band no hands or feet, conical head
	3.0	B	-	some lines	two deep eyes + 3 smaller ones, 4 dots in penis; toes on feet, stumpy arms
45-3-39	4.5	D	birdlike	linear	bird like with long rayed headdress. No foot visible. Waistband + linear infill
45-3-56	5.0	D	birdlike	linear	bird-like with long pointed headdress, 2 eyes + mouth. No foot visible
	4.5	D	birdlike	linear	bird like with rayed headdress, 4 eyes Waistband + 2 rows infill
45-3-99	8.3	B	-	lines dots	twisted perspective; penis to one side
45-3-110	3.2	D	birdlike	linear dots	leg with human foot, head twisted perspective w headdress. 2 eyes

Site Number	Max length (m)	Type Biaime/ Daramulan	Animal features	Infill	Other features
45-3-168	5.7	D	bird head	lines on neck	emu like; 2 legs w feet and penis to side
45-3-228	6.0	B	long ears	lines	buttocks shown + 2 legs, enormous ears, breasts? under arms, + 5 eyes
45-3-232	5.6	B	bird head	lines	no arms, twisted perspective head, headdress
45-3-954	3.5	B	-	heavy linear	feet, fingers 1 hand, no penis or headdress, 2 eyes
45-3-1289	6.5	B	-	lines	eyes, headdress
45-6-42	4.7	D	-	lines	2 penises, eyes, headdress
45-6-44	4.6	B	-	lines	large head, no neck; 1 leg added; toes 1 foot only; single line headdress, 3 eyes
45-6-85	5.2	D	snake head	-	snake + other head with headdress. Human foot
45-6-284	3.4	B	-	linear	7 dots on face no ears; fingers + toes; fringed body infill, big penis; holding boomerang assoc. with long trail mundoes
45-6-290	5.3	B	-	lines	2 eyes, no feet
	6.0	D	snake/ bird	-	1 leg, sinuous neck, arm + axe added, profile leg
	6.0	B/D	snake head	-	2 legs, sinuous neck, 1 arm with axe, 2 legs with feet (1 added) penis added
45-6-312	5.0	B	pointy ears	linear design	fingers toes, facial features, holding a number of material objects, body design.
45-6-313	4.5	B	-	lines	very long arms and legs foot on one leg only, fingers, toes
45-6-315	6.0	B	-	lines	fingers + toes, waist + arm bands, small penis, looped headdress; associated with shields
45-6-316	5.3	B	-	many lines	horizontal + vertical infill; fingers on arms and legs, no neck, facial features, penis
	4.7	B	-	many lines	horizontal + vertical infill finger like appendages on arms and legs, no neck, facial features include an inverted 'smile', penis with infill
45-6-323	5.0	B	-	lines	hands + feet with fingers + toes, barbed arrow in side, crossed body lines, penis
45-6-324	6.3	D	bird like	lines	emu like with human foot, arm added
45-6-346	4.2	B	-	lines	rayed headdress, girdle + arm bands, 2 eyes
	3.2	B	-	lines	rayed headdress, girdle + arm bands, body design, 1 foot
45-6-412	6.7	D	-	lines	male with key shaped headdress
	5.5+	D	-	-	female, pointed breast, 5 eyes
45-6-434	5.7	B	-	lines	waist + armbands, penis, feet with toes, headdress, holding fish
45-6-436	9.3	D	-	lines	leg added on to penis, arms added onto head; headdress
45-6-890	6.2	D	bird head	lines	foot with toes, arm + hand with fingers added. Birdlike head with beak. Long meandering penis, 6 eyes, dot on heel

They focus of both culture hero forms in the *Guringai* area, with relatively minor examples in northerly and westerly language areas, suggests design (and social) contact between these three language areas, with design focus stemming from the *Guringai*.

A similar pattern was demonstrated by the shields. The distribution of these same motifs in the shelter art component is compared below, to determine whether these patterns hold for both media.

Table 11.11: Daramulan and Biaime types, language areas and drainage basins.

Area	Daramulan	%	Biaime	%	Total	%
Darkingung U McDonald			1	100	1	2.8
Darkingung C McDonald			4	100	4	11.1
Darkingung Mangrove	1	100			1	2.8
Darug Cattai			1	100	1	2.8
Darug Berowra	1	50	1	50	2	5.6
Guringai Mooney	1	50	1	50	2	5.6
Guringai Brisbane Waters	3	75	1	25	4	11.1
Guringai Berowra	3	60	2	40	5	13.9
Guringai Cowan	5	31.3	10*	62.5	16	44.4
Totals	14	(38.9)	21	(58.3)	36	100.1

*one of these culture hero is half *Daramulan* and half *Biaime*.

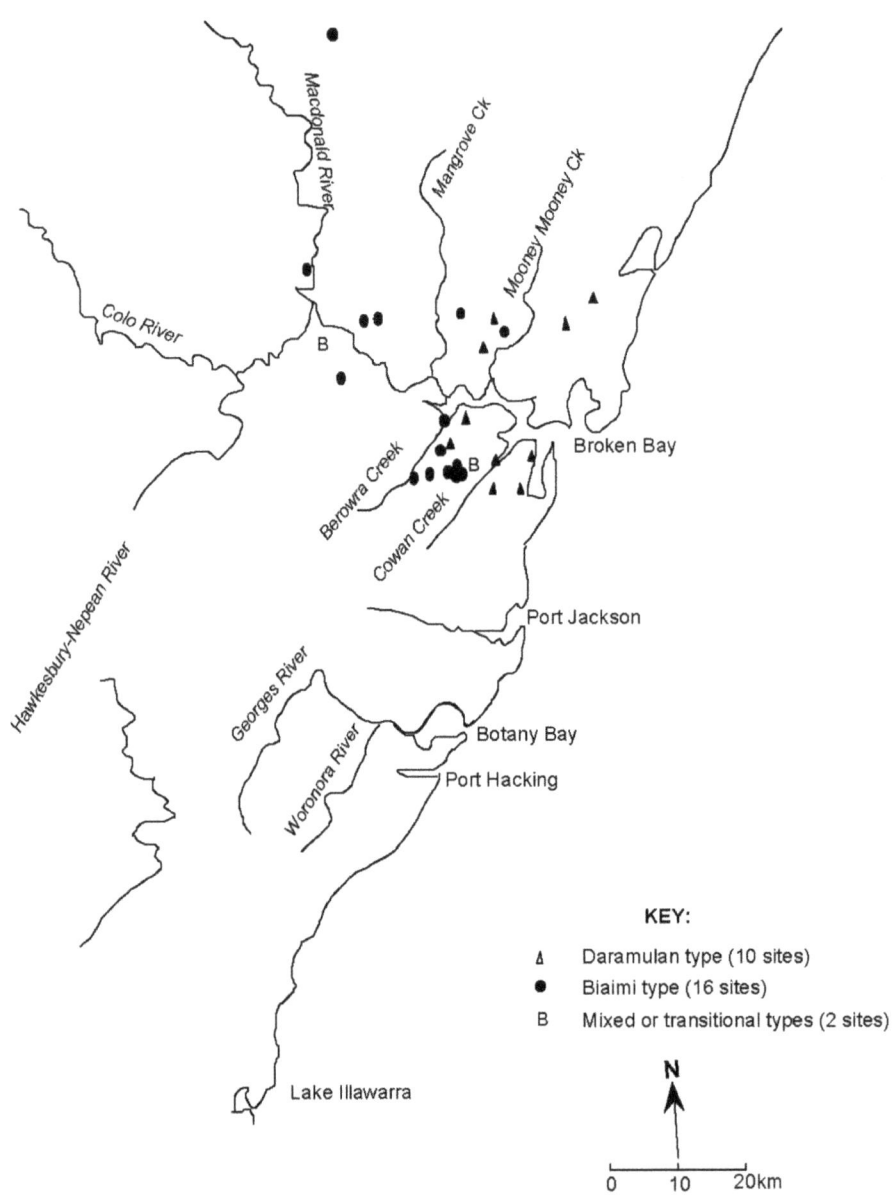

Figure 11.32: Distribution of Daramulan and Biaime type engraved motifs.

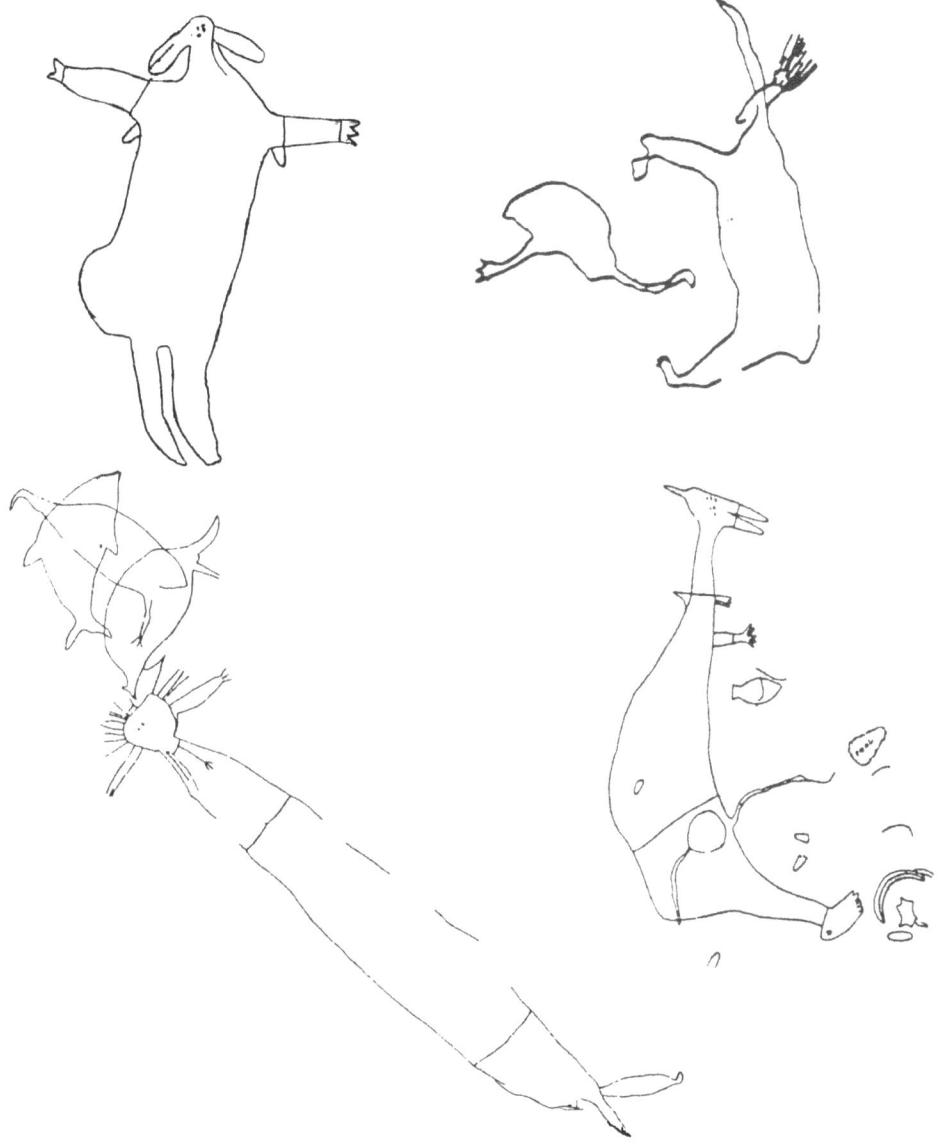

Figure 11.33: Engraved Culture Heroes. Daramulan motifs which appears to have been altered over time (top to bottom, L to R; after McCarthy 1956: group 3, Figure V; McCarthy 1954a: Figure 7C; Campbell 1899: Plate XXV, Fig 4; McCarthy 1956: Group 3, Figure V).

Engraving sites: conclusions

While a cohesive style region, these analyses have identified stylistic variability in the engraved art of the Sydney region. This is distinctive only in the south of the Basin, where the previously identified style boundary of the Georges River is confirmed. In most parts of the region stylistic variation is clinal. It can be explained in terms of defined language areas, and can also be identified as localised variability based on drainage basins.

The *Darkingung* and *Guringai* language areas have the most homogenous engraving assemblages closely followed by the *Darug*. The engravings in the *Tharawal* area are the least homogenous. The *Guringai* sites consistently demonstrate the highest levels of homogeneity, but also provide evidence for internal variability: the Cowan sites are less homogenous than those from the Pittwater drainage basin. In this central part of the region there is a clinal increase in variability as one moves west, away from the coast.

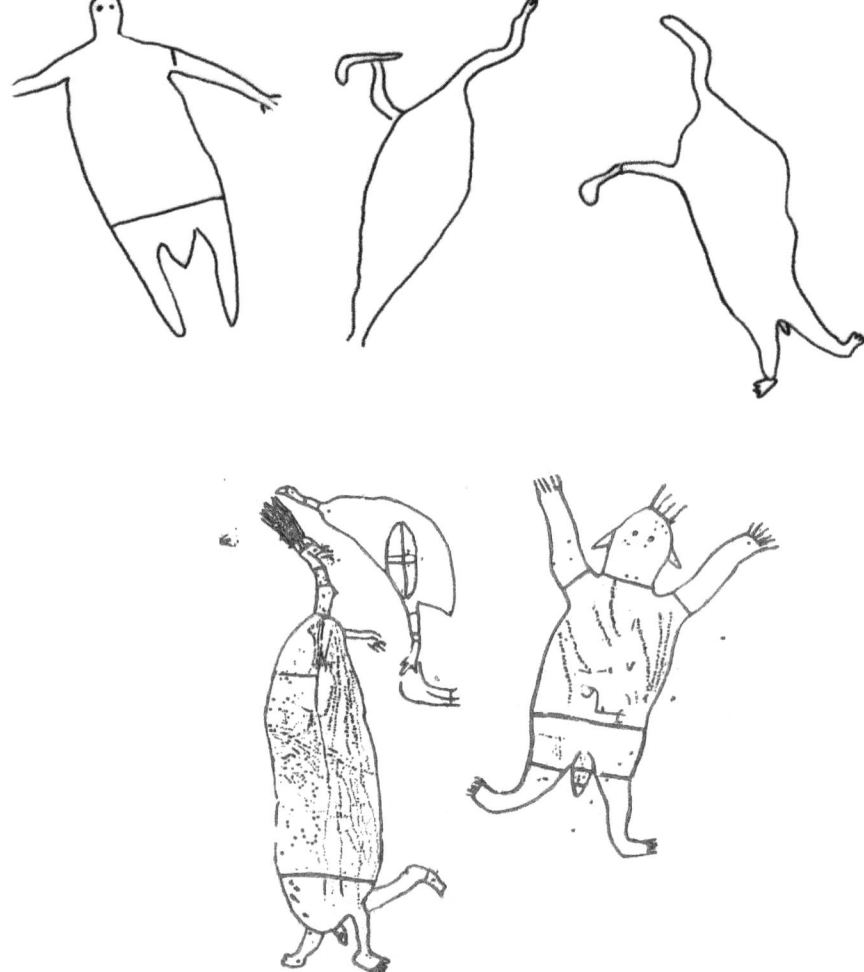

Figure 11.34: Paired and/or transitional *Daramulan* and *Biaime* motifs. Note that *Daramulan* below (on #45-2-16) has an altered outline also (top; from McCarthy and Hansen 1960: Figure 1; bottom; from McDonald 1986a: Figures 3 and 8).

Analysis of the *Tharawal* sites reveals an assemblage which dissimilar from all others. The sites are significantly more heterogeneous and the outlier foci indicate a different thematic focus. Comparison of the two drainage basins with good samples here suggests differences between the coastal and more inland sub-groups.

The results demonstrate a complex network of stylistic variability as defined by motif preference. In some language areas, sites show levels of internal cohesion. Several of the proposed language boundaries are supported by these analyses. Berowra Creek could be the boundary between the *Darug* and *Guringai* groups – the art on either sides of this showing varying design focus. The northern boundary for *Tharawal* is also supported by these analyses with the identified style boundary at the Georges River.

The analysis of rare motifs confirms the localised character of engraving themes around the Basin. Culture heroes are focussed in the western part of the *Guringai* territory and reveal a design link with the *Darkingung* - confirming the CA results.

The study of composition on several rare motifs also revealed design contact between *Guringai, Darkingung* and *Darug* language areas. The source of this contact appears to stem from the *Guringai*. Design variability on shield motifs is extraordinarily diverse in the *Guringai* area, with both less motifs and a marked decrease in design options being practised outside this language area. These results will be discussed further in chapter 13.

12

REGIONAL SYNCHRONIC VARIATION: SHELTER ART

Introduction

This chapter looks at regional stylistic variability in the shelter art assemblage. As with the engraved assemblage, the aim was to identify broad scale patterns. These analyses explored both motif depiction and technical variation. These results are compared with those achieved for the engraving sites in chapter 11.

Quantitative analyses here again focused on motif combinations, not on qualitative aspects of motif classes. The overall aim was to be able to compare the engraving and shelter art assemblages - despite their technical differences.

The approach and methodology for both art components were detailed in chapter 11. The classification system, original data and computer bivariate plots supporting the CA were provided in the original thesis[37] (McDonald 1994a: Appendices 1, 5, 6 and 7).

The CA results for both motif and technique are presented in terms of drainage basins. The three different locations investigated with the engraving assemblages, were again subject to more detailed analysis of the CA results.

The distribution of uncommon motifs is explored with this component also. Profile people, culture heroes, items of material culture and complex-non-figurative (CXNF's) appear to be good indicators of localised cultural choices being made by pigment artists. Contact motifs are again investigated (McDonald 2008).

As with the engraved component, small scale qualitative analyses of motif depiction and preference were undertaken. To enable comparison between the two media, shields and culture heroes were again the focus of this analysis.

Correspondence Analysis (CA): regional data, results and interpretation

Basic statistical information about motif and technical information (i.e. average assemblage size, motif frequencies, colour usage etc.) was presented in chapter 5. The multivariate technique (CA) used here allows quantified statements to be made about the regional homogeneity as well as demonstrating what variables distinguish sites (i.e. the sources of variance within the data base).

The same procedures were followed with this medium as were described previously for the engraving assemblage. Motif variables were the same as those used for the engraving sites with two additional motifs (Table 12.1).

Analysis commenced with unmodified count information (29 motif variables). This taxonomy was then reduced to seven clumped taxa (Table 12.1) and the CA was run using binary data. The CA of the shelter art's technique variable comprised 546 sites, while the CA of motif variables (which excluded sites with only unidentified motifs) involved 439 sites.

Motif

The first two components account for 54.4% of the variance in the sample. The scree slope plot (Wright 1992) demonstrates that the first two components describe considerable structure in the data (Figure 12.1). The plot of these first two co-ordinates reveals that the data is discriminated well by five of the seven variables (Figure 12.2a). Variables 1 and 4 are close to the origin and

[37]These are not included here as these represent many hundreds of pages of close-typed computer files.

Table 12.1: Shelter Art sites: Clumped motif variables used in Correspondence Analysis.

Variable No.	Motif/Variable description
1	anthropomorphic
2	terrestrial
3	birds
4	marine
5	material objects
6	tracks
7	other

Clumped variable 1 includes individual variables 1 - 5; 2 = v 6-8; 3 = v 9,10; 4 = v 11-14; 5 = v 15-18; 6 = v 21-25; 7 = v 26-29 (see Table 5.3). Unidentifiable motifs have been excluded from this level of analysis.

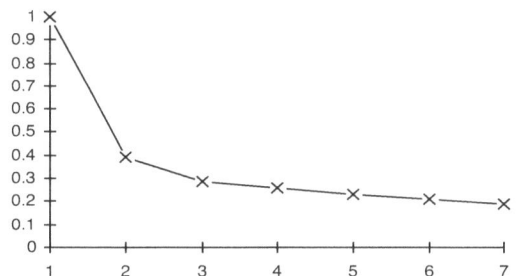

Figure 12.1: CA results. Scree slope plot of the latent roots showing that the variance is well accounted for by the first two components.

play little part in distinguishing between the sites.

In the first co-ordinate, variables 2 and 6 exhibit an inverse relationship with variables 3, 5 and 7 (Table 12.1). Sites which contain a combination of birds, material objects and/or other motifs are likely to be very different to sites which contain land animals and/or hand stencils (i.e. tracks). In the second co-ordinate, variables 2 and 6 are inversely related (although to a lesser extent than demonstrated by the first co-ordinate).

The distribution of motif variables on the graph is generally mirrored by the distribution of sites (see Figure 12.2b). Sites which are poorly discriminated by their motif assemblage lie close to the origin. Sites which are well discriminated on the basis of their motifs, are distributed around the graph, their position being determined by the motifs present in their assemblages, e.g. sites which have predominantly hand stencils are located in the negative quadrant: assemblages which contain lots of birds are located in the positive quadrant (Figure 12.3). Based on this analysis, it can be stated that the shelter art component is relatively homogeneous, with no major internal groupings.

Given the size of the data base, the usefulness of Figure 12.2b for detailed interpretation is low. Thus, the results were again replotted using various subdivisions of the data. All sub-plots are based on exactly the same results, but the smaller sample sizes enable more detailed interpretation.

The data were first subdivided into regional sub-groups which could be interpreted geographically. These groups were based on the AHIMS site identification number, which is in turn based on map sheet location (Table 12.2, see Figure 136). Groups 1 - 7 are directly comparable to the similarly numbered engraving Groups. Group 8 is to the west of Group 6 - an area where no engraving sites are recorded.

While an arbitrary division of the sample, this method achieved good control was on general east-west and north-south divisions in the data (McDonald 1985a). All bivariate plots from the CA results were presented (McDonald 1994a: Appendix 6, Figures A6:17 - 26). These are summarised here.

Table 12.2: Analytical grouping of shelter art sites according to AHIMS Id. numbers. Groups used in regional CA analysis.

Group	Map number	1:250,000/1:100,000	Sample size
Group 1	37 - 5 -	Singleton/Howes Valley	30 sites
	45 - 1 -	Sydney/Wallerawang	
Group 2	45 - 2 -	Sydney/St Albans	66 sites
Group 3	45 - 3 -	Sydney/Gosford	144 sites
Group 4	45 - 5 -	Sydney/Penrith	17 sites
Group 5	45 - 6 -	Sydney/Sydney	107 sites
Group 6	52 - 2 -	Wollongong/Wollongong	171 sites
Group 7	52 - 3 -	Wollongong/Port Hacking	6 sites
Group 8	52 - 1 -	Wollongong/Burragorang	5 sites

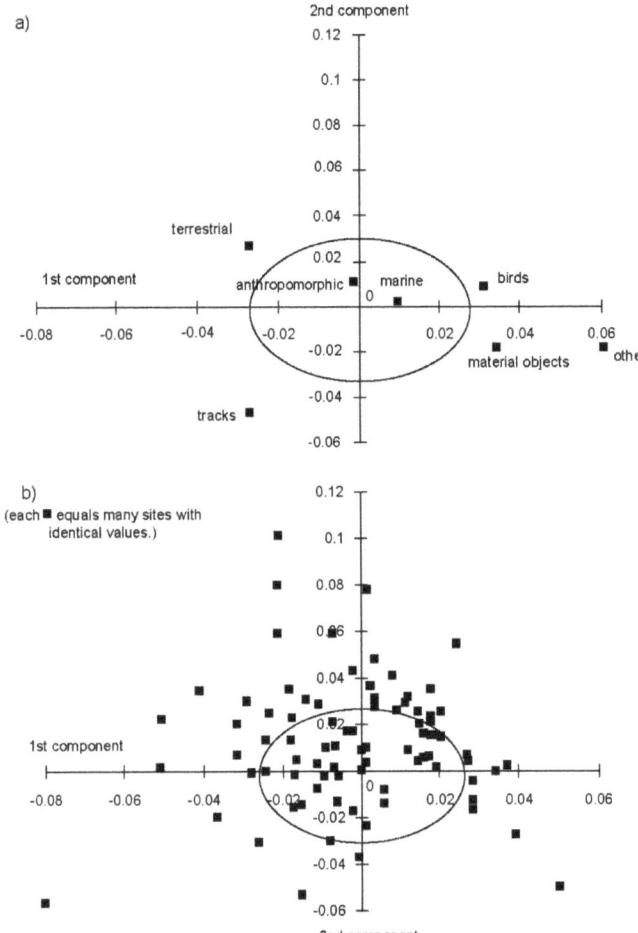

Figure 12.2: CA results, shelter motifs. Component scores a) motifs and b) sites.

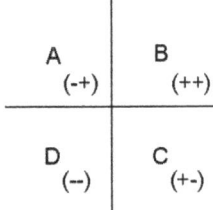

Figure 12.3: Quadrant identification used in interpreting the shelter art CA results.

While no distinctive groupings occur within the data base, that there are style clines across the Sydney Basin. These can be identified in the bivariate plots by comparing the degree(s) of homogeneity in each group, and by the presence and distribution of outlier sites, relative to the origin. The same technique was used for the engraving assemblage. The number of sites within a defined and consistent radius of the origin was noted[38], the percentage of 'common' sites was calculated and outlier sites in each area were identified. This was a necessary step given the disparate sample sizes.

The distribution of the outlier sites in particular quadrants (Figure 12.3) was investigated as variations in this result across the region enable more specific statements on localised variability. This analysis demonstrates the presence and nature of localised variability in assemblage content across the region.

Technique

The technique classification was initially devised for the *Rock Art Project* (McDonald 1985a, and see McDonald 1988a). It includes a combination of technique (variables 4-8, 16), form (variables 1-3, 9-11) and colour (variables 12-15: Table 5.4).

---

[38] The same consistent radius as used for the engraving sites was employed here. This has been drawn on each of the bivariate plots as an heuristic device.

Table 12.3: Shelter Art sites: technique variables used in the CA.

Variable No.	Technique description
1	outline
2	infill/solid
3	outline and infill
8	stencil
12	black pigment
13	white pigment
14	red pigment
15	yellow pigment
16	engraving (scratched, pecked)

Given the interdependence of some of these 18 technical variables (i.e. a depictive motif is described by a combination of all variables excluding #8); the technique taxonomy was reduced to include only unlinked variables (Table 12.3). The form variables (#'s 1-3) provide information which can be directly compared with the engraving assemblage.

The first two CA co-ordinates account for 64.4% of the variance in the data set and account well for its structure (Figure 12.4). A significant amount of the variance is accounted for by the first component. The bivariate plot of the first two co-ordinates reveals good discrimination between sites in the region (Figure 12.5).

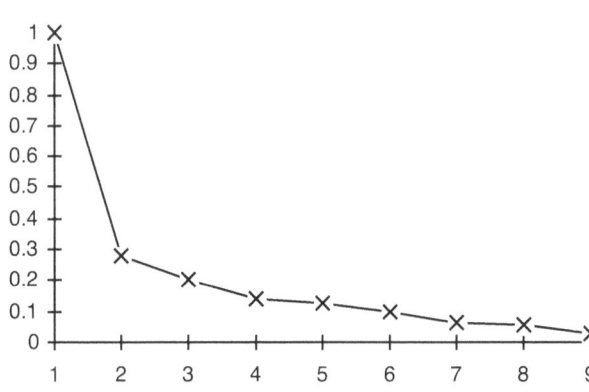

Figure 12.4: CA results Technique. Plot of the latent roots demonstrating that the variance in the data set is well accounted for by the first two components.

In the first component, variables 8, 12 and 13 (stencils, white pigment and yellow pigment) are inversely related to variables 1, 3 and 12 (black pigment, outline and outline and infill motifs). In the second co-ordinate, variable 16 (engravings) exhibits a strong positive value, while all other variables have a weak negative value. Archaeologically, these results indicate that sites which have large numbers of stencils and/or white pigment present are different to assemblages with black outline and infilled motifs. This dichotomy between white stencils and black drawings is a good summation of regional characteristics.

The bivariate plot for the distribution of sites (Figure 12.5b) reveals a solid clustering around the origin with sites being pulled out along the first co-ordinate. The majority of sites in the region are relatively homogeneous, but the variables used identify structure in the data and hence sources of variability amongst the assemblages. The regional sample was again too large to allow meaningful interpretation. Again, the results were replotted on the basis of the broad geographic sub-divisions described above (see Table 12.3).

These CA results describe how sites vary according to technique across the region. The relative homogeneity of the defined groups and the focus of each group's outlier sites was again the basis for interpreting varying levels of technical similarity/diversity.

Regional Comparison

Both motif and technique analyses indicate a core of more homogeneous sites in the centre-west of the Sydney Basin (Figure 12.6 and Figure 136). This homogeneous (core) focus is not in the same location that was found for the engraved assemblage. The sites from Groups 2 and 3 are the

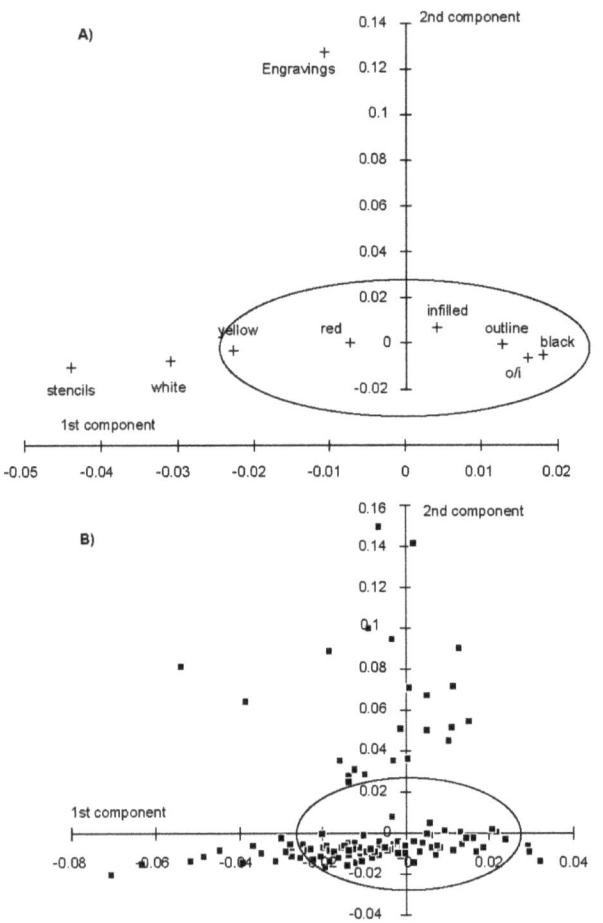

Figure 12.5: CA Results for Technique. Bivariate plot of component and eigen scores. A) variables and B) sites.

least diverse in the region. Sites from Groups 1 and 5 and 6 (and the small sample in Group 7) demonstrate considerable diversity in subject preference and techniques used.

And the technique variables reveal different levels of homogeneity than do the motif variables. The variability demonstrated by technique variables is less than that demonstrated by motif variables showing that there is a higher degree of technical homogeneity across the region than there is similarity in subject preference.

Subtle differences between the sites in the different groups are shown by the outlier sites. These indicate a preference in some areas of the Basin for hand stencilling (Groups 1, 2 and 5), while in other areas (Groups 3 and 6), the drawing of land animals in black pigment is most common.

Language Areas

As was done with the engraving sites, the shelter art assemblages were divided into language areas based on Capell's boundaries (Figure 3.1).

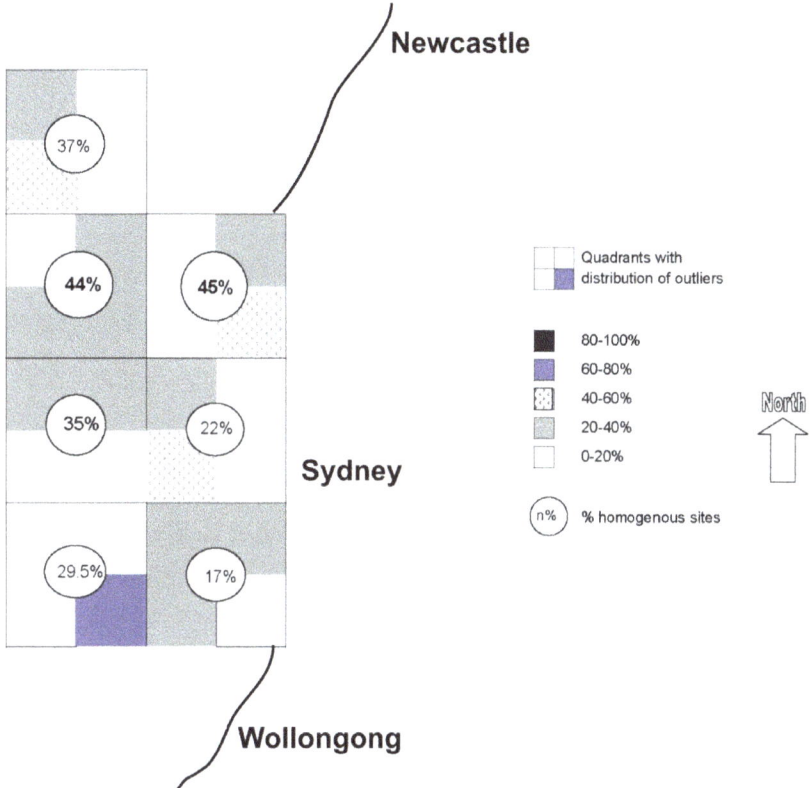

Figure 12.6: Shelter Art, Motif. Percentages of homogeneous shelter art sites in each analytical group.

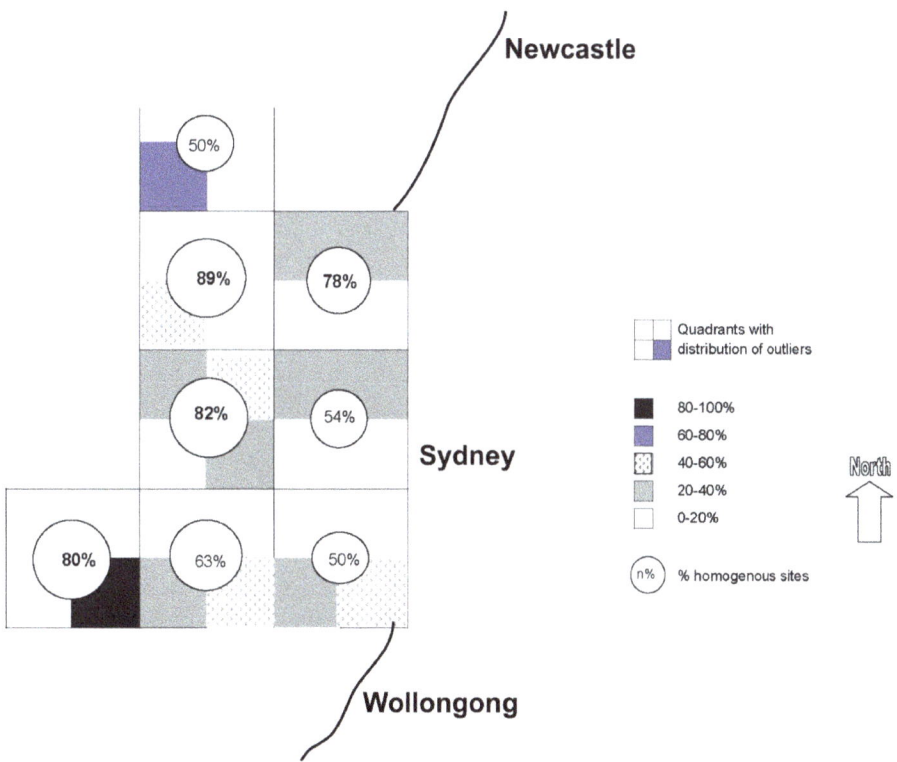

Figure 12.7: Shelter Art, Technique. Percentage of homogeneous shelter art sites in each analytical group.

Motif and Technical Variation across the Basin

Basic assemblage details (motif and technique) and the CA results are described for each of the language areas.

Much of the shelter art assemblage is unidentifiable due to poor preservation, *ad hoc* drawing activity, superimpositionning and the requirements of the classification system. The technique variables recorded included unidentifiable motifs, which results in a more accurate picture of technical ranges used across the region. Because of this, there is not always a correlation between motif proportions and technique characteristics: i.e., while hands might dominate the identifiable motifs, stencilling does not necessarily dominate in technique.

Darkingung Language Area

This group of inland shelter art sites represents the largest sample in the region (190 sites). While 7,725 motifs were recorded here, only 4,972 motifs could be classified: 30% of the assemblage comprises unidentified motifs. The largest recorded assemblages in the region are located here (average 41 motifs/site) as are the two biggest known assemblages: Swinton's with 857 motifs and Yengo 1 with 505 motifs. The average site size is quite high (34 motifs/site) even if these two sites are excluded.

The predominant motif is the hand (54%: Figure 12.8). The focus of the depictive motifs (Figure 12.9) is on macropods (19%) followed by anthropomorphs (16%) and other land animals (10%).

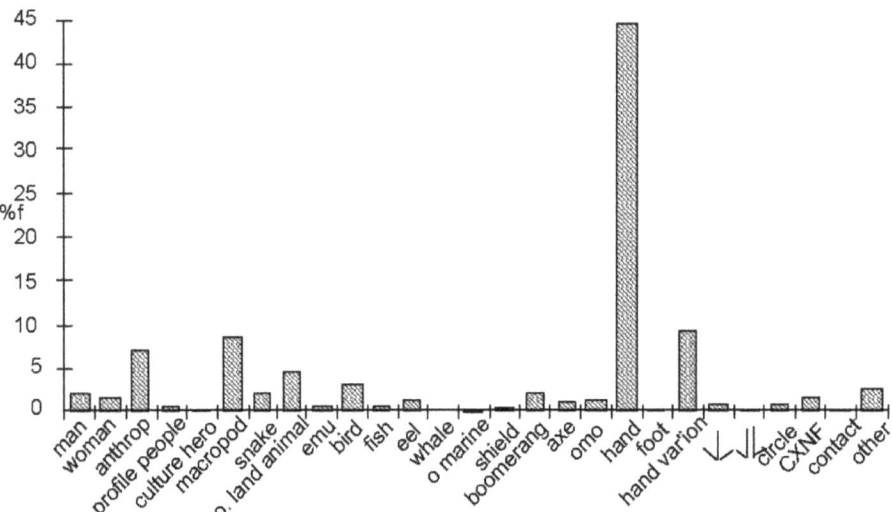

Figure 12.8: *Darkingung* **Language Area, Motif Assemblage (excluding unidentified motifs).**

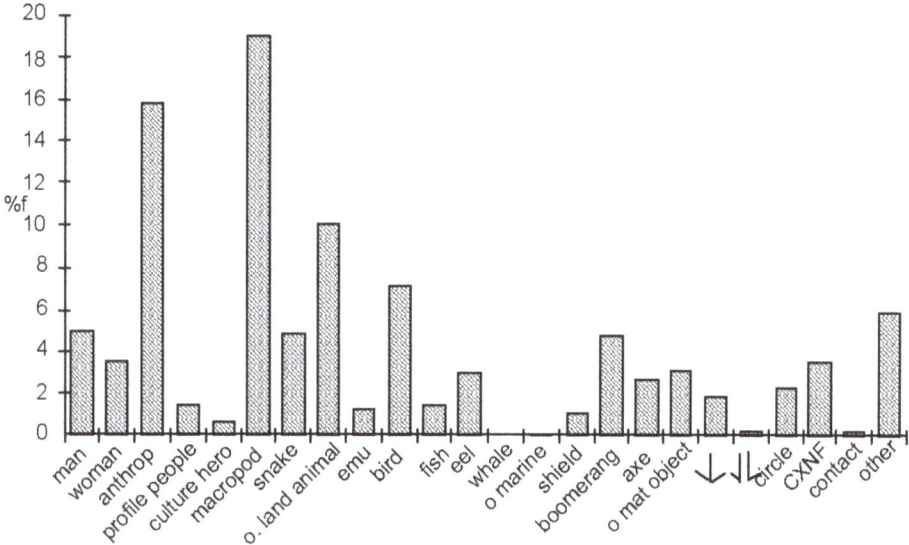

Figure 12.9: Darkingung Language Area: depictive motifs.

The whale is the only motif category not represented here. Non gendered anthropomorphs are the most commonly depicted human figures, followed by men and then women. Profile depictions are fairly uncommon. Most of the pigment culture heroes are located in this area. Boomerangs are the most commonly depicted material culture items. Despite the fact that hands dominate the recognisable motifs here, dry pigment (drawing) is the most commonly employed technique. Stencilling is common. Wet pigment (painting) is more common here than in any other language area (Figure 12.10).

Infilled motifs are slightly more common than the other two forms although all three are roughly equivalent (Figure 12.11).

The clear colour preference in this area is white pigment, followed by black, red and yellow (Figure 12.12).

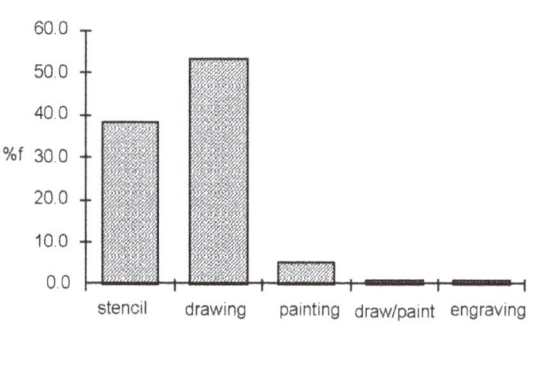

Figure 12.10: *Darkingung* **Language Area. Techniques employed.**

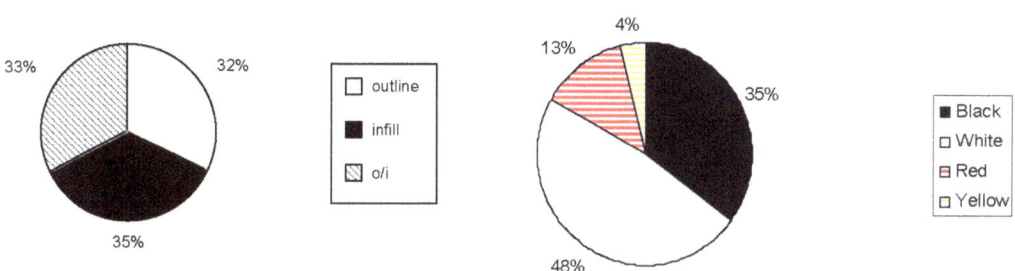

Figure 12.11: *Darkingung* **Language Area depictive motifs. Form.**

Figure 12.12: *Darkingung* **Language Area. Colour usage.**

Darug Language Area

This language area is also located inland and sites were split into two groups - north and south of the Cumberland Plain, testing the posited language boundaries here. In the northern area a total of 1,297 motifs were recorded from 36 sites. Only 851 of these were recognisable (34.4% unidentifiable). Average site size here is quite large (36 motifs/site). Hand stencils (including variations) dominate this assemblage (Figure 12.13).

'Other' dominates the depictive motifs followed by birds and other land animals. Human figures are again focused on non-gendered anthropomorphs, and there is a greater emphasis on profile figures than in the *Darkingung* assemblage (Figure 12.14).

Two culture heroes (at two sites) are located in this area. Drawing is the most commonly used technique followed by stencilling. The other technical options are less common (Figure 12.15).

The three defined forms are relatively evenly distributed (Figure 12.16) with infilled motifs slightly more common than the other two forms

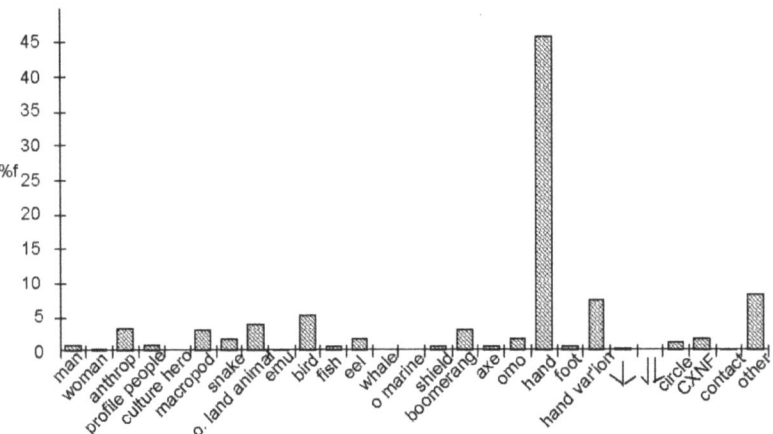

Figure 12.13: *Darug* **(North) Language Area. Motif assemblage.**

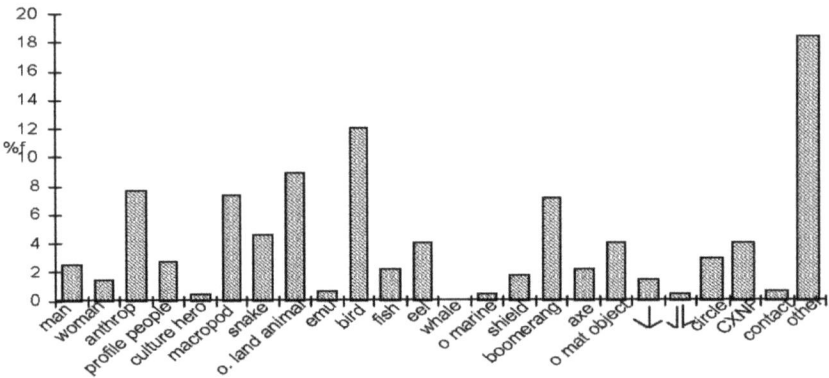

Figure 12.14: *Darug* **(North) Language Area. Depictive Motifs.**

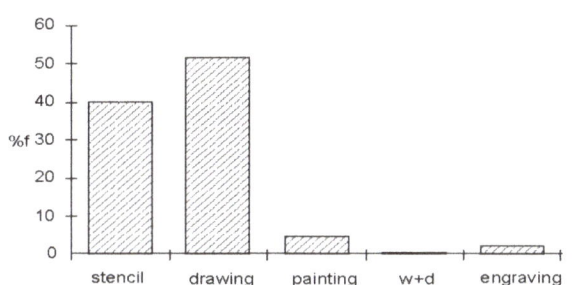

Figure 12.15: *Darug* (North) Language Area. Technical options employed.

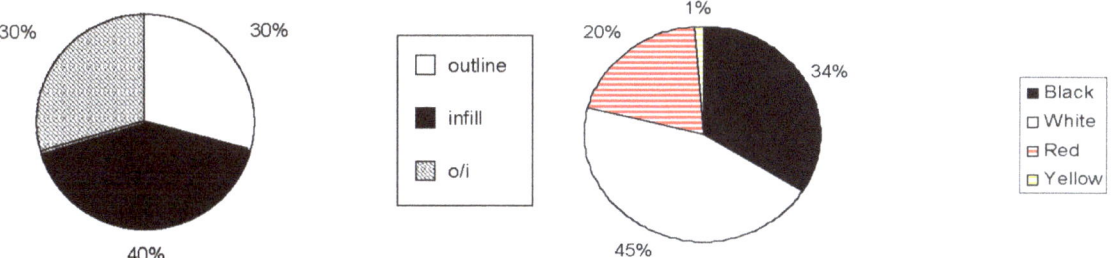

Figure 12.16: *Darug* (North) Language Area depictive motifs. Form.

Figure 12.17: *Darug* (North) Language Area. Colour preferences.

Colour preference in this area is again for white pigment, followed by black, red and yellow (Figure 12.17). Red is more common and white less dominant than in the *Darkingung* sample.

The southern *Darug* sample comprised 90 sites with 1,613 motifs. Only 722 of these were recognisable (55.2% unidentified). The average site size here is 17.9 motifs/site.

Hands again dominate but less so than in the preceding groups (Figure 12.18). Unlike the northern *Darug* sample, the dominant depictive motifs are other land animals followed by macropods and anthropomorphs. Culture heroes and women are present but extremely rare. 'Other material objects' are the most commonly depicted material culture items, followed by boomerangs (Figure 12.19).

Drawing is the most common technique. Stencilling is much less common, while the remaining technical options are uncommon or non-existent (Figure 149). Technical variability is much more limited in this *Darug* group compared with its northern counterpart. Outline and infilled motifs are more common here than outline only motifs. Infilled forms are quite rare (Figure 12.21).

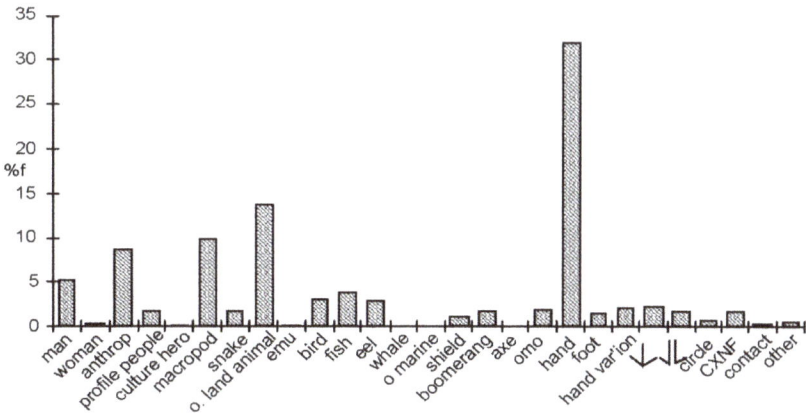

Figure 12.18: *Darug* (South) Language Area. Motif Assemblage.

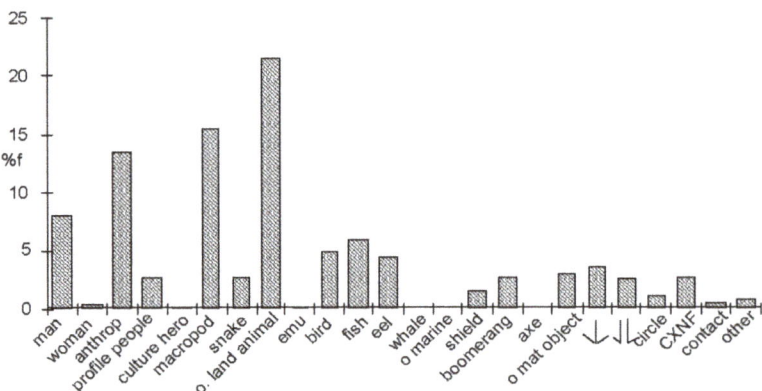

Figure 12.19: *Darug* (South) Language Area. Depictive Motifs.

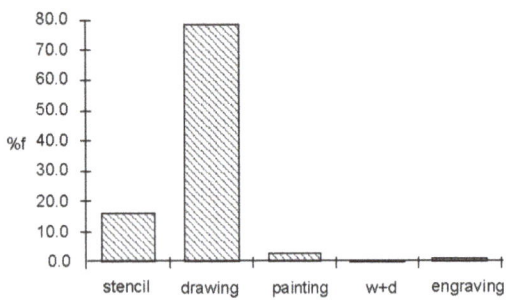

Figure 12.20: *Darug* (South) Language Area. Technical options employed.

Colour preferences here are very different to the more northerly groups. Black is the preferred colour followed by red, white and yellow (Figure 12.22).

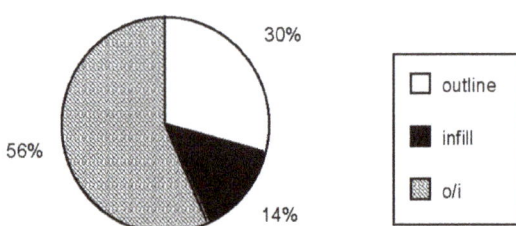

Figure 12.21: *Darug* (South) Language Area. Form.

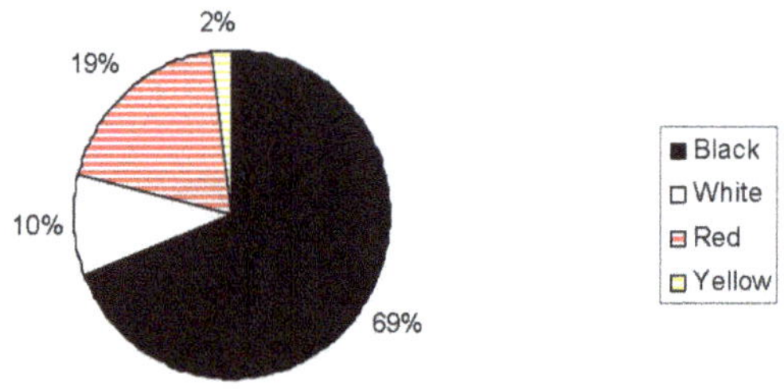

Figure 12.22: *Darug* (South) Language Area. Colour preferences.

Guringai Language Area

This sample represents the most northerly coastal group in the region. A total of 1,504 motifs were recorded from 78 sites here. Just over 38% of this assemblage was unidentifiable; 930 motifs were classifiable. The average site size here is 19.3 motifs/site, considerably smaller than the northern inland groups.

Hands dominate this group (Figure 12.23), while fish and macropods are co-dominant in the depictive assemblage (Figure 12.24). There are no culture heroes or profile people amongst the anthropomorphic figures here. Boomerangs are the most frequently depicted material objects followed by shields.

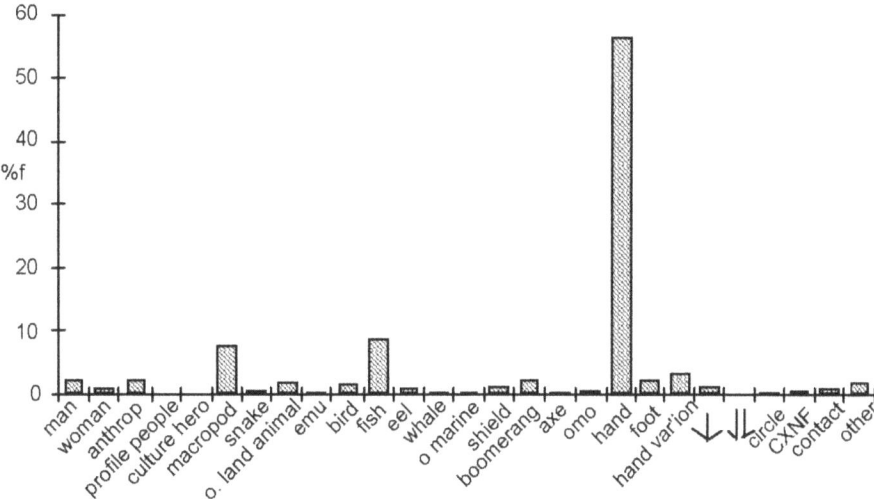

Figure 12.23: *Guringai* **Language Area. Motif Assemblage.**

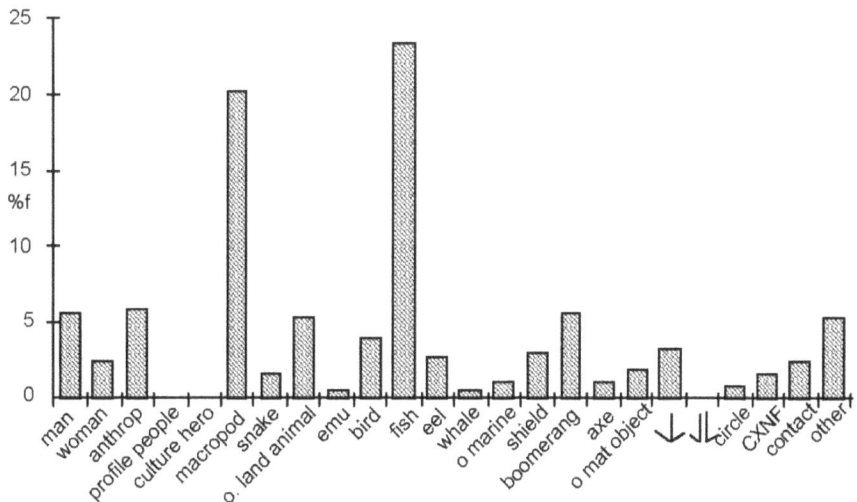

Figure 12.24: *Guringai* **Language Area. Depictive Motifs.**

Drawing and stencilling are co-dominant techniques (Figure 12.25). Painting is relatively common. Outline and infilled motifs are the preferred form followed by infilled and outline-only motifs (Figure 12.26). The preferred colour in this area is black, followed by white and red. Yellow is rarely used (Figure 12.27).

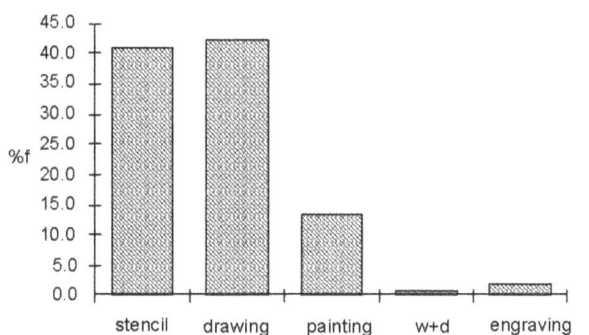

Figure 12.25: *Guringai* **Language Area. Technical options employed.**

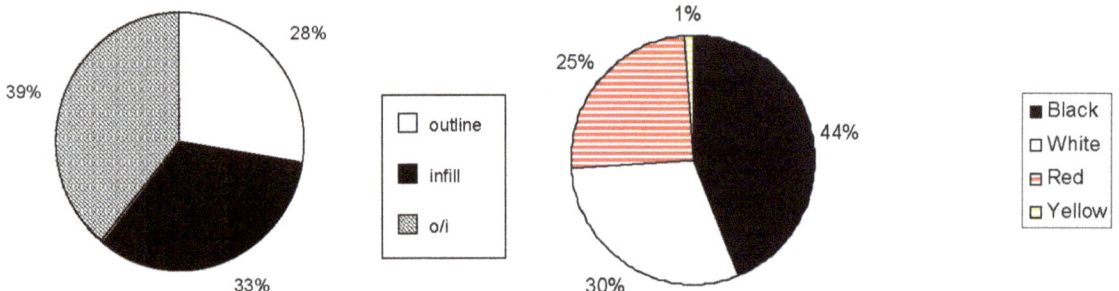

Figure 12.26: *Guringai* Language Area depictive motifs. Form.

Figure 12.27: *Guringai* Language Area. Colour preferences.

Sydney (*Eora*) Language Area

This language group is located, south of the *Guringai* and Port Jackson and north of the Georges River. This group has the smallest sample size (five sites) because of the focus in this area of European settlement, and because of the Cumberland Plain.

A total of 65 motifs were recorded here (averaging 13 motifs/site). Relatively few of the motifs (11%) were unidentifiable: 58 were recognisable.

Hand stencils dominate this assemblage, while the depictive focus is on fish and other marine animals (Figure 12.28, Figure 12.29). There is a much reduced motif classification for this area probably as a result of sample size.

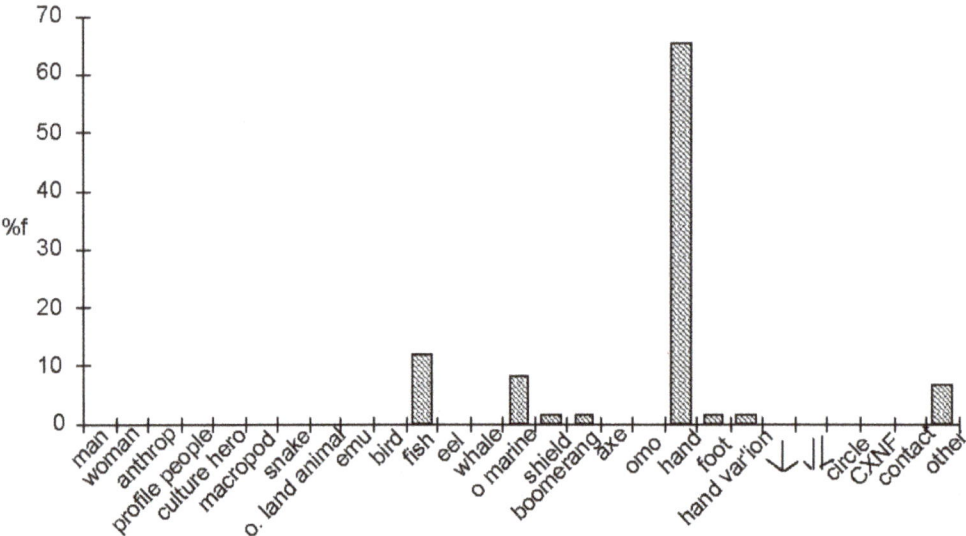

Figure 12.28: *Eora* Language Area. Motif Assemblage.

Chapter 12: Regional synchronic variation: shelter art

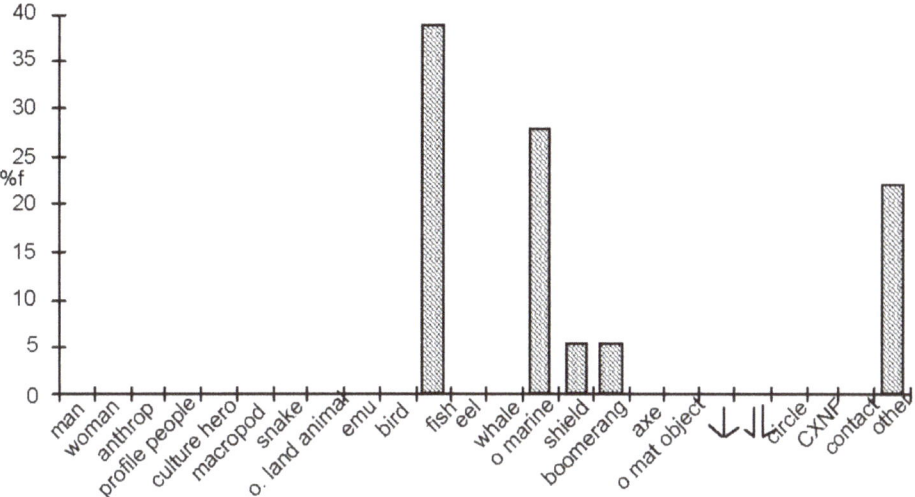

Figure 12.29: *Eora* **Language Area. Depictive Motifs.**

Stencilling dominates this assemblage, with drawing the only other recorded technique (Figure 12.30). Small sample size again makes these observations tentative. There is a preference for outline motifs followed by outline and infilled forms. No infilled-only forms were recorded in this area (Figure 12.31). The preferred colour in this area is white, followed by red, black and yellow (Figure 12.32).

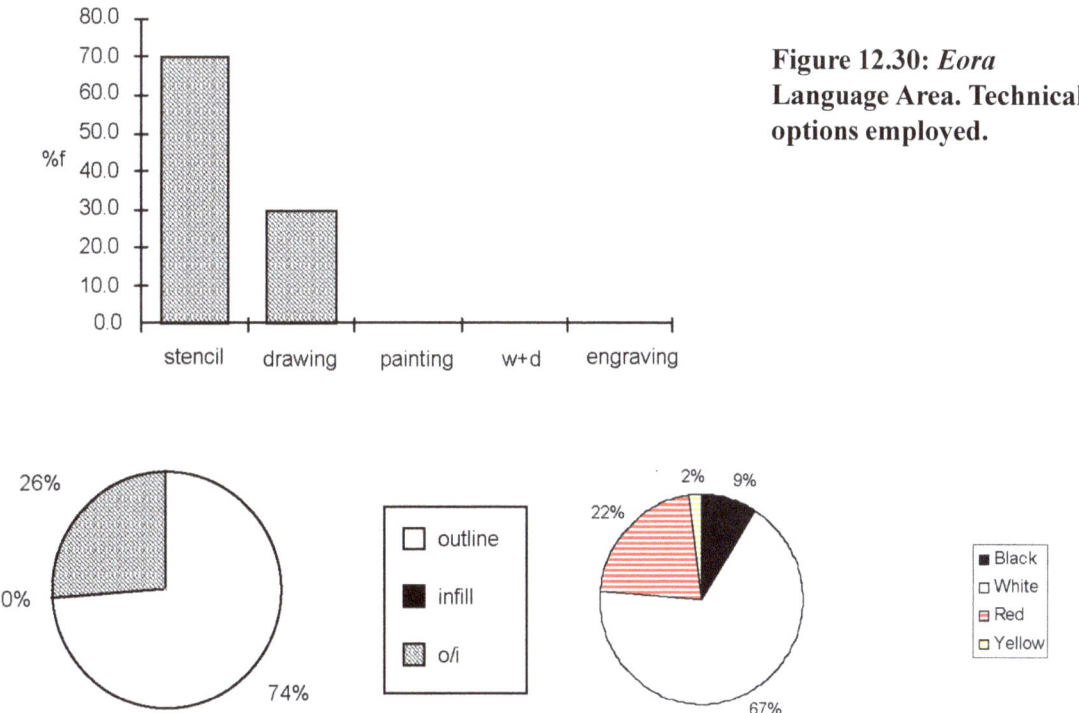

Figure 12.30: *Eora* **Language Area. Technical options employed.**

Figure 12.31: *Eora* **Language Area depictive motifs. Form.**

Figure 12.32: *Eora* **Language Area. Colour preferences.**

Tharawal Language Area

This area is also located largely on the coast, south of the Georges River. A total of 2,387 motifs were recorded here from 99 sites. A high proportion (58%) of this assemblage is indecipherable; there were 1,005 recognisable motifs. The sites here are of average size (24.1 motifs/site).

terra australis 27

305

This is the only area in the region where hands do not predominate the recognisable motifs (Figure 12.33). Macropods and hands are co-dominant. When hands are excluded, macropods dominate the depictive motifs followed by other land animals, anthropomorphs and birds (Figure 12.34). Profile anthropomorphs occur here as commonly as they do in the *Darug* (north and south) assemblages. Women and culture heroes, however, are extremely rare.

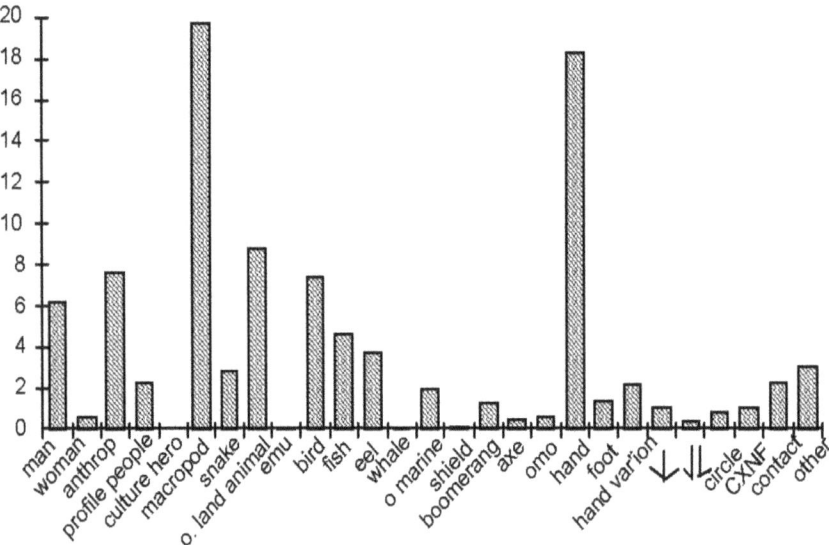

Figure 12.33: *Tharawal* **Language Area. Motif Assemblage.**

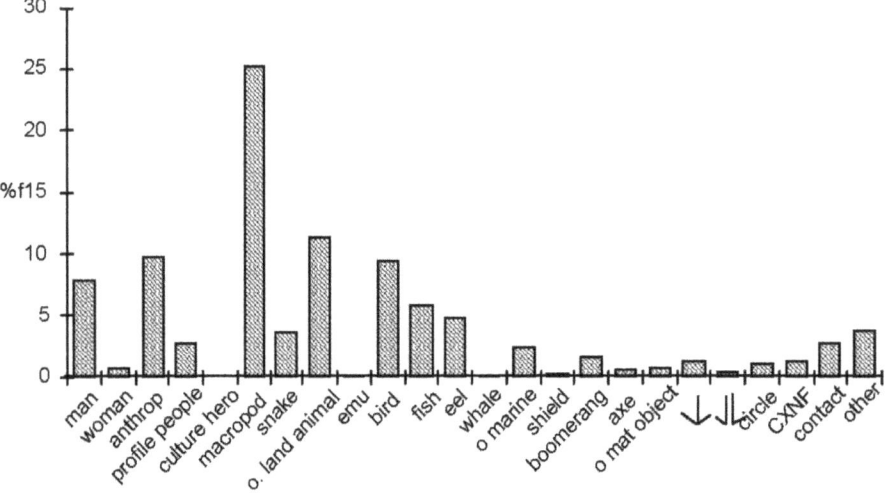

Figure 12.34: *Tharawal* **Language Area. Depictive Motifs.**

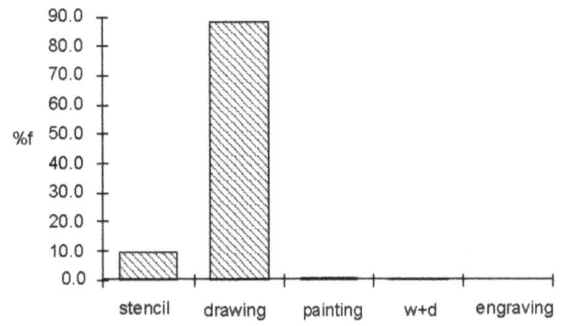

Figure 12.35: *Tharawal* **Language Area. Technical options employed.**

The drawing technique dominates this assemblage, with stencilling relatively uncommon (Figure 164). Painting (alone and in combination) is rare - as is engraving. Outlined motifs and outlined and infilled forms are the most common, while infilled only forms are less common (Figure 12.36). Black is the predominant colour, followed by red, white and yellow (Figure 12.37).

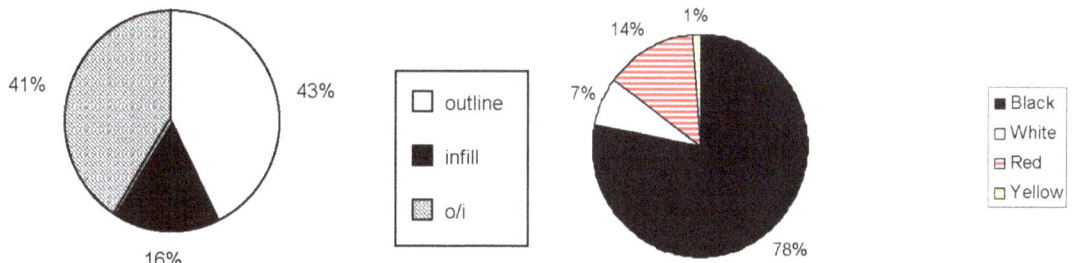

Figure 12.36: *Tharawal* **Language Area depictive motifs. Form.**

Figure 12.37: *Tharawal* **Language Area. Colour preferences.**

Summary

Pigment art assemblages in the *Tharawal* language area are clearly differentiated from all other language areas by the relative absence of hand stencils. There is a predominance of black pigment and scarcity of white pigment in this area, which further reinforces these differences. All other areas have varying levels of similarity and dissimilarity in their motif assemblages.

Despite their locations on the coast, there is no clear focus on marine animals in the *Guringai* or the *Tharawal* areas. Several motif classes are 'missing' from the central area of the Basin but present the northern and southern parts of the region. Profile people are common in the *Darkingung* and northern *Darug* sites and the southern *Darug* and *Tharawal* sites. None of these motifs are found in the *Guringai* or Sydney areas.

Viewing the Darug sites north and south of the Cumberland Plain separately revealed that some bedrock design notions (Sackett 1990) transcend the geographic distance between these two assemblages, while others do not. Both sets of *Darug* sites have a predominance of hand stencils with other land animals, macropods and anthropomorphs dominating (compared with *Tharawal* sites that have few hand stencils and are clearly dominated by macropods). Distance has created some differences in assemblage characteristics between the two sets of *Darug* sites. Birds, 'other' and boomerang motifs dominate in the northern group, while these elements are less important in the southern group. The schematic peculiarity of these southern sites (i.e. the use of four leg on terrestrial animals and two legs on birds, cf. two and one used, respectively, to the north) occurs in both the southern *Darug* and *Tharawal* sites, but not the northern *Darug* sites. This aspect has not been investigated in detail here.

Unidentified motifs were used in this analysis because of the technical information they provide. Without exception this category dominates all shelter art assemblages.

Colour usage in the different language areas reveals definite cultural preferences across the region. This preference does not reflect availability of resources. Charcoal is universally available. White pigment derives from pipeclay (kaolin) commonly found in creeklines around the region and would require only a local knowledge to procure. Red and yellow pigments derive from ironstone bedding within the sandstone formation. While requiring local knowledge to procure, these colours are ubiquitous in their distribution.

In the south of the region there is a definite preference for black pigment and a lesser focus on stencilling. In the north of the region there is a definite focus on white pigment. While this reflects the dominance of stencilling, there are also large numbers of white drawings and paintings in this area. This colour dominance supports a model of contact between the Hunter Valley (where white is prevalent) and this part of the Sydney region (Moore 1981; and see the Mount Yengo excavation report).

In the *Guringai* area, while black dominates, there is a much more use of red and white. Red is commonly used for stencilling here as well as for drawing.

Yellow is only rarely used in all language areas, although it is used more frequently in the *Darkingung* area. There are many sites with yellow stencils in this area, but relatively few drawings and paintings (except in the Warre Warren area: McDonald 1988a).

Correspondence Analysis, Language Areas and Drainage Basins

While trends in the motif assemblages and technical options across the Basin are quite clear, the CA results were used to interpret the significance of compositional differences and technical emphases in the different areas. Language areas and internal drainage basins were analysed, first for motif and then technique.

A total of 25 drainage basins with art sites were defined. The codes used here are the same as those used for the engraved assemblage (Table 12.4 and Table 12.5). The sample sizes here vary markedly compared with the engraved component. The largest sample of shelter sites derives from the *Darkingung* language area. This distribution reflects the work done in the Mangrove Creek Catchment (Attenbrow 1981, 1987, Gunn 1979, McDonald 1988a) and more broadly for the Rock Art Project (McDonald 1987, 1990a).

Table 12.4: Shelter Art sites (motif): Language areas, codes and sample sizes.

Language Group	Code	No. of sites
Darkingung	1	190
Guringai	2	78
Sydney	3	5
Darug	4	97
Tharawal	5	99

As for the engraved component's analysis, the following sample areas were used. These explored:

1) Intra-language area patterning within the *Darkingung* language area (drainage basins 1, 5 and 6);

2) East-west patterning across the proposed *Guringai/Darug* language boundary south of the Hawkesbury River (drainage basins 10 - 13); and,

3) The east-west patterning across the proposed *Tharawal/Darug* language boundary (drainage basins 18 - 21) south of the Georges River.

Shelter Art Motifs

The CA sample for the motif analysis comprised 469 sites. Sites with only unidentifiable motifs were excluded from these analyses. The CA results are shown here, with the sites plotted in their respective drainage basins and language areas.

The bivariate plots show both language areas and drainage basins. Core homogeneity is indicated and the distribution of outlier sites is shown using the quadrant method described.

1) Darkingung Language Area (drainage basins 1, 5 and 6)

This group of sites is north of the Hawkesbury River and includes on the major drainage basins of the Macdonald River and Mangrove Creek. The Upper Macdonald and central Macdonald were distinguished by their position relative to the Bala Range at its centre. The Bala Range forms part of a documented access route (Mathews 1899) along the Boree Track and what is now the Putty Road.

Table 12.5: Shelter Art sites (motif): Drainage Basins, Language Areas and Sample sizes.

Drainage Basin	Basin Code	Lang. Group	No. of sites
Upper Macdonald	1	1	19
Wollombi	2	1	9
Wyong	3	1/2	4/3
Colo	4	1	17/1
Central Macdonald	5	1	54
Mangrove Creek	6	1	84
Mooney Mooney	7	1/2	1/1
Brisbane Waters	8	2	5
Kurrajong	9	1/4	2/3
Cattai	10	1/4	-/6
Berowra	11	2/4	12/20
Cowan	12	2	29
Pittwater	13	2	8
Middle Harbour	14	2	14/1
Lane Cove	15	2/4	6/2
Port Jackson	16	3	4
Botany Bay	17	3	1
Port Hacking	18	5	5
Woronora	19	5	55
Mill/Williams	20	4/5	17/1
Georges	21	4/5	33/7
Nepean	22	4	1
Burragorang/Blue Mtns	23	4*/5	15/10
Cataract	24	4?/5	-/14
Avon/Cordeaux	25	4?/5	-/7

*May be mixture of *Darug* and *Gandangara* Language areas

Darkingung

The 157 sites in this group are homogeneous, with a heavy emphasis on anthropomorphs, terrestrial animals and birds and stencilled hands and weapons (Figure 12.38).

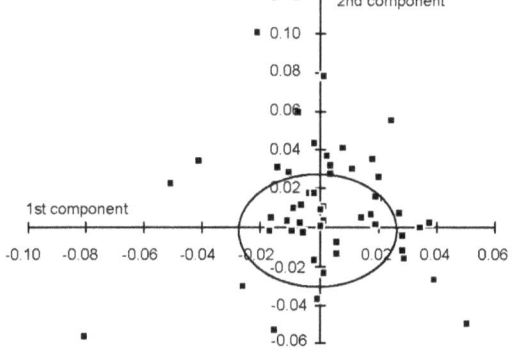

Figure 12.38: *Darkingung* **Language Area. Bivariate plot of CA scores: motifs.**

The three drainage basins show consistency in their core homogeneity (Figure 12.39), but clinal variation in motif focus.

Upper Macdonald

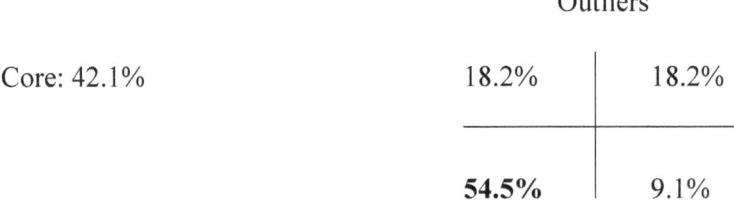

This group of 19 sites is relatively homogeneous with a strong emphasis on hands. Most of the sites in the negative quadrant comprise hand-only sites.

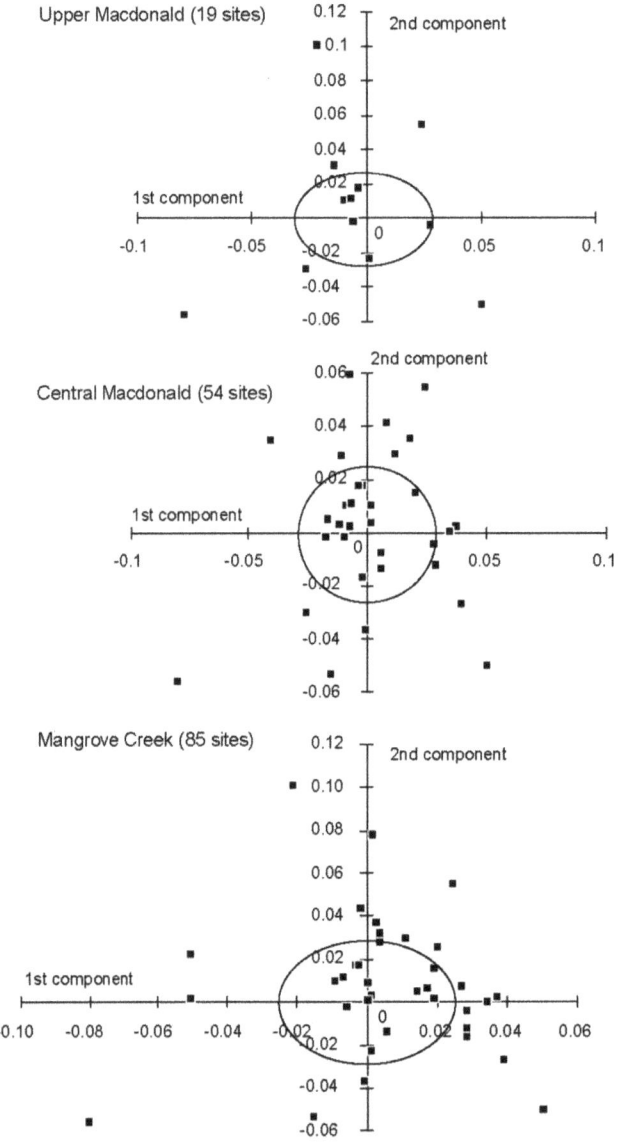

Figure 12.39: *Darkingung* **Language Area: Motif. Bivariate plots for the three drainage basin groupings.**

Central Macdonald

	Outliers	
Core: 44%	10%	22.3%
	33.3%	**33.3%**

This group of 54 sites is also quite homogenous, with the outlier focus on the negative side of the second component. Again there is a strong emphasis on hand stencils and on anthropomorphic figures. Macropods and other land animals figure strongly, as do birds and other material objects. Many sites contain complex-non-figurative motifs.

Mangrove Creek

	Outliers	
Core: 45%	8.5%	25.5%
	12.2%	**53.2%**

This group of sites has a similar degree of homogeneity to the Central Macdonald sites, with a decreasing emphasis on hands (i.e. a shift in focus to quadrant C). Anthropomorphs, terrestrial animals, material objects and birds figure strongly. Eels and fish also occur frequently. Other motifs occur quite often in quadrant C sites and many of these sites include small numbers of hand and weapon stencils.

2) East-west patterning Guringai/Darug language boundary (drainage basins 10 - 13)

The boundary between these two language areas is Berowra Creek. Both banks of this estuarine waterway were surveyed for the Rock Art Project (McDonald 1990b). For the purposes of testing this defined boundary, the sites are divided according to their location on left or right bank of Berowra Creek. This analysis indicates that the sites on either side of the creek have motif differences (Figure 12.40).

The *Darug* sites are more homogeneous than the *Guringai* sites. There is also a change in focus between the outlier sites in the two areas, with more human figures, land animals and birds occurring in the former and more hands occurring in the latter. There is considerable variability within these groups, based on drainage basins.

Darug

	Outliers	
Core: 46%	14.3%	14.3%
	29.6%	**42.8%**

Guringai

	Outliers	
Core: 29%	22.8	8.6%
	51.4%	17.1%

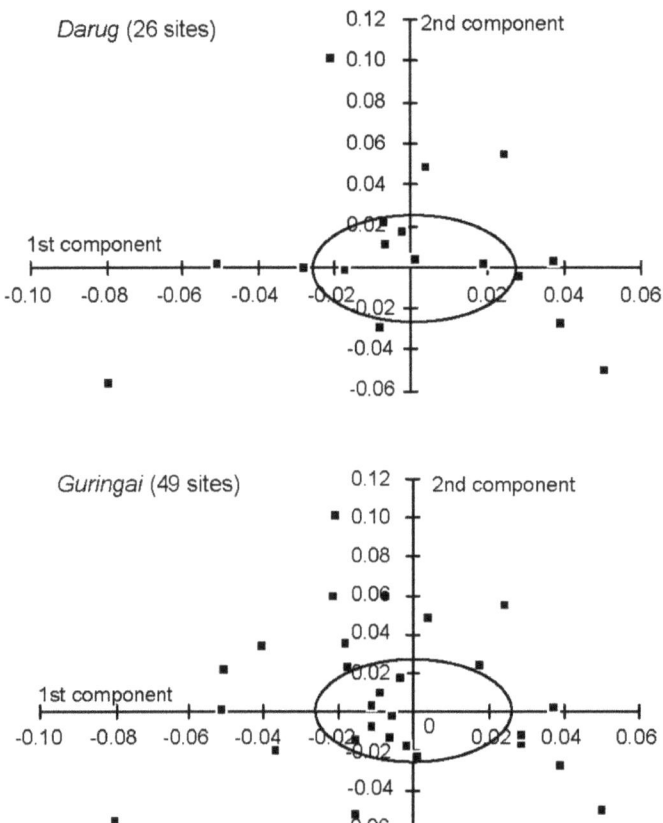

Figure 12.40: *Darug* and *Guringai* **Language Area: Motif. CA sccores.**

Cattai:

	Outliers	
Core: 50%	33.3%	**66.6%**
	0%	0%

This area has only six sites, and its results are thus treated tentatively. Three sites (50%) are in the core zone and all of the outlier sites are on the positive side of the 2nd component (and contain other land animals and other material objects).

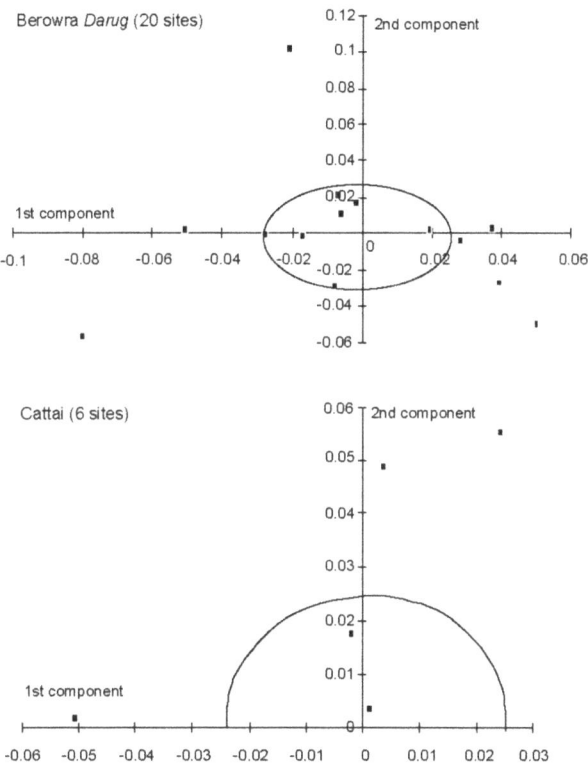

Figure 12.41: *Darug* drainage basins: motifs. Bivariate plot of CA scores.

Darug **Berowra:**

	Outliers	
Core: 45%	9.1%	0%
	36%	**55%**

This group of 20 sites are on the left bank of Berowra Creek. These sites are quite homogenous (Figure 12.41), with the main outlier focus in quadrant C and a minor focus in quadrant D. Quadrant C sites have a focus on terrestrial and anthropomorphic depictions. There are lots of sites with single macropods. Stencils (hand and weapons), land animals and eels dominate quadrant D sites.

Guringai **Berowra:**

	Outliers	
Core: 25%	22%	11%
	33%	**33%**

This group of 12 sites is considerably less homogeneous than those on the western bank of Berowra Creek (Figure 12.42). The focus of its outlier sites is also different: more on the negative side of the second component with hand stencils and eels (quadrant D), and anthropomorphs, macropods and other land animals (quadrant C).

Cowan

	Outliers	
Core: 34.5%	26%	11%
	53%	11%

This group of 29 sites is quite heterogeneous but is focussed on the negative side of the first component. The main compositional focus here is on hands, hand variations and fish (quadrant D) and on marine, terrestrial and other material objects (quadrant A).

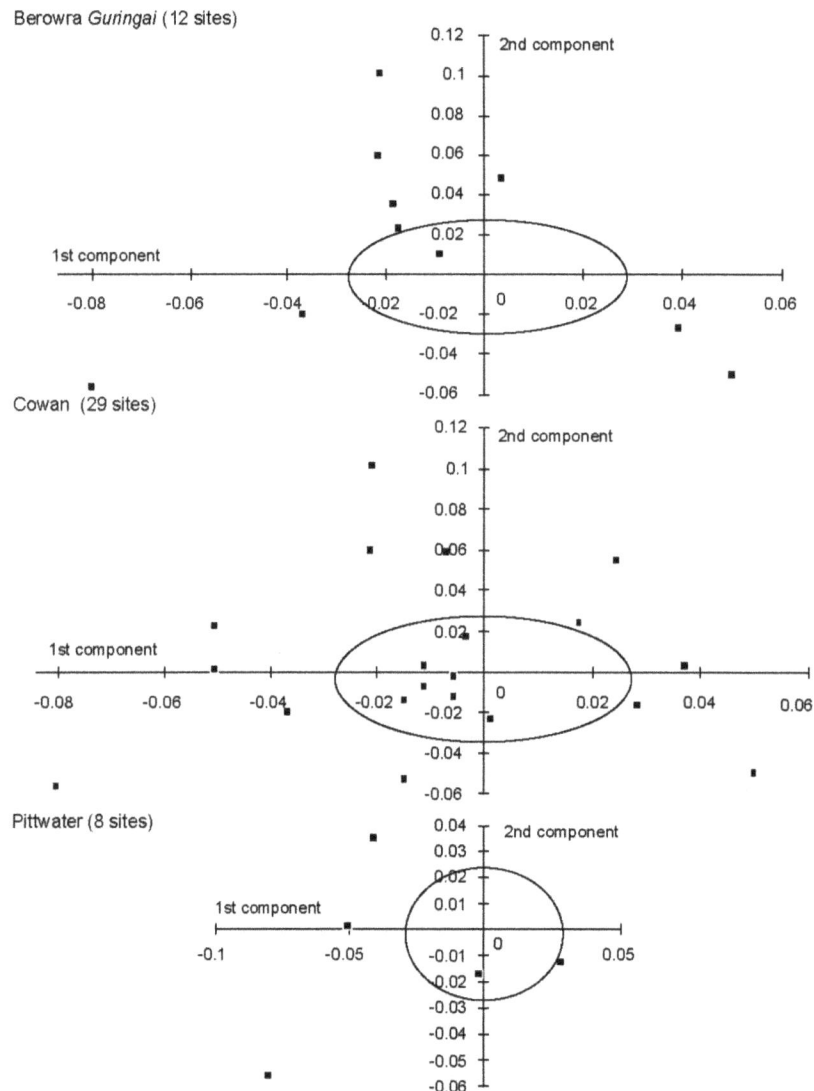

Figure 12.42: *Guringai* **drainage basins: motif. Bivariate plot of CA scores.**

Pittwater

	Outliers	
Core: 12%	14%	0%
	71%	14%

Chapter 12: Regional synchronic variation: shelter art

This small group of eight sites is the least homogenous of all those analysed in this area. The focus of this group is heavily on hands and other motifs. The Great Mackerel site with a large assemblage of mainly hand and material object stencils is the outlier site in quadrant A.

3) The Tharawal language area (Basins 18 - 21)

This group involves 118 sites south of the style boundary at the Georges River. These sites fall within Capell's designated *Tharawal* and *Darug* language group areas. The shelter art sites are mainly from the Georges River and Woronora catchments, unlike the engraving sites - which have a coastal focus.

Tharawal

	Outliers	
Core: 33.8%	6.7%	13.3%
	15.5%	**64.4%**

This group of sites is one of the least homogenous of those analysed according to language area (Figure 12.43).

Darug

	Outliers	
Core: 28%	5.5%	11.1%
	19.4%	**63.9%**

This group of sites is the least homogenous of those analysed according to language area. There are a few sites with hand stencils in this area, and a definite focus here is on animals and birds.

Darug Georges River

	Outliers	
Core: 27.3%	8.3%	16.7%
	16.7%	**58.3%**

The group of sites in this drainage basin is one of the least homogenous analysed. There are a few sites with hand stencils, but a definite focus on terrestrial animals and birds.

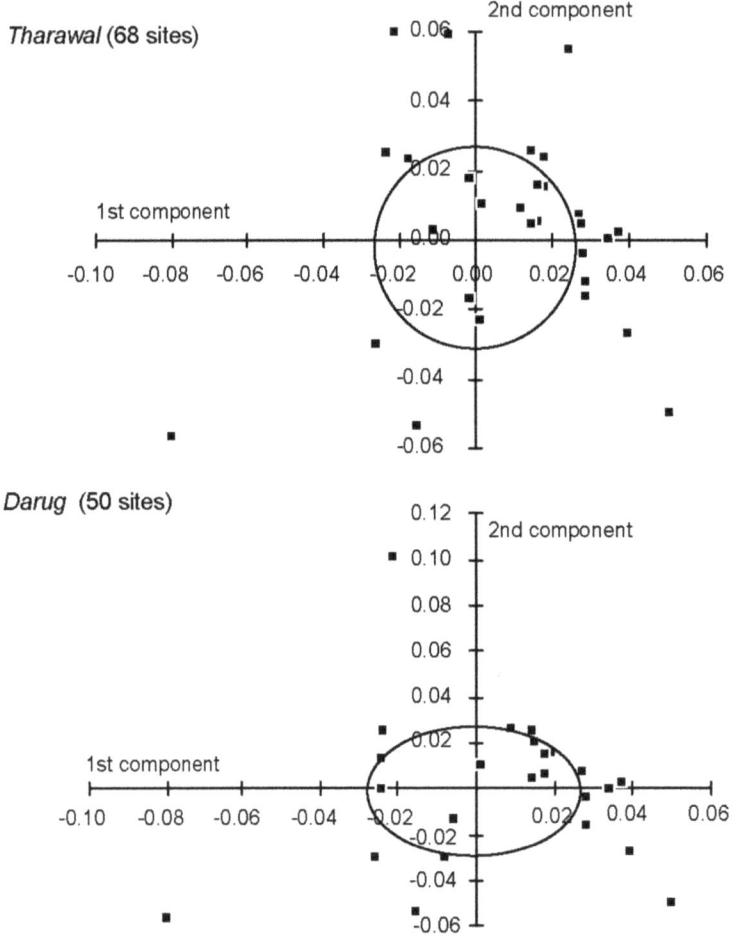

Figure 12.43: *Tharawal* and *Darug* language areas: motif. Bivariate plot of CA scores.

Darug Mill and Williams

This group of sites is slightly more homogenous than its neighbouring *Darug* drainage basin but with less diversity in motif preference. There is a definite focus on terrestrial animals, anthropomorphs and birds.

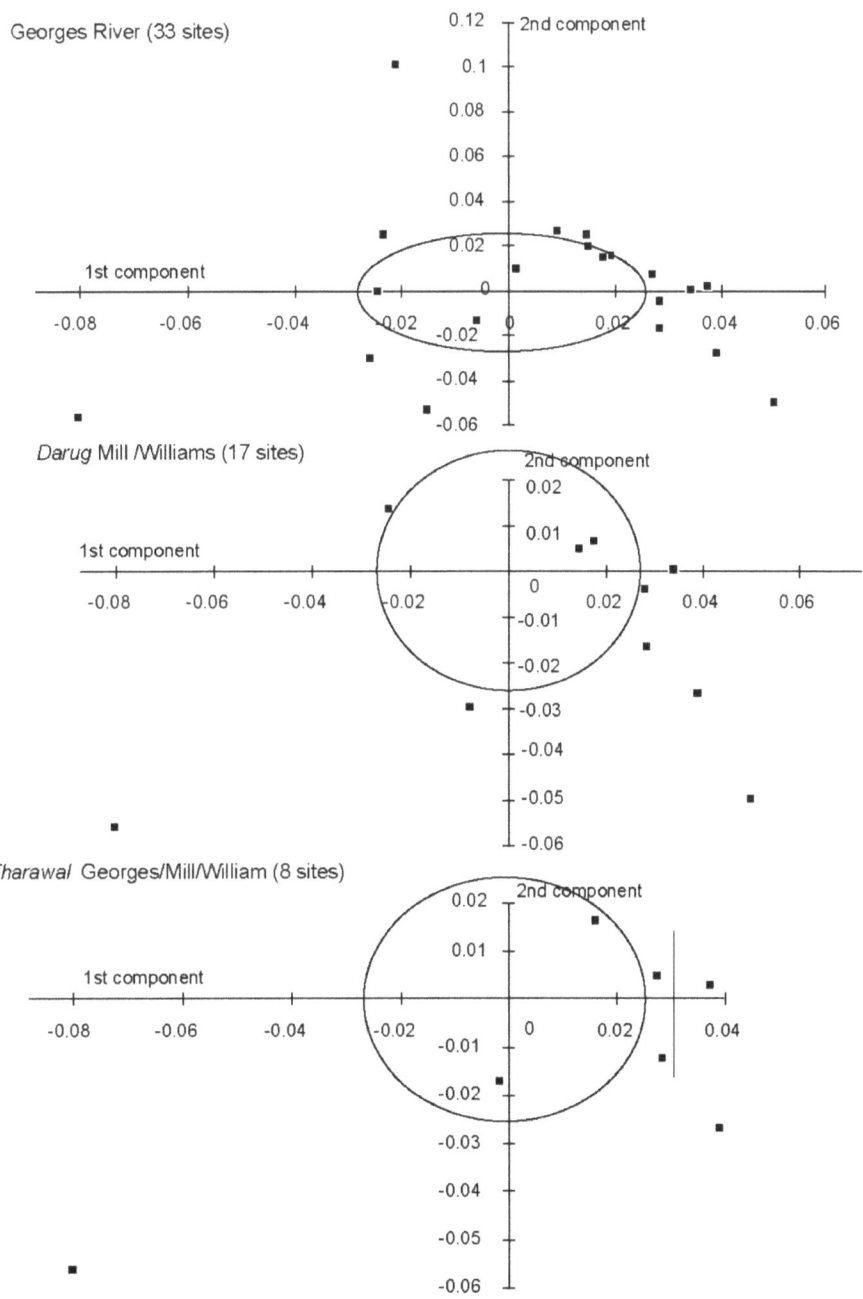

Figure 12.44: *Tharawal* and *Darug* drainage basins: motif. Bivariate plot of CA scores.

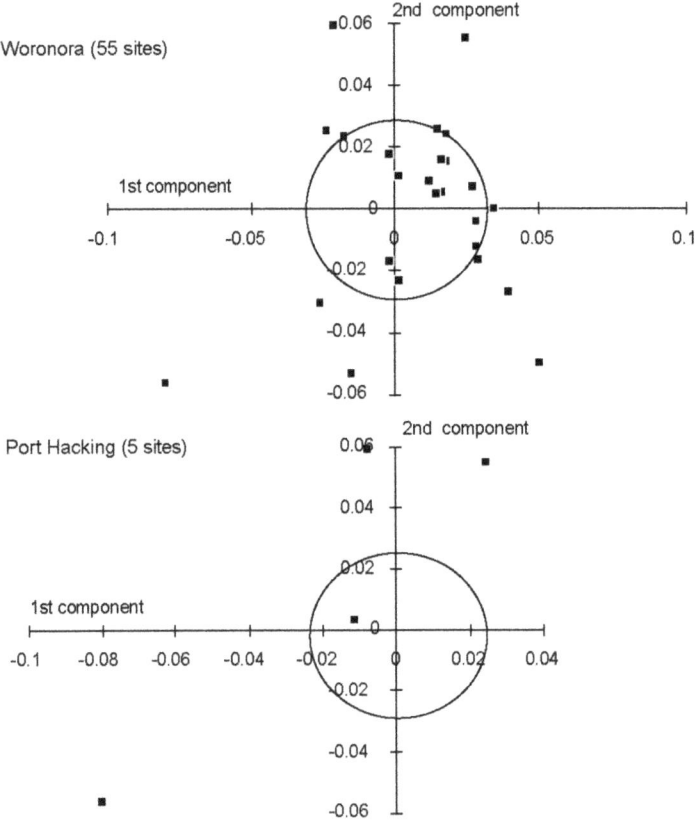

Figure 12.45: *Tharawal* drainage basins: motif. Bivariate plot of CA scores.

Tharawal **George, Mill and Williams**

	Outliers	
Core: 37.5%	0%	20%
	20%	**60%**

There are too few sites here to meaningfully discuss these results (these are included in the larger language group discussion).

Woronora

	Outliers	
Core: 34.5%	5.6%	11.1%
	11.1%	**72.2%**

This group of sites is more homogenous than the neighbouring *Darug* drainage basins. While there is slightly more diversity in motif preference here, the focus is definitely on terrestrial animals, anthropomorphs and birds.

Port Hacking

	Outliers	
Core: 20%	25%	25%
	50%	0%

This group is highly heterogeneous and has a different outlier focus to the other *Tharawal* sites. While several sites have hand stencils, there is a definite focus on fish and other marine depictions. The very small number of sites (n=5) makes conclusions regarding this area difficult.

Summary

Exploring how pigment motifs vary according to defined drainage basins and language areas has again revealed a mosaic of stylistic heterogeneity. The regional core of stylistically homogeneous sites appears to be in the *Darkingung* and northern *Darug* areas. The *Guringai* and southern *Darug* sites are the least homogenous, while the *Tharawal* sites are different again.

Subdividing the language areas into drainage basins provided further insight into localised variability. In the *Darkingung* area, all three drainage basins reveal very similar levels of homogeneity. There is clinal variation here motif preference with a focus on hands in the Upper Macdonald; hand stencils, terrestrial animals and birds in Central Macdonald; and anthropomorphs, terrestrial animals, birds and then hand stencils in Mangrove Creek.

The purported *Darug/Guringai* language boundary south of the Hawkesbury River was investigated. As with the engraved component, a strong separation between sites on either side of Berowra Creek was discovered, supporting the presence of this linguistic boundary. As was also found in the engraving assemblage, similarities between the *Darug* Berowra sites and the *Darkingung* Mangrove Creek sites are striking.

The southern *Darug* and *Tharawal* sites are also highly heterogeneous. Both southern *Darug* drainage basins demonstrate consistently high levels of heterogeneity and similar motif preferences. The *Tharawal* sites however are the most varied of the southern drainage basin with a focus on macropods, other land animals and birds.

Shelter Art Technique

All 564 shelter art sites were used for these analyses. The same drainage basin and language area divisions are used in these analyses as described above.

1) Darkingung language group (drainage basins 1, 5 and 6).

Darkingung

	Outliers	
Core: 68.9%	**32.1%**	17.9%
	17.9%	**32.1%**

The sites in this group are relatively homogeneous (Figure 12.46) with a dual emphasis in the outlier sites on engraved motifs and black outlined and infilled motifs (quadrants A and C). There is clinal variation in techniques used between the upper Macdonald and Mangrove Creek groups.

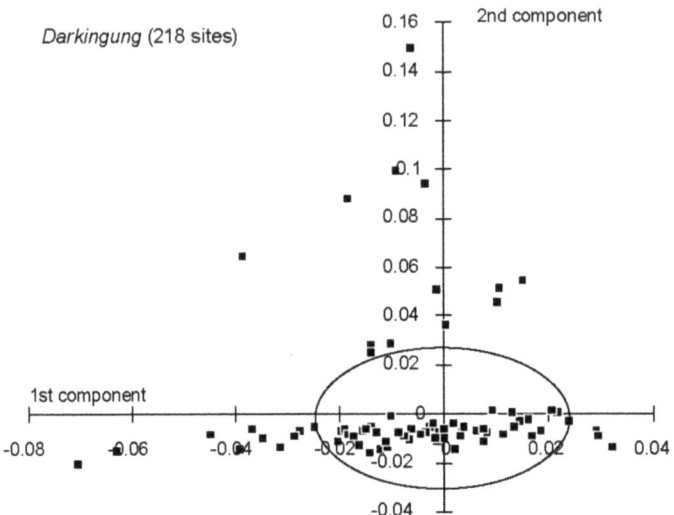

Figure 12.46: *Darkingung* **language area: technique. Bivariate plot of CA scores.**

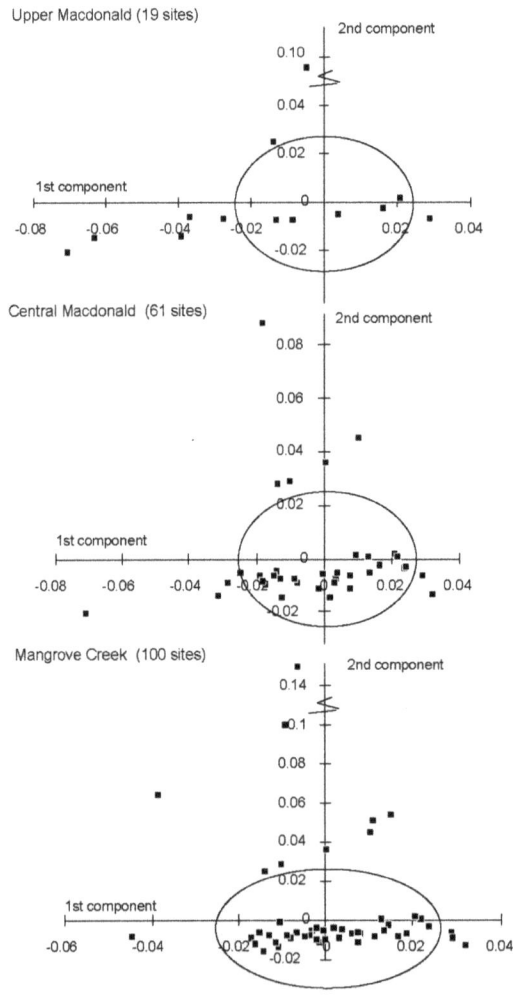

Figure 12.47: *Darkingung* **drainage basins: technique. Bivariate plot of CA scores.**

Chapter 12: Regional synchronic variation: shelter art

Upper Macdonald

Outliers

Core: 42.1%	36.4%	0
	54.5%	9.1%

This group is relatively homogeneous with the main emphasis on white stencils. A number of sites in this group have engraved motifs.

Central Macdonald

Outliers

Core: 78.7%	**30.8%**	**30.8%**
	23.1%	15.4%

This group of sites is very homogenous with the technical emphasis in outlier sites on engravings, white pigment and stencils.

Mangrove Creek

Outliers

Core: 68%	**31.3%**	18.8%
	3.0%	**46.9%**

These sites are relatively homogeneous but with a decreased emphasis on stencils and white pigment, and increased use of black and red pigments. A number of sites in this group have engraved motifs (quadrant A).

2) East-west patterning Guringai/Darug language boundary (drainage basins 10 - 13)

This analysis of technique yet again indicates that the sites on either side of the creek are different.

Darug

Outliers

Core: 70.4%	12.5%	12.5%
	0	**75%**

Guringai

                                    Outliers

         Core: 52.5%   20.7% | 13.8%
                                -------+-------
                                51.7% | 13.8%

The *Darug* sites are much more homogeneous than the *Guringai* sites (Figure 12.48). There is a change in outlier focus between the two areas, with more black pigment in the *Darug* sites and more stencils and white pigment in the *Guringai* sites. The drainage basins indicate a complex mosaic of technical options being used. Engraved motifs occur in both areas, but less so in the Darug assemblage.

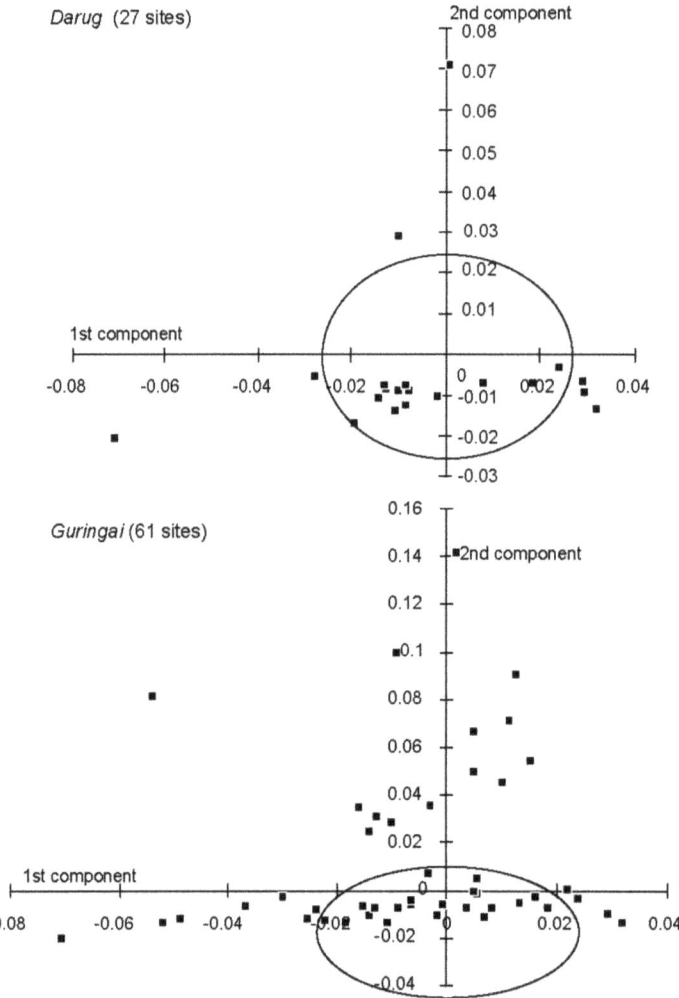

Figure 12.48: *Darug* and *Guringai* **language areas: technique. Bivariate plot of CA scores.**

Cattai

	Outliers	
Core: 83.3%	0%	**100%**
	0%	0%

This area has a very low number of sites and thus the results are treated tentatively. Five of the six sites are in the core zone and the only outlier site is in quadrant B.

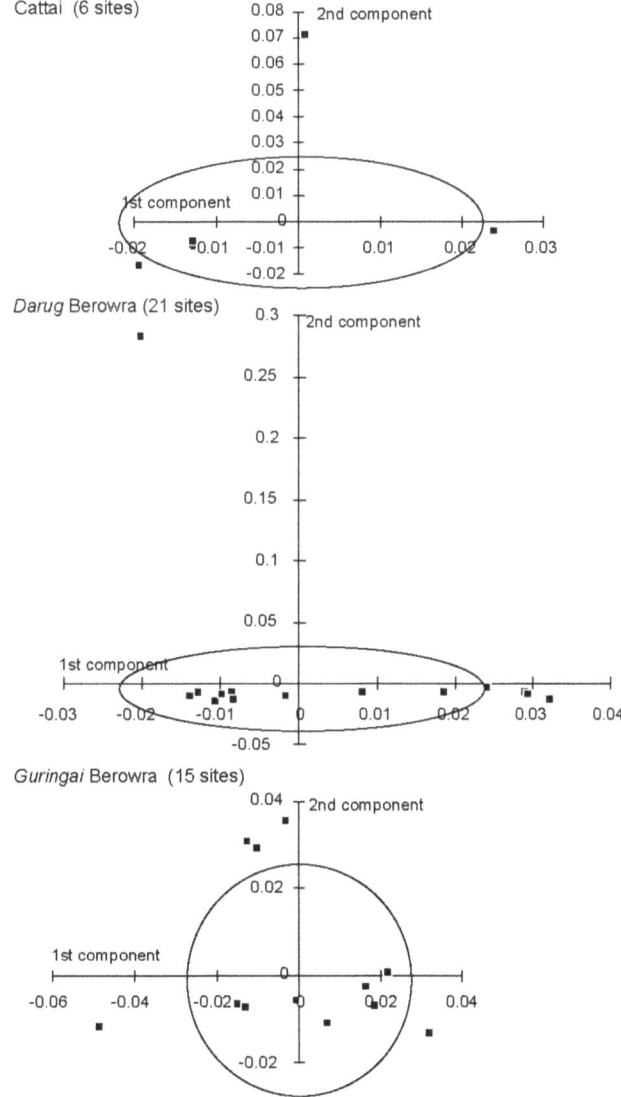

Figure 12.49: Cattai and Berowra drainage basins: technique. Bivariate plot of CA scores.

Darug **Berowra**

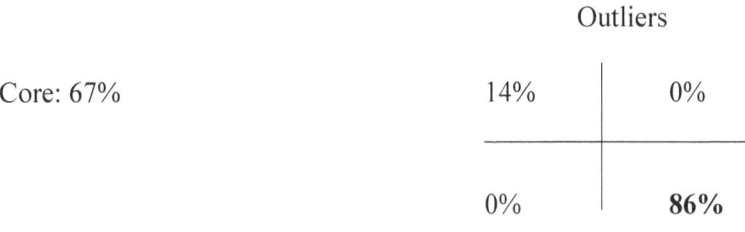

	Outliers	
Core: 67%	14%	0%
	0%	**86%**

This group of 21 sites are on the left bank of Berowra Creek. These sites are also very homogenous although less than the Cattai sites. The main outlier focus is in quadrant C (i.e. black pigment). The single outlier site in quadrant A has an engraved motif.

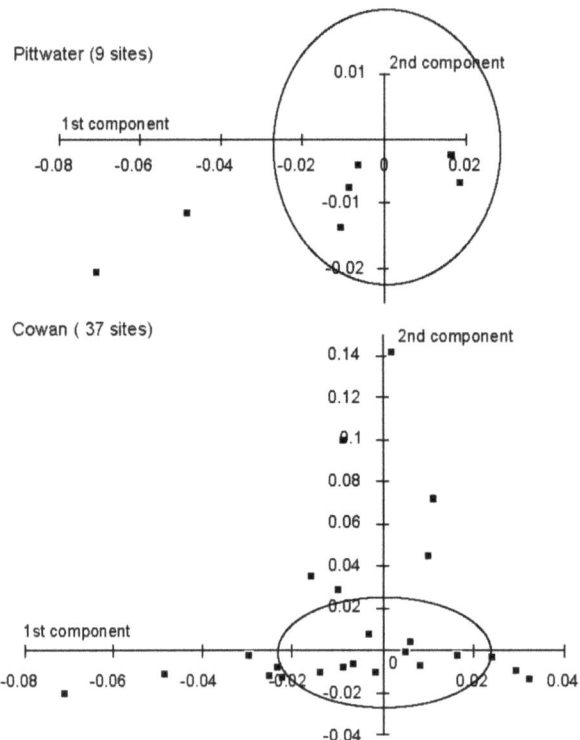

Figure 12.50: Pittwater and Cowan Drainage basins: technique. Bivariate plot of CA scores.

Guringai **Berowra**

	Outliers	
Core: 60%	**50%**	0%
	16.7%	33.3%

This group of 15 sites is slightly less homogeneous than those on the western side of this drainage basin but the technical emphasis of its outlier sites is completely different. There are many more stencils here and a number of sites with engraved motifs (quadrant A).

Cowan

	Outliers	
Core: 48.6%	15.8%	21.1%
	52.6%	10.5%

This group is quite heterogeneous and is focussed in the D quadrant with a minor focus in the B quadrant. The technical emphasis here is on white and red stencils (quadrant D). A number of sites have engraved motifs (quadrant A).

Pittwater

Outliers

Core: 55.6%	0%	0%
	100%	0%

This group is relatively homogenous although the small sample size here (n=9) is noted. The technical emphasis in this group is on red and white stencils. There are no sites with engraved motifs.

3) Tharawal and Darug language areas (Basins 18 - 21)

This group of 152 sites south of the Georges River span the designated boundary between the *Tharawal* and *Darug* language groups.

Tharawal

Outliers

Core: 61.4%	0%	2.9%
	9.8%	**88.2%**

This group of sites is technically homogenous with a focus on black drawings. A few outlier sites have red and white stencils. There are no engraved motifs in this area.

Darug

Outliers

Core: 64.1%	0%	0%
	9.7%	**91.3%**

This group of sites is fairly homogenous with a technical emphasis on black drawings. A few outlier sites have stencils (and red and white pigment). There are no engraved motifs. The technical emphases and core homogeneity in these two areas are very similar (Figure 12.51). A language boundary between these groups is not supported.

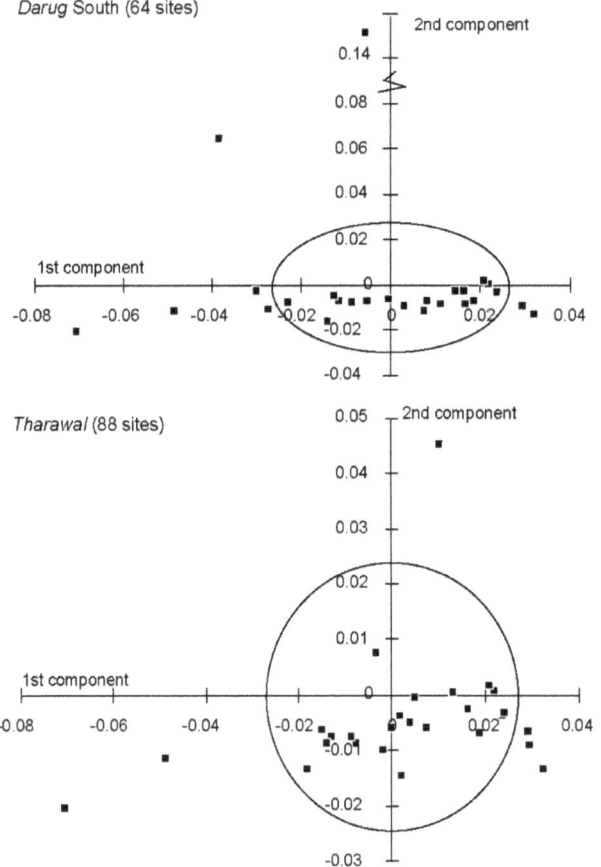

Figure 12.51: *Darug* and *Tharawal* Language Areas: technique. CA scores.

Darug Georges River

	Outliers	
Core: 63.4%	0%	10%
	7%	**93%**

The technical variables here are relatively homogenous with a definite outlier focus on black drawings.

Darug Mill and Williams

	Outliers	
Core: 65%	0%	0%
	13%	**88%**

These sites are relatively homogeneous and like the other southern *Darug* group has an outlier focus on black drawings. One outlier site has numerous red and white stencils.

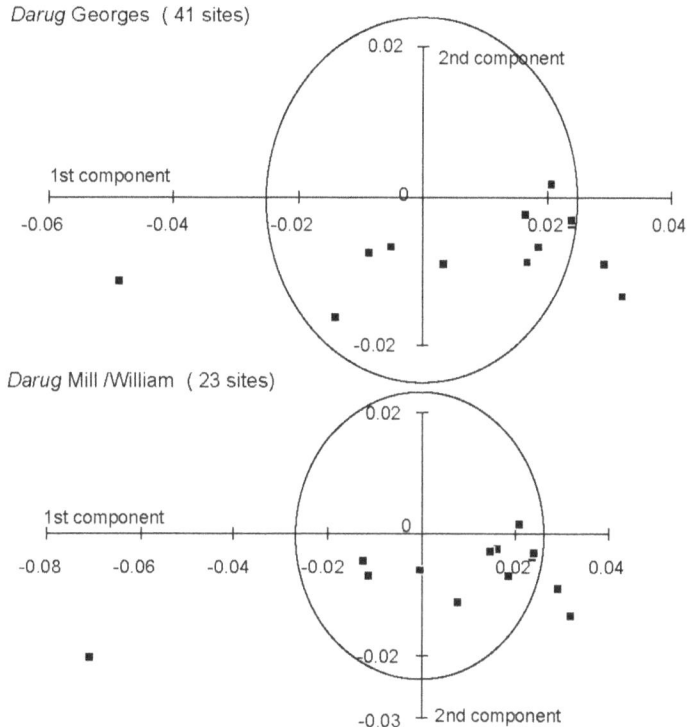

Figure 12.52: *Darug* Georges and Mill/Williams drainage basins. Bivariate plot of CA scores.

Tharawal **George, Mill and Williams**

	Outliers	
Core: 57%	0%	0%
	0%	**100%**

This small sample of sites here are less homogeneous but the outlier emphasis is still on black drawings.

Woronora

	Outliers	
Core: 63%	0%	4%
	4%	**92%**

This group is also quite homogeneous but again there is a strong focus on black drawings. A few sites have stencils only (including yellow ones: quadrant D) and several have engravings.

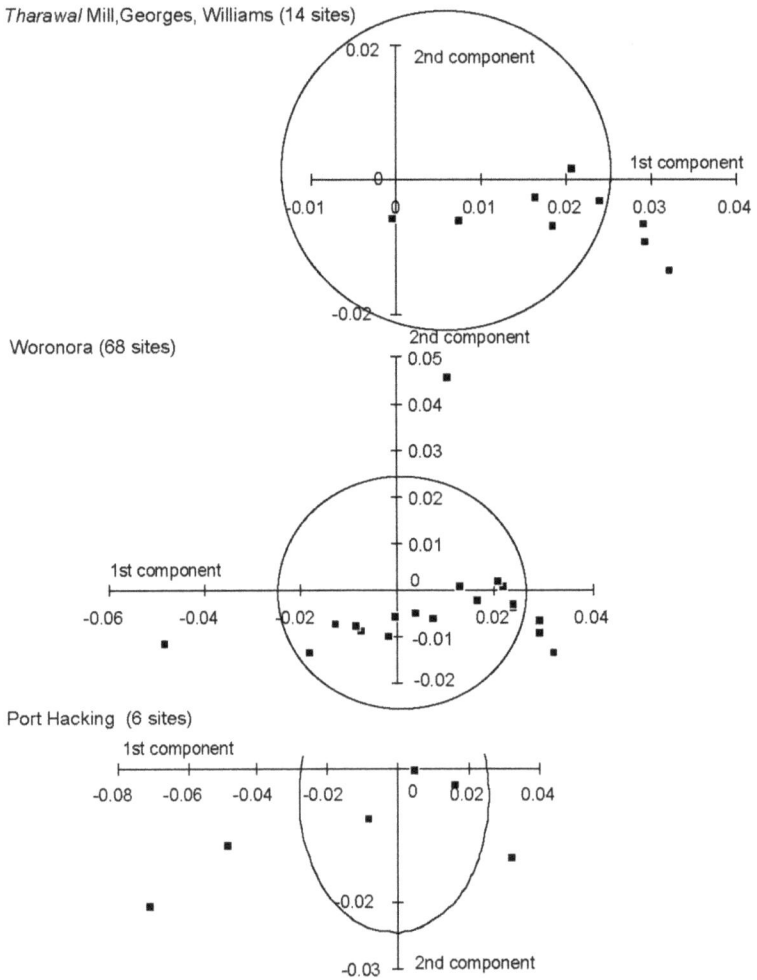

Figure 12.53: *Tharawal* drainage basins: technique. Bivariate plot of CA scores.

Port Hacking

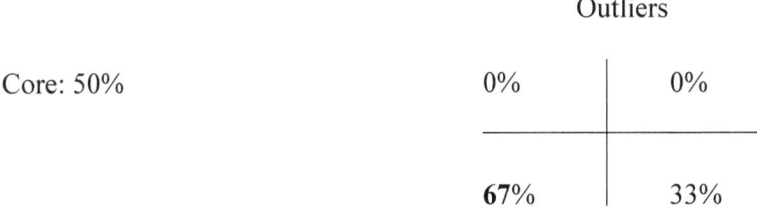

This group is less homogeneous than the other *Tharawal* sites, but the small sample size makes conclusions difficult. This group has a different outlier focus to other *Tharawal* groups. The two outlier sites in quadrant D contain stencils in red and white pigment.

Summary

The shelter sites demonstrate more homogeneity on the basis of technique variables than was found with motif preference. The *Darkingung* and northern *Darug* sites are the most homogeneous. The southern *Darug* and *Tharawal* sites demonstrate relatively high and similar levels of technical variability. The *Tharawal* and southern *Darug* sites show consistent levels of homogeneity. The *Guringai* sites are the most heterogeneous.

A general trend, from north to south, is demonstrated in the use of white pigment and charcoal. In all but *Darkingung* and *Guringai* areas the emphasis is on black drawings. In the *Guringai* area the use of red pigment influence the technical diversity. In the *Darkingung* area, engraved motifs play a part in the local diversity.

Internal variability was demonstrated across the different drainage basins. This was marked in the *Darkingung* sample, with the Upper Macdonald sites being the most heterogeneous in the region but the Central Macdonald group being one of the most homogeneous. This disparity in technique is marked, particularly in light of the highly consistent motif homogeneity and outlier foci in these two locations.

There is less disparity displayed by sites on either side of Berowra Creek using technique variables, although the outlier focus on either side of the creek is markedly different. Again, the *Darug* Berowra sites are similar to the Mangrove creek sites in levels of overall homogeneity - although more engraved motifs are found in the latter.

Rare Motifs

Rare and unique motifs were analysed to establish their geographic distributions in the hope that this would elucidate localised stylistic traits. Analysis concentrated on non-economic motifs in an effort to reduce environmental influences.

The number of times that any individual motif occurred at any shelter site in the region demonstrated some interesting results (Table 12.6; Figure 12.54).

Table 12.6: Shelter Art Motif totals. Maximum motif incidence, number of sites in the region with motif present, and *%f* of sites with motif.

Motif	Total	Max incidence	Sites with Motif present	% of Sites with Motif
Man	244	34	86	16.0
Woman	104	7	50	9.1
Anthropomorph	552	26	154	29.0
Profile Person	81	9	31	5.6
Culture Hero	18	6	9	1.6
Macropod	803	45	219	40.0
Snake	176	22	84	15.0
Other Land Animal	473	17	161	29.0
Emu	36	6	21	3.8
Other Bird	320	38	98	19.0
Fish	206	21	77	14.0
Eel	155	16	75	14.0
Whale	3	1	3	0.5
Other Marine	34	5	20	3.6
Shield	54	7	34	6.2
Boomerang	183	33	67	12.0
Axe	82	14	29	5.3
Other material object	117	12	50	9.1
Hand	3,588	417	206	37.4
Foot	69	9	36	6.5
Hand variation	609	79	86	16.0
Bird track	89	13	39	7.1
Roo track	24	5	12	2.18
Circle	81	30	24	4.4
CXNF	126	12	59	11.0
contact	45	22	16	2.9
other	266	43	63	11.0

The most frequently depicted motifs did not always occur at the most site locations. Macropod (40%) are found at most sites in the region followed by hands (37.4%), other land animals and anthropomorphs (c.29%). Unlike the engraved *mundoes* which are a common motif but found at relatively few sites, hand stencils are ubiquitous. While not having the same prevalence as hand stencils, macropods occur widely in pigment sites given their overall numerical contribution (9.4% of the identifiable motifs).

The analysis of the maximum number of times that a particular motif occurs in a site shows that there are some shelter sites in the region with large numbers of certain motifs (i.e. 417 hands at Yengo 1; 45 macropods at #45-3-917, etc.: see Table 12.6). The geographic distribution of sites with some of the rarer motifs was plotted, as were the language groups into which these fall (Table 12.7).

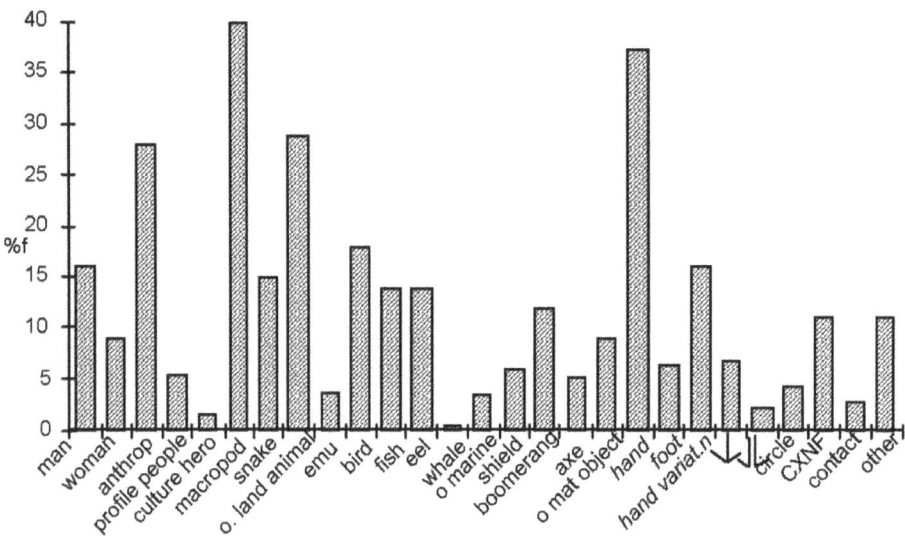

Figure 12.54: Occurrences (%f) of sites with particular motifs.

The distribution of the rare motifs were also plotted (Figure 12.55 to Figure 12.58). These illustrate the focus for particular motifs in some areas and the relative absence of these same motifs from other areas. Women and snake motifs represent a proportionally higher contribution to the *Darkingung* sites, while profile people occur much more commonly than elsewhere in the *Tharawal* sites. Similarly, feet and contact motifs occur much more frequently in the *Guringai* sample than elsewhere, as do kangaroo tracks in the *Darug* assemblage and other marine animals in the Sydney assemblage.

'Other marine' themes occur relatively infrequently in the *Darkingung* and *Darug* sites. All anthropomorphic depictions are either rare or completely absent from the *Guringai* sample. Culture heroes and snakes are absent or extremely rare in the *Tharawal* sites.

An approximate randomization method (Noreen 1989, Wright 1992) was used on the Table 12.7 to test the statistical significance of these differences. Taking a probability level of 0.05 as statistically significant, the following significant differences are identified between language areas (Table 12.7).

There are statistically significant differences between the *Darkingung* language area and all other language areas. This is least strongly demonstrated between the *Darkingung* and *Guringai* groups (Table 12.8).

There are also statistically significant differences between the *Guringai* sites and those from the *Darug* and *Tharawal* areas. The most significant of these is between the *Guringai* and *Tharawal* groups. Interestingly the only significant difference recorded for the *Eora* sites is with the *Darkingung* sites (sampling is likely to be implicated in this result). There is **not** a significant difference between the *Darug* and *Tharawal* assemblages.

Table 12.7: Rare Shelter Art Motifs: Distribution per Language Area. Statistically significant results in red/bold.

Motif	Number (and %) of Sites with motif in each Language Area									
	Darkingung		*Guringai*		*Eora*		*Darug*		*Tharawal*	
Woman	**34**	**68**	**4**	**8**	0	0	6	12	6	12
Profile person	14	45	0	0	0	0	6	19	**11**	**35**
Culture hero	6	67	0	0	0	0	3	33	0	0
Snake	**15**	**71**	**2**	**0**	0	0	3	14	1	4.8
Other marine	**2**	**10**	3	15	**3**	**15**	1	5	**11**	**55**
Shield	18	53	6	18	1	2.9	7	21	**2**	**5.9**
Axe	19	66	3	10	0	0	5	17	2	6.9
Roo tracks	**3**	**25**	0	0	0	0	**5**	**42**	**4**	**33**
Bird tracks	16	41	5	13	0	0	12	31	6	15
foot	12	33	**10**	**28**	1	2.8	9	25	4	11
CXNF	37	63	3	5.0	0	0	10	17	9	15
contact	7	44	**4**	**25**	0	0	3	19	2	13
Total motif sample	7725	52.9	1504	10.3	65	0.4	2910	20.0	2387	16.4

Table 12.8: Shelter art motifs. Significant values achieved for rare motifs in the five language areas.

Language Areas compared		Significance value
Darkingung	*Guringai*	.004
Darkingung	*Eora*	<.001
Darkingung	*Darug*	<.001
Darkingung	*Tharawal*	<.001
Guringai	*Darug*	.030
Guringai	*Tharawal*	<.001

These analyses confirm general stylistic clines across the region as well as significant localised differences in the use of rare motifs. A comparison of these differences with those found in the engraved motif assemblage will contribute to the understanding of stylistic patterning in the region (chapter 3).

Composition

Here the compositional details of two rare motif categories (shields and culture heroes) are analysed. These same two motifs were analysed in the engraved assemblage. The pigment motifs demonstrate less compositional 'rigour' than the engraved component. Officer (1984) argued that this was due to pigment art being 'less culturally fettered'. The fact that the techniques used are usually freehand drawing no doubt also contributes to a greater flexibility in the graphic vocabulary used in this art form.

Shields

The 34 sites with 54 painted, drawn or stencilled shields were analysed (Figure 12.56). It was hoped that this analysis of pigment shield designs may contribute further to determining the interrelatedness of contacts around the region - as shown by engraved shields (see chapter 11).

Much less design structure was identified with the pigment shield motifs. The vast majority (83%) consist of either a simple outline or solid internal infill. In one site (Figure 5.21) there is a stencilled parrying shield (with no decorative infill).

In the Warre Warren area there are several examples with internal designs that correspond to the classification developed for the engraved shields (e.g. Figure 5.18). In all, a total of ten shields, from seven sites, can be classified using the design categories defined. Only three identified forms occur: types 2B, 2C and 2E (Figure 11.30). All three of these were recorded in one site in the Warre Warren area (#45-3-970), while three other examples of type 2C were recorded at another Warre Warren site (#45-3-1602), on the Colo River (#45-2-292) and at Manly (#45-6-1262).

Two forms unidentified in the engraving assemblage were recorded at two sites in Middle Harbour and Lane Cove, these being a double longitudinal line (2E) and a single chevron design, from site #52-2-453 in the *Tharawal* Avon/Cordeaux drainage basin.

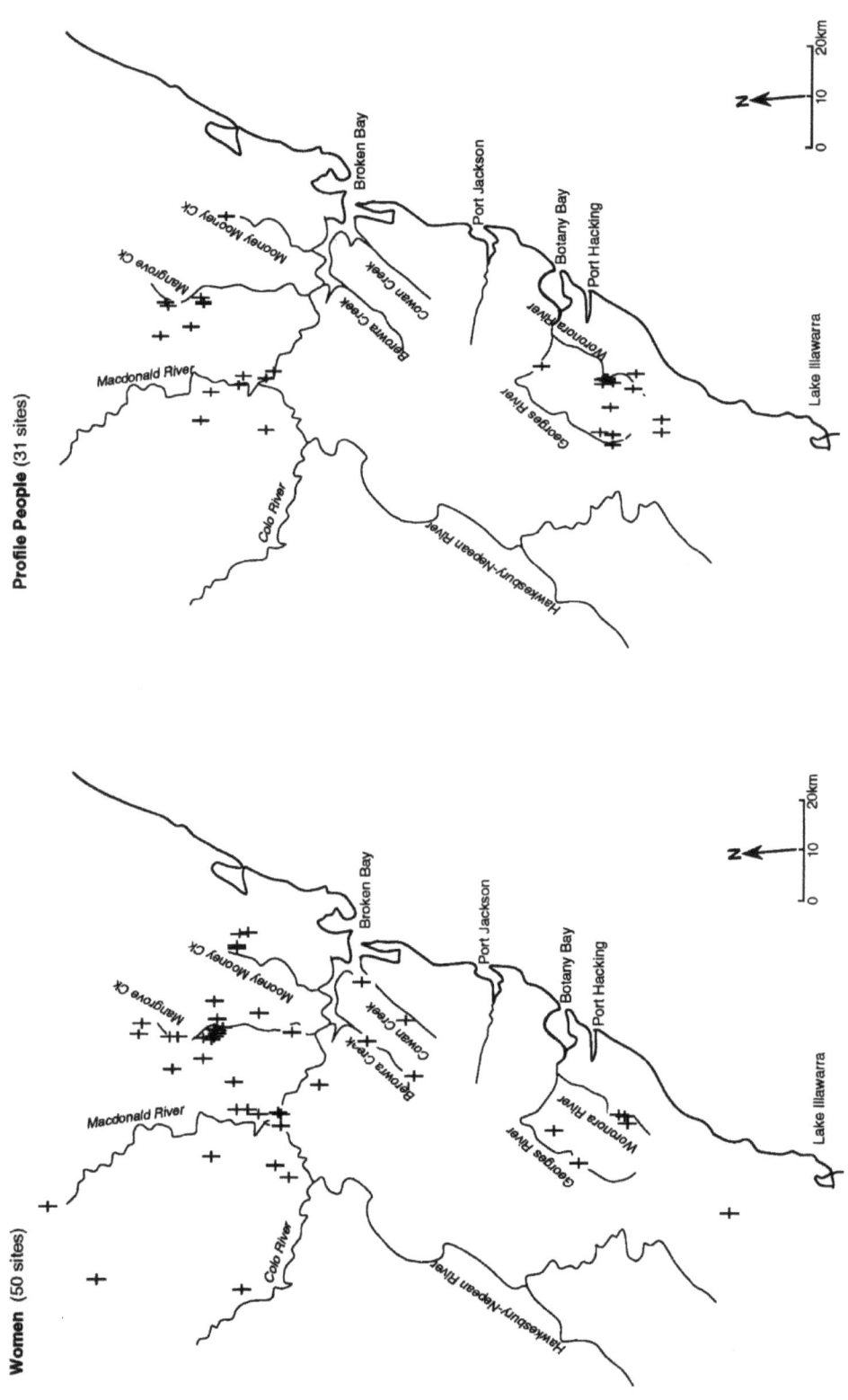

Figure 12.55: Pigment art sites. Distribution of shelters with women and profile people motifs.

Chapter 12: Regional synchronic variation: shelter art

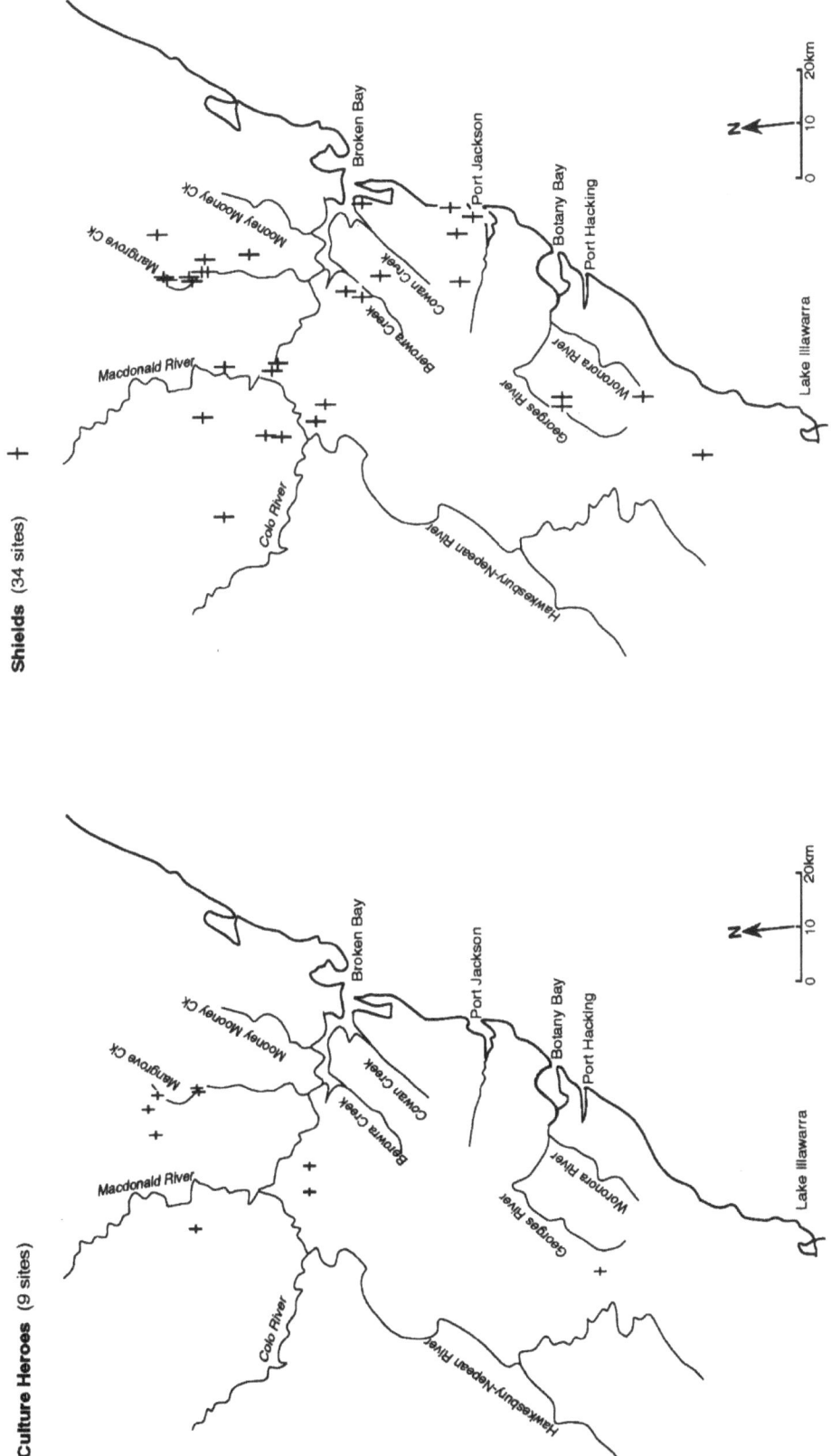

Figure 12.56: Pigment art sites. Distribution of shelters with culture heroes and shield motifs.

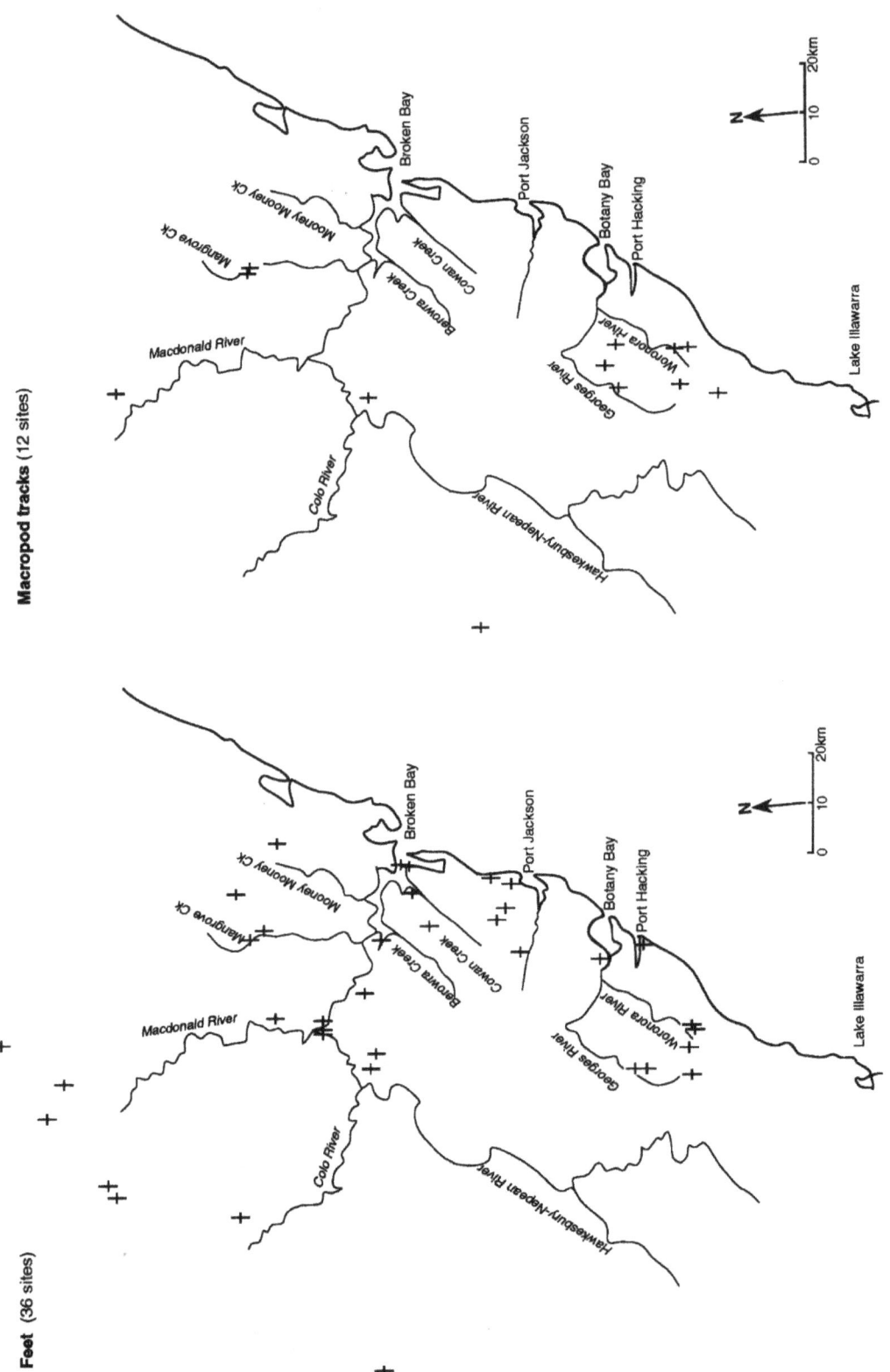

Figure 12.57: Pigment art sites. Distribution of shelters with human feet and macropod track motifs.

Chapter 12: Regional synchronic variation: shelter art

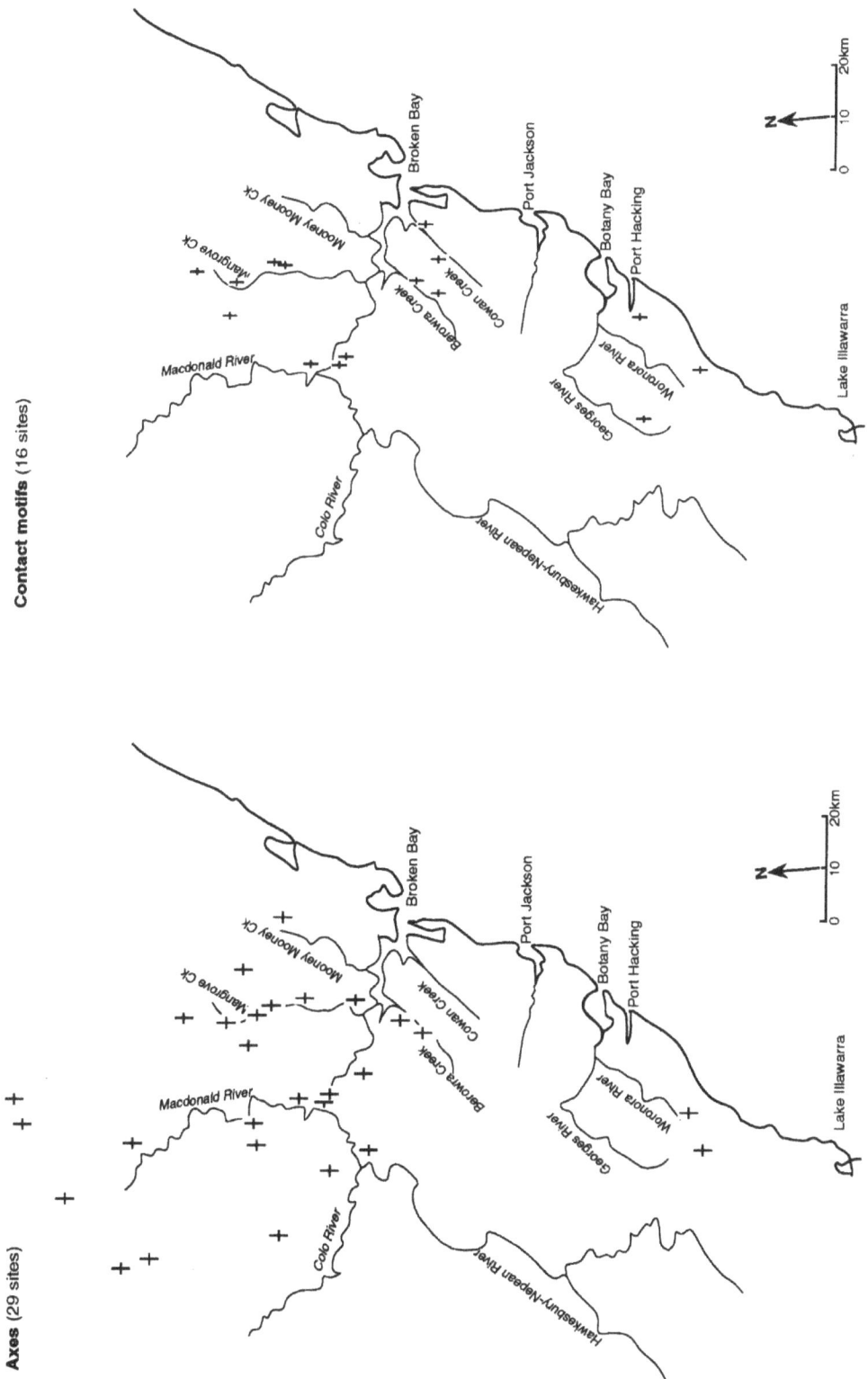

Figure 12.58: Pigment art sites. Distribution of shelters with axes and contact motifs.

The small sample of shields with internal designs makes a detailed distribution analysis redundant. The relative absence of shield designs, in comparison with the engraved component, is interesting in itself. It reinforces the general contention that this is a less stylistically constrained medium. While the sample size of pigment shield designs is very small, certain similarities and differences with the engraved assemblage's results can be made.

- The double (horizontal) cross design (2C) was the most common engraved variety, and 40% of the pigmented shields with designs were also of this type;

- The pigment shields with designs were located mainly in the *Darkingung* area with one only in the *Guringai* area. This is in stark contrast to the definite engraved shield design focus in the *Guringai* area;

- The only pigment shield with a St George cross (Type 2B) design was located in the *Darkingung* language area, Mangrove Creek catchment. In this same site two other design forms (2C and 2E) were also found. Type 2B was concentrated in the *Guringai* and *Darug* language areas with the engraved shields;

- The one engraved shield in *Tharawal* language area had no internal design; the drawn version has a quite distinctive chevron design, not recorded elsewhere in the pigment assemblage and not in the engraved assemblage (however, see Collins 1798[1975]: Appendix VI; Plates 5 and 6);

- No pigment shields had the diamonds design element at either end of the shield. This engraved design element was concentrated in the Cowan and Middle Harbour catchments. The pigment shield assemblage (in Middle Harbour and Lane Cove catchments) included a design which did not occur in the engraved component - a double longitudinal line design.

Culture Heroes

Only 18 pigment culture heroes occur in nine sites across the region (Figure 12.56). The discussion here is restricted to those 17 motifs from eight sites from the north of the region[39] (Table 12.9). In the shelter art component, this motif's form and distribution is significantly more restricted than is found in the engraved component. All of the pigment culture heroes are of the *Biaime* type. The analysis of this rare motif's distribution indicates that it occurs only in *Darkingung* and *Darug* language areas. The engraved culture hero focus is in the *Guringai* language area. **No** pigment culture heroes were recorded in the *Guringai* area.

This motif is found amongst a range of assemblage sizes, including in the largest recorded in the region (Swinton's). At four sites, these motifs occur in isolation. In two sites they are paired male and female. In one site (45-2-189: Sim 1969) there are six pigment culture heroes.

The two engraved forms of culture hero are the *Daramulan* and the *Biaime* types. In the engraved component, excessive size was a significant aspect of the classification. Decorative infill and/or the presence of therianthropic characteristics were also considered important. Larger size is a consistent criterion in this medium also, although the pigment forms are generally smaller than their engraved counterparts. The mean size of the pigment culture heroes is 1.25m (standard dev. 0.6) and the largest of these motifs is c.3.0m long (at Canoelands: Clegg 1977, McCarthy 1961).

---

[39]One culture hero was recorded by the Sydney Prehistory Group (1983) south of the Georges River. This was counted during the Rock Art Project (McDonald 1985a). This reanalysis of motifs necessitated further inspection of the original scale recordings, held by NPWS (in the AHIMS Sites Register). The original SPG drawings went missing from the Sites Register during the move from the city to Hurstville (Ian Johnson, then NPWS Sites Registrar, pers. comm.). The motif from site #52-2-23 could not be reanalysed for this analysis.

Size appears to be due in part to the smaller 'canvas' size available in shelter sites. The consistent size range demonstrated by these motifs and the fact that these are indeed larger than the majority of anthropomorphic figures depicted in shelter art sites[40] suggests that relative scale is still a consideration in this motif's graphic 'vocabulary' or schemata. This motif is also differentiated from plain anthropomorphs on the basis of infilled decoration and attachments (e.g. the double and single horned anthropomorphs in the Mangrove and Mogo Creek catchments).

Table 12.9: Shelter Art sites. Culture heroes: compositional details.

Site	Max length (m)	Animal features	Headdress	Horns	Other features	Colour	Technique
45-2-189	1.25	-	√	-	redrawn	Black, white + red	Drawn outline/infill
	1.15	-	√	-	redrawn, 3 white eyes	Black, white + red	Drawn outline/infill
	1.0	-	√	-	redrawn, white eyes	Black, white + red	Drawn outline/infill
	2.1	-	√	-	head and arms only, white eyes	White + red	Drawn outline/infill
	0.9	-	-	-	4 white eyes holding spear + bag	Black, white + red	Drawn outline/infill
	1.8	-	√	-	6 white eyes	Black, white + red	Drawn outline/infill
45-3-252	1.1	-	-	√	late in sequence	Red	Drawn outline/infill
45-3-317	1.4	-	-	√	female white eyes	Red + White	Drawn + painted outline/infill
	1.5	-	-	√	male white eyes	Red + White	Drawn + painted outline/infill
	0.95	-	-	√	female white eyes	Red + White	Drawn + painted outline/infill
45-3-568	3.0	√	-	-	several production episodes ?	Red, yellow, black + white	Drawn outline/infill
45-3-794	0.7	-	-	-	male	Black	Drawn outline/infill
	0.75	-	-	-	female	Black + engraved	Drawn outline/infill
45-3-814	1.7	-	-	√		Red	Drawn outline/infill
45-3-1136	1.2	-	-	-		Red	Drawn outline/infill
45-3-1602	c. 1.0	-	-	√	lower half weathered	White	Painted outline
	1.15	-	√	-		Red + White	Drawn outline/infill

The distribution of the horned anthropomorph culture hero form is extremely localised. Eight of the culture heroes (47%) are in the Mangrove Creek catchment and five of these are the red horned anthropomorph form (Figure 5.14)[41]. A sixth red horned anthropomorph is located near Mogo Creek less than 10km west of Mangrove Creek. A pair of black culture heroes in the Mangrove Creek catchment (at #45-3-794) has a very similar morphology (minus the horns).

[40] This has not been quantified on a large scale. A limited analysis of the 33 anthropomorphic figures (men, women, profile people and anthropomorphs) at the Swinton's site indicated that the mean size for these was 0.46m (st dev. 0.2). The culture hero at this site was 1.1m long.

[41] The distribution of this motif is highly suggestive of the *Darkingung* mythical creature called *Ghindaring*. This malevolent creature was said to inhabit the rocky places on sides of mountains and have a body 'with a red glow like burning coals' (Mathews 1904: 345).

Only one of the pigment culture heroes is therianthropic. The Canoelands culture hero has a macropod-like head (i.e. in profile) with otherwise anthropomorphic characteristics (albeit seven fingers on one hand).

These motifs have been produced in range of colours, although red is the most common. Most of these motifs are drawn, but several also have painted elements. Monochrome is less common than bichrome and there is the one polychrome example (at Canoelands: Table 12.9). One black example also includes some abraded lines.

This analysis provides contradictory evidence to the engraved component. The most obvious difference in these results is in the more restricted geographic distribution of the pigment form motif. There is almost no overlap in the distribution of this motif in the two different art components (cf. Figure 11.26 and Figure 12.56).

The absence of pigment *Daramulan*-type culture heroes is very interesting. This however does correlate with the distribution of this form generally: with the exception of one *Daramulan* found at Maroota, all of the *Daramulan*-type engraved culture heroes are in *Guringai* territory.

Shelter art sites: conclusions

The analysis of this art component included a general investigation of regional characteristics including technical options and motif focus.

Colour usage across the region reveals definite stylistic preferences which are not related to pigment availability. In the south of the region there is a definite preference for black pigment. In the north of the region there is a definite preference for white pigment. While this reflects the prevalence of stencilling as a technique - white drawings and paintings are also common. The dominance of white colour use in the north supports a model of contact between this part of the Sydney region and the Hunter Valley.

The CA of motif and technique indicates that there is a core of greater stylistic homogeneity in the centre-west of the Sydney Basin. From this core central area the surrounding art demonstrates increasing stylistic variability. This core style area is not the same as that identified for the engraved assemblage. And technique variables reveal a slightly different core focus pigment motifs. The techniques used in pigment art across the region are relatively homogenous.

While these analyses have demonstrated localised stylistic variability across the Sydney region, only in one area is this variability distinctive enough to identify a style boundary.

Stylistic variation can be explained in terms of the defined language areas. The core of stylistically homogeneous shelter art sites occurs in the *Darkingung* and northern *Darug* language areas. The *Guringai* and southern *Darug* sites are the least homogenous, while the *Tharawal* sites are also relatively heterogeneous.

The proposed location of the *Darug/Guringai* language boundary is supported by shelter motifs, with a strong separation between sites on either side of Berowra Creek. There is less disparity shown when technique variables are considered, although the technical choices demonstrated (i.e. outlier foci) on either side of the creek are markedly different.

The investigation of *Darug* sites north and south of the Cumberland Plain revealed that some 'bedrock design notions' (Sackett 1990) transcend the distance between these two assemblages. Both sets of *Darug* sites have a predominance of hand stencils with macropods, other land animals and anthropomorphs dominating. Distance, however, has created some fundamental differences between the *Darug* sites on either side of the Cumberland Plain. These are most obvious in terms of schema. The southern sites have a schematic preference for the use of four legs on terrestrial animals and two legs on birds (compared with two and one, respectively, in the north). This schematic vocabulary is present in the southern *Darug* and *Tharawal* sites, but not the northern *Darug* sites. This stylistic convention does not support the proposed boundary between the *Darug* and *Tharawal* language groups. The pigment art suggests that there was more social interaction between the southern *Darug* group and *Tharawal* speakers than there was between southern and northern *Darug* members.

The analysis of rare motifs demonstrated significant differences between the two art components. The amount of information-laden detail in the *engraved* shields, and the distribution of engraved shield designs were not mirrored by similar design detail or distribution patterns in the pigment shields. Culture heroes also showed different distribution patterns and a much more restricted design vocabulary.

These results support a model for the expression of very different social behaviours in the two art contexts.

13

DREAMTIME SUPERHIGHWAY: MODELLING A REGIONAL STYLE

This research has explored the interrelatedness of prehistoric art and other archaeological evidence in the Sydney region. Sources of stylistic variability in open engraving sites and sheltered pigment sites have been investigated. Diachronic and synchronic variations have been scrutinized, as have the effects of art medium and site location. The contemporaneity of art and occupation deposits in shelter sites was examined as a means of testing assumptions about the age of the art. This work was also important for developing models about how pigment and engraved art may have functioned across this area. A model to explain changing social interactions across the region which is founded in information exchange theory has been developed. The model adopts the view that style is a means of non-verbal communication used to negotiate identity (Wiessner 1990). It is proposed that the rock art in the Sydney region functioned as a prehistoric information superhighway. Through stylistic behaviour, groups around the region, who were not in constant verbal contact with each other, were able to communicate important social messages and demonstrate both broad-scale group cohesion and within-group distinctiveness.

Art characteristics in two contexts

Engraving sites

The engraved site sample was 717 sites containing 7,804 motifs. The average engraving assemblage contains 11 motifs. The majority of sites have less than eight motifs. Most engraving sites are located on open horizontal expanses of sandstone. A small number of engraving sites are found on vertical boulders close to major waterways.

The topographic location of engraving sites is more diverse than previous work suggested. Roughly half of the engraving sites are on ridgelines with almost as many in the hillslopes zone.

The predominant motif in the region is the human foot/track (*mundoe*), followed by fish, macropods, men and bird tracks. The predictive power of the classificatory distinction between Panaramitee and Simple Figurative assemblages, based on track proportions (Maynard 1976), is diminished by this result – although human tracks are not common in Panaramitee sites.

Shelter art sites

The shelter art sample comprised 546 sites containing 14,424 motifs. The average shelter art assemblage contains 26 motifs. Most sites, however, contain less than ten motifs.

The topographic location for this art context is more focussed than the engraved component. The majority of shelter art sites are located on hillslopes. The remainder are equally distributed between ridgetop and valley bottom locations.

The predominant identifiable pigment motif in the region is the hand (stencil). Macropods, anthropomorphs and land animals dominate the depictive motifs. The pigment art is generally in a fairly poor state of preservation. This, combined with its more *ad hoc* production, meant that only 60% of all pigment motifs in the region could be classified to motif type.

Most shelter art sites also have surface evidence for occupation, although this is difficult to quantify based on site records. Excavation programmes around the region directed at testing PADs (Potential Archaeological Deposits) have revealed that a large proportion of sites with no surface manifestations are indeed occupation sites (e.g. Attenbrow 2004, JMcD CHM 2005a).

General comparisons of the two contexts

A comparison of the two art components reveals the following points about the nature of these extensive regional art bodies:

1) The two art bodies represent different manifestations of the same art tradition – although they demonstrate inherently distinctive traits due to their different repertoires of technical options;

2) Shelter assemblages are generally larger than open engraving assemblages;

3) There are striking similarities in the motif preferences demonstrated by the two components with several major differences (e.g. marine depictions dominate the engraving assemblage, but these themes are relatively absent from the shelter art assemblage);

4) Stylistic clines and boundaries are demonstrated by both art components, and there is considerable congruence in the location of these;

    A previously identified style boundary within the region (McMah 1965, McDonald 1985a) is confirmed at the Georges River. The location of this boundary is based on the presence/ absence and proportions of different motifs and on overriding schematic differences on either side of this river;

5) Shelter art sites are present in great numbers across the entire Hawkesbury sandstone landscape, while the distribution of engraving sites is more restricted. There is a dense core of engraving sites in the central coastal area of the Basin. Engraving sites decline in frequency towards the north-west, although assemblage sizes in this area are very large. To the south of the Basin, particularly south of the Georges River, the number and size of engraving sites diminish.

Systematic survey in areas with otherwise low numbers of engravings has revealed a real absence of this site type in parts of the region (e.g. Attenbrow 1987, Attenbrow and Negerevich 1984, McDonald 1988a, 1990b).

The distribution of engraving sites, particularly the apparent focus in the *Guringai* area and relative paucity south of the identified style boundary may be indicative of different social mechanisms operating across the region.

Art production in shelter locations is ubiquitous across the region. Again this suggests that these two art contexts provided the opportunity for distinctly different symbolic behaviours. All local groups across the region used pigment art to negotiate identity. This type of behaviour occurred on both sides of the style boundary. While different stylistic behaviours are indicated by this boundary, it would appear that the behaviour - or social context - of this type of art production, transcends this boundary.

Contemporaneity of art and deposit

The contemporaneity of art and occupation deposit was demonstrated at three shelter art sites. At all three sites, it could be argued that the main phase of pigment art production coincided with the most intensive period of shelter occupation. In multi-phased art sites, earlier low-intensity occupation could be argued to have an artistic component, also of low intensity. Proving this earlier association was more difficult, however.

Prior to this study, more than 30 shelter sites with art had been excavated across the region. While only one of these had been excavated to provide a context for the art, an analysis of these shelters demonstrates broad contemporaneity of art and domestic activities across the region. Significantly this analysis showed that the patterns of occupation in shelter art sites mirrored the indices exhibited by occupation sites generally. This result suggests there are no intrinsic differences in the nature of occupation in shelters with or without art. Only the UDM shelter indicates that art production in sheltered locations may have continued once shelters were abandoned as places of occupation – and this evidence is inconclusive.

Pigment art produced in occupation shelters would have been accessible to and viewed by undifferentiated parts of the community engaged in general domestic activities. This art was fulfilling a function very different from art being produced in a context where there was a restricted viewing public. This argument is developed further below.

The analyses of contemporaneity also demonstrate evidence for diachronic change in art production in the region throughout its occupation. The clearest evidence for this is with pecked intaglio motifs in shelter sites, interpreted here as 'residual Panaramitee' (see Mt Yengo discussion). While most of these engravings are not in datable contexts, the Yengo 1 excavation supports the contention that this art style predates the majority of the art and occupation in the Sydney region. A similar pattern has been demonstrated in numerous other Australian art regions (e.g. central, south-eastern and northern Queensland, central Australia, and elsewhere.).

Only a small number of shelter sites contain residual Panaramitee engravings. The evidence for an earlier, low-density artistic tradition predating the main occupation and artistic period of the region matches the evidence for low-density early occupation of rockshelters and open locations (e.g. Attenbrow 2004; JMcD CHM 2005a, 2005b, 2005c). There would appear to be a continuing tradition over time for art to be produced in shelters that are also occupied.

In some parts of the region, there are more sites where earlier, residual Panaramitee art was produced. There are 55 shelter sites with engravings amongst their assemblages in the north of the region, but only two such sites to the south. Almost half of the northern shelters contain miniature-Sydney or incised engravings. Most of the sites with pecked circles and animal tracks are located in the Macdonald River and Mangrove Creek drainage basins. This suggests that early occupation of the region may have been focussed in these major tributaries of the Hawkesbury River (Bowdler 1977)[42]. The majority of the earliest occupation sites in the region are in similar contexts (e.g. Attenbrow 2004; Kohen *et al*. 1984; JMcD CHM 2005a, b and c).

Diachronic change in the shelter art

Diachronic variability was explored using 65 shelter art sites in the Mangrove Creek Valley. On the basis of superimpositionning, motif preference and multivariate analyses of motif and technical variables, three phases of art production were discerned. Based on broadly similar patterns identified in a number of sites in different locations, this sequence is extrapolated to the broader region:

---

[42] The two southern sites with engraved macropod and bird tracks are located in the Georges and Wollondilly River catchments, the latter being in the upper reaches of the Hawkesbury-Nepean. There are also two sites with engraved bird tracks in the Blue Mountains, around the headwaters of the Grose Valley (another tributary of the Hawkesbury River).

Sydney Basin Art Phase 1   pecked engravings of mostly tracks and circles.

Sydney Basin Art Phase 2   red paintings, red hand stencils, and possibly white hand stencils (the latter two do not co-occur).

Sydney Basin Art Phase 3   a proliferation of techniques and colour usage, perhaps starting with plain dry black and dry red motifs and then developing into a range of paints, dry bichromes, stencils of varying colours, polychromes and incised motifs. Outline-only motifs end the sequence in many shelters, although contact motifs include white stencils and drawn red and white outlined and infilled forms.

Across the region, during Phase 3, there is often localised variation in technique proportions, motif preference and timing, as identified in the synchronic analyses.

A chronology for the Mangrove Creek art sequence was based on associated dates in particular shelter sites. This was initially correlated with the UMCC lithic phases, and based on associations in certain sites. The art sequence was assigned to broader cultural phases which seemed to be the most judicious calibration of the art with the stone tool phases.

Art Phase 1	Pre- Bondaian	> 4,000 years BP (minimum)
Art Phase 2	Early Bondaian	> 4,000 years BP
Art Phase 3	Middle to Late Bondaian	4,000 BP - European contact

Difficulties in accurately contextualising this chronology arise because of the scarcity of sites with art in dateable contexts, and because of inconsistencies in dating the UMCC Stone Tool Phases in particular sites. A correlation of Art Phases with broader Lithic Phases appears to be the most judicious way of categorising this material. It is argued that the main art production period in most sites is contemporaneous with the most intensive period of stone tool production within shelters; i.e. the Middle Bondaian. Art production continued into the late Bondaian, however, and indeed up until contact.

The production of the art in shelter contexts continued without appreciable stylistic change from the Middle to late Bondaian. The significance of this finding is discussed, below, with reference to the model for stylistic behaviour developed for the region.

The diachronic analyses indicated that appreciable changes in style did not occur over the main period of art production. Thus, synchronic variability should not be unduly affected by diachronic 'noise'. As indicated above, 'older' art sites, as with the earlier occupation evidence, appear to be focussed on tributaries of the Hawkesbury-Nepean. There seems to have been a rapid decline in art production in the post-contact period. The cultural upheaval which occurred with the European invasion does not appear to have been given voice in the art.

Synchronic variability: both art components

Engravings

Multivariate analyses indicate that the Sydney region's engravings are relatively homogeneous and that there were no distinctive internal divisions. This result is as would be expected in a regional art body. The CA results do demonstrate, however, that stylistic variability can be discerned across the region. In only one area is this variability significant enough to suggest a style boundary. In other parts of the region, stylistic changes are clinal. The CA demonstrates that sites in the central coastal area have the greatest stylistic homogeneity. As one travels out from this central coastal area, stylistic heterogeneity increases. The sites at the very north and south of the region are the most heterogeneous found in the region.

Broad geographic trends in style were identified, and the assemblage was investigated in terms of contact language areas and drainage basins. These analyses focussed on three areas:

- the north-western inland zone, in *Darkingung* country;

- the central coast and inland south of the Hawkesbury River, crossing the *Guringai* and *Darug* language boundary; and,

- the south of the Georges River, in *Tharawal* country.

The *Darkingung* and *Guringai* language areas have the most homogenous engraving assemblages, followed by the *Darug*. The *Tharawal* engravings are the least homogenous.

The analyses based on motif classes have demonstrated thematic differences across the region, reinforcing the idea that a significant degree of stylistic information can be provided by this approach. This broader level of analysis was complemented by an investigation of rare motifs and 'compositional details' (Sackett 1990). More specific levels of information were sought to determine the degree of *vernacular* variability.

The analysis of rare motifs confirmed the localised character of the engraving assemblages in different parts of the Basin. It also confirmed the separation between the *Tharawal* and all other language groups. Sites with culture heroes, profile anthropomorphs, axes and contact motifs occur in the west of *Guringai* territory and to a lesser extent in *Darkingung* territory. These motifs' distributions suggest a design 'link' between these two language areas, confirming the CA results. These motifs demonstrate aggregational contact (Conkey 1980) between these language groups.

Compositional details also revealed design contact between *Guringai*, *Darkingung* and *Darug* language areas and confirmed the dissimilarity of the *Tharawal* engravings. Design variability on shield motifs was extraordinarily diverse in the *Guringai* area, with both fewer motifs and a marked decrease in design options being found outside this language area. Again, graphic vocabulary stems from the *Guringai* area, and design influence flows from this area into the other areas.

These findings have significance in light of ethnohistoric evidence for the inequality of groups around Port Jackson area. The *Guringai* speakers to the north of Sydney were noted to be socially influential:

> (by the) influence of their numbers and muscular appearance ... there is no doubt of their decided superiority over all tribes with whom we are acquainted. (Collins 1798[1975]:453)

Shelter Art Sites

The analysis of this art context initially commenced with an overview of regional technical options and motif foci. Again broadly similar regional characteristics were identified as were varying stylistic preferences across the region. The variability demonstrated by technique variables is generally less than that demonstrated by motif variables.

Colour usage in the different language areas reinforces differences in stylistic preference across the region. In the south of the region there is a definite preference for black drawings. In the centre of the region there is a co-dominance of red, black and white pigments. In the north of the region there is a definite focus on white stencils, white drawings and white paintings. This dominance of white supports a model of contact between the Hunter Valley and this part of the Sydney region.

In most parts of the region, stylistic changes are clinal. The degree of stylistic homogeneity is generally less in this component than in the engraving component. The *Tharawal* language area is clearly differentiated from all other language areas by its relative absence of hand stencils.

Variation can be explained in terms of the defined language areas, and there are varying degrees of stylistic variability in each. The analysis of sites according to defined drainage basins and language areas confirmed a mosaic of stylistic variability.

The stylistic core of shelter sites based on motif and technique occurs in the *Darkingung* and northern *Darug* areas. The *Guringai* and southern *Darug* sites are the least homogenous, while the *Tharawal* sites similarly demonstrate high levels of stylistic heterogeneity. The location of the *Darug/Guringai* language boundary was supported by shelter motifs, and by outlier focus based on technique.

In the *Darkingung* area, the motifs from three drainage basins were revealed as having very similar levels of homogeneity, with clinal variation in motif preference. This same sample revealed quite marked variability in terms of technique.

The southern *Darug* and *Tharawal* sites are highly heterogeneous in terms of motif themes while the techniques used in these adjoining language areas are very similar. The sites in these areas do not support a boundary between these language groups, and there are strong associations demonstrated by shelter art sites between the southern *Darug* and *Tharawal*. These include the schematic differences described above, and the fact that neither is associated with an extensive engraved component.

The analysis of rare pigment motifs demonstrated significant differences between this and the engraved components. Design contact and influence identified in the engraved shields were not matched by similar design detail in the sheltered component and nor were the ornate and impressive engraved culture heroes (found mainly in the *Guringai* area) matched either in extent or in the range of variation in the shelter art component. These results support a model for the expression of very different social behaviours in the two art contexts.

Social context and stylistic information

Shelters versus engravings

In order to investigate the stylistic variability in these two art contexts, it was necessary to consider the potential for heterogeneity in each. If the basic potential for variability exists in either medium, then variability demonstrated needs to be greater than the intrinsic potential for variability for significance to be claimed.

The motif classes used for the two assemblages were almost identical, with two extra motifs being counted in the shelter art assemblage. It was possible that the very similar motif classifications would make it difficult to differentiate the potential of either medium for greater heterogeneity (e.g. Gamble 1982 cf. Soffer 1987).

One potential source of greater heterogeneity in the shelter art assemblage is assemblage size. Shelter art assemblages are considerably larger than their open engraved components. The general statistics for site size (Table 13.1) indicate that the potential for heterogeneity of motif assemblages is far greater for shelter art sites than it is for engraving sites. The variability inherent in shelter art sites is three times as great as that demonstrated by the engraving sites. When looking at the numbers of motifs used in the different media, however, this variability is substantially decreased (Table 13.2). While the maximum number of motifs present at any one shelter site is greater than found with engraving sites, the standard deviations for the two are very similar.

Motif occurrence (maximum number of times a motif is present) and the motif emphasis (percentage of sites at which motifs have been depicted) both indicate that while there are different foci in the two media, both show a general consistency of motif use. No particular emphasis is shown for particular motifs in either context. *Mundoes*, fish and macropods are the most commonly depicted motifs in the engraved assemblage. These occur in 22% (157), 35% (250) and 34.5% (247) of engraving sites (Table 11.5). The most commonly depicted motifs in the shelter art assemblage are hands and macropods. These occur in 37% (206) and 40% (219) of pigment art sites (Table 12.6).

Table 13.1: Statistics for regional assemblage sizes, both art components.

	Shelter Art sites	Engraving sites
Minimum	1	1
Maximum	857	174
Mean	26.4	10.9
Stand Dev.	55.4	18.0

Table 13.2: Statistics for total number of motif categories recorded in both art components across the region.

	Shelter Art sites	Engraving sites
Minimum	1	1
Maximum	24	19
Mean	4.5	3.4
Stand Dev.	3.8	4.0

The potential for heterogeneity can be calculated based on the motif classifications used and the maximum number of times that motifs occurred in the two assemblages (C. Smith 1989: 126). The formula for potential combinations is:

$$F(n) = \frac{n!}{r!(n-r)!}$$

where 'n' equals the number of elements available in the system, and 'r' equals the number of elements combined.

For the engraving sites, 27 variables were used with no site having more than 19 different motifs. With the shelter art sites, 29 variables were used with no sites having more than 24 variables.

Engraving sites:

$$n = 27$$
$$r = 19$$
$$F(n) = \frac{27!}{19!(8)!}$$
$$= 2{,}220{,}075$$

Shelter art sites:

$$n = 29$$
$$r = 24$$
$$F(n) = \frac{29!}{24!(5)!}$$
$$= 118{,}755$$

The potential for heterogeneity (in terms of motif combinations) is significantly greater for engraving sites than for shelter art sites.

The multivariate analyses undertaken for this research shows that both Sydney art components exhibit synchronic stylistic variability in each of the language areas and drainage basins. The degree of stylistic heterogeneity was found to be <u>greater</u> in the shelter art component than is found in the engraving component - which is not the result which would be expected, given the calculated baseline potential to be heterogeneous.

The engraving component is more homogenous than the shelter art (a statistically significant difference) and this has equally significant implications for a cultural interpretation.

These varying levels of stylistic heterogeneity are seen to reveal different levels of transmission of social information. The greater stylistic homogeneity in the engraved medium demonstrates larger scale group cohesion. The more stylistically heterogeneous pigment sites demonstrate localised group identifying behaviour. This argument is supported by the archaeological data based on the principles of information exchange theory (Wobst 1977;

Wiessner 1989, 1990). The nature of participants (producers and viewers) in art production and social context is important in developing this argument.

The uneven distribution of engraving sites around the region suggests that the social cohesion provided by this medium was unequal – especially compared to the messaging potential provided by the ubiquitous shelter art medium. The *Guringai* territory with its highly homogenous engraving sites and design influence would appear to be the core area of social cohesion. The *Guringai* influence diminishes to the north-west of the region and is non-existent to the south of the region. The relative absence of engravings south of the Georges River style boundary indicates either truncation of broad-scale group cohesion (i.e. a different culture area) with alternative social mechanisms perhaps operating in this southern area.

Public vs Private engravings

Social context based on levels of visibility was investigated using engraving sites in the *Guringai* language area. The art sites in the foreshore resource zone at the bottom of steep cliffs and the intermediate hillslopes provide a different social context to the open engraving sites on the less economically-productive plateaux above.

Bowdler demonstrated the potential of shell middens to inform upon 'dynamic relationships in the economic sphere of cultural systems' (Bowdler 1976: 248). Since then, the contribution of women to the formation of middens has been widely recognised, although with few exceptions (e.g. McDonald 1992b; Meehan 1982, 1988) Australian analyses have continued to focus on a range of unengendered topics such as dietary estimates, changing resource structures and the exploitation of these (e.g. Bailey 1975; Beaton 1985; Mackay and White 1987; Przywolnik 2005; Sullivan 1987), or on dietary reconstructions and the question of intensification (e.g. Hughes and Lampert 1982, Woods 1989).

The sites around the foreshore are close to estuarine resources and most would have been accessed either by canoe or on foot. The estuarine foreshores in the *Guringai* area are characterised by extensive and often continuous open middens as well as middens in shelters. Many of these are associated with rock art, both the pigment form and the unusual vertical engraving type described earlier. The foreshore zone was the focus for women's shellfish collecting and men fishing. Ethnohistoric reports indicate that women beached their canoes and ate their catch on shore (Collins 1798[1975]; Tench 1793(1961]).

The art around the foreshore informs us of very different artistic behaviours from art sites which are viewed by fewer people or by people operating in a less casual context. The art produced in this highly visible context should reflect group identifying (bounding) behaviour while that occurring in socially less visible locations should demonstrate bonding (Wobst 1977, Wiessner 1990). The art produced around the foreshore should function much as in shelter art sites, since the audience to this art's production includes the entire group in their daily economic round.

The analysis of open ridgetop engravings on horizontal platforms and foreshore engravings on vertical boulders was done within a single language area. The overall motif emphases (i.e. graphic vocabulary) in these locations were similar but stylistic expression was not. CA results of the engraving sites in the two contexts confirm that the engraving sites located around the foreshores are stylistically more heterogeneous than those occurring on the plateaus.

The presence and use of the vertical engraving sites around the foreshore zones, in a different local social context, provides supplementary evidence for bounding behaviour in the region. The analysis of selected stylistic variability contextualised by social visibility confirms a number of predictions about style being used as non-verbal communication.

A model for social and territorial interaction across the region
The Sydney region is a fertile coastal zone with varying localised biomass (coastal, estuarine, hinterland). In terms of foraging and social interaction, the environmental and seasonal conditions

were fairly predictable. Owing to these relatively predictable resources and the likely social pressure induced by large population sizes especially post-sea level rise, well-developed group identity and formalised group membership would be expected as a mechanism to control access to resources and a prescribed territory.

In a relatively non-stratified society, such as in the Sydney region at contact, significant within-group differentiation on grounds other than age or gender would not be expected. The Sydney region may have presented a closed and cohesive cultural system to the groups outside its boundaries, but within this system interactive networks were relatively open. The fact that there were overriding cultural similarities among the groups across the region suggests that while bounding behaviour was practised, the lack of inter language-group cultural distinctiveness resulted in low levels of inter-group differentiation. Different local groups did exist across the region, notably as expressed in language, but the societal differences between them were less than the overriding similarities of the larger cultural bloc.

While the population density was generally high, there was not fierce competition over resources as these were distributed across the entire region. Social interaction within and across linguistic boundaries was observed at European contact. Organised social events, e.g. initiation ceremonies and dances, as well as the exploitation of windfall resources such as feasts on beached whales, resulted in aggregations of large numbers of people of mixed language groups. There is no ethnographic evidence for a rigid demarcation of territorial boundaries – particularly on the resource-rich coastal strip. This territorial organisation, and the overriding regional cultural similarities, initially suggested that the spatial organisation of art traits would not demonstrate smaller-scale boundary maintenance (Wobst 1977).

Previous social reconstructions tend to suggest that a number of dialect groups existed in each of the four identified language groups (Kohen and Lampert 1988). The foraging areas of local groups probably varied according to local biomass, but were relatively small. The analysis of drainage basins undertaken in this work attempted to achieve meaningful subdivisions within the larger language areas, and indeed, successfully demonstrated patterning in the art within language areas. This patterning may well represent the distribution of smaller dialect groups, and suggests that conscious group territoriality operated on a scale smaller than the language area.

Archaeological reconstruction suggests considerable change in settlement patterns over the period of the region's occupation. By 5,000 years ago, after sea-level stabilisation, occupation was well established within all habitats, but widely ramified social change is likely to have taken place owing to increased population pressure. Habitation indices and the interrelationship of art suggest that these patterns can now be expanded beyond technological description of the Eastern Regional Sequence (Hiscock and Attenbrow 2003).

Detailed work in the Mangrove Creek valley built on Attenbrow's (1987, 2004) model. The analyses undertaken here reinforced the variability inherent at the local level. Extrapolation of these patterns to the regional level is possible based on extensive work which has been completed over the last decade, particularly in open sites on the Cumberland Plain.

This research and subsequent analyses suggests that the timing for the transition between the Middle and Late Bondaian requires modification. The introduction of fishhooks c.1,000 BP and a number of more recent Middle Bondaian dates (e.g. Yengo 1 >540 BP: UDM c.1,200 BP and Loggers at c.780 BP) suggest that the changeover between these phases may have been later than the 1,600 BP as initially posited (Attenbrow 1987; 2004). A transition which coincides with the introduction of fish-hooks on the coastal strip and with the declining shelter occupation rates in the most recent millennium would be a more parsimonious treatment of the various strands of archaeological data, and thus a date of c.1,000 years is proposed for this changeover. The following dated phases are recognised across the Sydney Region during which time settlement patterns and signalling behaviours are expected to have fluctuated.

Pre Bondaian   30,000 years ago to 8,000BP

During the Pleistocene, groups appear to have been highly mobile, travelling considerable distances between sites. At this time, the focus of stone acquisition was on the Hawkesbury-Nepean River gravels. The cores and tools which people carried were quite large, but they used the stone sparingly, leaving few artefacts behind, and rarely discarding their cores (which acted as portable quarries). Rock art production focused on iconic designs found broadly across the continent and art reinforced broad-scale social networks.

Early Bondaian 8,000 years to c.4,000 years BP

Rising sea levels, stabilising after 6,000 BP, forced groups previously occupying the drowning coastal plain inland. Population densities across the region were still relatively low. Use of rock shelters was increasing or at the very least artefact discard increased so as to be archaeologically visible. Backed artefacts were introduced into the stone tool repertoire and in some areas were produced intensively. The focus of lithic sourcing shifted from the known gravel beds on the Hawkesbury-Nepean to more localised resources. While iconic engravings continued to be produced, along with transitional forms, the increased population pressures in the later part of this phase saw the early development of Sydney style figurative pigment art and open engraved art (Art Phase 2). The additional components of this art form appear when social networks become more powerful, partly necessitated by increasing social stresses resulting from sea level rise and decreasing (or shifting) territorial ranges.

Middle Bondaian   c.4,000 years to c.1,000 years BP

A dramatic increase in population densities occurred during this time period. The use of rock shelters for habitation increased and there was a conspicuous increase in the use of these locations for artefact manufacture and discard. An increased population necessitated social mechanisms to mediate uncontrolled and possible conflict-marked interactions. Evidence for increasing cultural control is the death by ritual spearing of the Narrabeen man around 3,700 cal BP. Other evidence for increasing social prescription included a proliferation of symbolic behaviour, particularly that which demonstrated local group social affiliation. Symbolic behaviour probably took many forms including body decoration and scarification, as well as the use of decorated portable material culture. The pigment and engraved art of the region developed and flourished in this milieu of escalating sociality.

Late Bondaian c.1,000 years to European contact

There is no firm evidence for a population decline during this period, but there are indications of changing social organisation and lithic strategies. Settlement indices show that rockshelters continued to be used but that occupation and artefact deposition rates dropped in these locations. It is argued that as a result of changes to the social system (Walters 1988) the focus shifted at this time to open camp site locations. This focus on open sites is supported by the establishment dates for open middens along the south-east coast generally and by myriad open sites on the Cumberland Plain (chapter 4).

The ethnohistoric evidence supports this recent habitation focus in open 'villages ... on the sea coast' (Beaglehole 1955, Collins 1802[1975]: 460, Phillip 1789[1970], Tench 1793[1961]). The archaeological and ethnohistoric evidence suggest that over the last millennium, occupation patterns involved a move away from shelters as a primary focus for habitation. The ethnohistoric literature also supports a model which encompasses larger territorial groupings, i.e. larger than single residence groups. Camping behaviour and spatial relations indicate that most rockshelters do not have a large enough floor area to accommodate a group larger than a single residence group.

The move out of shelters did not occur because the symbolic importance of these sites increased but was a pragmatic reaction to increased spatial requirements of larger social grouping. The increasing size of residence groups over time can be explained in terms of the shift to the dual social system (Hamilton 1980, Walters 1988).

Technological changes - particularly the introduction of fish hooks - suggest that there may have been an increase and a change in the nature of socially-shared food procurement activities (male hunting parties and women fishing). Such changes may have increased the reliability of food supplies, and supported increasing group sizes (Bowdler 1981). The change from family groups to gender-based foraging groups would not have been permanent social arrangements – but probably reflected daily and seasonal pursuits. The use of shelters on an *ad hoc* basis would have continued and increased mobility (compared to the preceding phase) is not suggested by the habitation indices (cf. Attenbrow 2004). The continued, but lower intensity use of shelters throughout the last millennium can be explained in terms of the patterns of aggregation and dispersal observed at contact. During times of seasonal abundance, groups lived in large, semi-permanent open 'villages'. In times of resource stress these larger groups dispersed and continued to exploit their range of already-established rockshelters, in their smaller family or gender-based hunting and fishing groups.

Social interaction throughout this period would have necessitated use of mechanisms to control social interaction and enforce larger-scale group cohesion. Symbolic behaviour, throughout the last millennium, would have continued to be an important facilitator of interaction. It is possible that many engraving sites were produced during this late period – particularly in the resource rich coastal strip. Increasing social complexity would also have produced the required conditions (Wiessner 1984, 1989) for demonstrations of group cohesion: fear, real or potential inter-group competition and aggression, and a need for co-operation to achieve certain goals.

On the periphery of the region there is evidence that this type of social cohesion is reduced. The engraved assemblages show both a marked decline in numbers and increased levels of heterogeneity. The relative paucity of engraving sites, and high levels of heterogeneity south of the Georges River style boundary, suggest that enforcing group cohesion in this part of the Sydney culture bloc was not as important.

The ubiquitous shelter art sites consistently demonstrate higher levels of heterogeneity across the region and particularly at its periphery - supporting this model. Higher levels of stylistic homogeneity in this medium at the margins of the region indicate that territorial bounding behaviour at the culture area's periphery was of increased importance. A model for this territorial interaction is suggested by these art analyses (Figure 13.1).

The concept of social context has been extremely important in developing this argument. So has the notion that style is a means of non-verbal communication used to negotiate identity (Wiessner 1990). Shelter art sites, with their demonstrable public context, and vertical engraving sites in similarly public economical resource zones both demonstrate higher levels of stylistic heterogeneity. Stylistic heterogeneity is argued here to be a marker of localised group identifying behaviour. Higher levels of stylistic homogeneity in the engraving medium demonstrate a mechanism for increased broad-scale group cohesion. Discerning the social contexts of these different locations has been crucial to understanding the social messaging potential of these different art contexts. Complex patterns of variability across the region in both contexts demonstrate the nature of ongoing contacts between language groups, as well as areas where the tensions resulting from these contacts were greatest (i.e. between language groups and at the periphery of the culture bloc).

The rock art in the Sydney region functioned as a prehistoric information superhighway. Through stylistic behaviour, groups around the region, who were not in constant verbal contact with each other, were able to communicate important social messages and demonstrate both broad-scale group cohesion and within-group distinctiveness. Throughout the Sydney region people signalled information about themselves using symbolic and iconographic signatures. This socially mediated their interactions, creating buffers against inter-group conflict during several phases of substantial change in their society, settlement patterns and territoriality.

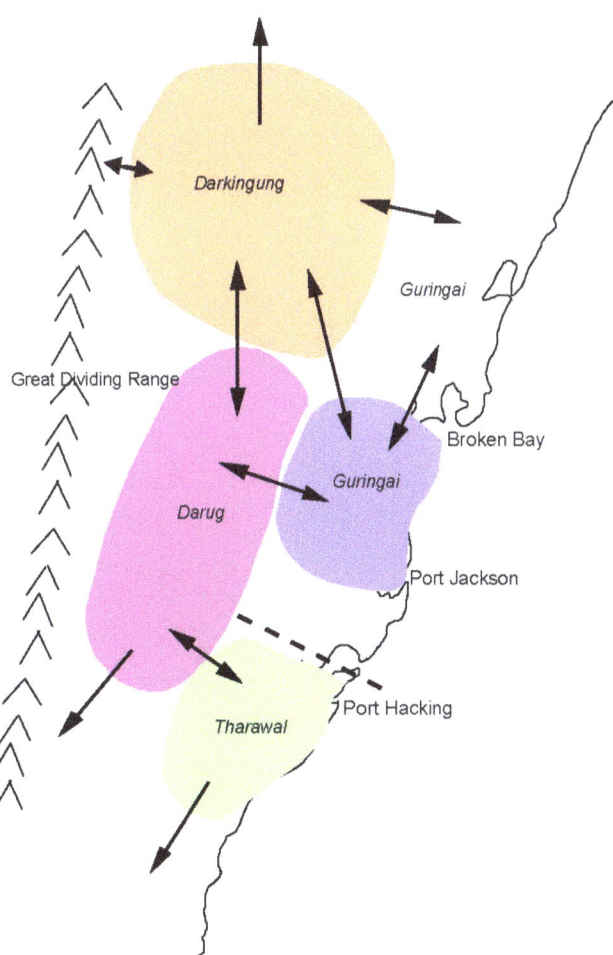

Figure 13.1: Model for territorial organisation and interaction across the Sydney region.

14

REFERENCES

Allen, G.R., Midgley, S.H., and Allen, M. 2002. *Field Guide to the Freshwater Fishes of Australia*. Western Australian Museum. Perth.

Angas, G.F. 1847. *Savage Life and Scenes in Australia and New Zealand: being an artist's impression of countries and people at the antipodes.* 2nd ed. Smith, Elder and Co, London. Reprinted 1969. Australiana Facsimile Editions No. 184. Libraries Board of South Australia.

Aplin, K. 1981. Faunal Remains from Archaeological Sites in Mangrove Creek Catchment. In V. Attenbrow (ed.) Mangrove Creek Dam: Salvage Excavation Project. Volume 2. Report to NSW National Parks and Wildlife Service (NPWS) on behalf of the Public Works Department. Report held by NSW NPWS.

Aplin, K. 1985. Report on Faunal Remains from site M11. In M. Koettig 1985 Archaeological investigation of three sites on Upper Mill Creek, near Lucas Heights, Sydney; pp. 82-86. Unpublished report to MWDA. Report held NSW NPWS.

Aplin, K. and Gollan, K. 1982. Faunal Remains from Archaeological Sites in the Mangrove Creek Dam Catchment II: The Deep Creek Fauna. Appendix 3 in V.J. Attenbrow, Mangrove Creek Dam: Salvage Excavation Project. 2 Volumes. Report to NSW National Parks and Wildlife Service (NPWS) on behalf of the Public Works Department.

Ashton, N.M. 1983. Spatial patterning in the Middle-Upper Palaeolithic transition. *World Archaeology* 15(2):224-234.

Attenbrow, V.J. 1981. Mangrove Creek Dam: Salvage Excavation Project. Volumes 1 and 2. Report to NSW NPWS on behalf of PWD. Report held NSW NPWS.

Attenbrow, V.J. 1982. The Archaeology of Upper Mangrove Creek: Research in Progress. In S. Bowdler (ed.) *Coastal Archaeology in Eastern Australia*, pp.67-79. Department of Prehistory, Research School of Pacific Studies, Australian National University, Canberra.

Attenbrow, V.J. 1987. The Upper Mangrove Creek Catchment: A study of quantitative changes in the archaeological record. Unpublished PhD Thesis, University of Sydney, Sydney.

Attenbrow, V.J. 1988. Research into the Aboriginal occupation of the Hunter's Hill Municipality. Report to Hunter's Hill Municipal Council. Report held NSW NPWS.

Attenbrow, V.J. 1990. The Port Jackson Archaeological Project: report on Stage I. Report to Australian Institute of Aboriginal and Torres Strait Islander Studies and Local Aboriginal Land Council.

Attenbrow, V.J. 1992. Port Jackson Archaeological Project - Stage II. Report to the Australian Institute of Aboriginal and Torres Strait Islander Studies.

Attenbrow, V. 2002. *Sydney's Aboriginal Past: investigating the archaeological and historical records*. University of New South Wales Press Ltd, Sydney.

Attenbrow, V. 2003. Habitation and land use patterns in the Upper Mangrove Creek catchment, New South Wales central coast, Australia. *Australian Archaeology* 57:20-31.

Attenbrow, V. 2004. *What's Changing: Population Size or Land use patterns? The archaeology of Upper Mangrove Creek, Sydney Basin.* Terra Australis 21, Pandanus Books, Research School of Pacific and Asian Studies, Australian National University, Canberra.

Attenbrow, V.J. and Negerevich, T. 1981. Lucas Heights Waste Depot proposed extensions: Aboriginal sites in the Barden's Creek Valley. Report to MWDA. Report held NSW NPWS.

Attenbrow, V.J. and Negerevich, T. 1984. The assessment of sites. Lucas Heights waste disposal depot: a case study. In S. Sullivan and S. Bowdler (eds), *Site Survey and Significance assessment in Australian Archaeology,* pp.136-151. Canberra: Research School of Pacific Studies, Australian National University.

Bahn, P.G. 1986. No sex, please, we're Aurignacians. *Rock Art Research* 3(2):99-105.

Bahn, P.G. and Vertut, J. 1988. *Images of the Ice Age.* Windward, England.

Bailey, G.N. 1975. The Role of Molluscs in Coastal Economies: the results of midden analysis in Australia. *Journal of Archaeological Science* 2:45-62.

Baker, N. 1992. Technological structure and organisation in the Hunter Valley: technological analysis of artefacts within the Narama Coal Lease, Hunter Valley, NSW. Appendix 2 to E. Rich (1992) Narama Salvage Project, Lower Bayswater Creek, Hunter Valley, NSW, pp. 1-98. Brayshaw McDonald Pty Ltd report to Envirosciences and the Narama Joint Venture.

Baker, R.G.V. and Haworth, R.J. 2000. Smooth or oscillating late Holocene sea-level curve? Evidence from the palaeozoology of foxed biological indicators in East Australia and beyond. *Marine Geology* 163(104):367-386.

Backhouse, J. 1843. *A Narrative of a Visit to the Australian Colonies.* Hamilton, Adams and Co., London.

Balme, J., L. Dagg, M. David, I. Davidson and J. Ross 2001. Archaeological investigations: Parklea, NSW. Salvage Excavations at PK/CD1+2 and PK/CD4+6. Volume 3. Report prepared for Rouse Hill (Stage 1) Pty Ltd.

Barker, B. 2004. *The Sea People: Late Holocene maritime specialisation in the Whitsunday Islands, central Queensland.* Terra Australis 20, Pandanus Books, Research School of Pacific and Asian Studies, Australian National University, Canberra.

Barrallier, F. 1802. *Journal of the expedition into the interior of New South Wales.* Printed in 1975, Marsh Walsh Publishing, Melbourne.

Barratt, G. 1981. *The Russians at Port Jackson 1814-1822.* Australian Institute of Aboriginal Studies, Canberra.

Barrington, G. 1802. *The History of New South Wales, including Botany Bay, Port Jackson, Parramatta, Sydney and all its Dependencies.* Paternoster Row, London.

Baxter, M.J. 1994. *Exploratory Multivariate Analysis in Archaeology.* Edinburgh University Press, Edinburgh.

Basedow, H. 1904. Anthropological notes made on the South Australian Government Northwest prospecting Expedition 1903. *Transactions of the Royal Society of South Australia* 28:12-53.

Beadle, N.C.W., Evans, O.D. and Carolin, R.C. 1986. *Flora of the Sydney Region.* Reeds Book Pty Ltd., N.S.W.

Beaglehole, J.C. (ed.) 1955. *The journals of Captain James Cook on his voyage of discovery. The voyage of the Endeavour 1768-1771.* Cambridge: Hakluyt Society.

Beaton, J.M. 1985. Evidence for a Coastal Occupation Time-Lag at Princess Charlotte Bay (North Queensland) and Implications for Coastal Colonisation and Population Growth Theories for Aboriginal Australia. *Archaeology in Oceania* 20(1):1-20.

Beaton, J.M. 1991a. Excavations at Rainbow Cave and Wanderer's Cave: Two Rockshelters in the Carnarvon Range, Queensland. *Queensland Archaeological Research* 8:3-32.

Beaton, J.M. 1991b. Cathedral Cave: A rockshelter in Carnarvon Gorge, Queensland. *Queensland Archaeological Research* 8:33-84.

Beck, W., D.J. Donohue, A.J.T. Jull, G. Burr, W.S. Broecker, G. Bonani, J. Hajdas and E. Malotki 1998. Ambiguities in direct dating of rock surfaces using radiocarbon measures. *Science*, 280: 2132-2135.

Bednarik, R.G. 1988. Comment on FD McCarthy's Rock art sequences: a matter of clarification. *Rock Art Research* 5(1): 35-38.

Bednarik, R. 1996. Only time will tell: A review of the methodology of direct rock art dating. *Archaeometry* 38:1-13.

Begouen, R. and J. Clottes 1985. L'art mobilier des Magdaleniens. *Archeologia* 207, novembre:40-49.

Bell, D. 1983. *Daughters of the Dreaming.* McPhee Gribble/George Allen and Unwin, Melbourne.

Benson, D.H. 1979. Vegetation survey of Upper Mangrove Creek, Wyong. Unpublished report to Public Works Department NSW.

Benzecri, J.-P. 1992. *Correspondence Analysis Handbook.* Marcel Dekker, Inc. New York, Basel, Hong Kong.

Bonhomme Craib and Associates 1999. Archaeological salvage of site RSI (NPWS # 45-5-982), Regentville, Western Sydney. Unpublished report to TransGrid. Report held NPWS NSW.

Berndt, C.H. 1986. Digging sticks and spears: or, a two-sex model. In F. Gale (ed.) *Woman's role in Aboriginal society*, pp. 64-84. 3rd edition, Australian Institute of Aboriginal Studies, Canberra. 1st edition, 1978.

Bird, C.F.M. and Frankel, D. 1991. Chronology and explanation in western Victoria and south-east South Australia. *Archaeology in Oceania* 26:1-16.

Bolviken, E., Helskog, E. Helskog, H. 1982. Correspondence analysis: an alternative to principal components. *World Archaeology* 14(1):41-60.

Boot, P. 1996. Aspects of prehistoric change in the south coast hinterland of New South Wales. In S. Ulm, I. Lilley and A. Ross (eds) *Australian Archaeology '95: Proceedings of the 1995 AAA Annual Conference. Tempus Volume 6*: 63-79, UQ, Brisbane.

Bowdler, S. 1971. Ball's Head: The excavation of a Port Jackson rock shelter. *Records of the Australian Museum* 28:117-128.

Bowdler, S. 1976. Hook, Line and Dilly Bag: an interpretation of an Australian coastal shell midden. *Mankind* 10(4):248-58.

Bowdler, S. 1977. The coastal colonisation of Australia. In J. Allen, J. Golson and R. Jones (eds) *Sunda and Sahul: Prehistoric Studies in Southeast Asia, Melanesia and Australia*, pp.205-246. Academic Press, London.

Bowdler, S. 1981. Hunters in the highlands: Aboriginal adaptations in the eastern Australian uplands. *Archaeology in Oceania* 16(2):99-111.

Bowdler, S. 1984a. Hunter *Hill, Hunter Island*. Terra Australis 8, Department Prehistory, Research School of Pacific Studies, Australian National University, Canberra.

Bowler, J.M., Johnston, H., Olley, J.M., Prescott, J.R., Roberts, R.G., Shawcross, W. and Spooner, N.A. 2003. New ages for human occupation and climatic change at Lake Mungo, Australia. *Nature* 421: 837-840.

Bradley, W. 1786-92. *A voyage to New South Wales; the Journal of LT. William Bradley RN of HMS Sirius.* Ms.A3631, Mitchell Library, Sydney. Published 1969 by William Dixon Foundation, Publication No. 11, Ure Smith Pty Ltd.

Branagan, D.F., Herbert, C. and Langford-Smith, T. 1976. *The Sydney Basin: an outline of the geology and geomorphology.* Science Press, Sydney.

Brayshaw, H.C. 1986. *Aborigines of the Hunter Valley: a study of colonial records.* Bicentennial Publication No. 4, Scone and Upper Hunter Historical Society, NSW.

Brockwell, C.J. 1989. Archaeological investigations of the Kakadu wetlands, Northern Australia. Unpublished MA Thesis, Australian National University, Canberra.

Bridges, B. 1970. Pemulwuy: a "Noble Savage". *Notes of the Royal Australian Historical Society* 88:3-4.

Bureau of Meteorology 1975. Rainfall figures for District 61. Australian Govt. Publishing Service, Canberra.

Bureau of Meteorology 1979. Climatic survey: Sydney Region 5; New South Wales. Australian Govt. Publishing Service, Canberra.

Butlin, N.G. 1983. *Our original aggression: Aboriginal populations of south-eastern Australia 1788-1850.* George, Allen and Unwin, Sydney.

Byrnes, J. 1982. Origin of Silcrete in the Cumberland Basin. Unpublished Petrological Report 82/17, Geological Survey of N.S.W. Department of Mineral Resources.

Callaghan, M. 1980. Some previously unconsidered environmental factors of relevance to South Coast prehistory. *Australian Archaeology* 11:43-47.

Campbell, J. 2002. *Invisible Invaders: Smallpox and Other Diseases in Aboriginal Australia 1780-1880.* Melbourne: Melbourne University Press.

Campbell, W.D. 1899. Aboriginal carvings of Port Jackson and Broken Bay. *Memoirs of the Geological Society of NSW, Ethnological Series 1.* Department of Mines and Agriculture.

Capell, A. 1970. Aboriginal languages in the south central coast, NSW: Fresh discoveries. *Oceania* 41:20-27.

Cartailhac, E. and Breuil, H. 1906. La caverne d'Altimira a Santillane, pres Santander (Espagne). Imp. de Monaco.

Chaloupka, G. 1977. Aspects of the chronology and schematisation of two prehistoric sites on the Arnhem Land Plateau. In P. Ucko (ed.) *Form in Indigenous Art*, pp. 243-59. Australian Institute of Aboriginal Studies, Canberra.

Chaloupka, G. 1985. Chronological sequence of Arnhem Land Plateau rock art. In R. Jones (ed.) *Archaeological Research in Kakadu National Park*, pp. 269-80. Australian National Parks and Wildlife Service Special Publication No 13, Canberra.

Chaloupka, G. 1994. *Journey in Time: the world's longest continuing art tradition.* Reed, Australia.

Chappell, J.M.A. 1982. Sea levels and sediments: some features of the context of coastal archaeological sites in the Tropics. *Archaeology in Oceania* 17(2):69-78.

Chappell, J.M.A. 1983. Sea level changes, 0 to 40KA. In J.M.A. Chappell and D. Grindrod (eds) *Climanz. Proceedings of the First Climanz Conference*, pp.121-122. Department of Biogeography and Geomorphology, Research School of Pacific Studies, Australian National University, Canberra.

Chappell, J. 2003. The LGM - Longer, colder and harder. Paper presented at AAA conference, Jindabyne, in session called The Last Glacial Maximum in Australia – Implications for Understanding Human Occupation.

Chappell, J.M.A. and Thom, B.G. 1977. Sea levels and coasts. In J. Allen, J. Golson and R. Jones (eds) *Sunda and Sahul*, pp. 275-291. Academic Press, Sydney.

Chippindale, C. and Taçon P.S.C. 1998b. The many ways of dating Arnhem Land rock-art, northern Australia. In C. Chippindale and Taçon P.S.C. (eds) *The Archaeology of Rock-art,* pp.90-111. Cambridge University Press, UK.

Clegg, F. 1990. *Simple Statistics: A Course Book for the Social Sciences.* Cambridge University Press, Cambridge.

Clegg, J.K. 1977. A method of resolving problems which arise from style in art. In P.J. Ucko (ed.) *Form in Indigenous Art*, pp.260-276. Australian Institute of Aboriginal Studies, Canberra.

Clegg, J.K. 1978. *Mathesis pictures; Mathesis words*. Unpublished MA Thesis, University of Sydney, Sydney.

Clegg, J.K. 1979. *Milligan's: The State at 1/xii/79.* Unpublished Ms. Copy held on file NSW NPWS.

Clegg, J.K. 1981. *Notes towards Mathesis Art*. Clegg Calendars, Balmain.

Clegg, J.K. 1987. Style and Tradition at Sturt's Meadows. *World Archaeology* 19(2):236-255.

Clendinnen, I. 2003. *Dancing with Strangers.* Melbourne, Australia.

Clottes, J. and J. Courtin 1996. *The Cave beneath the Sea: Palaeolithic Images at Cosquer.* Harry N. Abrahams, New York.

Coates, I. 1999. Lists and letters: an analysis of some exchanges between British museums, collectors and Australian Aborigines (1895-1910). Unpublished PhD Thesis, Department of Anthropology and Prehistory, The Faculties, Australian National University, Canberra.

Cole, N.A. and Trezise, P.J. 1992. Laura engravings: a preliminary recording of the Amphitheatre Site. In J.J. McDonald and I.P. Haskovec (eds) *State of the Art: Regional rock art studies in Australia and Melanesia,* pp.83-88. AURA Publication No.6, Melbourne.

Cole, N.A., Watchman, A. and Morwood, M.J. 1994. Chronology of Laura Rock Art. In M.J. Morwood and D.R. Hobbs (eds) *Quinkin Prehistory: the archaeology of Aboriginal art in S.E. Cape York Peninsula*, pp.147-160. Tempus Volume 3. University of Queensland, Brisbane.

Collins, D. 1798. *An Account of the English Colony in New South Wales.* Volume I. London: Cadell and Davies. Republished 1975, B.H. Fletcher (ed.) A.H. and A.W. Reed, Sydney (RAHS in assoc.).

Collins, D. 1802. *An Account of the English Colony in New South Wales.* Volume II. London: Cadell and Davies. Republished 1975, B.H. Fletcher (ed.) A.H. and A.W. Reed, Sydney; (RAHS in assoc.).

Collins, D. 1804. *An Account of the English Colony in New South Wales.* Whitcombe and Tombs Ltd., Melbourne, 1910.

Conkey, M.W. 1978. Style and Information in Cultural Evolution: towards a predictive model for the Palaeolithic. In C.L. Redman, M.J. Berman, E.V. Curtvin, W.T. Langhorne, N.M. Vergassi and J.C. Wauser (eds) *Social Archaeology: beyond subsistence and dating*, pp.61-85. Academic Press, New York.

Conkey, M.W. 1980. The Identification of Hunter-Gatherer Aggregation sites - the case of Altamira. *Current Anthropology* 21(5):609-630.

Conkey, M.W. 1987. Interpretative problems in Hunter-Gatherer Regional Studies: some thoughts on the European Upper Palaeolithic. In O. Soffer (ed.) *The Pleistocene Old World: Regional Perspectives*, pp.63-77. Plenum Press. New York.

Conkey, M.W. and Hastorf, C.A. 1990. Introduction. In Conkey, M.W. and C.A. Hastorf (eds) *The uses of style in archaeology*, pp.1-4. New Directions in Archaeology, Cambridge University Press, Cambridge.

Cook, N., Davidson, I. and Sutton, S. 1990. Why are so many ancient rock paintings red? *Australian Aboriginal Studies* 1:30-32.

Corkill, T. 1999. Here and there: links between stone sources and Aboriginal archaeological sites in Sydney, Australia. Masters of Philosophy thesis, Department of Archaeology, University of Sydney.

Corkill, T. 2005. Sourcing stone from the Sydney region: a hatchet job. *Australian Archaeology* 60:41-50.

Cox, J., Maynard, L. and Megaw, J.V.S. 1968. The Excavation of a rock shelter at Audley, Royal National Park, NSW. *Anthropology and Physical Anthropology in Oceania* III:94-104.

Curson, P. 1985. All Dead ! All Dead ! In *Times of Crisis - epidemics in Sydney 1788 - 1900*, pp. 41-53. University of Sydney Press.

David, B. 1991. Fern Cave, rock art and social formations: Rock art regionalization and demographic models in south-eastern Cape York Peninsula. *Archaeology in Oceania* 26:41-57.

David, B. 1992. Analysing space: investigating context and meaning in the rock paintings of the Chillagoe-Mungana limestone belt of north Queensland. In J. McDonald and I. P. Haskovec (eds) *State of the Art: Regional rock art studies in Australia and Melanesia*, pp.159-163. AURA Publication No.6, Melbourne.

David, B. 1994. *A space-time odyssey: rock art and regionalisation in north Queensland prehistory.* Unpublished PhD Thesis, University of Queensland, Brisbane.

David, B. 2002. Landscapes, rock art and the Dreaming: an archaeology of preunderstanding. Leicester University Press, London.

David, B., B. Barker and I.J. McNiven (eds) 2006. *The social archaeology of Australian Indigenous Societies.* Aboriginal Studies Press, Canberra.

David, B. and Cole, N. 1990. Rock art and inter-regional interaction in northeast Australian prehistory. *Antiquity* 64:788-806.

David, B., R.A. Armitage, M. Hyman, M.W. Rowe, and E. Lawson 1999. How old is North Queensland Rock-Art? A review of the evidence with new AMS determinations. *Archaeology in Oceania* 34:103-20.

Davidson, D.S. 1938. An ethnic map of Australia. *Proceedings of the American Philosophical Society* 79:649-79.

Dawes, W. 1790. Languages of the Port Jackson Aborigines. Unpublished manuscript.

d'Errico, F. 1994. Birds of Cosquer Cave: the Great Ark (*Pinguinus impennis*) and its significance during the Upper Palaeolithic. *Rock Art Research* 11(1):45-57.

Dickens, J. 1992. "Shielding the Dreaming": a Formal Analysis of Central Australian Shield Designs. Unpublished BA(Hons) Thesis, Australian National University, Canberra.

Dodson, J.R. and Hope, G. 1983. Southwestern Victoria and the Coorang, 9000-5000 BP. In J.M.A. Chappell and D. Grindrod (eds) *Climanz. Proceedings of the First Climanz Conference* pp.74-75. Department of Biogeography and Geomorphology, Research School of Pacific Studies, Australian National University, Canberra.

Dragovich, D. 1986. Minimum age of some desert varnish near Broken Hill, NSW. *Search*, 17 5/6, May/June:149-150.

Edwards, R. 1971. Art and Aboriginal prehistory. In D.J. Mulvaney and J. Golson (eds) *Aboriginal man and environment in Australia*, pp.356-367. Australian National University Press, Canberra.

Elkin, A.P. 1946. The Wollombi and Parish of Wollombi: a history. Manuscript, University of Sydney, Sydney.

Elkin, A.P. 1949. The origin and interpretation of petroglyphs in south-east Australia. *Oceania* 20:119-57.

Enright, W.J. 1901. Aboriginal districts and notes. *Science of Man* 4(5):80-81.

Fletcher-Jones, N.A. 1985. Across the First Frontier: the behavioural ecology of the Sydney Region Aborigines. Unpublished PhD Thesis, University of Durham, United Kingdom.

Flood, J. 1976. Man and ecology in the highlands of south-eastern Australia: a case study. In N. Peterson (ed.) *Tribes and Boundaries in Australia*, pp. 30-49. Social Anthropology Series No. 10, Australian Institute of Aboriginal Studies, Canberra. Humanities Press Inc, New Jersey, USA.

Flood, J. 1980. *The Moth Hunters: Aboriginal Prehistory of the Australian Alps.* Australian Institute of Aboriginal Studies, Canberra.

Flood, J. 1987. Rock Art of the Koolburra Plateau, north Queensland. *Rock Art Research* 4(2):91-126.

Flood, J.M and Horsfall, N. 1986. Excavation of Green Ant and Echidna Shelters, Cape York Peninsula. *Queensland Archaeological Research* 3:4-64.

Foley, R. 1981. A model of regional archaeological structure. *Proceedings of the Prehistoric Society* 47:1-17.

Ford, J. 2006. Painting contact: characterising the paints of the South Woronora Plateau Rock Art Assemblage, Wollongong, NSW. Unpublished BA(Hons) Thesis, Australian National University, Canberra.

Forge, A. 1991. Hand stencils: rock art or not art. In P. Bahn and A. Rosenfeld (eds) *Rock Art and Prehistory: papers presented to symposium G of the AURA Congress, Darwin 1988*, pp.39-44. Oxbow Monograph 10, Oxford.

Franklin, N. 1984. Of !Macropods and !Men: an analysis of the Simple Figurative styles. Unpublished BA(Hons) Thesis, University of Sydney, Sydney.

Franklin, N. 1988. Comment on F.D. McCarthy's Rock Art Sequences: a matter of clarification. *Rock Art Research* 5(1):24-28.

Franklin, N. 1991. Explorations of the Panaramitee style. In Bahn, P. and A. Rosenfeld (eds) *Rock Art and Prehistory: papers presented to symposium G of the AURA Congress, Darwin 1988*, pp. 120-135. Oxbow Monograph 10, Oxford.

Franklin, N. 2004. *Explorations of Variability in Australian Prehistoric Rock Engravings* BAR - S1318. Oxford.

Frost, R., David, B. and Flood, J.M. 1992. Pictures in transition: discussing the interaction of visual forms and symbolic contents in Wardaman rock pictures. In M.J. Morwood and D.R. Hobbs (eds) *Rock Art and Ethnography*, pp.27-32. Occasional AURA Publication No.5, Melbourne.

Galt-Smith, B. 1997. Motives for Motifs: Identifying Aggregation and dispersion settlement patterns in the rock art assemblages of Central Australia. Unpublished BA Honours Thesis: University of New England, Armidale.

Gamble, C.S. 1982. Interaction and alliance in Palaeolithic society. *Man* 17(1): 92-107.

Geib, P.R., Fairley, H.C. and Bungart, P.W. 1986. Archaeological survey in the Glen Canyon National Recreation Area: Year 1 Descriptive Report, 1984-1985. Northern Arizona University Archaeological Report No 999. Flagstaff.

Geib, P.R. and Fairley, H.C. 1992. Radiocarbon dating of Fremont Anthropomorphic Rock Art in Glen Canyon, South-central Utah. *Journal of Field Archaeology* 19:155-168.

Gibbs, M. and P. Veth 2002. Ritual Engines and the Archaeology of Territorial Ascendancy. In S. Ulm, C. Westcott, J. Reid, A. Ross, I. Lilley, J. Pragnell and L, Kirkwood (eds) *Barriers, Borders, Boundaries: Proceedings of the 2001 Australian Archaeological Association Annual Conference* Tempus 7, pp.11-19. Brisbane: University of Queensland.

Gillespie, R. 1991. The Australian marine shell correction factor. *Quaternary Dating Workshop 1990*. Department of Biogeography and Geomorphology, Research School of Pacific Studies, Australian National University, Canberra.

Glover, E. 1974. Report on the excavation of a second rock shelter at Curracurrang Cove, NSW. In J.V.S. Megaw, (ed.) *The Recent Archaeology of the Sydney District: Excavations 1964-1967*, pp.13-18. Australian Institute of Aboriginal Studies, Canberra.

Godden Mackay Logan Pty Ltd and Austral Archaeology Pty Ltd 1997 Prince of Wales Project 1995. Randwick Destitute Children's Asylum Cemetery. Archaeological investigation Volume 2 –Part 3 – Aboriginal Archaeology. Report prepared for the South Eastern Sydney Area Health Service, Heritage Council of NSW and NSW Department of Health.

Godwin, L. 1990. Inside information: settlement and alliance in the late Holocene of north-eastern New South Wales. Unpublished PhD Thesis, University of New England, Armidale.

Gonzales Ethegaray, J. 1974. Pinturas y grabados de la ceuva de las Chimeneas (Peunte Viesgo, Santander). *Monografias de Artes Rupestre, Arte Paleolitico No. 2,* Barcelona.

Gorecki, P.P. 1988. Hunters and shelter - the need for ethnoarchaeological data. In B. Meehan and R. Jones (eds) *Archaeology with ethnography: an Australian perspective*, pp.159-170. Department of Prehistory, Research School of Pacific Studies, Australian National University, Canberra.

Gould R.A. 1977. Puntutjarpa Rockshelter and the Australian Desert Culture. *Anthropological Papers of the American Museum of Natural History 54*(1).

Gunn R.B. 1979. Report on the Aboriginal Rock Art of the Upper Mangrove Creek Catchment. Report to NSW NPWS.

Gunn R.B. 1983. Aboriginal rock art of the Grampians. *Records of the Victorian Archaeological survey no 16*. Ministry for Conservation, Melbourne.

Gunn, R. G. 1995. Regional patterning in the Aboriginal rock art of central Australia: a preliminary report. *Rock Art Research* 12:117-127.

Gunn, R. G. 2000. Central Australian Rock Art: second report. *Rock Art Research* 17(2):111-26.

Gunn, R. G. 2003. Arrernte rock-art: interpreting physical permanence in a changing social landscape. *Australian Aboriginal Studies* 2003/1:52-73.

Gunnerson, J.H. 1969. The Fremont culture. *Papers of the Peabody Museum 59(2)*. Harvard University, Cambridge, Massc.

Gunson, N. (ed.) 1974. *Australian Reminiscences and papers of L.E. Threlkeld: missionary to the Aborigines 1824 - 1859*. Volumes I and II. Australian Aboriginal Studies No. 40. Australian Institute of Aboriginal Studies, Canberra.

Hamilton, A. 1978. Dual social systems: technology, labour and women's secret rites in the eastern Western Desert of Australia. Unpublished paper presented to the International Conference on Hunters and Gatherers, Paris, June 1978. Ms. held AITSIS, Canberra

Hamilton, A. 1980. Dual social systems: technology, labour and women's secret rites in the eastern Western Desert of Australia. *Oceania* 51:4-19.

Hart, C.W.M. 1970. Fieldwork among the Tiwi, 1928-1929. In G.D. Spendler (ed.) *Being an anthropologist - fieldwork in eleven cultures,* pp. 142-163. Holt, Rinehart and Winston, New York.

Haskovec, I.P. 1992. Mt Gilruth revisited. *Archaeology in Oceania* 27(2):61-74.

Haworth, R. J., R. G. V. Baker and P. G. Flood. 2002. Predicted and observed Holocene sea-levels on the Australian coast: what do they indicate about hydro-isostatic models in far-field sites? *Journal of Quaternary Science* 17:581-591.

Hatte, L. 1992. 'Boring coastal stuff': Rock images of the Townsville district, north Queensland. In J.J. McDonald and I. P. Haskovec (eds) *State of the Art: Regional rock art studies in Australia and Melanesia*, pp.71-75. Occasional AURA Publication No.6, Melbourne.

Head, J. 1991. The radiocarbon dating of freshwater and marine shell. *Quaternary Dating Workshop 1990*: Department of Biogeography and Geomorphology, Research School of Pacific and Asian Studies, Australian National University, Australian National University, Canberra.

Hiatt, L.R. 1962. Local organisation among the Australian Aborigines. *Oceania* 32:267-86.

Hiatt, L.R. 1965. *Kinship and Conflict: a study of an Aboriginal community in northern Arnhem Land.* Australian National University Press, Canberra.

Higgs, S. 2003. Emu Bums and Second Generation Knobs: a taxonomic look at some Sydney region rock engravings. Unpublished BA (Hons) Thesis, University of Sydney, Sydney.

Hiscock, P. 1981. Comments on the use of chipped stone artefacts as a measure of 'intensity of site usage'. *Australian Archaeology* 13(2):30-34.

Hiscock, P. 1986. Technological change in the Hunter Valley and its implications for the interpretation of Late Holocene change in Australia. *Archaeology in Oceania* 21(1):40-50.

Hiscock, P. 1993. Bondaian technology in the Hunter Valley, New South Wales. *Archaeology in Oceania* 28(2):65-76.

Hiscock, P. 1994. Technological responses to risk in Holocene Australia. *Journal of World Prehistory* 8(3):267-291.

Hiscock, P. 2003. Quantitative exploration of size variation and the extent of reduction in Sydney Basin assemblages: a tale from the Henry Lawson Drive rockshelter. *Australian Archaeology* 57:64-74.

Hiscock, P. and Attenbrow, V. 2003. Early Australian implement variation: a reduction model. *Journal of Archaeological Science* 30:239-249.

Hiscock, P. and Attenbrow, V. 2005. *Australia's Eastern Regional Sequence Revisited: technology and change at Capertee3.* BAR International Series 1397, Hadrien Books, Oxford.

Howitt, A.W. 1904. *The Native tribes of south-eastern Australia.* Macmillan, London.

Hughes, P.J. and Djohadze, J. 1980. Radiocarbon dates from archaeological sites on the south coast of New South Wales and the use of depth age curves. *Occasional Papers in Prehistory No. 1*, Dept. of Prehistory, Research School of Pacific Studies, Australian National University, Canberra.

Hughes, P.J. and Lampert, R.J. 1982. Prehistoric population change in southern coastal New South Wales. In S. Bowdler (ed.) *Coastal Archaeology in Eastern Australia*, pp.16-28. Department of Prehistory, Research School of Pacific Studies, Australian National University, Canberra.

Hughes, P.J. and Sullivan, M.E. 1979. A Geoarchaeological Investigation of Sandstone shelters in the Mangrove Creek Catchment, NSW. Appendix 2 In V. Attenbrow, Mangrove Creek Dam: Salvage Excavation Project. Volume 2. Report to NSW National Parks and Wildlife Service (NPWS) on behalf of the Public Works Department. Report held by NSW NPWS.

Hunter, J. 1793. *An historical journal of the transactions at Port Jackson and Norfolk Island.* London: J Stockdale. Published 1968 Australian Facsimile Editions No. 148, Libraries Board of South Australia, Adelaide.

Hyman, M. and M.W. Rowe 1997. Plasma Chemical Extraction and AMS Dating of Rock Paintings. *American Indian Rock Art*, Vol. 23, edited by S.M. Freers, pp.1-9. American Rock Art Research Association, San Miguel, CA.

HRNSW 1893. *Historic Records of New South Wales.* Mitchell Library.

James, R.E. 1993. Stones, samples and stories we tell. Unpublished BA (Hons) Thesis, University of New England, Armidale.

Johnson, I. 1979. The getting of data: a case study from the recent industries of Australia. Unpublished PhD thesis, Australian National University, Canberra.

Jo McDonald CHM Pty Ltd 1997a. Archaeological salvage of Site RM1 at Richmond, NSW: Test and Salvage excavation report. Report to Restifa and Partners on behalf of Woolworth's.

Jo McDonald CHM Pty Ltd 1997b. Interim heritage management report: ADI Site St Marys. Test excavation report (2 Volumes). Report to Lend Lease - ADI Joint Venture in response to the Section 22 Committee Interim Report.

Jo McDonald CHM Pty 1998a. Archaeological salvage of site WGO3-2 (NPWS # 45-5-971) at Wattle Grove, NSW. Report prepared for Delfin Management Services Pty Ltd.

Jo McDonald CHM 1998b. Archaeological Test Excavations at Site RWP-1 (NPWS 45-5-2404) Richmond Water Reuse Project, Richmond, NSW. Report to Sydney Water.

Jo McDonald CHM 1998c. Archaeological salvage excavation of an Aboriginal site, corner of Baker and George Street, Windsor, NSW.

Jo McDonald CHM Pty Ltd 1999. Test excavation of PAD5 (RH/SP9) and PAD31 (RH/CC2) for the Rouse Hill (Stage 2) Infrastructure Project at Rouse Hill and Kellyville, NSW. Report to RHIC.

JMcD CHM Pty Ltd. 2002. Archaeological excavations at Windsor Road, Kellyville, NSW (Site RH/CD12: NPWS # 45-5-962): Archaeological salvage programme prior to residential development. Report prepared for Australand Holdings.

JMcD CHM Pty Ltd. 2003. Archaeological Salvage Excavations at the Proposed Xavier College: Site ADI 47+48 (NPWS # 45-5-1048), Ninth Avenue, Llandilo, NSW. Report prepared for PMDL on behalf of the Catholic Education Office, Diocese of Parramatta, NSW.

Jo McDonald CHM Pty Ltd. 2004a. Archaeological Salvage Excavation at Conrad Road PADI (Site RH/SP 13J) Rouse Hill, NSW. Report prepared for Craig and Rhodes Pty Ltd on behalf of Australand Holdings Ltd.

Jo McDonald CHM Pty Ltd. 2004b. Archaeological Test Excavation at 1 Smith Street (Site SSP1) Parramatta, NSW. Report prepared for Casey and Lowe Pty Ltd on behalf of Sydney Water.

Jo McDonald CHM Pty Ltd. 2005a. *Salvage excavation of six sites along Caddies, Second Ponds, Smalls and Cattai Creeks in the Rouse Hill Development Area, NSW.* Australian Archaeological Consultancy Monograph Series, Volume 1, AACAI, St Lucia, Queensland.

Jo McDonald CHM Pty Ltd. 2005b. Archaeological salvage excavation of site CG1 (NPWS # 45-5-2648), at the corner of Charles and George Streets, Parramatta, NSW. Report prepared for Meriton Apartments Pty Ltd.

Jo McDonald CHM Pty Ltd. 2005c. Archaeological testing and salvage excavations at Discovery Point in the former grounds of Tempe House, NSW. Report to Australand Holdings Pty Ltd.

Jo McDonald CHM Pty Ltd. 2005d. Archaeological salvage excavation of eight archaeological landscapes in the Second Ponds Creek Valley, Rouse Hill Development Area, NSW. Report to RHI and Landcom. Three Volumes.

Jo McDonald CHM Pty Ltd. 2005e. Archaeological Salvage Excavation of Site RTA-G1: 109-113 George Street Parramatta, NSW. Report to Landcom.

Jo McDonald CHM Pty Ltd. 2005f. Salvage Excavation of Human Skeletal Remains at Ocean and Octavia Streets, Narrabeen (Site #45-6-2747). Report to Energy Australia.

Jo McDonald CHM Pty Ltd. 2006a. Archaeological Salvage Excavation of Site CG3: 101a-105 George Street, Parramatta, NSW. Report to Rahi Developments Ltd.

Jo McDonald CHM Pty Ltd. 2006b. Archaeological Salvage Excavation of the Colebee Release Area, Schofields, NSW. Report prepared for Medallist Golf Holdings Pty Ltd. Three Volumes.

Jo McDonald CHM Pty Ltd. 2006c. Archaeological salvage of a stormwater easement and testing within the State Heritage Register Conservation Precinct at the former Tempe House, Discovery Point. Report to Australand Holdings Pty Ltd.

Jo McDonald CHM Pty Ltd. 2007. Salvage Excavation of Four Archaeological Sites in the Caddies Creek Precinct, Rouse Hill Regional Centre, NSW. Report to Lend Lease GPT (Rouse Hill) Pty Ltd.

Jones, R. 1977. The Tasmanian Paradox. In R.V.S. Wright (ed.) *Stone tools as cultural markers: change evolution and complexity*, pp.189-204. Australian Institute of Aboriginal Studies, Humanities Press Inc., New Jersey.

Kelleher, M. 2003 Archaeology of sacred space: the spatial nature of religious behaviour in the Blue Mountains National Park, Australia. Two volumes. Unpublished PhD Thesis, University of Sydney, Sydney.

Kershaw, P., P.G. Quilty, B. David, S. van Huet, and A. McMinn 2000. Palaeobiogeography of the Quaternary of Australasia. In A.J. Wright, G.C. Young, J.A. Talent and R.J. Laurie (eds) *Palaeobiogeography of Australasian Faunas and Floras*, pp. 471-515. Memoir 23 of the Association of Australasian Palaeontologists, Canberra.

Keyser, J.D. 2001. Relative Dating methods. In D. Whitley (ed.) *Handbook of Rock Art Research, pp.116-138*. Altamira Press, Walnut Creek, California, USA.

Koettig, M.K. 1976. Rising damp. Unpublished MA Thesis, University of Sydney, Sydney.

Koettig, M.K. 1985. Archaeological investigation of three sites on Upper Mill Creek, near Lucas Heights, Sydney. Unpublished report to MWDA. Report held NSW NPWS.

Koettig, M.K. 1992. Salvage excavations of Aboriginal sites on the Camberwell Lease. Unpublished report to Camberwell Coal P/L. Report held NSW NPWS.

Koettig, M.K. and McDonald, J.J 1984. Archaeological survey for Aboriginal sites in the Upper Mill Creek Area: an alternative site for the Lucas Heights Waste Disposal Depot. Unpublished report to MWDA. Report held NSW NPWS.

Kohen, J.L. 1986. Prehistoric settlement in the western Cumberland Plain: resources, environment and technology. Unpublished PhD Thesis, School of Earth Sciences, Macquarie University, Sydney.

Kohen, J.L. 1988. The Dharug of the western Cumberland Plain: ethnography and demography. In B. Meehan and R. Jones (eds) *Archaeology with ethnography: an Australian perspective*, pp.238-250. Dept of Prehistory, Research School of Pacific Studies, Australian National University, Canberra.

Kohen, J. 1993. *The Darug and Their Neighbours: The Traditional Aboriginal Owners of the Sydney Region*. Blacktown: Darug Link in association with Blacktown and District Historical Society.

Kohen, J.L. and Lampert, R.J. 1988. Hunters and Fishers in the Sydney Region. In D.J. Mulvaney and P.J. White (eds) *Australians to 1788. Australians: a Historical Library*, pp.342-365. Fairfax, Syme and Weldon Associates, Broadway, Australia.

Kohen, J.L., Stockton, E. and Williams, M. 1984. Shaw's Creek K2 Rockshelter: a prehistoric occupation site in the Blue Mountains piedmont, eastern NSW. *Archaeology in Oceania* 19(2):57-73.

Koettig, M. and P. Hughes 1995 Test excavation at RS1 at Regentville near Penrith on the Cumberland Plain, Sydney. Report to Pacific Power.

Konecny, T. 1981. What could it all mean: Conclusions on an enigmatic subject. Unpublished BA(Hons) Thesis, University of Sydney.

Kroebber, A.L. 1939. Cultural and natural areas of native North America. *University of California publications in American archaeology and ethnology* 48:1-242.

Lampert, R.J. 1966. An excavation at Durras North, NSW. *Archaeology and Physical Anthropology in Oceania* 1(2):84-118.

Lampert, R.J. 1971a. *Burrill Lake and Currarong: Coastal Sites in Southern New South Wales.* Terra Australis 1, Department Prehistory, Research School of Pacific Studies, Australian National University, Canberra

Lampert, R.J. 1971b. Coastal Aborigines of south-eastern Australia. In D.J. Mulvaney and J. Golson (eds) *Aboriginal man and environment in Australia*, pp.114-132. ANU Press, Canberra.

Lampert, R.J. 1988. Aboriginal life around Port Jackson, 1788-92. In B. Smith and A. Wheeler (eds) *The Art of the First Fleet and other early Australian Drawings*, pp.19-69. Oxford University Press in association with the Australian Academy of the Humanities and the British Museum (Natural History), Melbourne.

Lampert, R.J. and Hughes, P.J. 1974. Sea level change and Aboriginal coastal adaptations in southern NSW. *Archaeology and Physical Anthropology in Oceania* IX(3):226-235.

Lampert, R.J. and Konecny, T. 1989. Aboriginal spears of Port Jackson type discovered - a bicentennial sequel. *Antiquity* 63:137-41.

Lampert, R.J. and Turnbull, G.E. 1970. The Manufacture of shell fish-hooks on the south coast of NSW. *Mankind* 7(4):312-3.

Lawrence, R.J. 1968. *Aboriginal habitat and economy.* Department of Geography Occasional Papers No 6, School of General Studies, Australian National University, Canberra.

Layton, R. 1989. The political use of Australian Aboriginal body painting and its archaeological implications. In I. Hodder (ed.) *The Meanings of Things: material culture and symbolic expression*, pp.1-11. Unwin Hyman, London.

Layton, R. 1992a. *Australian Rock Art: a new synthesis.* Cambridge University Press, Cambridge.

Layton, R. 1992b. The role of ethnography in the study of Australian rock art. In M.J. Morwood and D.R. Hobbs (eds) *Rock Art and Ethnography,* pp.7-9. Occasional AURA Publication No.5, Melbourne.

Leroi-Gourhan, A. 1968. *The Art of Prehistoric Man in Western Europe.* Thames and Hudson, London.

Lewis, D. 1988. *The rock paintings of Arnhem Land, Australia: Social, Ecological and Material Culture change in the post-glacial period.* BAR International Series 415.

Lipe, W.D. 1970. Anasazi communities in the Red Rock Plateau, south-eastern Utah. In W.A. Longacre (ed.) *Reconstructing Pueblo Societies*, pp.84-139. University of New Mexico Press, Albuquerque.

Lister, F.C. 1964. *Kaiparowits Plateau and Glen Canyon Prehistory: an interpretation based on ceramics.* University of Utah Anthropological Papers 71. Salt Lake City.

Lorblanchet, M. 1977. From naturalism to abstraction in European prehistoric rock art. In P.J. Ucko (ed.) *Form in Indigenous Art*, pp.44-56. Australian Institute of Aboriginal Studies, Canberra.

Lourandos, H. 1985. Intensification and Australian Prehistory. In T.D. Price, and J.A. Brown (eds) *Prehistoric Hunter-Gatherer: the emergence of cultural complexity*, pp.385-423. Academic Press.

Lourandos, H. 1997. *Continent of hunter-gatherers: new perspectives in Australian Prehistory*, Cambridge University Press, Cambridge.

Lourandos, H. and Ross, A. 1994. The great 'intensification debate': its history and place in Australian Archaeology. *Australian Archaeology* 39:54-63.

Loy, T.H., R. Jones, D.E. Nelson, B. Meehan, J. Vogel, J. Southon, and R. Cosgrove 1990. Accelerator radiocarbon dating of human blood proteins in pigments from Late Pleistocene Art Sites in Australia. *Antiquity* 64:110-16.

MacIntosh, N.W.G. 1965. Dingo and Horned Anthropomorph in an Aboriginal Rock Shelter. *Oceania* 36(2):85-101.

Mackay, R. and White, J.P. 1987. Musseling in on the NSW Coast: a comment. *Archaeology in Oceania* 22:107-111.

McBryde, I. 1974. *Aboriginal Prehistory in New England*. Sydney University Press, Sydney.

McBryde, I. 1976. Subsistence patterns in New England prehistory. *University of Queensland Occasional Papers in Anthropology* 6:48-68.

McBryde, I. 1979. Ethnohistory in an Australian context: Independent discipline or convenient data quarry? *Aboriginal History* 3(2):128-151.

McBryde, I. 1989. *Guests of the Governor - Aboriginal Residents of the First Government House.* Friends of the First Government House Site, Sydney.

McCarthy, F.D. 1939. The Aboriginal rock engravings of the Sydney District. *Australian Museum Magazine* 6:401-9.

McCarthy, F.D. 1940. Aboriginal Australian material culture: causative factors in its composition. Parts 1 and 2. *Mankind* 2(8):241-169 and *Mankind* 2(9):294-320.

McCarthy, F.D. 1944. Records of rock engravings of the Sydney District Numbers 7-20. *Mankind* 3(6):161-71.

McCarthy, F.D. 1946. Records of rock engravings of the Sydney District Numbers 21-32. *Mankind* 3(8):217-225.

McCarthy, F.D. 1947a. Records of rock engravings in the Sydney District: 33-37. *Mankind* 3(9):266-272.

McCarthy, F.D. 1947b. Records of rock engravings in the Sydney District: 38. *Mankind* 3(11):322-9.

McCarthy, F.D. 1948. The Lapstone Creek excavation: Two culture periods revealed in eastern NSW. *Records Australian Museum* 22:1-34.

McCarthy, F.D. 1949. Records of rock engravings of the Sydney District Numbers 39-40 *Mankind* 4(2):61-7.

McCarthy, F.D. 1954a. Records of rock engravings of the Sydney District Numbers 41-55. *Mankind* 5(1):8-32.

McCarthy, F.D. 1954b. Aboriginal rock engravings near Main Roads in the County of Cumberland. *Main Roads* XIX(4), June:105-112.

McCarthy, F.D. 1956. Rock engravings of the Sydney-Hawkesbury District (Part 1): Flat Rocks Ridge; a Daruk ceremonial ground. *Records Australian Museum* 24(5):37-58.

McCarthy, F.D. 1957. Theoretical considerations of Australian Aboriginal Art. *Journal and Proceedings of the Royal Society of NSW* 91:3-22.

McCarthy, F.D. 1959a. Rock engravings of the Sydney-Hawkesbury District (Part 2): some important ritual groups in the County of Cumberland. *Records Australian Museum* 24(14):203-15.

McCarthy, FD 1959b. Cave art of the Conjola District, New South Wales. *Records Australian Museum* 24(13):191-202.

McCarthy, F.D. 1960. The cave paintings of Groote Eylandt and Chasm Island. In C.P. Mountford, (ed.) *Records of the American-Australian Expedition to Arnhem Land*, pp. 297-414. Vol. 2, Anthropology and Nutrition. Melbourne University Press.

McCarthy, F.D. 1961. A remarkable ritual gallery of cave paintings in eastern NSW. *Records Australian Museum* 25(7):115-20.

McCarthy, F. D. 1962. The rock engravings at Port Hedland, North-Western Australia. *Kroeber Anthropological Society Papers* 26:1-73. University of California, Berkeley.

McCarthy, F.D. 1964. The archaeology of the Capertee Valley, NSW. *Records Australian Museum* 26(6):197-264.

McCarthy, F.D. 1976. *Rock Art of the Cobar pediplain.* Regional Research Studies 7. Australian Institute of Aboriginal Studies, Canberra.

McCarthy, F.D. 1978. New Light on the Lapstone Creek Excavation. *Australian Archaeology* 8:49-60.

McCarthy, F.D. 1979. *Australian Aboriginal Rock Art.* (4th ed.; 3rd ed. 1967) Australian Museum, Sydney.

McCarthy, F.D. 1983. Catalogue of rock engravings in the Sydney-Hawkesbury District, N.S.W. 3 Volumes; Edited by E.A. Higgenbotham. NSW NPWS, Sydney.

McCarthy, F.D. 1988. Rock Art Sequences: a matter of clarification. (with comments by Clegg, David, Franklin, McDonald, Maynard, Moore, Morwood, Rosenfeld and Bednarik: with author's reply). *Rock Art Research* 5(1):16-42.

McCarthy, F.D. and Hansen, F. 1958. Records of rock engravings of the Sydney District Numbers 56-71. *Mankind* 5(5):200-208.

McCarthy, F.D. and Hansen, F. 1960. Records of rock engravings of the Sydney District Numbers 72-102. *Mankind* 5(9):385-400.

McCarthy, F.D. and MacIntosh 1962. The archaeology of Mootwingee, western NSW. *Records Australian Museum* 25(13):249-298.

McDonald, J.J. 1983. The identification of species in a Panaramitee style engraving site. In M. Smith (ed.) *Archaeology at ANZAAS*, pp.236-48. West Australian Museum, Perth.

McDonald, J.J. 1985a. Sydney Basin Aboriginal Heritage Study: Rock Engravings and shelter Art Sites. Stage I. Unpublished report to NSW NPWS. Report held at NSW NPWS.

McDonald, J.J. 1985b. An excavation at Cherrybrook, Site #45-6-1649. Report to the Metropolitan Water Sewerage and Drainage Board. Unpublished report held at NSW NPWS.

McDonald, J.J. 1986a. Maroota Historic Site: Archaeological survey. Unpublished Report to NSW NPWS. Report held at NSW NPWS.

McDonald, J.J. 1986b. Preliminary archaeological reconnaissance of the proposed Schofields Regional Depot, Plumpton, NSW. Unpublished report to Metropolitan Waste Disposal Authority. Report held at NSW NPWS.

McDonald, J.J. 1987. Sydney Basin Aboriginal Heritage Study: Shelter Art Sites. Stage II. Unpublished report to NSW NPWS. Report held at NSW NPWS.

McDonald, J.J. 1988a. *The proposed Warre Warren Aboriginal Place, McPherson State Forest: Archaeological Investigation.* The Forestry Commission of NSW, Sydney.

McDonald, J.J. 1988b. Comment on FD McCarthy's 'Rock Art Sequences: a matter of clarification'. *Rock Art Research* 5(1):28-30.

McDonald, J.J. 1990a. Sydney Basin Aboriginal Heritage Study: Rock Engravings and Shelter Art Sites. Stage III. Unpublished report to NSW NPWS. Report held at NSW NPWS.

McDonald, J.J. 1990b. Review of Brayshaw (1982) Archaeological survey in West Menai Urban Release Area. Unpublished report to Travers Morgan on behalf of the Department of Housing. Report held by NSW NPWS.

McDonald, J.J. 1991. Archaeology and Art in the Sydney Region: Context and theory in the analysis of a dual medium style. In P. Bahn and A. Rosenfeld (eds) *Rock Art and Prehistory: papers presented to symposium G of the AURA Congress, Darwin 1988*, pp.78-85. Oxbow Monograph 10, Oxford.

McDonald, J.J. 1992a. *The Archaeology of the Angophora Reserve rockshelter: helping the police with their enquires.* Environmental Heritage Monograph Series No. 1, NSW NPWS.

McDonald, J.J. 1992b. The Great Mackerel Rockshelter excavation: Women in the archaeological record? *Australian Archaeology* 35:32-50.

McDonald, J.J. 1992c. Rock Art in the Sydney Region: synchronic and functional variation in a dual-media, regional art body. In J.J. McDonald and I. Haskovek (eds) *State of the Art: Regional Rock Art Studies in Australia and Melanesia*, pp. 20-36. AURA Publication No. 6, Melbourne.

McDonald, J.J. 1993a. The depiction of species in macropod track engravings at an Aboriginal art site in western New South Wales. *Records of the Australian Museum, Supplement* **17**:105-116.

McDonald, J.J. 1993b. On a clear day you can see Mt Yengo: or, Investigating the archaeological manifestations of culturally significant foci in the prehistoric landscape. In J. Steinbring A. Watchman, P. Faulstich and P. Taçon (eds) *Time and Space: Dating and spatial consideration in rock art research*, pp.1-18. AURA Publication No. 8. ARARA, Melbourne.

McDonald, J.J. 1993c. Excavation of two sites on Eastern Creek (PB1 and PB2), Doonside. Unpublished report to Blacktown City Council. Report held NSW NPWS.

McDonald, J.J. 1994a. Dreamtime Superhighway: An analysis of Sydney Basin rock art and prehistoric information exchange. Unpublished PhD Thesis, Australian National University, Canberra.

McDonald, J.J. 1994b. Everywhere you go you know he's been there before: The influence of Fred McCarthy on subsequent rock art research in the Sydney Basin. In M.E. Sullivan, S. Brockwell and A. Webb (eds) *Archaeology in the North: Proceedings of the 1993 Australian Archaeology Association Conference*, pp. 40-56. North Australian Research Unit, Australian National University, Darwin.

McDonald, J.J. 1995. Looking for a woman's touch: indications of gender in shelter sites in the Sydney Basin. In Balme, J. and Beck, W. (eds) *Gendered Archaeology: the second Australian Women in Archaeology Conference*, pp.92-96. ANH Publications, RSPAS, Australian National University, Canberra.

McDonald, J.J. 1996. The conservation of landscapes - a strategic approach to cultural heritage management. In S. Ulm, I. Lilley and A. Ross (eds) Australian Archaeology '95: Proceedings of the 1995 AAA Annual Conference. *Tempus Volume 6:*113-121, UQ, Brisbane.

McDonald, J.J. 1998a. Shelter art in the Sydney Basin - a space-time continuum: exploring different influences on stylistic change. In C. Chippindale and P. Taçon (eds) *Rock Art: New Directions in Archaeology,* pp.319-335. Cambridge University Press.

McDonald, J.J. 1998b. Beyond hook, line and dillybag: Gender, economics and information exchange in prehistoric Sydney. In Casey, M., D. Donlon, J. Hope and S. Welfare (eds) *Redefining Archaeology: Feminist Perspectives,* pp. 96-104. ANH Publications, Australian National University, Canberra.

McDonald, J.J. 1999. Bedrock notions and isochrestic choice: evidence for localised stylistic patterning in the engravings of the Sydney region. *Archaeology in Oceania* 34(3):145-160.

McDonald, J.J. 2000a. Media and social context: influences on information exchange networks in prehistoric Sydney. *Australian Archaeology* 51:54-63.

McDonald, J.J. 2000b. Archaeology, rock art, ethnicity and Native Title. In I. Lilley (ed.) *Native Title and Transformation of Archaeology in the Postcolonial World.* Oceania Monograph 50, pp.54-77. University of Sydney, Sydney.

McDonald, J.J. 2000c. The dating dilemma: problems in interpreting charcoal AMS dates from the Sydney region. In Ward, G.K. and C. Tuniz (eds.) *Advances in dating Australia rock markings: papers from the First Australian Rock-Picture Dating Workshop*, pp.90-94. AURA Publication No. 10. ARARA, Melbourne.

McDonald, J.J. 2005a. Australia: 'Rock Art Capital of the World'. In T. Murray (ed.) *Archaeology in Australia,* pp.98-115. Australian Scholarly Publishing, Melbourne.

McDonald, J.J. 2005b. Archaic faces to headdresses: The changing role of the rock art across the arid zone. In Veth, P., P. Hiscock and M. Smith (eds.) *Desert peoples: archaeological perspectives*, pp.116-141. Blackwell Studies in Global Archaeology, Boston.

McDonald, J.J. 2005c. Archaeology in the De Rose Native Title Claim. In Harrison, R. J. McDonald and P. Veth (eds) *Archaeology and Native Title. Australian Aboriginal Studies* 5:43-62.

McDonald, J.J. 2006. 'Rock-Art' In J. Balme and A. Patterson (eds) *Archaeology in Practice: a student's Guide to Archaeological Analyses*, pp.59-96. Blackwell Publishing Ltd, UK, USA, Australia.

McDonald, J.J. 2007. Dampier Archipelago: motifs and trait analysis in the assessment of scientific values of a rock art province. In *Rock art in the frame of Cultural Heritage of Humankind. Proceedings of the XXII Valcamonica Symposium 2007*, pp. 321-326. UNESCO, ICOMOS-CAR, UISPP, IIMP and the Centro Camuno di Studi Preistorica.

McDonald, J.J. 2008. Rock art and cross-cultural interaction in Sydney: how did both sides perceive the other? In Veth, P.M., P. Sutton and M. Neale (eds.) *Strangers on the Shore: Early Coastal Contacts with Australia*, pp. 46-89. National Museum of Australia Press, Canberra.

McDonald, J.J., D. Donlon, J.H. Field, R.L.K. Fullagar, J. Brenner-Coltrain, P. Mitchell and M. Rawson 2007. The first archaeological evidence for death by spearing in Australia. *Antiquity* 81:877-885.

McDonald, J., Mitchell, P., and Rich, E. 1996. A further investigation of site RS1 (# 45-5-892) at Regentville, Mulgoa Creek, western Sydney, NSW. Unpublished Pacific Power report to TransGrid.

McDonald, J.J., K.C. Officer, D. Donahue, T. Jull, J. Head and B. Ford 1990. Investigating AMS: dating prehistoric rock art in the Sydney Sandstone Basin, NSW. *Rock Art Research* 7(2):83-92.

McDonald, J.J. and Rich, E. 1994. The Discovery of a heat-treatment pit on the Cumberland Plain, western Sydney. *Australian Archaeology* 38: 46-48.

McDonald, J.J., E. Rich, and H. Barton 1994. The Rouse Hill Infrastructure Project (Stage 1) on the Cumberland Plain, western Sydney. In M.E. Sullivan, S. Brockwell and A. Webb (eds) *Archaeology in the North: Proceedings of the 1993 Australian Archaeology Association Conference*, pp.259-293. North Australian Research Unit, Australian National University, Darwin.

McDonald, J.J. and Smith, L. 1984. Archaeological survey of proposed (F3) Freeway Extensions: Wahroonga - Berowra. Unpublished report to Department of Main Roads. Report held at NSW NPWS.

McDonald, J. and Steelman, K. 2008. Rock art dating results from the Calvert and Carnarvon Ranges, Western Desert. Report to the Australian Institute of Aboriginal and Torres Strait Islander Studies, Canberra.

McDonald, J.J. and P.M. Veth 2005. S16 Report: Rock art dating work in the Calvert and Carnarvon Ranges, Western Desert. Unpublished report to Department of Indigenous Affairs, WA.

McDonald, J.J. and P.M. Veth 2006a. Rock art and social identity: a comparison of graphic systems operating in arid and fertile environments in the Holocene. In I. Lilley (ed.) *Archaeology of Oceania: Australia and the Pacific Islands*, pp.96-115. Blackwell Studies in Global Archaeology, Oxford.

McDonald, J.J. and P.M. Veth 2006b. Dampier Archipelago: additional analysis of scientific values, in support of National Heritage Listing. Unpublished Report to Department of Environment and Heritage.

McDonald, J.J. and P.M. Veth 2007. Pilbara and Western Desert Rock Art: Style graphics in arid landscapes. In *Rock art in the frame of Cultural Heritage of Humankind. Proceedings of the XXII Valcamonica Symposium 2007*, pp. 327-334. UNESCO, ICOMOS-CAR, UISPP, IIMP and the Centro Camuno di Studi Preistorica.

McDonald, J.J. and P.M. Veth 2008. Rock Art of the Western Desert and Pilbara: pigment dates provide new perspectives on the role of art in the Australian arid zone. *Australian Aboriginal Studies* 2008/1 (in press).

McMah, L. 1965. A quantitative analysis of the Aboriginal rock carvings in the District of Sydney and the Hawkesbury River. Unpublished BA(Hons) Thesis, University of Sydney, Sydney.

McNiven, I. 1999 Fissioning and regionalisation: the social dimensions of changes in Aboriginal use of the Great Sandy Region, southeast Queensland. In Hall, J. and McNiven, I.J. (eds) *Australian Coastal Archaeology*, pp.157-168. ANH Publications, Department of Archaeology and Natural History, Australian National University, Canberra.

Mann, J.F. 1885. Notes on the Aborigines of Australia. *Proceedings of the Geographic Society of Australasia; NSW Vic Branches* I.

Markgraf, V. and H.F. Diaz 2000. The past ENSO Record: a synthesis. In H.F. Diaz and V. Markgraf (eds) *El Nino and the Southern Oscillation. Multiscale variability and Global and regional impacts*, pp.465-88. Cambridge: Cambridge University Press.

Marshack, A. 1972. *The Roots of Civilisation*. McGraw-Hill, New York.

Mary Dallas Consulting Archaeologists 2002. Aboriginal Archaeological Assessment of Lot 8 DP586986, Captain Cook Drive, Kurnell. Unpublished report prepared for R.W. Corkery and Co. Pty Ltd. on behalf of Rocla Ltd.

Mathews, R.H. 1893. Rock paintings of the Aborigines in caves on Bulgar Creek, near Singleton. *Journal of the Royal Society of NSW* 27:353-58.

Mathews, R.H. 1895a. The rock paintings and carvings of the Australian Aborigines. *Journal of the Anthropological Institute of Great Britain and Ireland* 25:145-63.

Mathews, R.H. 1895b. Aboriginal rock paintings and carvings. *Proceedings of the Royal Society of Victoria* III:137-73.

Mathews, R.H. 1895c. The Aboriginal rock pictures of Australia. *Transactions and Proceedings of the Royal Geographic Society of Australasia* X:46-70.

Mathews, R.H. 1895d. Australian rock pictures. *American Anthropologist VIII*:268-78.

Mathews, R.H. 1896a. Australian ground and tree drawings. *American Anthropologist* IX(2):33-49.

Mathews, R.H. 1896b. The rock pictures of the Australian Aborigines. *Proceedings and Transactions of the Queensland Royal Geographic Society* XI:86-105.

Mathews, R.H. 1896c. The rock paintings and carvings of the Australian Aborigines. *Journal of the Anthropological Institute of Great Britain and Ireland* XXV:145-63.

Mathews, R.H. 1896d. The Keeparra ceremony of initiation. *Journal of the Anthropological Institute of Great Britain and Ireland XXVI*:320-40.

Mathews, R.H. 1897a. Australian rock carvings. *Proceedings of the American Philosophical Society* 36:195-208.

Mathews, R.H. 1897b. Rock carvings and paintings of the Australian Aborigines. *Proceedings of the American Philosophical Society* 36:466-78.

Mathews, R.H. 1897c. The Burbung of the Darkinung Tribe. *Proceedings of the Royal Society of Victoria* 10:1-12.

Mathews, R.H. 1897d. Rock carving by the Australian Aborigines. *Proceedings of the Royal Society of Queensland* XII:97-8.

Mathews, R.H. 1897e. Totemic divisions of some Australian tribes *Journal of the Royal Society of NSW* 31:154-76.

Mathews, R.H. 1898a. The rock paintings and carvings of the Australian Aborigines (Part II). *Journal of the Anthropological Institute* XXVII:532-541.

Mathews, R.H. 1898b. Initiation ceremonies of Australian tribes. *Proceedings of the American Philosophical Society* 38:54-73.

Mathews, R.H. 1899. Rock Carvings of the Australian Aborigines. *Proceedings and Transcripts of the Royal Geographical Society of Australia* XIV:9-11.

Mathews, R.H. 1900. Marriage and Descent among the Australian Aborigines. *Journal and Proceedings of the Royal Society of NSW* 34:120-35.

Mathews, R.H. 1901. The Thurrawal Language. *Journal and Proceedings of the Royal Society of NSW 35*:127-60.

Mathews, R.H. 1903. Languages of the Kamilaroi and other Aboriginal tribes of NSW. *Journal of the Anthropological Institute of Great Britain and Ireland* 33:259-83.

Mathews, R.H. 1904. Ethnological notes on the Aboriginal tribes of NSW and Victoria. *Journal and Proceedings of the Royal Society of NSW* 38:203-381.

Mathews, R.H. 1908. Some mythology of the *Gundungarra* tribe, New South Wales. *Ziet f Ethnologia* 40:291-310.

Mathews, R.H. 1910. Some rock engravings of the Aborigines of New South Wales. *Journal and Proceedings of the Royal Society of NSW* 44:401-405.

Mathews, R.H. and Enright, B.A. 1895. Rock Paintings and Carvings of the Aborigines of NSW. *Report of the Australasian Association for the Advancement of Science* VI:624-637.

Mathews, R.H. and Everitt, M.M. 1900. The organisation, language and initiation ceremonies of the Aborigines of the south-east coast of NSW. *Journal and Proceedings of the Royal Society of NSW* 34:262-81.

Maynard, L. 1976. An archaeological approach to the study of Australian rock art. Unpublished MA Thesis, University of Sydney, Sydney.

Maynard, L. 1977. Classification and Terminology in Australian rock art. In P.J. Ucko (ed.) *Form in Indigenous Art*, pp.385-402. Australian Institute of Aboriginal Studies, Canberra.

Maynard, L. 1979. The archaeology of Australian Aboriginal art. In S.M. Mead (ed.) *Exploring the Visual Art of Oceania*, pp.83-110. University Press of Hawaii, Honolulu.

Meehan, B. 1982. *Shell bed to shell midden.* Australian Institute of Aboriginal Studies, Canberra.

Meehan, B. 1988. The 'dinnertime camp'. In Meehan, B. and R. Jones (eds) *Archaeology with ethnography: an Australian perspective,* pp.171-181. Department of Prehistory, Research School of Pacific Studies, Australian National University, Canberra.

Meehan, B. and R. Jones (eds) 1988. *Archaeology with ethnography: an Australian perspective.* Department of Prehistory, Research School of Pacific Studies, Australian National University, Canberra.

Megaw, J.V.S. 1967. Archaeology, Art and Aborigines: a survey of historical sources and later Australian prehistory. *Journal and Proceedings of the Royal Australian Historic Society* 53:277-94.

Megaw, J.V.S. 1968. A dated cultural sequence for the south Sydney region of NSW. *Current Anthropology* 9(4):325-329.

Megaw, J.V.S. 1969. Captain Cook and bone barbs at Botany Bay. *Antiquity* 43(171):213-216.

Megaw, J.V.S. 1974. The recent archaeology of the South Sydney district - a summary. In J.V.S. Megaw (ed.) *The Recent Archaeology of the Sydney District: Excavations 1964-1967,* pp. 35-38. Australian Institute of Aboriginal Studies, Canberra, ACT.

Megaw, J.V.S. 1993. Something old, something new: further notes on the Aborigines of the Sydney District as represented by their surviving artefacts, and as depicted in some early European representations. *Records of the Australian Museum, Supplement* 17:25-44.

Megaw, J.V.S. and Roberts, A. 1974. The 1967 excavations at Wattamolla Cove, Royal National Park, NSW. In J. Megaw, (ed.) *The Recent Archaeology of the Sydney District: Excavations 1964-1967*, pp.1-12. Australian Institute of Aboriginal Studies, Canberra, ACT.

Miller, R. 1983. Bull Cave: Its relevance to the prehistory of the Sydney Region. Unpublished BA(Hons) Thesis, Department of Anthropology, University of Sydney, Sydney.

Moore, D. 1970. Results of an Archaeological survey of the Hunter River Valley, NSW. Part I: The Bondaian industry of the upper Hunter and Goulburn River Valleys. *Records of the Australian Museum* 28(2):54-64.

Moore, D. 1977. The hand stencil as a symbol. In Ucko, P.J. (ed.) *Form in Indigenous Art.*, pp. 318-24. Australian Institute of Aboriginal Studies, Canberra.

Moore, D. 1981. Results of an archaeological survey of the Hunter River Valley, NSW. Part II: Problems with the lower Hunter and contacts with the Hawkesbury Valley. *Records of the Australian Museum* (9):388-442.

Morphy, H. 1977. Too many meanings: an analysis of the artistic system of the Yolngu of north-east Arnhem Land. Unpublished PhD Thesis, Australian National University, Canberra.

Morphy, H. 1989. On representing ancestral beings. In H. Morphy (ed.) *Animals into Art*, pp.144-160. Unwin Hyman, London.

Morris, B. 1978. Colonial Occupation of NSW: the Aboriginals' experience. Unpublished BA Honours Thesis, Macquarie University, Sydney.

Morwood, M.J. 1979. Art and Stone: towards a prehistory of Central Western Queensland. Unpublished PhD Thesis, Australian National University, Canberra.

Morwood, M.J. 1984. The prehistory of the Central Queensland highlands. In Wendorf, F. and A. Close (eds) *Advances in World Archaeology*, pp.325-379. Volume 3. Academic Press, New York.

Morwood, M.J. 1986. The archaeology of art: excavations at Maidenwell and Gatton shelters, SE Queensland. *Queensland Archaeological Research* 3:88-132.

Morwood, M.J. 1987. The archaeology of social complexity in south-eastern Queensland. *Proceedings of the Prehistoric Society* 53:337-350.

Morwood, M.J. 1988. Comment on FD McCarthy's Rock Art Sequences: a matter of clarification. *Rock Art Research* 5(1):32-33.

Morwood, M.J. 1992. Changing art in a changing landscape: a case study from the upper Flinders region of the north Queensland highlands. In J.J. McDonald and I.B. Haskovec (eds) State *of the Art: Regional Rock Art Studies in Australia and Melanesia*, pp.60-71. AURA Publication No. 6, Melbourne.

Morwood, M.J. 2002. *Visions from the Past: the archaeology of Australian Aboriginal Art.* Allen and Unwin, Australia.

Morwood, M.J. and Hobbs, D.R. (eds) 1994. *Quinkin Prehistory: the archaeology of Aboriginal art in S.E. Cape York Peninsula.* Tempus Volume 3. University of Queensland, Brisbane.

Mountford, C.P. and Edwards, R. 1963. Rock engravings of Panaramitee Station, north-eastern South Australia. *Transactions of the Royal Society of South Australia* 86:131-146.

Nanson, G.C., Young, R.W., and Stockton, E.D. 1987. Chronology and palaeoenvironment of the Cranebrook Terrace (near Sydney) containing artefacts more than 40,000 years old. *Archaeology in Oceania* 22(2):72-78.

NSW National Parks and Wildlife Service 1996, *Ku-ring-gai Chase National Park*. NSW National Parks and Wildlife Service, Hurstville.

Nobbs, M.F. and Dorn, R.I. 1988. Age determinations for rock varnish formation within petroglyphs. *Rock Art Research* 5(2):108-146.

Nelson, D.E. (ed.) 2000. *The beeswax art of northern Australia.* Simon Fraser University, Burnaby.

O'Connell, J.F. and Allen, J. 2004. Dating the colonization of Sahul (Pleistocene Australia–New Guinea): a review of recent research. *Journal of Archaeological Science* 31:835-853.

Officer, K.L.C. 1984. From Tuggerah to Dharawal: variation and function within a regional art style. Unpublished BA (Hons) Thesis, Australian National University, Canberra.

Officer, K.L.C. 1992. The edge of the sandstone: what makes style change. In J.J. McDonald and I. Haskovec (eds) *State of the Art: Regional Rock Art Studies in Australia and Melanesia*, pp.6-14. AURA Publication No. 6, Melbourne.

Officer, K.L.C. 1993. Style and Graphic: an archaeological model for the analysis of rock art. Unpublished PhD Thesis, Department of Prehistory and Anthropology, Australian National University, Canberra.

Peterson, N. 1971. Open sites and the ethnographic approach to the archaeology of Hunter-Gatherers. In D.J. Mulvaney and J. Golson (eds) *Aboriginal man and environment in Australia*, pp.239-248. ANU Press, Canberra.

Peterson, N. 1973. Campsite location amongst Australian Hunter-Gatherers: archaeological and ethnographic evidence for a key determinant. *Archaeology and Physical Anthropology in Oceania* 8(3):173-92.

Peterson, N. 1976. The natural and cultural areas of Aboriginal Australia: a preliminary analysis of population groupings with adaptive significance. In N. Peterson (ed.) *Tribes and Boundaries in Australia*, pp.50-71. Social Anthropology Series No. 10, Australian Institute of Aboriginal Studies, Canberra. Humanities Press Inc, New Jersey, USA.

Peterson, N. 1985. *Australian Territorial Organisation: a band perspective.* University of Sydney, Sydney.

Peterson, N. 1986. The importance of women in determining the composition of residential groups in Aboriginal Australia. In F. Gale (ed.) *Woman's role in Aboriginal society*, pp.16-27. 3rd edition Australian Institute of Aboriginal Studies, Canberra. 1st edition, 1978.

Phillip, A. 1789. *The Voyage of Governor Phillip to Botany Bay*: with contributions from other officers of the first fleet and observations on affairs of the time by Lord Auckland. John Stockdale, London. Reprinted 1970, Angus and Robertson; in association with RAHS.

Phillip, A. 1791. *Extracts of letters from Arthur Phillip, Esq. Governor of New South Wales to Lord Sydney.* Debrett, London. Reprinted 1963 Australiana Facsimile Editions No 15, Public Library of South Australia, Adelaide.

Poiner, G. 1971. The Process of the Year. Unpublished BA(Hons) Thesis, University of Sydney, Sydney.

Poiner, G. 1976. The process of the year among Aborigines of the Central and South Coast of New South Wales. *Archaeology and Physical Anthropology in Oceania* 11(3):186-206.

Przywolnik, K. 2005. Long term transitions in Hunter-Gatherers of coastal Northwestern Australia. In Veth, P., P. Hiscock and M. Smith (eds) *Desert peoples: archaeological perspectives*, pp.177-205. Blackwell Studies in Global Archaeology, Boston.

Quinnell. M.C. 1975. Aboriginal rock art in Carnarvon Gorge, south-central Queensland. *Anthropological Society of Queensland Newsletter* 69:2-10.

Radcliffe-Brown, A. 1930. The rainbow serpent myth in south-eastern Australia. *Oceania* 1:242-247.

Radcliffe-Brown, A.R. 1931. The social organization of Australian tribes, Part III. *Oceania* 1:426-456.

Rich, E. 1992. Narama Salvage Project, Lower Bayswater Creek, Hunter Valley, NSW. Unpublished report prepared for Brayshaw McDonald Pty Ltd on behalf of Envirosciences and the Narama Joint Venture. Report held NSW NPWS.

Rich, E. 1993. Archaeological investigation of site RC-PAD and MF2 Mt Flora near Mittagong, NSW. Unpublished report to Mitchell McCotter Pty Ltd and CSR Readymix. Report held NSW NPWS.

Rich, E. and J. McDonald 1995. Archaeological salvage of site WH3 (# 45-5-965): Project 12603, Cowpastures Road, West Hoxton, NSW. Unpublished report to Lean Lackenby and Hayward on behalf of Landcom, Parramatta.

Roberts, R., G.L. Walsh, A. Murray, J. Olley, R. Jones, M. Morwood, C. Tuniz, E. Lawson, M. MacPhail, D. Bowdery, and I. Naumann 1997. Luminescence dating of Rock Art and Past Environments using Mud-wasp nests in Northern Australia. *Nature* 387:696-99.

Robertson, G. R. 2005. Backed Artefact Use in Eastern Australia: a Residue and Use-wear Analysis. Unpublished PhD Thesis, University of Queensland, Brisbane.

Rosenfeld, A. 1991. Panaramitee: dead or alive? In P. Bahn, and A. Rosenfeld (eds) *Rock Art and Prehistory: papers presented to symposium G of the AURA Congress, Darwin 1988*, pp.136-144. Oxbow Monograph 10, Oxford.

Rosenfeld, A. 1993. A review of the evidence for the emergence of rock art in Australia. In M.A. Smith, M. Spriggs and B. Frankhouser (eds) *Sahul in Review: Pleistocene Archaeology in Australia, New Guinea and Island Melanesia*, pp.71-80. Occasional Papers in Prehistory, No. 24. Research School of Pacific Studies, Australian National University, Canberra.

Rosenfeld, A. 2002. Rock art as an indicator of changing social geographies in central Australia. In David, B. and M. Wilson (eds) *Inscribed landscapes: marking and making places*, pp.61-78. Honolulu: University of Hawaii Press.

Rosenfeld, A., Horton, D., and Winter, J. 1981. *Early Man in North Queensland.* Terra Australis 6. Department of Prehistory, Research School of Pacific Studies, Australian National University, Canberra.

Rosenfeld, A. and C.E. Smith 1997. Recent developments in radiocarbon and stylistic methods of dating rock art. *Antiquity* 71:405-411.

Rosenfeld, A. and M.A. Smith 2002. Rock-Art and the History of Puritjarra Rock Shelter, Cleland Hills, Central Australia. *Proceedings of the Prehistoric Society* 68:103-124.

Ross, A. 1976. Inter-tribal contacts: What the First Fleet saw. Unpublished BA (Hons) Thesis, University of Sydney, Sydney.

Ross, A. 1984. If There Were Water: Prehistoric Settlement Patterns in the Victorian Mallee. Unpublished PhD Thesis, School of Earth Sciences, Macquarie University, Sydney.

Ross, A. 1985. Archaeological Evidence for Population Change in the Middle to Late Holocene in Australia. *Archaeology in Oceania* 20(3):81-89.

Ross, A. 1988. Tribal and linguistic boundaries: a reassessment of the evidence. In G. Aplin, (ed.) *A Difficult Infant: Sydney before Macquarie*, pp.42-53. University of New South Wales Press, Sydney.

Ross, A. 1990. Aboriginal life on the Lower Hawkesbury at the time of European settlement. In J. Powell, and L. Banks, (eds) *Hawkesbury River History: Governor Phillip, Exploration and early settlement*, pp.31-42. Dharug and Hawkesbury Historical Society.

Ross, J. 1997. Painted relationship: an archaeological analysis of a distinctive anthropomorphic rock art motif in northwestern central Queensland. Unpublished BA (Hons) Thesis, University of New England, Armidale.

Ross, J. 2003. Rock Art, Ritual and Relationships: an archaeological analysis of rock art from the central Australian arid zone. Unpublished PhD Thesis, University of New England, Armidale.

Ross, J. 2005. Rock art of the Red Centre. In M. Smith and P. Hesse (eds) *23°S: Archaeology and Environmental History of the Southern Deserts,* pp.217-230. National Museum of Australia Press, Canberra.

Roth, W.E. 1909 North Queensland Ethnography: Fighting Weapons. *Records of the Australian Museum*, Vol. VII (4):189-211. Edited by R. Etheridge. Bulletin No.13.

Rowe, M.M 2001. Dating by AMS radiocarbon analysis. In D. Whitley (ed.) *Handbook of Rock Art Research,* pp.139-166. Altamira Press, Walnut Creek, California, USA.

Roy, R.S. and Thom, B.G. 1981. Late Quaternary marine deposition in New South Wales and Southern Queensland - an evolutionary model. *Journal of the Geological Society of Australia* 28:471-489.

Sackett, J.R. 1990. Style and ethnicity in archaeology: the case for isochrestism. In M. Conkey and C. Hastorf (eds) *The uses of style in archaeology,* pp.32-43. New Directions in Archaeology, Cambridge University Press, Cambridge.

Sanders, J., Bedvard M., Leahy, B., Robinson, M. and Sheringham, P. 1988. Preliminary report on the vegetation of Yengo National Park and Parr State Recreation Area. Unpublished report to NSW NPWS.

Schaasma, P. 1985. Form, Content and Function: Theory and Method in North American Rock Art Studies. In M. Schiffer (ed.) *Advances in Archaeological Method and Theory*, pp.237-277. Volume 8. Academic Press. New York.

Schrire, C. 1972. Ethno-archaeological models and subsistence behaviour in Arnhem land. In D. Clarke (ed.) *Models in Archaeology*, pp.653-669. Methuen, London.

Schrire, C. 1982. *The Alligator Rivers - Prehistory and Ecology in western Arnhem Land.* Terra Australis 7. Department of Prehistory, Research School of Pacific Studies, Australian National University, Canberra.

Scott, W. 1929 (1982). *The Port Stephens Blacks.* Cambridge Press, Newcastle. First published 1929 by The Chronicle Office, Dungog, NSW.

Sefton, C. 1988. Prehistory in the Illawarra. Unpublished MA Thesis, University of Sydney, Sydney.

Shawcross, W. 1967. An investigation of Prehistoric Diet and Economy on a Coastal Site at Galatea Bay, New Zealand. *Proceedings of the Prehistory Society* 7:107-131.

Shennan, S. 1988. *Quantifying Archaeology.* Edinburgh University Press, Edinburgh.

Sim, I.M. 1963a. Records of rock engravings of the Sydney District Numbers 111-127. *Mankind* 6(1) May:19-26.

Sim, I.M. 1963b. Record of rock engravings of the Sydney District. *Mankind* 6(2):55-64.

Sim, I.M. 1965. Records of rock engravings of the Sydney District Numbers 138-145. *Mankind* 6(6):275-287.

Sim, I.M. 1966a. *Rock engravings of the Macdonald River District, NSW.* Occasional Papers in Aboriginal Studies No 7. Australian Institute of Aboriginal Studies, Canberra.

Sim, I.M. 1966b. Records of Rock engravings of the Sydney District Numbers 146-162: Gosford District, Part I. *Mankind* 6(8):347-354.

Sim, I.M. 1969. A gallery of cave art in the Macdonald River District, NSW. *Archaeology and Physical Anthropology in Oceania* 4(2):144-79.

Smith, C.E. 1991. Colonising with style: reviewing the nexus between rock art, territoriality and the colonisation and occupation of Sahul. *Australian Archaeology* 34:34-42.

Smith, C.E. 1994. Situating style: an ethnoarchaeological study of social and material context in an Australian Aboriginal artistic system. Unpublished PhD Thesis, University of New England, Armidale.

Smith, L-J. 1983. What's in the Size of a Macropod? A Study of variance in prehistoric pictures from the Mangrove Creek Area. Unpublished BA(Hons) Thesis, University of Sydney, Sydney.

Smith, L-J. 1986. Artefact analysis of a 3,450 year old open site on the Cumberland Plain, NSW. *Australian Archaeology* 23:11-25.

Smith, L-J. 1989. Aboriginal Site Planning Study in the Sydney Basin: The Cumberland Plain, Stage 1. Unpublished report to the NSW National Parks and Wildlife Service. Report held by NSW NPWS.

Smith, L-J., Rich, E. and Hesp, P. 1990. Aboriginal sites on Kurnell Peninsula: a management study. Unpublished report to the NSW NPWS and the Australian Heritage Commission.

Smith, M.A. 1982. Devon Downs reconsidered: changes in site use at Lower Murray Valley rockshelter. *Archaeology in Oceania* 17(3):109-116.

Smith, M. A., Fankhauser, B. and Jercher, M. 1998 The changing provenance of red ochre at Puritjarra rockshelter, central Australia: late Pleistocene to present. *Proceedings of the Prehistoric Society* 64:275-292.

Soffer, O. 1987. Upper Palaeolithic connubia, refugia, and the archaeological record from Eastern Europe. In O. Soffer (ed.) The *Pleistocene old world: regional perspectives*, pp.333-348. Plennum Press, New York.

Spate, W. and Jennings, A. 1983. The stability of sandstone rock surfaces at Milligan's, near Woy Woy. Unpublished Ms held by the authors.

Stanner, W.E.H. 1965. Aboriginal territorial organisation: estate, range, domain and regime. *Oceania* 36(1):1-26.

Steele, D. 1994. Yengo 1: Report on the Excavated Bone Assemblage. In J. McDonald 1994c: Appendix 1A. Report lodged with NPWS, Hurstville.

Stockdale, J. (ed.) 1789. *The voyage of Governor Phillip to Botany Bay, with an account of the establishment of the Colonies of Port Jackson and Norfolk Island* (with contributions by Lieutenants Shortland, Watts, Ball and Captain Marshall). Reprinted in 1970. J. Stockdale, London.

Stockton, E.D. and Holland, W. 1974. Cultural sites and their environment in the Blue Mountains. *Archaeology and Physical Anthropology in Oceania* 9:36-65.

Sullivan, M.E. 1987. The recent prehistoric exploitation of edible mussel in Aboriginal shell middens in southern NSW. *Archaeology in Oceania* 22(3):97-106.

Sutton, P. 1995. Atomism versus collectivism: The problem of group definition in native title cases. In J. Fingleton and J. Finlayson (eds.) *Anthropology in the native title era: proceedings of a workshop*, pp.1-10. Workshop conducted by the Australian Anthropological Society and the Native Titles Research Unit, Australian Institute of Aboriginal and Torres Strait Islander Studies.

Sydney Prehistory Group 1983. *In Search of the Cobrakall: a survey of sites in the Campbelltown area south of Sydney*. Parts I and II. NSW NPWS.

Talbot, R.K. and J.D. Wilde 1989. Giving form to the formative: shifting settlement patterns in the eastern Great Basin and northern Colorado Plateau. *Utah Archaeology 1:3*-18.

Taçon, P.S.C. 1989. From Rainbow Snakes to X-ray fish: the nature of the recent rock painting tradition in western Arnhem Land, Australia. Unpublished PhD Thesis, Australian National University, Canberra.

Taçon, P.S.C. and Chippindale, C. 1993. Two old painted panels from Kakadu: variation and sequence in Arnhem Land Rock Art. In J. Steinbring, A. Watchman, P. Faulstich and P. Taçon (eds) *Time and Space: Dating and spatial consideration in rock art research*, pp.32-56. AURA Publication No. 8. ARARA, Melbourne.

Taçon, P.S.C., W. Brennan, S. Hooper, M. Kelleher and D. Pross 2005. Greater Wollemi: a new Australian rock art area bordering Sydney. *International Newsletter on Rock Art* 43:1-6.

Taçon, P.S.C., M. Kelleher, W. Brennan, S. Hooper and D. Pross 2006. Wollemi petroglyphs, NSW: an unusual assemblage with rare motifs. *Rock Art Research* 23(2):227-238

Taylor, L. 1987. 'The same but different': Social reproduction and innovation in the art of the Kunwinjku of western Arnhem Land. Unpublished PhD Thesis, Australian National University, Canberra.

Taylor, L. 1989. Seeing the 'inside': Kunwinjku paintings and the symbols of the divided body. In H. Morphy (ed.) *Animals into Art,* pp.371-389. Unwin Hyman, London.

Tench, W. 1789. *A narrative of the expedition to Botany Bay.* London: Debrett.

Tench, W. 1793. *A complete account of the settlement at Port Jackson.* G. Nichol, Pall Mall.

Tench, W. 1789 1793. *Sydney's first four years: being a reprint of A narrative of the expedition to Botany Bay and A complete account of the settlement at Port Jackson.* Reprinted in 1961. Angus and Robertson in association with RAHS.

Thorley, P. 1999. Regional Archaeological research in the Palmer River catchment. *Australian Aboriginal Studies* 2:62-8.

Threlkeld, L.E. 1892. *An Australian language as spoken by the Awabakal: the people of Awaba and Lake Macquarie: being an account of language, traditions and customs.* Re-arranged, condensed and edited by J Fraser. Government Printer, Sydney.

Tindale, N.B. 1940. Results of the Harvard-Adelaide Universities anthropological expedition 1938-1939. Distribution of Australian Tribes: a field survey. *Transactions of the Royal Society of South Australia* 64:140-231.

Tindale, N.B. 1974. *Aboriginal Tribes of Australia - their terrain, environmental controls, distribution, limits, and proper names.* University of California Press, Berkley.

Tracey, R. 1974. Three minor sites near Curracurrang Cove with a preliminary note on a rock shelter at Newport, NSW. In J.V.S. Megaw (ed.) *The Recent Archaeology of the Sydney District: Excavations 1964-1967,* pp.19-27. Australian Institute of Aboriginal Studies, Canberra, ACT.

Troy, J. 1990. *Australian Aboriginal contact with the English language in New South Wales: 1788 to 1845.* Pacific Linguistic Series B-103. Research School of Pacific Studies, Australian National University, Canberra.

Troy, J. 1994 The Sydney language. In N. Theiberger and W. McGregor (eds) *Macquarie Aboriginal Words,* pp.61-78. Macquarie Dictionary, Macquarie University, North Ryde, Sydney.

Veth, P. 2000. Origins of the western desert language: convergence in linguistic and archaeological space and time models. *Archaeology in Oceania* 35(1):11-19.

Veth, P. 2006. Social dynamism in the archaeology of the Western Desert. In David, B., B. Barker and I.J. McNiven (eds) *The social archaeology of Australian Indigenous Societies,* pp.242-253. Aboriginal Studies Press, Canberra.

Veth, P. and McDonald, J.J. 2004. Can archaeology be used to address the principles of Exclusive Possession in Native Title? In R. Harrison and C. Williamson (eds) *After Captain Cook: The Archaeology of the Recent Indigenous past in Australia,* pp.121-132. Altamira Press, CA, USA.

Vinnicombe, P. 1980. Prediction and Predilection. A study of Aboriginal sites in the Gosford-Wyong Region. Report to NSW NPWS.

Vinnicombe, P. 1984. Single sites or site complexes? A case study from north of the Hawkesbury, NSW. In S. Sullivan and S. Bowdler (eds) *Site survey and Significance assessment in*

Australian Archaeology, pp.107-118. Department of Prehistory, Research School of Pacific Studies, Australian National University, Canberra.

Walsh, G.L. 1979. Mutilated hands or signal stencils? A consideration of irregular hand stencils from central Queensland. *Australian Archaeology* 9:33-41.

Walsh, G.L. 1994. *Bradshaws and the ancient rock art of north-west Australia.* Bradshaw Foundation, Santa Fe.

Walters, I. 1988. Fish hooks: Evidence for dual social systems in south-eastern Australia? *Australian Archaeology* 27:98-114.

Ward, I.A.K., R.L.K. Fullagar, T. Boer-Mah, L.M. Head, P.S.C. Tacon and K. Mulvaney 2006 Comparison of sedimentation and occupation histories inside and outside rockshelters, Keep River region, north-western Australia. *Geoarchaeology* 21(1): 1-27.

Watchman, A. 1992. Doubtful dates for Karolta Engravings. *Journal of the Australian Institute of Aboriginal and Torres Strait Islander Studies* 1:51-55.

Watchman, A. 1993a. Perspectives and potentials for absolute dating rock paintings. *Antiquity* 67:58-65.

Watchman, A. 1993b. Evidence of a 25,000-year-old pictograph in northern Australia. *Geoarchaeology* 8(6):465-473.

Watchman, A. 1994. Geological note: Radiocarbon dating fatty acids in Holocene siliceous rock surface accretions. *Australian Journal of Earth Sciences* 41:179-180.

Watchman, A. and N. Cole 1993. Accelerator radiocarbon dating of plant fibre binders in rock paintings from northeast Australia. *Antiquity* 67:355-358.

Watchman, A. and R. Lessard 1993. Focused laser extraction of carbonaceous substances for AMS Radiocarbon dating. In J. Steinbring, A. Watchman, P. Faulstich and P. Taçon (eds) *Time and Space: Dating and spatial consideration in rock art research*, pp.74-77. AURA Publication No. 8. ARARA, Melbourne.

Watchman, A., G.L. Walsh, M.J. Morwood, and C. Tuniz 1997. AMS Radiocarbon Age estimates for early rock paintings in the Kimberley, N.W. Australia: preliminary results. *Rock Art Research* 14:18-26.

Watling, T. 1794. *Letters from an exile at Botany Bay.* Printed by Ann Bell, Penrith, Sydney.

White, J. 1790. *Journal of a voyage to New South Wales.* London. Reprinted 1962 (ed. A.H. Chisolm). Sydney: Angus and Robertson.

White, J.P. and O'Connell, J.F. 1982. *A Prehistory of Australia, New Guinea and Sahul.* Academic Press.

White, J.P. and Weineke, C. 1975. Henry Lawson Drive Rockshelter: Excavation report. Unpublished report to NSW NPWS. Report held NSW NPWS.

Wiessner, P. 1983. Style and information in Kalahari San projectile points. *American Antiquity* 48:253-76.

Wiessner, P. 1984. Reconsidering the behavioural basis for style: a case study among the Kalahari San. *Journal of Anthropological Archaeology* 3:190-234.

Wiessner, P. 1985. Style or isochrestic variation? a reply to Sackett. *American Antiquity* 50:12-24.

Wiessner, P. 1989. Style and changing relations between individual and society. In Hodder I. (ed.) *The Meanings of Things: material culture and symbolic expression*, pp.56-63. Unwin Hyman, London.

Wiessner, P. 1990. Is there a unity to style? In M. Conkey, and C. Hastorf (eds) *The uses of style in archaeology*, pp.105-112. New Directions in Archaeology, Cambridge Uni. Press, Cambridge.

Williams, E. 1985. Wet Underfoot? Earth Mound Sites and the Recent Prehistory of South western Victoria. Unpublished PhD Thesis, Australian National University, Canberra.

Wobst, H.M. 1977. Stylistic Behaviour and Information Exchange. In Cleland, C.E. (ed.) *For the Director: Research Essays in honour of JB Griffen*, pp.317-342. Anthropological Papers, Museum of Anthropology, University of Michigan, No 61, Ann Arbor.

Wobst, H.M. 1983. Palaeolithic archaeology - some problems with form, space and time. In G. Bailey (ed.) *Hunter-Gatherer economy in Prehistory: a European perspective*, pp.220-225. Advances in Archaeological Method and Theory. Cambridge University Press, Cambridge.

Wood, V.K. 1989. Angophora Reserve rockshelter: a faunal analysis. Unpublished B.Litt Thesis, Australian National University, Canberra.

Worgan, G.B. 1788. *Journal of a First Fleet Surgeon: GB Worgan; Surgeon of the Sirius.* based on his journal and a letter to his brother Richard. Ms C830 Mitchell Library; Published 1978, by the William Dixon Foundation, Library Council of NSW and Library of Australian History.

Worms, E.A. 1954. Prehistoric petroglyphs of the upper Yule River, north-western Australia. *Anthropos* 49:1067-1088.

Wright, R.V.S. 1971. Prehistory in the Cape York Peninsula. In D.J. Mulvaney and J. Golson (eds) *Aboriginal man and environment in Australia,* pp.133-140. Australian National University Press, Canberra.

Wright, R.V.S. 1992. Doing Multivariate Archaeology and Prehistory: handling large data sets with MV-Arch. Department of Anthropology, University of Sydney, Sydney.

www.ingramcontent.com/pod-product-compliance
Lightning Source LLC
Chambersburg PA
CBHW041545220526
45473CB00015B/2965